Universitext

Universitext is a series of textbooks that provides material from a wide variety of mathematical disciplines at master's level and beyond. The books, often well class-tested by their author, may have an informal, personal, even experimental approach to their subject matter. Some of the most successful and established books in the series have evolved through several editions, always following the evolution of teaching curricula, into very polished texts.

Thus as research topics trickle down into graduate-level teaching, first textbooks written for new, cutting-edge courses may make their way into Universitext.

For further volumes:
http://www.springer.com/series/223

Universitext

Universitext is a series of textbooks that presents material from a wide variety of mathematical disciplines at master's level and beyond. The books, often well class-tested by their author, may have an informal, personal, even experimental approach to their subject matter. Some of the most successful and established books in the series have evolved through several editions, always following the evolution of teaching curricula, into very polished texts.

Thus as research topics trickle down into graduate-level teaching, first textbooks written for new, cutting-edge courses may make their way into *Universitext*.

For further volumes:
http://www.springer.com/series/223

Andreas E. Kyprianou

Fluctuations of Lévy Processes with Applications

Introductory Lectures

Second Edition

Andreas E. Kyprianou
Department of Mathematical Sciences
University of Bath
Bath, UK

ISSN 0172-5939 ISSN 2191-6675 (electronic)
Universitext
ISBN 978-3-642-37631-3 ISBN 978-3-642-37632-0 (eBook)
DOI 10.1007/978-3-642-37632-0
Springer Heidelberg New York Dordrecht London

Library of Congress Control Number: 2013958153

Mathematics Subject Classification (2010): 60G50, 60G51, 60G52

Printed on acid-free paper

Springer is part of Springer Science+Business Media (www.springer.com)

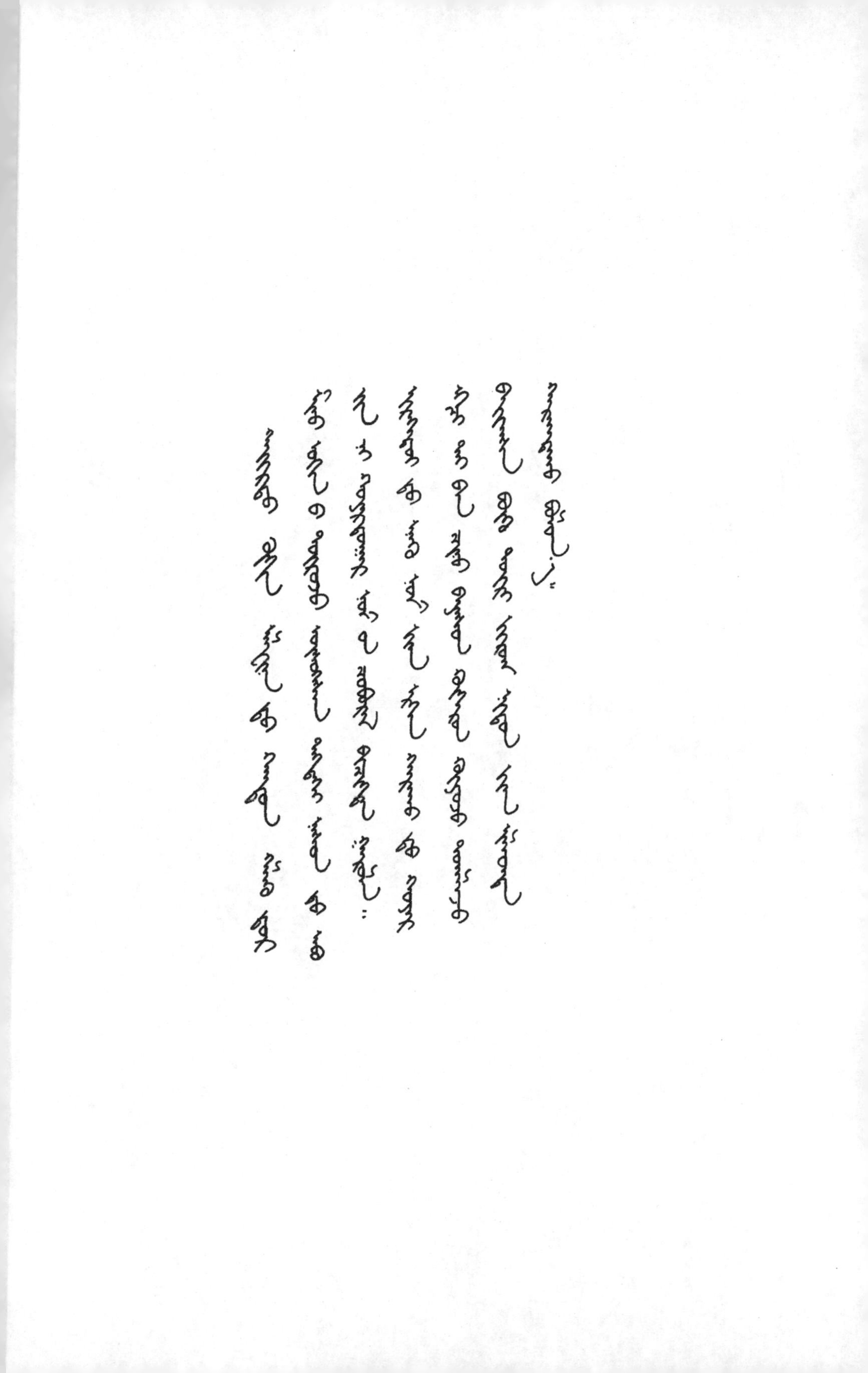

[illegible]
[illegible]
[illegible]
[illegible]
[illegible]
[illegible]
[illegible]
[illegible]

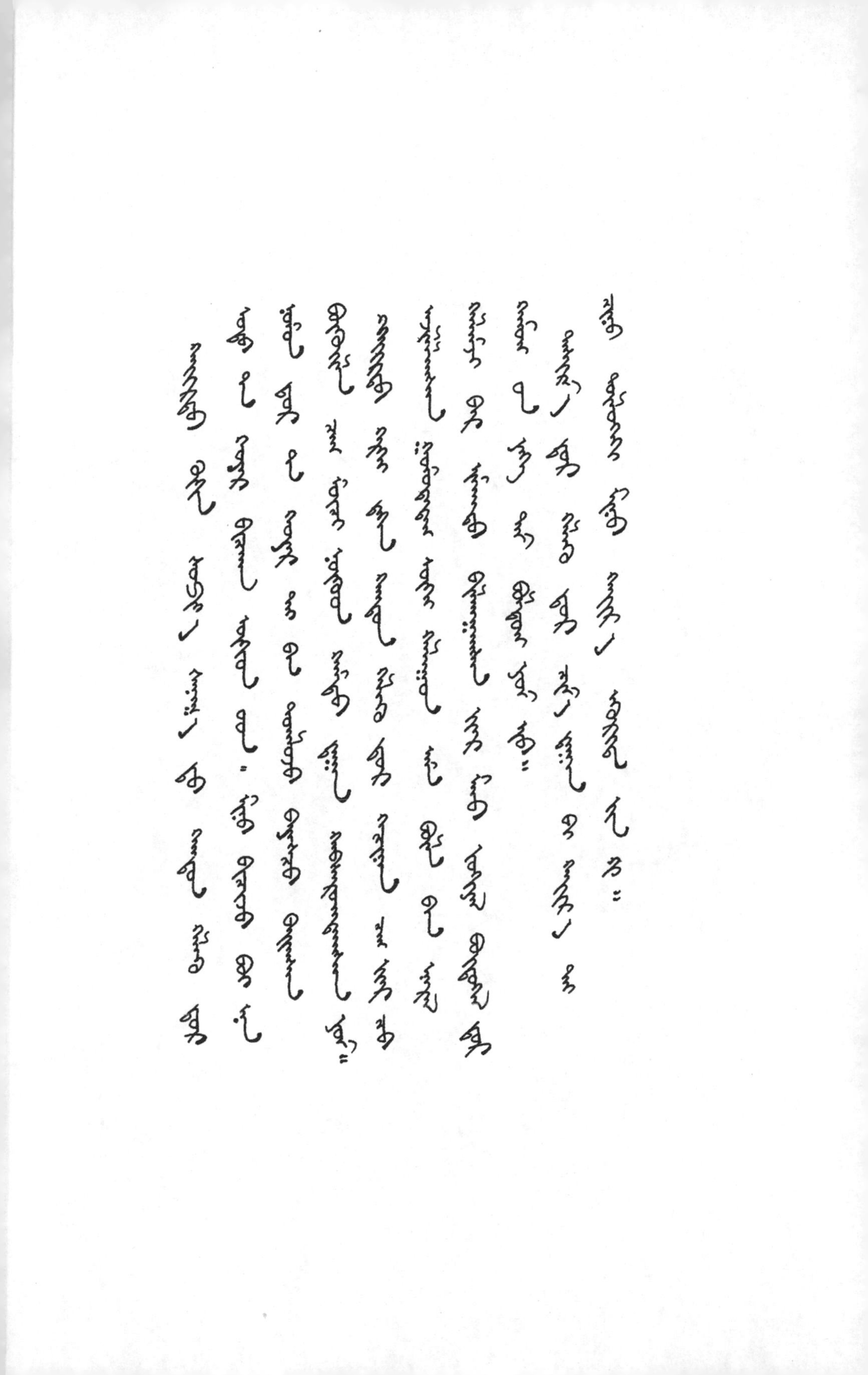

[illegible]
[illegible]
[illegible]
[illegible]
[illegible]
[illegible]
[illegible]
[illegible]
[illegible]
[illegible]

Preface to the Second Edition

For the second edition, I have made a number of typographic, historical and mathematical corrections to the original text. I am deeply grateful to many people who have been kind enough to communicate some of these corrections to me. In this respect, I would like to mention the following names, again, in alphabetical order: Hansjoerg Albrecher, Larbi Alili, Sandra Palau Calderon, Loïc Chaumont, Ron Doney, Leif Döring, Irmingard Eder, Janos Engländer, Her Majesty Queen Elizabeth II, Clément Foucart, Hans Gerber, Sasha Gnedin, Martin Herdegen, Friedrich Hubalek, Lyn Imeson, Robert Knobloch, Takis Konstantopoulos, Alexey Kuznetsov, Eos Kyprianou, Ronnie Loeffen, Juan Carlos Pardo, Pierre Patie, José-Luis Tripitaka Garmendia Pérez, Victor Rivero, Antonio Elbegdorj Murillo Salas, Paavo Salminen, Uwe Schmock, Renming Song, Matija Vidmar, Zoran Vondraček, Long Zhao and Xiaowen Zhou. I must give exceptional thanks to my four current Ph.D. students, Maren Eckhoff, Marion Hesse, Curdin Ott and Alex Watson, who diligently organised themselves to give an extremely thorough read of the penultimate draft of this document. Likewise, Erik Baurdoux and Kazutoshi Yamazaki deserve exceptional thanks for their meticulous proof-reading of substantial parts of the text. In particular, Erik must be commended for his remarkable stamina and ability to spot the most subtle of errors.

The biggest thanks of all however *must* go to the mighty Nick Bingham who committed himself to reading the entire book from cover to cover. Aside from errors of a mathematical and historical nature, he uncovered untold deficiencies in my use of the English language.[1,2] Sincerely, thank you, Nick, for having the patience to fight your way through both my grammar and punctuation and to teach me by example.

I have also included some additional material which reflects some of the many developments that have occurred in the theory and application of Lévy processes

[1]Many thanks to Erik Baurdoux who ironically pointed out that, in the penultimate draft of this manuscript, even the original version of the sentence referred to by this footnote was a grammatical mess.

[2]Erik also took issue with the wording in footnote 1 above.

since the last edition, and which I believe are accessible at the level that I originally pitched this book. Within existing chapters, I have included new material on the theory of special subordinators and I have updated the discussion on particular examples of Wiener–Hopf factorisations. I have also included three new chapters. One chapter concerns the theory of scale functions and another their use in the theory of ruin. Finally, the third new chapter addresses the theory of positive self-similar Markov processes. Another notable change to the book is that the full set of solutions at the back has been replaced by a more terse set of hints. This follows in response to the remarks of several colleagues who have used the book to teach from, as well as using the exercises as homeworks. Finally, the title of the book has also changed. Everyone hated the title of the first edition, most of all me. Within the constraints of permuting the original wording, I am not sure that the new title is a big improvement.

The final big push to finish this second edition took place during my six-month sabbatical as a guest at the Forschungsinstitut für Mathematik, ETH Zürich. I am most grateful to Paul Embrechts and the FIM for the invitation and for accommodating me so comfortably.

Once again, by way of a new inscription, special thanks go to Jagaa, Sophia, Sanaa and, the new addition to the family, little Alina (although she is not so little any more as it took me so long to get through the revision in the end).

Zürich, Switzerland — Andreas E. Kyprianou
December 2012

Preface to the First Edition

In 2003, I began teaching a course entitled *Lévy processes* on the Amsterdam-Utrecht masters programme in stochastics and financial mathematics. Quite naturally, I wanted to expose my students to my own interests in Lévy processes; that is, the role that certain subtle behaviour concerning their fluctuations plays in explaining different types of phenomena appearing in a number of classical models of applied probability. Indeed, recent developments in the theory of Lévy processes, in particular concerning path fluctuations, have offered the clarity required to revisit classical applied probability models and improve on well-established and fundamental results.

Whilst teaching the course, I wrote some lecture notes which have now matured into this text. Given the audience of students, who were either engaged in their "afstudeerfase"[1] or just starting a Ph.D., these lecture notes were originally written with the restriction that the mathematics used would not surpass the level that they should, in principle, have reached. Roughly speaking, that means the following: having experience to the level of third year or fourth year university courses delivered by a mathematics department on

- foundational real and complex analysis,
- basic facts about L^p spaces,
- measure theoretic probability theory,
- elements of the classical theory of Markov processes, stopping times and the strong Markov property,
- Poisson processes and renewal processes,
- Brownian motion as a Markov process and
- elementary martingale theory in continuous time.

For the most part, this affected the way in which the material is handled when compared with the classical texts and research papers from which almost all of the results and arguments in this text originate. A good example of this is the conscious exclu-

[1] The afstudeerfase is equivalent to the typical European masters-level programme.

sion of calculations involving the master formula for the Poisson point process of excursions of a Lévy process from its maximum.

There are approximately 80 exercises, which are also pitched at a level appropriate to the aforementioned audience. Indeed, several of the exercises have been included in response to some of the questions that have been asked by students themselves, concerning curiosities of the arguments given in class. Arguably some of the exercises are quite long. These exercises reflect some of the other ways in which I have used preliminary versions of this text. A small number of students in Utrecht also used the text as an individual reading/self-study programme contributing to their "kleine scripite" (extended mathematical essay) or "onderzoekopdracht" (research project). In addition, some exercises were used as (take-home) examination questions. The exercises in the first chapter are, in particular, designed to show the reader that the basics of the material presented thereafter is already accessible assuming basic knowledge of Poisson processes and Brownian motion.

There can be no doubt, particularly to the more experienced reader, that the current text has been heavily influenced by the outstanding books of Bertoin (1996a) and Sato (1999), especially the former which also takes a predominantly pathwise approach to its content. It should be reiterated however that, unlike these two books, this text is *not* intended as a research monograph nor as a reference manual for the researcher.

Writing of this text began whilst I was employed at Utrecht University in the Netherlands. In early 2005, I moved to a new position at Heriot–Watt University in Edinburgh, and then, in the final stages of completion of the book, to the University of Bath. Over a period of several months my presence in Utrecht was phased out and my presence in Edinburgh was phased in. Along the way, I passed through the Technical University of Munich and the University of Manchester. I should like to thank these four institutes and my hosts for giving me the facilities necessary to write this text (mostly time and a warm, dry, quiet room with an ethernet connection). I would especially like to thank my colleagues at Utrecht for giving me the opportunity and environment in which to develop this course, Ron Doney, during his two-month absence, for lending me the key to his office, thereby giving me access to his book collection whilst mine was in storage, and Andrew Cairns for arranging to push my teaching duties into 2006, thereby allowing me to focus on finalising this text.

Let me now thank the many, including several of the students who took the course, who have made a number of remarks, corrections and suggestions (major and minor) which have helped to shape this text. In alphabetical order these are: Larbi Alili, David Applebaum, Johnathan Bagley, Erik Baurdoux, M.S. Bratiychuk, Catriona Byrne, Zhen-Qing Chen, Gunther Cornelissen, Irmingard Eder, Abdelghafour Es-Saghouani, Serguei Foss, Uwe Franz, Shota Gugushvili, Thorsten Kleinow, Paweł Kliber, Claudia Klüppelberg, V.S. Korolyuk, Ronnie Loeffen, Alexander Novikov, Zbigniew Palmowski, Goran Peskir, Kees van Schaik, Sonja Scheer, Wim Schoutens, Budhi Arta Surya, Enno Veerman, Maaike Verloop and Zoran Vondraček. In particular, I would also like to thank Peter Andrew, Jean Bertoin, Nick Bingham, Ron Doney, Niel Farricker, Alexander Gnedin, Amaury Lambert, Antonis Papapantoleon and Martijn Pistorius who rooted out many errors from extensive sections of the text and provided valuable criticism. Antonis Papapantoleon very kindly produced some simulations of the paths of Lévy processes which have been included in Chap. 1. I am most grateful to Takis Konstantopoulos who read through earlier drafts of the entire text in considerable detail, taking the time to discuss with me at length many of the issues that arose. The front cover was produced in consultation with Hurlee Gonchigdanzan and Jargalmaa Magsarjav. All further comments, corrections and suggestions on the current text are welcome.

Finally, the deepest gratitude of all goes to Jagaa, Sophia and Sanaa for whom the special inscription is written.

Edinburgh, UK Andreas E. Kyprianou
June 2006

Contents

Chapter 1
Lévy Processes and Applications

In this chapter, we define and characterise the class of Lévy processes. To illustrate the variety of processes captured within the definition of a Lévy process, we explore briefly the relationship between Lévy processes and infinitely divisible distributions. We also discuss some classical applied probability models, which are built on the strength of well-understood path properties of elementary Lévy processes. We hint at how generalisations of these models may be approached using more sophisticated Lévy processes. At a number of points later on in this text, we handle these generalisations in more detail. The models we have chosen to present are suitable for the course of this text as a way of exemplifying fluctuation theory but are by no means the only applications.

1.1 Lévy Processes and Infinite Divisibility

Let us begin by recalling the definition of two familiar processes, a Brownian motion and a Poisson process.

A real-valued process, $B = \{B_t : t \geq 0\}$, defined on a probability space $(\Omega, \mathcal{F}, \mathbb{P})$ is said to be a Brownian motion if the following hold:

(i) The paths of B are $\mathbb{P}$-almost surely continuous.
(ii) $\mathbb{P}(B_0 = 0) = 1$.
(iii) For $0 \leq s \leq t$, $B_t - B_s$ is equal in distribution to B_{t-s}.
(iv) For $0 \leq s \leq t$, $B_t - B_s$ is independent of $\{B_u : u \leq s\}$.
(v) For each $t > 0$, B_t is equal in distribution to a normal random variable with zero mean and variance t.

A process valued on the non-negative integers, $N = \{N_t : t \geq 0\}$, defined on a probability space $(\Omega, \mathcal{F}, \mathbb{P})$, is said to be a Poisson process with intensity $\lambda > 0$ if the following hold:

(i) The paths of N are $\mathbb{P}$-almost surely right-continuous with left limits.
(ii) $\mathbb{P}(N_0 = 0) = 1$.

A.E. Kyprianou, *Fluctuations of Lévy Processes with Applications*, Universitext,
DOI 10.1007/978-3-642-37632-0_1, © Springer-Verlag Berlin Heidelberg 2014

(iii) For $0 \leq s \leq t$, $N_t - N_s$ is equal in distribution to N_{t-s}.
(iv) For $0 \leq s \leq t$, $N_t - N_s$ is independent of $\{N_u : u \leq s\}$.
(v) For each $t > 0$, N_t is equal in distribution to a Poisson random variable with parameter λt.

On first encounter, these processes would seem to be considerably different from one another. Firstly, Brownian motion has continuous paths whereas a Poisson process does not. Secondly, a Poisson process is a non-decreasing process, and thus has paths of bounded variation over finite time horizons, whereas a Brownian motion does not have monotone paths and, in fact, its paths are of unbounded variation over finite time horizons.

However, when we line up their definitions next to one another, we see that they have a lot in common. Both processes have right-continuous paths with left limits, both are initiated from the origin and both have stationary and independent increments. We may use these common properties to define a general class of one-dimensional stochastic processes, which are called *Lévy processes.*

Definition 1.1 (Lévy Process) A process $X = \{X_t : t \geq 0\}$, defined on a probability space $(\Omega, \mathcal{F}, \mathbb{P})$, is said to be a Lévy process if it possesses the following properties:

(i) The paths of X are $\mathbb{P}$-almost surely right-continuous with left limits.
(ii) $\mathbb{P}(X_0 = 0) = 1$.
(iii) For $0 \leq s \leq t$, $X_t - X_s$ is equal in distribution to X_{t-s}.
(iv) For $0 \leq s \leq t$, $X_t - X_s$ is independent of $\{X_u : u \leq s\}$.

Unless otherwise stated, from now on, when talking of a Lévy process, we shall always use the measure $\mathbb{P}$ (with associated expectation operator $\mathbb{E}$) to be implicitly understood as its law.[1] We shall also associate to X the filtration $\mathbb{F} = \{\mathcal{F}_t : t \geq 0\}$, where, for each $t \geq 0$, $\mathcal{F}_t$ is the natural enlargement of the sigma-algebra generated by $\{X_s : s \leq t\}$. (See Definition 1.3.38. of Bichteler (2002) for a detailed description of what this means.) In particular, this assumption ensures that, for each $t \geq 0$, $\mathcal{F}_t$ is complete with respect to the null sets of $\mathbb{P}|_{\mathcal{F}_t}$ and there is right-continuity, in the sense that $\mathcal{F}_t = \bigcap_{s>t} \mathcal{F}_s$.[2]

The term "Lévy process" honours the work of the French mathematician Paul Lévy who, although not alone in his contribution, played an instrumental role in

---

[1] We shall also repeatedly abuse this notation throughout the book as, on occasion, we will need to talk about a Lévy process, X, referenced against a random time horizon, say $\mathbf{e}$, which is independent of X and exponentially distributed. In that case, we shall use $\mathbb{P}$ (and accordingly $\mathbb{E}$) for the product law associated with X and $\mathbf{e}$.

[2] Where we have assumed natural enlargement here, it is commonplace in other literature to assume that the filtration $\mathbb{F}$ satisfies "*les conditions habituelles*". In particular, for each $t \geq 0$, $\mathcal{F}_t$ is complete with respect to all null sets of $\mathbb{P}$. This can create problems, for example, when looking at changes of measure (as indeed we will in this book). The reader is encouraged to read Warning 1.3.39. of Bichteler (2002) for further investigation.

bringing together an understanding and characterisation of processes with stationary independent increments. In earlier literature, Lévy processes can be found under a number of different names. In the 1940s, Lévy himself referred to them as a sub-class of *processus additifs* (additive processes), that is, processes with independent increments. For the most part, however, research literature through the 1960s and 1970s refers to Lévy processes simply as *processes with stationary independent increments.* One sees a change in language through the 1970s and by the 1980s the use of the term "Lévy process" had become standard.

From Definition 1.1 alone it is difficult to see just how rich the class of Lévy processes is. The mathematician de Finetti (1929) introduced the notion of *infinitely divisible* distributions and showed that they have an intimate relationship with Lévy processes. It turns out that this relationship gives a reasonably good impression of how varied the class of Lévy processes really is. To this end, let us now devote a little time to discussing infinitely divisible distributions.

Definition 1.2 We say that a real-valued random variable, Θ, has an infinitely divisible distribution if, for each $n = 1, 2, \ldots$, there exists a sequence of i.i.d. random variables $\Theta_{1,n}, \ldots, \Theta_{n,n}$ such that

$$\Theta \stackrel{d}{=} \Theta_{1,n} + \cdots + \Theta_{n,n},$$

where $\stackrel{d}{=}$ is equality in distribution. Alternatively, we could have expressed this relation in terms of probability laws. That is to say, the law μ of a real-valued random variable is infinitely divisible if, for each $n = 1, 2, \ldots$, there exists another law μ_n of a real-valued random variable such that $\mu = \mu_n^{*n}$. (Here μ_n^{*n} denotes the n-fold convolution of μ_n.)

In view of the above definition, one way to establish whether a given random variable has an infinitely divisible distribution is via its characteristic exponent. Suppose that Θ has characteristic exponent $\Psi(u) := -\log \mathbb{E}(\mathrm{e}^{\mathrm{i}u\Theta})$, defined for all $u \in \mathbb{R}$. Then Θ has an infinitely divisible distribution if, for all $n \geq 1$, there exists a characteristic exponent of a probability distribution, say Ψ_n, such that $\Psi(u) = n\Psi_n(u)$, for all $u \in \mathbb{R}$.

The full extent to which we may characterise infinitely divisible distributions is described by the characteristic exponent Ψ and an expression known as the Lévy–Khintchine formula.

Theorem 1.3 (Lévy–Khintchine formula) *A probability law, μ, of a real-valued random variable is infinitely divisible with characteristic exponent Ψ,*

$$\int_{\mathbb{R}} \mathrm{e}^{\mathrm{i}\theta x} \mu(\mathrm{d}x) = \mathrm{e}^{-\Psi(\theta)}, \quad \textit{for } \theta \in \mathbb{R},$$

if and only if there exists a triple (a, σ, Π), where $a \in \mathbb{R}$, $\sigma \in \mathbb{R}$ and Π is a measure concentrated on $\mathbb{R}\backslash\{0\}$ satisfying $\int_{\mathbb{R}}(1 \wedge x^2)\Pi(\mathrm{d}x) < \infty$, such that

$$\Psi(\theta) = \mathrm{i}a\theta + \frac{1}{2}\sigma^2\theta^2 + \int_{\mathbb{R}}\big(1 - \mathrm{e}^{\mathrm{i}\theta x} + \mathrm{i}\theta x \mathbf{1}_{(|x|<1)}\big)\Pi(\mathrm{d}x),$$

for every $\theta \in \mathbb{R}$. Moreover, the triple (a, σ^2, Π) is unique.

Definition 1.4 The measure Π is called the Lévy (characteristic) measure.

The proof of the Lévy–Khintchine characterisation of infinitely divisible random variables is quite lengthy and we choose to exclude it in favour of moving as quickly as possible to fluctuation theory. The interested reader is referred to Lukacs (1970) or Sato (1999) to name but two of many possible references.

A special case of the Lévy–Khintchine formula was established by Kolmogorov (1932) for infinitely divisible distributions with second moments. However, it was Lévy (1934a 1934b) who gave a complete characterisation of infinitely divisible distributions and, in doing so, he also characterised the general class of processes with stationary independent increments. Later, Khintchine (1937) and Itô (1942) gave further simplification and deeper insight to Lévy's original proof. All of this was integrated in Lévy's book of 1948 (with second edition in 1965); cf. Lévy (1948).

Let us now discuss in greater detail the relationship between infinitely divisible distributions and processes with stationary independent increments.

From the definition of a Lévy process, we see that, for any $t > 0$, X_t is a random variable belonging to the class of infinitely divisible distributions. This follows from the fact that, for any $n = 1, 2, \ldots$,

$$X_t = X_{t/n} + (X_{2t/n} - X_{t/n}) + \cdots + (X_t - X_{(n-1)t/n}), \tag{1.1}$$

together with the facts that X has stationary independent increments and that $X_0 = 0$. Suppose, now, that we define, for all $\theta \in \mathbb{R}$, $t \geq 0$,

$$\Psi_t(\theta) = -\log \mathbb{E}\big(\mathrm{e}^{\mathrm{i}\theta X_t}\big).$$

Then using (1.1) twice, we have, for any two positive integers m, n, that

$$m\Psi_1(\theta) = \Psi_m(\theta) = n\Psi_{m/n}(\theta).$$

Hence, for any rational $t > 0$,

$$\Psi_t(\theta) = t\Psi_1(\theta). \tag{1.2}$$

If t is an irrational number, then we can choose a decreasing sequence of rationals $\{t_n : n \geq 1\}$ such that $t_n \downarrow t$ as n tends to infinity. Almost sure right-continuity of X implies right-continuity of $\exp\{-\Psi_t(\theta)\}$ (by dominated convergence) and hence (1.2) holds for all $t \geq 0$.

In conclusion, any Lévy process has the property that, for all $t \geq 0$,

$$\mathbb{E}(\mathrm{e}^{\mathrm{i}\theta X_t}) = \mathrm{e}^{-t\Psi(\theta)}, \tag{1.3}$$

where $\Psi(\theta) := \Psi_1(\theta)$ is the characteristic exponent of X_1. Moreover, the latter has an infinitely divisible distribution.

Definition 1.5 In the sequel, we shall also refer to $\Psi(\theta)$ as the characteristic exponent of the Lévy process.

It is now clear that each Lévy process can be associated with an infinitely divisible distribution. What is not clear is whether given an infinitely divisible distribution, one may construct a Lévy process X, such that X_1 has that distribution. This issue is dealt with by the following theorem, which gives the Lévy–Khintchine formula for Lévy processes.

Theorem 1.6 (Lévy–Khintchine formula for Lévy processes) *Suppose that $a \in \mathbb{R}$, $\sigma \in \mathbb{R}$ and Π is a measure concentrated on $\mathbb{R}\backslash\{0\}$ such that $\int_{\mathbb{R}}(1 \wedge x^2)\Pi(\mathrm{d}x) < \infty$. From this triple, define for each $\theta \in \mathbb{R}$,*

$$\Psi(\theta) = \mathrm{i}a\theta + \frac{1}{2}\sigma^2\theta^2 + \int_{\mathbb{R}} \big(1 - \mathrm{e}^{\mathrm{i}\theta x} + \mathrm{i}\theta x \mathbf{1}_{(|x|<1)}\big)\Pi(\mathrm{d}x).$$

Then there exists a probability space, $(\Omega, \mathcal{F}, \mathbb{P})$, on which a Lévy process is defined having characteristic exponent Ψ.

The proof of this theorem is rather complicated, but very rewarding as it also reveals much more about the general structure of Lévy processes. Later, in Chap. 2, we will prove a stronger version of this theorem, which also explains the path structure of the Lévy process in terms of the triple (a, σ, Π).

1.2 Some Examples of Lévy Processes

To conclude our introduction to Lévy processes and infinite divisible distributions, let us proceed to some concrete examples. Some of these will also be of use later to verify certain results from the forthcoming fluctuation theory we will present.

1.2.1 Poisson Processes

For each $\lambda > 0$, consider a probability distribution μ_λ which is concentrated on $k = 0, 1, 2, \ldots$ such that $\mu_\lambda(\{k\}) = \mathrm{e}^{-\lambda}\lambda^k/k!$, that is to say, the Poisson distribution.

An easy calculation reveals that

$$\sum_{k\geq 0} \mathrm{e}^{\mathrm{i}\theta k}\mu_\lambda(\{k\}) = \mathrm{e}^{-\lambda(1-\mathrm{e}^{\mathrm{i}\theta})}$$

$$= [\mathrm{e}^{-\frac{\lambda}{n}(1-\mathrm{e}^{\mathrm{i}\theta})}]^n.$$

The right-hand side is the characteristic function of the sum of n independent Poisson variables, each of which has parameter λ/n. In the Lévy–Khintchine decomposition, we see that $a = \sigma = 0$ and $\Pi = \lambda\delta_1$, where δ_1 is the Dirac measure supported on $\{1\}$.

Recall that a Poisson process, $\{N_t : t \geq 0\}$, is a Lévy process such that, for each $t > 0$, N_t is Poisson distributed with parameter λt. From the above calculations, we have

$$\mathbb{E}(\mathrm{e}^{\mathrm{i}\theta N_t}) = \mathrm{e}^{-\lambda t(1-\mathrm{e}^{\mathrm{i}\theta})}$$

and hence its characteristic exponent is given by $\Psi(\theta) = \lambda(1 - \mathrm{e}^{\mathrm{i}\theta})$, for $\theta \in \mathbb{R}$.

1.2.2 Compound Poisson Processes

Suppose now that N is a Poisson random variable with parameter $\lambda > 0$ and that $\{\xi_i : i \geq 1\}$ is a sequence of i.i.d. random variables (independent of N) with common law F which has no atom at zero. By first conditioning on N, we have for $\theta \in \mathbb{R}$,[3]

$$E(\mathrm{e}^{\mathrm{i}\theta\sum_{i=1}^N \xi_i}) = \sum_{n\geq 0} E(\mathrm{e}^{\mathrm{i}\theta\sum_{i=1}^n \xi_i})\mathrm{e}^{-\lambda}\frac{\lambda^n}{n!}$$

$$= \sum_{n\geq 0}\left(\int_{\mathbb{R}} \mathrm{e}^{\mathrm{i}\theta x} F(\mathrm{d}x)\right)^n \mathrm{e}^{-\lambda}\frac{\lambda^n}{n!}$$

$$= \mathrm{e}^{-\lambda\int_{\mathbb{R}}(1-\mathrm{e}^{\mathrm{i}\theta x})F(\mathrm{d}x)}. \tag{1.4}$$

We see from (1.4) that distributions of the form $\sum_{i=1}^N \xi_i$ are infinitely divisible with triple $a = -\lambda\int_{0<|x|<1} xF(\mathrm{d}x)$, $\sigma = 0$ and $\Pi(\mathrm{d}x) = \lambda F(\mathrm{d}x)$, for $x \neq 0$. If F consists of an atom of unit mass at 1, then we have simply a Poisson distribution. Note also that if we allow the distribution F to have an atom at zero, then the expression in the exponent on the right-hand side of (1.4) remains the same. Moreover, straightforward computations show that we may interpret it as corresponding to the characteristic exponent of a compound Poisson process with arrival rate $\lambda(1 - F(\{0\}))$ and jump distribution $F(\mathrm{d}x)/(1 - F(\{0\}))$, for $x \in \mathbb{R}\backslash\{0\}$.

---

[3]Here and throughout the remainder of the book, we use the convention that, for any $n = 0, 1, 2, \ldots$, $\sum_{n+1}^n \cdot = 0$.

Suppose now that $\{N_t : t \geq 0\}$ is a Poisson process with intensity $\lambda > 0$ and consider a compound Poisson process $\{X_t : t \geq 0\}$ defined by

$$X_t = \sum_{i=1}^{N_t} \xi_i, \quad t \geq 0.$$

Using the fact that N has stationary independent increments together with the mutual independence of the random variables $\{\xi_i : i \geq 1\}$, by writing

$$X_t = X_s + \sum_{i=N_s+1}^{N_t} \xi_i,$$

for $0 \leq s < t < \infty$, it is clear that X_t is the sum of X_s and an independent copy of X_{t-s}. Right-continuity and left limits of the process $\{N_t : t \geq 0\}$ also ensure right-continuity and left limits of X. In conclusion, compound Poisson processes are Lévy processes. From the calculations in the previous paragraph, for each $t \geq 0$, we may substitute N_t for the variable N_1 to discover that the Lévy–Khintchine formula for a compound Poisson process takes the form $\Psi(\theta) = \lambda \int_{\mathbb{R}}(1 - \mathrm{e}^{\mathrm{i}\theta x})F(\mathrm{d}x)$. Note in particular that the Lévy measure of a compound Poisson process is always finite with total mass equal to the rate λ of the underlying process N.

Compound Poisson processes provide a direct link between Lévy processes and random walks. Recall that a random walk is a discrete-time process of the form $S = \{S_n : n \geq 0\}$ where

$$S_0 = 0 \quad \text{and} \quad S_n = \sum_{i=1}^{n} \xi_i, \quad \text{for } n \geq 1. \tag{1.5}$$

A compound Poisson process is nothing more than a random walk whose jumps have been spaced out in time with independent and exponentially distributed inter-arrival periods.

1.2.3 Linear Brownian Motion

Take the probability law

$$\mu_{s,\gamma}(\mathrm{d}x) := \frac{1}{\sqrt{2\pi s^2}} \mathrm{e}^{-(x-\gamma)^2/2s^2} \mathrm{d}x,$$

supported on $\mathbb{R}$, where $\gamma \in \mathbb{R}$ and $s > 0$. This is the well-known Gaussian distribution with mean γ and variance s^2. It is well known that

$$\begin{aligned}\int_{\mathbb{R}} \mathrm{e}^{\mathrm{i}\theta x} \mu_{s,\gamma}(\mathrm{d}x) &= \mathrm{e}^{-\frac{1}{2}s^2\theta^2 + \mathrm{i}\theta\gamma} \\ &= \left[\mathrm{e}^{-\frac{1}{2}(\frac{s}{\sqrt{n}})^2\theta^2 + \mathrm{i}\theta\frac{\gamma}{n}}\right]^n,\end{aligned}$$

showing, again, that it is an infinitely divisible distribution, this time with $a = -\gamma$, $\sigma = s$ and $\Pi = 0$.

We immediately recognise the characteristic exponent $\Psi(\theta) = s^2\theta^2/2 - \mathrm{i}\theta\gamma$ as that of a scaled Brownian motion with linear drift (otherwise referred to as *linear Brownian motion*),

$$X_t := sB_t + \gamma t, \quad t \geq 0,$$

where $B = \{B_t : t \geq 0\}$ is a standard Brownian motion. It is a trivial exercise to verify that X has stationary independent increments with continuous paths as a consequence of the fact that B does.

1.2.4 Gamma Processes

For $\alpha, \beta > 0$, define the gamma-(α, β) distribution by its associated probability measure

$$\mu_{\alpha,\beta}(\mathrm{d}x) = \frac{\alpha^\beta}{\Gamma(\beta)} x^{\beta-1} \mathrm{e}^{-\alpha x} \mathrm{d}x,$$

concentrated on $(0, \infty)$. Note that when $\beta = 1$, this is the exponential distribution. We have

$$\int_0^\infty \mathrm{e}^{\mathrm{i}\theta x} \mu_{\alpha,\beta}(\mathrm{d}x) = \frac{1}{(1 - \mathrm{i}\theta/\alpha)^\beta} = \left[\frac{1}{(1 - \mathrm{i}\theta/\alpha)^{\beta/n}}\right]^n$$

and infinite divisibility follows. For the Lévy–Khintchine decomposition, we have $\sigma = 0$ and $\Pi(\mathrm{d}x) = \beta x^{-1}\mathrm{e}^{-\alpha x}\mathrm{d}x$, concentrated on $(0, \infty)$ and $a = -\int_0^1 x\Pi(\mathrm{d}x)$. However, this is not immediately obvious. The following lemma proves to be useful in establishing the above triple (a, σ, Π). Its proof is Exercise 1.3; see also Bingham (1975).

Lemma 1.7 (Frullani integral) *For all $\alpha, \beta > 0$ and $z \in \mathbb{C}$ such that*[4] $\Re z \leq 0$, *we have*

$$\frac{1}{(1 - z/\alpha)^\beta} = \exp\left\{-\int_0^\infty (1 - \mathrm{e}^{zx})\beta x^{-1}\mathrm{e}^{-\alpha x}\mathrm{d}x\right\}.$$

To see how this lemma helps, note that the Lévy–Khintchine formula for a gamma distribution takes the form

$$\Psi(\theta) = \beta \int_0^\infty (1 - \mathrm{e}^{\mathrm{i}\theta x})\frac{1}{x}\mathrm{e}^{-\alpha x}\mathrm{d}x = \beta \log(1 - \mathrm{i}\theta/\alpha),$$

[4]The notation $\Re z$ refers to the real part of z.

for $\theta \in \mathbb{R}$. The choice of a in the Lévy–Khintchine formula is the necessary quantity to cancel the term coming from $\mathrm{i}\theta x \mathbf{1}_{(|x|<1)}$ in the integral with respect to Π, in the general Lévy–Khintchine formula.

According to Theorem 1.6, there exists a Lévy process whose Lévy–Khintchine formula is given by Ψ, the so-called *gamma process*.

Suppose now that $X = \{X_t : t \geq 0\}$ is a gamma process. Stationary independent increments tell us that, for all $0 \leq s < t < \infty$, $X_t = X_s + \widetilde{X}_{t-s}$, where $\widetilde{X}_{t-s}$ is an independent copy of X_{t-s}. The fact that $\widetilde{X}_{t-s}$ is strictly positive with probability one (on account of it being gamma distributed) implies that $X_t > X_s$ almost surely. Hence a gamma process is an example of a Lévy process with almost surely non-decreasing paths (in fact its paths are strictly increasing). Another example of a Lévy process with non-decreasing paths is a compound Poisson process where the jump distribution F is concentrated on $(0, \infty)$. Note, however, that a gamma process is not a compound Poisson process, on two counts. Firstly, its Lévy measure has infinite total mass, unlike the Lévy measure of a compound Poisson process, which is necessarily finite (and equal to the arrival rate of jumps). Secondly, whilst a compound Poisson process with positive jumps does have paths which are almost surely non-decreasing, it does not have paths that are almost surely strictly increasing.

Lévy processes whose paths are almost surely non-decreasing (or simply non-decreasing for short) are called *subordinators*. We will return to a formal definition of this subclass of processes in Chap. 2.

1.2.5 Inverse Gaussian Processes

Suppose, as usual, that $B = \{B_t : t \geq 0\}$ is a standard Brownian motion. Define the first passage time

$$\tau_s = \inf\{t > 0 : B_t + bt > s\}. \tag{1.6}$$

This is the first time a Brownian motion with linear drift $b > 0$ crosses above level s. Recall that τ_s is a stopping time[5] for Brownian motion and, since Brownian motion has continuous paths, we know that $B_{\tau_s} + b\tau_s = s$ almost surely. From the strong Markov property, it is known that $\{B_{\tau_s+t} + b(\tau_s + t) - s : t \geq 0\}$ is equal in law to $\{B_t + bt : t \geq 0\}$ and hence, for all $0 \leq s < t$,

$$\tau_t = \tau_s + \widetilde{\tau}_{t-s},$$

where $\widetilde{\tau}_{t-s}$ is an independent copy of τ_{t-s}. This shows that the process $\tau := \{\tau_t : t \geq 0\}$ has stationary independent increments. Continuity of the paths of $\{B_t + bt : t \geq 0\}$ ensures that τ has right-continuous paths. Further, it is clear that τ

[5]We assume that the reader is familiar with the basic notion of a stopping time for a Markov process as well as the strong Markov property. Both will be dealt with in more detail for a general Lévy process in Chap. 3.

has almost surely non-decreasing paths, which guarantees its paths have left limits as well as being yet another example of a subordinator. According to its definition as a sequence of first passage times, τ is also the almost sure right inverse of the graph of $\{B_t + bt : t \geq 0\}$. From this, τ earns its name as the inverse Gaussian process.

According to the discussion following Theorem 1.3, it is now immediate that, for each fixed $s > 0$, the random variable τ_s is infinitely divisible. Its characteristic exponent takes the form

$$\Psi_s(\theta) = s\big(\sqrt{-2\mathrm{i}\theta + b^2} - b\big),$$

for all $\theta \in \mathbb{R}$, where Ψ_s corresponds to the triple $a = -2sb^{-1}\int_0^b (2\pi)^{-1/2}\mathrm{e}^{-y^2/2}\mathrm{d}y$, $\sigma = 0$ and

$$\Pi(\mathrm{d}x) = s\frac{1}{\sqrt{2\pi x^3}}\mathrm{e}^{-\frac{b^2 x}{2}}\,\mathrm{d}x,$$

concentrated on $(0, \infty)$. The law of τ_s can also be computed explicitly as

$$\mu_s(\mathrm{d}x) = \frac{s}{\sqrt{2\pi x^3}}\mathrm{e}^{sb}\mathrm{e}^{-\frac{1}{2}(s^2x^{-1}+b^2x)}\mathrm{d}x,$$

for $x > 0$. For the proof of these facts, see Exercise 1.6.

1.2.6 Stable Processes

Stable processes are the class of Lévy processes whose characteristic exponents correspond to those of stable distributions. Stable distributions were introduced by Lévy (1924, 1925) as a third example of infinitely divisible distributions after Gaussian and Poisson distributions. A random variable, Y, is said to have a *stable distribution* if, for all $n \geq 1$, it observes the distributional equality

$$Y_1 + \cdots + Y_n \overset{d}{=} a_n Y + b_n, \tag{1.7}$$

where $Y_1, \ldots, Y_n$ are independent copies of Y, $a_n > 0$ and $b_n \in \mathbb{R}$. By subtracting b_n/n from each of the terms on the left-hand side of (1.7) and then dividing through by a_n one sees, in particular, that this definition implies that any stable random variable is infinitely divisible. It turns out that $a_n = n^{1/\alpha}$, for $\alpha \in (0, 2]$; see Feller (1971), Sect. VI.1. In that case, we refer to the parameter α as the *stability index*. A smaller class of distributions are the *strictly stable distributions*. A random variable Y is said to have a strictly stable distribution if it observes (1.7) but with $b_n = 0$. In that case, we necessarily have

$$Y_1 + \cdots + Y_n \overset{d}{=} n^{1/\alpha} Y. \tag{1.8}$$

The case $\alpha = 2$ corresponds to zero mean Gaussian random variables and is excluded in the remainder of the discussion as such distributions have been dealt with in Sect. 1.2.3.

Stable random variables observing the relation (1.7) for $\alpha \in (0, 1) \cup (1, 2)$ have characteristic exponents of the form

$$\Psi(\theta) = c|\theta|^\alpha \left(1 - \mathrm{i}\beta \tan \frac{\pi\alpha}{2} \operatorname{sgn} \theta\right) + \mathrm{i}\theta\eta, \tag{1.9}$$

where $\beta \in [-1, 1]$, $\eta \in \mathbb{R}$ and $c > 0$. Stable random variables observing the relation (1.7) for $\alpha = 1$, have characteristic exponents of the form

$$\Psi(\theta) = c|\theta| \left(1 + \mathrm{i}\beta \frac{2}{\pi} \operatorname{sgn} \theta \log|\theta|\right) + \mathrm{i}\theta\eta, \tag{1.10}$$

where $\beta \in [-1, 1]$, $\eta \in \mathbb{R}$ and $c > 0$. Here, we work with the sign function, $\operatorname{sgn}\theta = \mathbf{1}_{(\theta>0)} - \mathbf{1}_{(\theta<0)}$. To make the connection with the Lévy–Khintchine formula, one needs $\sigma = 0$ and

$$\Pi(\mathrm{d}x) = \begin{cases} c_1 x^{-1-\alpha} \mathrm{d}x & \text{for } x \in (0, \infty) \\ c_2 |x|^{-1-\alpha} \mathrm{d}x & \text{for } x \in (-\infty, 0), \end{cases} \tag{1.11}$$

where $c = -(c_1 + c_2)\Gamma(-\alpha)\cos(\pi\alpha/2)$, $c_1, c_2 \geq 0$ and $\beta = (c_1 - c_2)/(c_1 + c_2)$ if $\alpha \in (0, 1) \cup (1, 2)$ and $c_1 = c_2$ if $\alpha = 1$. The choice of $a \in \mathbb{R}$ in the Lévy–Khintchine formula is then implicit. Exercise 1.4 shows how to make the connection between Π and Ψ with the right choice of a (which depends on α). Unlike the previous examples, the distributions that lie behind these characteristic exponents are *heavy tailed* in the sense that the tails of their distributions decay slowly enough to zero, so that they only have moments strictly less than α. The value of the parameter β gives an indication of asymmetry in the Lévy measure and likewise for the distributional asymmetry (although this fact is not immediately obvious). The densities of stable processes are known explicitly in the form of convergent power series. See Zolotarev (1986), Sato (1999) and Samorodnitsky and Taqqu (1994) for further details of all the facts given in this paragraph. With the exception of the defining property (1.8), we shall generally not need detailed information on distributional properties of stable processes in order to proceed with their fluctuation theory. This explains our reluctance to give further details here.

Two examples of the aforementioned power series that tidy up to more compact expressions are centred Cauchy distributions, corresponding to $\alpha = 1$, $\beta = 0$ and $\eta = 0$, and $\frac{1}{2}$-stable distributions, corresponding to $\alpha = 1/2$, $\beta = 1$ and $\eta = 0$. In the former case, $\Psi(\theta) = c|\theta|$, for $\theta \in \mathbb{R}$, and its law is given by

$$\frac{c}{\pi} \frac{1}{(x^2 + c^2)} \mathrm{d}x, \tag{1.12}$$

for $x \in \mathbb{R}$. In the latter case, $\Psi(\theta) = c|\theta|^{1/2}(1 - \mathrm{i}\operatorname{sgn}\theta)$ for $\theta \in \mathbb{R}$ and its law is given by

$$\frac{c}{\sqrt{2\pi x^3}} \mathrm{e}^{-c^2/2x} \mathrm{d}x.$$

Note that an inverse Gaussian distribution coincides with a $\frac{1}{2}$-stable distribution for $s = c$ and $b = 0$.

Suppose that $\mathcal{S}(c, \alpha, \beta, \eta)$ is the distribution of a stable random variable with parameters c, α, β and η. For each choice of $c > 0$, $\alpha \in (0, 2)$, $\beta \in [-1, 1]$ and $\eta \in \mathbb{R}$, Theorem 1.6 tells us that there exists a Lévy process with characteristic exponent given by (1.9) or (1.10), according to the choice of parameters. Further, from the definition of its characteristic exponent, it is clear that, at each fixed time, the α-stable process will have distribution $\mathcal{S}(ct, \alpha, \beta, \eta t)$.

In this text, we shall henceforth make an abuse of notation and refer to an α-stable process to mean a Lévy process based on a strictly stable distribution.

Strict stability means that the associated characteristic exponent takes the form

$$\Psi(\theta) = \begin{cases} c|\theta|^{\alpha}(1 - \mathrm{i}\beta \tan \frac{\pi\alpha}{2} \operatorname{sgn}\theta) & \text{for } \alpha \in (0,1) \cup (1,2) \\ c|\theta| + \mathrm{i}\eta\theta & \text{for } \alpha = 1, \end{cases} \tag{1.13}$$

where the parameter ranges for c, β and η are as above. The reason for the restriction to strictly stable distributions is that we will want to make use of the following fact. If $\{X_t : t \geq 0\}$ is an α-stable process, then from its characteristic exponent (or equivalently the scaling properties of strictly stable random variables), we see that, for all $\lambda > 0$, $\{X_{\lambda t} : t \geq 0\}$ has the same law as $\{\lambda^{1/\alpha} X_t : t \geq 0\}$.

1.2.7 Other Examples

There are many more known examples of infinitely divisible distributions (and hence Lévy processes). Of the many known proofs of infinitely divisibility for specific distributions, most of them are non-trivial, often requiring intimate knowledge of special functions. A brief list of such distributions might include generalised inverse Gaussian (see Good 1953 and Jørgensen 1982), truncated stable (see Tweedie 1984; Hougaard 1986; Koponen 1995; Boyarchenko and Levendorskii 2002a and Carr et al. 2003), generalised hyperbolic (see Halgreen 1979; Bingham and Kiesel 2004 and Eberlein 2001; Barndorff-Nielsen and Shephard 2001), Meixner (see Schoutens and Teugels 1998), Pareto (see Steutel 1970 and Thorin 1977a), F-distributions (see Ismail and Kelker 1979), Gumbel (see Johnson and Kotz 1970 and Steutel 1973), Weibull (see Johnson and Kotz 1970 and Steutel 1970), lognormal (see Thorin 1977b), Student t-distribution (see Grosswald 1976 and Ismail 1977), Lamperti-stable (see Caballero et al. 2010) and β-class (see Kuznetsov 2010a).

Despite being able to identify a large number of infinitely divisible distributions, and hence their associated Lévy processes, it is not clear at this point what the paths of Lévy processes look like. The task of giving a mathematically precise account of this lies ahead in Chap. 2. In the meantime, let us make the following informal remarks concerning paths of Lévy processes.

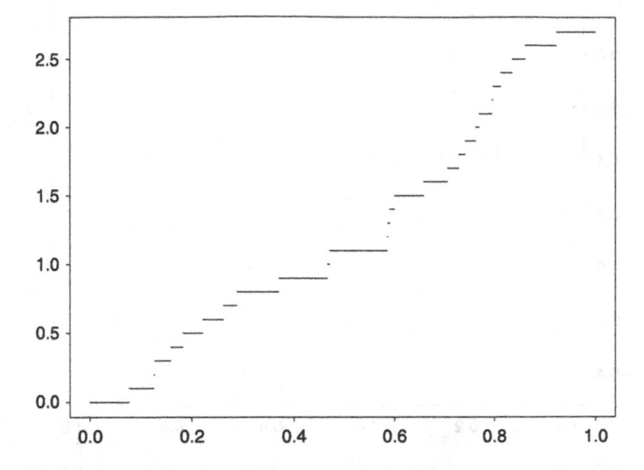

Fig. 1.1 A sample path of a Poisson process; $\Psi(\theta) = \lambda(1 - \mathrm{e}^{\mathrm{i}\theta})$ where λ is the jump rate.

Exercise 1.1 shows that a linear combination of a finite number of independent Lévy processes is again a Lévy process. It turns out that one may consider any Lévy process as an independent sum of a Brownian motion with drift and a countable number of independent compound Poisson processes with different jump rates, jump distributions and drifts. The superposition occurs in such a way that the resulting path remains almost surely finite at all times. Moreover, for each $\varepsilon > 0$, over all fixed time intervals, the process experiences at most a countably infinite number of jumps of magnitude ε or less with probability one, and an almost surely finite number of jumps of magnitude greater than ε. In this description, a necessary and sufficient condition for there to be an almost surely finite number of jumps over each fixed time interval is that the Lévy process is a linear combination of a Brownian motion with drift and an independent compound Poisson process. Depending on the underlying structure of the jumps and the presence of a Brownian motion in the described linear combination, a Lévy process will either have paths of bounded variation on all finite time intervals or paths of unbounded variation on all finite time intervals.

Below, we include six computer simulations to give a rough sense of what the paths of Lévy processes look like. Figures 1.1 and 1.2 depict the paths of a Poisson process and a compound Poisson process, respectively. Figures 1.3 and 1.4 show the paths of a Brownian motion and the independent sum of a Brownian motion and a compound Poisson process, respectively. Finally Figs. 1.5 and 1.6 show the paths of a variance-gamma process and a normal inverse Gaussian processes. Both are pure jump processes (no Brownian component as described above). Variance-gamma processes are discussed in more detail later in Sect. 2.7.3 and Exercise 1.5, normal inverse Gaussian processes are Lévy processes whose jump measure is given by $\Pi(\mathrm{d}x) = (\delta\alpha/\pi|x|)\exp\{\beta x\}K_1(\alpha|x|)\mathrm{d}x$, for $x \in \mathbb{R}$, where $\alpha, \delta > 0$, $\beta \leq |\alpha|$ and $K_1(x)$ is the modified Bessel function of the third kind with index 1 (the precise meaning of this is not worth the detail at this moment in the text). Both experience an infinite number of jumps over a finite time horizon. However, variance-gamma

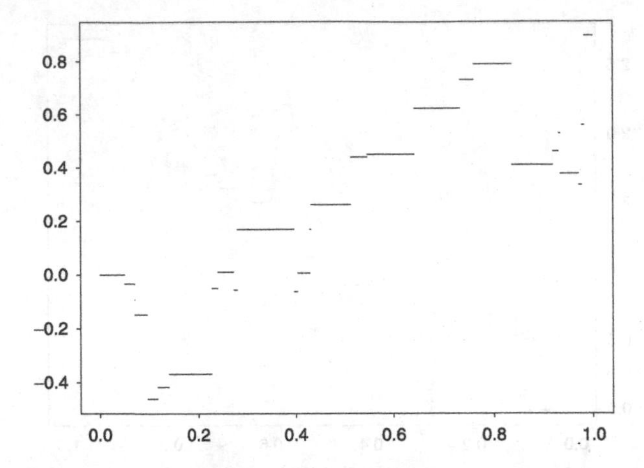

Fig. 1.2 A sample path of a compound Poisson process; $\Psi(\theta) = \lambda \int_{\mathbb{R}}(1 - e^{i\theta x})F(dx)$ where λ is the jump rate and F is the common distribution of the jumps.

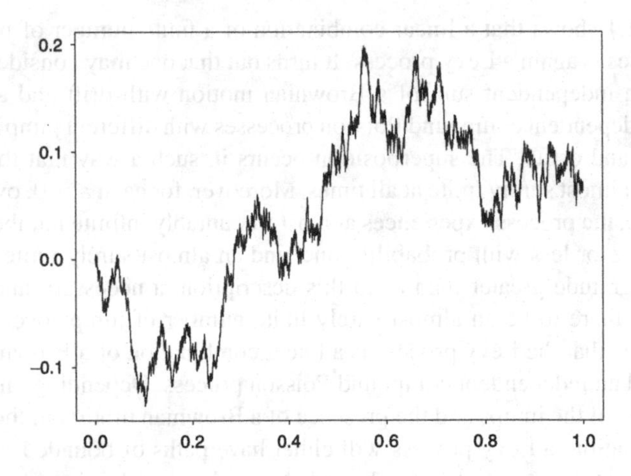

Fig. 1.3 A sample path of a Brownian motion; $\Psi(\theta) = \theta^2/2$.

processes have paths of bounded variation whereas normal inverse Gaussian processes have paths of unbounded variation. The reader should be warned that computer simulations can only depict a finite number of jumps in any given path. All figures were very kindly produced by Antonis Papapantoleon for the purpose of this text.

1.3 Lévy Processes and Some Applied Probability Models

In this section, we introduce some classical applied probability models, which are structured around basic examples of Lévy processes. This section provides a par-

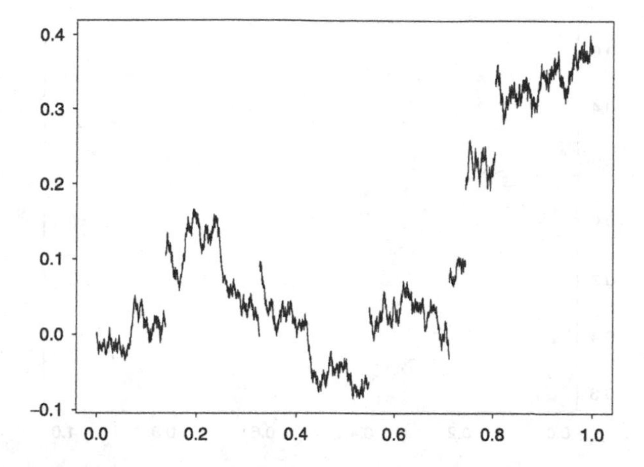

Fig. 1.4 A sample path of the independent sum of a Brownian motion and a compound Poisson process; $\Psi(\theta) = \theta^2/2 + \int_{\mathbb{R}}(1 - e^{i\theta x})F(dx)$.

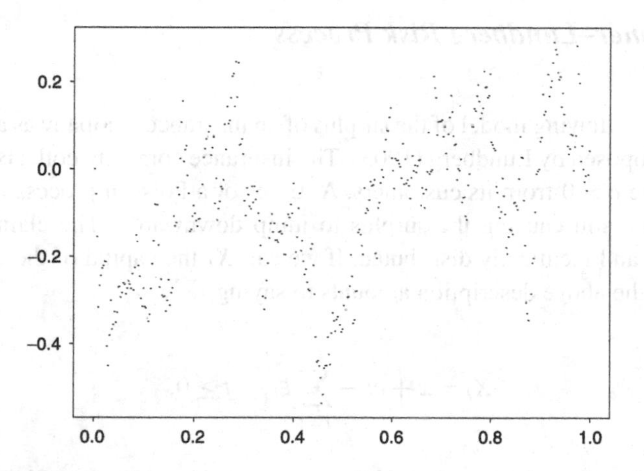

Fig. 1.5 A sample path of a variance-gamma processes. The characteristic exponent is given by $\Psi(\theta) = \beta \log(1 - i\theta c/\alpha + \beta^2\theta^2/2\alpha)$ where $c \in \mathbb{R}$ and $\beta > 0$.

ticular motivation for the study of the fluctuation theory that follows in subsequent chapters. (There are of course other reasons for wanting to study fluctuation theory of Lévy processes.) With the right understanding of the models given below, much richer generalisations may be studied. At different points later on in this text, we will return to these models and reconsider these phenomena in the light of the theory that has been presented along the way. In particular, all of the results either stated or alluded to below will be proved in greater generality in later chapters.

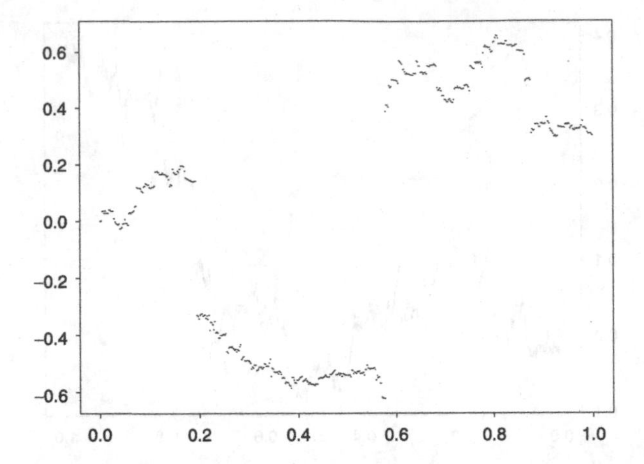

Fig. 1.6 A sample path of a normal inverse Gaussian process; $\Psi(\theta) = \delta(\sqrt{\alpha^2 - (\beta + \mathrm{i}\theta)^2} - \sqrt{\alpha^2 - \beta^2})$ where $\alpha, \delta > 0$, $|\beta| < \alpha$.

1.3.1 Cramér–Lundberg Risk Process

Consider the following model of the surplus of an insurance company as a process in time, first proposed by Lundberg (1903). The insurance company collects premiums at a fixed rate $c > 0$ from its customers. At times of a Poisson process, a customer will make a claim causing the surplus to jump downwards. The claim sizes are independent and identically distributed. If we call X_t the capital of the company at time t, then the above description amounts to saying,

$$X_t = x + ct - \sum_{i=1}^{N_t} \xi_i, \quad t \geq 0,$$

where $x > 0$ is the initial capital of the company, $N = \{N_t : t \geq 0\}$ is a Poisson process with rate $\lambda > 0$, and $\{\xi_i : i \geq 1\}$ is a sequence of positive, independent and identically distributed random variables, also independent of N. The process $X = \{X_t : t \geq 0\}$ is nothing more than a compound Poisson process with drift of rate c, initiated from $x \geq 0$. Denote its law by $\mathbb{P}_x$ and, for convenience, write $\mathbb{P}$ instead of $\mathbb{P}_0$.

Financial ruin in this model (or just *ruin* for short) will occur if the surplus of the insurance company drops below zero. Since this will happen with probability one if $\mathbb{P}(\liminf_{t\uparrow\infty} X_t = -\infty) = 1$, an additional assumption imposed on the model is that

$$\lim_{t\uparrow\infty} X_t = \infty. \tag{1.14}$$

A sufficient condition to guarantee (1.14) is that the distribution of ξ has finite mean, say $\mu > 0$, and that

$$\frac{\lambda\mu}{c} < 1,$$

the so-called *security loading condition (per unit time)*. Indeed, to see why, note that the Strong Law of Large Numbers for Poisson processes, which states that $\lim_{t\uparrow\infty} N_t/t = \lambda$, and the obvious fact that $\lim_{t\uparrow\infty} N_t = \infty$, together imply that

$$\lim_{t\uparrow\infty} \frac{X_t}{t} = \lim_{t\uparrow\infty} \left(\frac{x}{t} + c - \frac{N_t}{t} \frac{\sum_{i=1}^{N_t} \xi_i}{N_t} \right) = c - \lambda\mu > 0.$$

Hence, under the security loading condition it follows that ruin will occur with probability less than one. Fundamental quantities of interest in this model, when X drifts to infinity, are the distribution of the time to ruin and the deficit at ruin, otherwise identified as

$$\tau_0^- := \inf\{t > 0 : X_t < 0\} \text{ and } X_{\tau_0^-} \text{ on } \{\tau_0^- < \infty\}.$$

The following classic result links the probability of ruin to the conditional distribution

$$\eta(x) = \mathbb{P}\big(-X_{\tau_0^-} \leq x \,|\, \tau_0^- < \infty\big), \quad x \geq 0.$$

Theorem 1.8 (Pollaczek–Khintchine formula) *Suppose that $\lambda\mu/c < 1$. For all $x \geq 0$,*

$$1 - \mathbb{P}_x\big(\tau_0^- < \infty\big) = (1-\rho)\sum_{k\geq 0} \rho^k \eta^{*k}(x), \tag{1.15}$$

where $\rho = \mathbb{P}(\tau_0^- < \infty)$.

Formula (1.15) is not particularly explicit in the sense that it gives no information about constant ρ, nor about the distribution η. It turns out that these unknowns can be identified explicitly, as the next theorem reveals.

Theorem 1.9 *In the Cramér–Lundberg model* (*with $\lambda\mu/c < 1$*), *$\rho = \lambda\mu/c$ and*

$$\eta(x) = \frac{1}{\mu}\int_0^x \big[1 - F(y)\big]\mathrm{d}y, \tag{1.16}$$

where F is the distribution of ξ_1.

This result can be derived via a classical path analysis of random walks. Moreover, the aforesaid analysis gives some taste of general fluctuation theory for Lévy processes that we will spend quite some time on in this book. The proof of Theorem 1.9 can be found in Exercise 1.8.

The Pollaczek–Khintchine formula, together with some additional assumptions on F, gives rise to interesting asymptotic behaviour of the probability of ruin. Specifically, we have the following result.

Theorem 1.10 *If $\lambda\mu/c < 1$ and there exists a $\nu \in (0,\infty)$ such that $\mathbb{E}(\mathrm{e}^{-\nu X_1}) = 1$, then*

$$\mathbb{P}_x(\tau_0^- < \infty) \leq \mathrm{e}^{-\nu x},$$

for all $x > 0$. If further, the distribution of F is non-lattice, then

$$\lim_{x\uparrow\infty} \mathrm{e}^{\nu x}\mathbb{P}_x(\tau_0^- < \infty) = \left(\frac{\lambda\nu}{c-\lambda\mu}\int_0^\infty x\mathrm{e}^{\nu x}[1-F(x)]\mathrm{d}x\right)^{-1},$$

where the right-hand side should be interpreted as zero if the integral is infinite.

In the above theorem, the parameter ν is known as the *Lundberg exponent*. See Cramér (1994a, 1994b) for a review of these results.

In more recent times, some authors in this field have moved to working with more general classes of Lévy processes for which there are no positive jumps, in place of the Cramér–Lundberg process. See for example Huzak et al. (2004a, 2004b), Chan (2004) and Klüppelberg et al. (2004a, 2004b). It turns out that working with this class of Lévy processes preserves the idea that the surplus of the insurance company is the aggregate superposition of lots of independent claims, arriving sequentially through time, offset against a deterministic increasing process, corresponding to the accumulation of premiums, even when there are an almost surely infinite number of jumps downwards (claims) in any fixed time interval. We will provide a more detailed interpretation of this class in Chap. 2. In Chaps. 4 and 7, amongst other things, we will also re-examine the Pollaczek–Khintchine formula and the asymptotic probability of ruin, given in Theorem 1.10, in the light of these generalised risk models.

1.3.2 The $M/G/1$ Queue

Let us recall the classical definition of the $M/G/1$ queue. Customers arrive at a service desk according to a Poisson process and join a queue. Customers have service times that are independent and identically distributed. Once served, they leave the queue. The terminology $M/G/1$ refers to the fact that the arrival process is Markovian, the service times are General and there is 1 server.

At each time $t \geq 0$, the workload, W_t, is defined to be the time it will take a customer, who joins the back of the queue at that moment, to reach the service desk. That is to say, the amount of processing time remaining in the queue at time t. Suppose that at an arbitrary moment, which we shall call time zero, the server is not idle and the workload is equal to $w > 0$. On the event that t is before the first time the queue becomes empty, we have that

$$W_t = w + \sum_{i=1}^{N_t} \xi_i - t. \tag{1.17}$$

Here, as with the Cramér–Lundberg risk process, $N = \{N_t : t \geq 0\}$ is a Poisson process with intensity $\lambda > 0$ and $\{\xi_i : i \geq 0\}$ are positive random variables that are independent and identically distributed, with common distribution F and mean $\mu < \infty$. The process N models the arrivals of new customers and $\{\xi_i : i \geq 0\}$ are understood to be their respective service times. The negative unit drift simply corresponds to the decrease in time as the server deals with jobs at a constant rate. Thanks to the lack-of-memory property, once the queue becomes empty, the queue remains empty for an exponentially distributed period of time with parameter λ, after which, a new arrival causes a jump in W, which has distribution F. The process proceeds to evolve as the compound Poisson process, described above, until the queue next empties, and so on.

The workload is clearly not a Lévy process as it is impossible for $\{W_t : t \geq 0\}$ to decrease in value from the state zero, whereas it can decrease in value from any other state $x > 0$. However, it turns out that it is quite easy to link the workload to a familiar functional of a Lévy process, which is also a Markov process. Specifically, suppose we define $\{X_t : t \geq 0\}$ as equal to the same Lévy process describing the Cramér–Lundberg risk model, with $c = 1$ and $x = 0$. Then

$$W_t = (w \vee \overline{X}_t) - X_t, \quad t \geq 0,$$

where the process $\overline{X} := \{\overline{X}_t : t \geq 0\}$ is the running supremum of X. That is, $\overline{X}_t := \sup_{u \leq t} X_u$, $t \geq 0$. Whilst it is easy to show that the pair $(\overline{X}, X)$ is a Markov process, with a little extra work it can also be shown that W is a strong Markov process (this is dealt with in more detail in Exercise 3.2). Clearly then, under $\mathbb{P}$, the process W behaves like $w - X$ until the stopping time

$$\tau_w^+ := \inf\{t > 0 : X_t > w\}.$$

At the time τ_w^+, the process $W = \{W_t : t \geq 0\}$ first becomes zero in value. On account of the strong Markov property and the lack-of-memory property, W then remains zero for an interval of time, whose length is exponentially distributed with parameter λ. Note that during this interval of time, $w \vee \overline{X}_t = \overline{X}_t = X_t$. At the end of this so-called *idle period*, X makes another negative jump distributed according to F and, accordingly, W makes a positive jump with the same distribution, and so on, thereby matching the description of the evolution of W in the previous paragraph; see Fig. 1.7.

Note that this description still makes sense when $w = 0$, in which case, for an initial period of time, which is exponentially distributed, W remains equal to zero until X first jumps (corresponding to the first arrival in the queue).

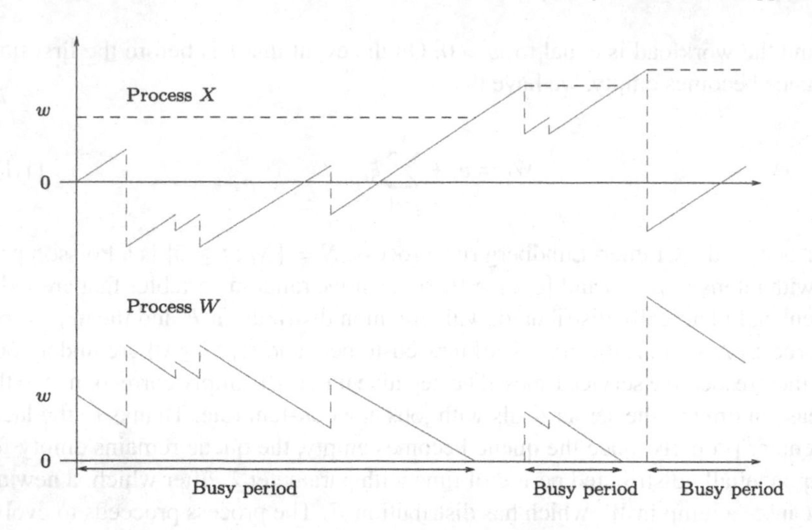

Fig. 1.7 Sample paths of X and W.

There are a number of fundamental points of interest concerning both local and global behavioural properties of the workload of the $M/G/1$ queue. Take, for example, the time it takes before the queue first empties, namely τ_w^+. It is clear from a simple analysis of the paths of X and W that τ_w^+ is finite with probability one, if the underlying process X drifts to infinity with probability one. With the help of the Strong Law of Large Numbers, it is easy to deduce that this happens when $\lambda\mu < 1$. Another common situation of interest in this model corresponds to the case that the server is only capable of dealing with a maximum workload of z units. The first time the workload exceeds the buffer level z,

$$\sigma_z := \inf\{t > 0 : W_t > z\},$$

therefore becomes relevant. In particular, one is interested in the probability of $\{\sigma_z < \tau_w^+\}$, which corresponds to the event that the workload exceeds the buffer level before the server can complete a busy period.

The following two theorems give some classical results concerning the idle time of the $M/G/1$ queue and the stationary distribution of the workload. Roughly speaking, they say that when there is heavy traffic ($\lambda\mu > 1$), eventually the queue never becomes empty, the workload grows to infinity and the total time that the queue remains empty is finite with a particular distribution. Further, when there is light traffic ($\lambda\mu < 1$), the queue repeatedly becomes empty and the total idle time grows to infinity, whilst the workload process converges in distribution. At the critical value $\lambda\mu = 1$, the workload grows to arbitrary large values but, nonetheless, the queue repeatedly becomes empty and the total idle time grows to infinity. Ultimately, all these properties are reinterpretations of the long-term behaviour of a special class of reflected Lévy processes.

Theorem 1.11 *Suppose that* $W = \{W_t : t \geq 0\}$ *is the workload of an* $M/G/1$ *queue with arrival rate* λ *and service distribution* F, *having mean* μ. *Define the total idle time*

$$I = \int_0^\infty \mathbf{1}_{(W_t=0)} \mathrm{d}t.$$

(i) *Suppose that* $\lambda\mu > 1$. *Let*

$$\psi(\theta) = \theta - \lambda \int_{(0,\infty)} (1 - \mathrm{e}^{-\theta x}) F(\mathrm{d}x), \quad \theta \geq 0,$$

and define θ^* *to be the largest root of the equation* $\psi(\theta) = 0$. *Then*[6]

$$\mathbb{P}(I \in \mathrm{d}x \,|\, W_0 = w) = (1 - \mathrm{e}^{-\theta^* w})\delta_0(\mathrm{d}x) + \theta^* \mathrm{e}^{-\theta^*(w+x)}\mathrm{d}x.$$

(ii) *If* $\lambda\mu \leq 1$ *then* I *is infinite with probability one.*

Note that the function ψ, given above, fulfils the relation $\psi(\theta) = \log \mathbb{E}(\mathrm{e}^{\theta X_1})$, for $\theta \geq 0$, and is called the *Laplace exponent* of the underlying Lévy process which drives the process W. It is easy to check, by differentiating it twice, that ψ is a strictly convex function. Moreover, it is zero at the origin and tends to infinity at infinity. Furthermore, under the assumption $\lambda\mu > 1$, $\psi'(0+) < 0$ and hence θ^* exists, is finite and is in fact the only solution to $\psi(\theta) = 0$, other than $\theta = 0$, in $[0, \infty)$.

Theorem 1.12 *Let* $W = \{W_t : t \geq 0\}$ *be the same as in Theorem* 1.11.

(i) *Suppose that* $\lambda\mu < 1$. *Then for all* $w \geq 0$ *the workload process has a stationary distribution,*

$$\lim_{t\uparrow\infty} \mathbb{P}(W_t \leq x \,|\, W_0 = w) = (1 - \rho) \sum_{k=0}^\infty \rho^k \eta^{*k}(x),$$

where

$$\eta(x) = \frac{1}{\mu} \int_0^x [1 - F(y)] \mathrm{d}y \quad and \quad \rho = \lambda\mu.$$

(ii) *If* $\lambda\mu \geq 1$ *then* $\limsup_{t\uparrow\infty} W_t = \infty$ *with probability one.*

---

[6] Following standard notation, the measure δ_0 is the *Dirac measure*, which assigns a unit atom to the point 0.

Some of the conclusions in the above two theorems can already be obtained with basic knowledge of compound Poisson processes. Theorem 1.11 is proved in Exercise 1.9 and gives some feeling for the fluctuation theory that will be touched upon later on in this text. The remarkable similarity between part (i) of Theorem 1.12 and the Pollaczek–Khintchine formula is of course no coincidence. Indeed, the fundamental principles that are responsible for these two results are embedded within a larger fluctuation theory for general Lévy processes. We will revisit Theorems 1.11 and 1.12 later, but for more general versions of the workload process of the $M/G/1$ queue, known as general storage models. Such generalisations involve working with a class of Lévy process that have no positive jumps (that is $\Pi(0,\infty)=0$) and defining, as before, $W_t=(w\vee\overline{X}_t)-X_t$. When there are an infinite number of jumps in each finite time interval, this process may be thought of as modelling a processor that deals with an arbitrarily large number of small jobs and occasional large jobs. The precise interpretation of such a generalised $M/G/1$ workload process and issues concerning the distribution of the busy period, the stationary distribution of the workload, time to buffer overflow and other related quantities, will be dealt with later on in Chaps. 2, 4 and 8.

1.3.3 Optimal Stopping Problems

A fundamental class of problems motivated by applications in physics, optimal control, sequential testing and economics (to name but a few) concerns optimal stopping problems of the form: Find $v(x)$ and a stopping time, τ^*, belonging to a specified family of stopping times, $\mathcal{T}$, such that

$$v(x)=\sup_{\tau\in\mathcal{T}}\mathbb{E}_x\big(\mathrm{e}^{-q\tau}G(X_\tau)\big)=\mathbb{E}_x\big(\mathrm{e}^{-q\tau^*}G(X_{\tau^*})\big), \tag{1.18}$$

for all $x\in\mathbb{R}$. Here, $X=\{X_t:t\geq 0\}$ is an $\mathbb{R}$-valued Markov process with probabilities $\{\mathbb{P}_x:x\in\mathbb{R}\}$ (with the usual understanding that $\mathbb{P}_x$ is the law of X given that $X_0=x$), $q\geq 0$ and $G:\mathbb{R}\to[0,\infty)$ is a function suitable to the application at hand. The optimal stopping problem (1.18) is not the most general class of such problems that one may consider but will suffice for the discussion at hand.

In many cases it turns out that the optimal strategy takes the form

$$\tau^*=\inf\big\{t>0:(t,X_t)\in D\big\},$$

where $D\subset[0,\infty)\times\mathbb{R}$ is a domain in time-space called the *stopping region*. Further, there are many examples within this class for which $D=[0,\infty)\times I$ where I is an interval or the complement of an interval. In other words an optimal strategy is the first hitting time of X into I,

$$\tau^*=\inf\{t>0:X_t\in I\}. \tag{1.19}$$

A classic example of an optimal stopping problem in the form (1.18), for which the solution agrees with (1.19), is the following, taken from McKean (1965). Find

$$v(x) = \sup_{\tau \in \mathcal{T}} \mathbb{E}_x\left(\mathrm{e}^{-q\tau}\left(K - \mathrm{e}^{X_\tau}\right)^+\right), \quad x \in \mathbb{R}, \tag{1.20}$$

where now $q > 0$, $\mathcal{T}$ is the family of stopping times with respect to the filtration $\mathcal{F}_t := \sigma(X_s : s \le t)$ and X is a linear Brownian motion, $X_t = \sigma B_t + \gamma t$, $t \ge 0$ (see Sect. 1.2.3). Note that we use here the standard notation $y^+ = y \vee 0$. This particular example, when seen in the right context, models the optimal time to sell a risky asset for a fixed value K when the value of the asset's dynamics are those of an exponential linear Brownian motion. Optimality in this case is determined via the expected discounted gain at the selling time. On account of the underlying source of randomness being Brownian motion and the optimal strategy taking the simple form (1.19), the solution to (1.20) turns out to be explicitly computable as follows.

Theorem 1.13 *The solution* (v, τ^*) *to* (1.20) *can be represented by*

$$\tau^* = \inf\{t > 0 : X_t < x^*\},$$

where

$$\mathrm{e}^{x^*} = K\left(\frac{\Phi(q)}{1 + \Phi(q)}\right),$$

$\Phi(q) = (\sqrt{\gamma^2 + 2\sigma^2 q} + \gamma)/\sigma^2$ *and*

$$v(x) = \begin{cases} (K - \mathrm{e}^x) & \text{if } x < x^* \\ (K - \mathrm{e}^{x^*})\mathrm{e}^{-\Phi(q)(x - x^*)} & \text{if } x \ge x^*. \end{cases}$$

The solution to this problem reflects the intuition that the optimal time to stop should be at a time when X is as negative as possible, taking into consideration that waiting too long to stop incurs an exponentially weighted penalty. Note that, in $(-\infty, x^*)$, the value function $v(x)$ is equal to the *gain function* $(K - \mathrm{e}^x)^+$ as the optimal strategy τ^* dictates that one should stop immediately here. A particular curiosity of the solution to (1.20) is the fact that at x^*, the value function v joins smoothly to the gain function. In other words,

$$v'(x^*-) = -\mathrm{e}^{x^*} = v'(x^*+).$$

A natural question, in light of the above optimal stopping problem, is whether one can characterise the solution to (1.20) when X is replaced by a general Lévy process. Indeed, if the same strategy of first passage below a specified level is still optimal, one is then confronted with needing information about the distribution of the overshoot of a Lévy process when first crossing below a barrier in order to compute the function v. This is also of interest if one would like to address the question as to whether the phenomenon of smooth fit is still to be found in the general Lévy process setting.

Later in Chap. 11, we give a brief introduction to some general principles appearing in the theory of optimal stopping and apply them to a handful of examples, where the underlying source of randomness is provided by a Lévy process. The first of these examples is the generalisation of (1.20), as mentioned above. All of the examples presented in Chap. 11 can be solved (semi-)explicitly thanks to a degree of simplicity in the optimal strategy, such as (1.19), coupled with knowledge of fluctuation theory of Lévy processes. In addition, through these examples, we will attempt to give some insight into how and when smooth pasting occurs as a consequence of a subtle type of path behaviour of the underlying Lévy process.

1.3.4 Continuous-State Branching Processes

Originating, in part, from the concerns of the Victorian British upper classes that aristocratic surnames were becoming extinct, the theory of branching processes now forms a cornerstone of classical applied probability. Some of the earliest work on branching processes dates back to Watson and Galton (1874). However, approximately 100 years later, it was discovered by Heyde and Seneta (1977) that the lesser known work of I.J. Bienaymé, dated around 1845, contained many aspects of the later-dated work of Galton and Watson. The *Bienaymé–Galton–Watson* process, as it is now known, is a discrete-time Markov chain with state space $\{0, 1, 2, \ldots\}$, described by the sequence $\{Z_n : n = 0, 1, 2, \ldots\}$, satisfying the recursion $Z_0 > 0$ and

$$Z_n = \sum_{i=1}^{Z_{n-1}} \xi_i^{(n)},$$

for $n = 1, 2, \ldots$, where $\{\xi_i^{(n)} : i = 1, 2, \ldots\}$ are independent and identically distributed on $\{0, 1, 2, \ldots\}$. We use the usual notation $\sum_{i=1}^{0}$ to represent the empty sum. The basic idea behind this model is that Z_n is the population count in the n-th generation and from an initial population Z_0 (which may be randomly distributed) individuals reproduce asexually and independently, with the same distribution of numbers of offspring. These reproductive properties are referred to as the *branching property*. Note that, as soon as $Z_n = 0$ it follows from the given construction that, for all $k = 1, 2, \ldots$, $Z_{n+k} = 0$. A particular consequence of the branching property is that, if $Z_0 = a + b$, then Z_n is equal in distribution to $Z_n^{(1)} + Z_n^{(2)}$, where $Z_n^{(1)}$ and $Z_n^{(2)}$ are independent with the same distribution as an n-th generation Bienaymé–Galton–Watson process initiated from population sizes a and b, respectively.

A mild modification of the Bienaymé–Galton–Watson process is to set it into continuous time by assigning life lengths to each individual, which are independent and exponentially distributed with parameter $\lambda > 0$. Individuals reproduce at their moment of death, in the same way as described previously for the Bienaymé–Galton–Watson process. If $Y = \{Y_t : t \geq 0\}$ is the $\{0, 1, 2, \ldots\}$-valued process describing the population size, then it is straightforward to see that the lack-of-memory property of the exponential distribution implies that, for all $0 \leq s \leq t$,

$$Y_t = \sum_{i=1}^{Y_s} Y_{t-s}^{(i)},$$

where, given $\{Y_u : u \leq s\}$, the variables $\{Y_{t-s}^{(i)} : i = 1, \dots, Y_s\}$ are independent, with the same distribution as Y_{t-s} conditional on $Y_0 = 1$. In that case, we may talk of Y as a continuous-time Markov chain on $\{0, 1, 2, \dots\}$, with probabilities, say, $\{P_y : y = 0, 1, 2, \dots\}$, where P_y is the law of Y under the assumption that $Y_0 = y$. As before, the state 0 is absorbing in the sense that, if $Y_t = 0$, then $Y_{t+u} = 0$ for all $u > 0$. The process Y is called the *continuous-time Markov branching process*. The branching property for Y may now be formulated as follows.

Definition 1.14 (Branching property) For any $t \geq 0$ and y_1, y_2 in the state space of $Y = \{Y_t : t \geq 0\}$, the random variable Y_t under $P_{y_1+y_2}$ is equal in law to the independent sum $Y_t^{(1)} + Y_t^{(2)}$, where the distribution of $Y_t^{(i)}$ is equal to that of Y_t under P_{y_i}, for $i = 1, 2$.

So far there appears to be little connection with Lévy processes. However, a remarkable time transformation shows that the path of Y is intimately linked to the path of a compound Poisson process whose jump distribution is supported in $\{-1, 0, 1, 2, \dots\}$ and which is stopped at the first instant that it hits zero. To explain this in more detail, let us introduce the probabilities $\{\pi_i : i = -1, 0, 1, 2, \dots\}$, where $\pi_i = P(\xi = i + 1)$ and ξ has the same distribution as the typical family size in the Bienaymé–Galton–Watson process. To avoid complications, let us assume that $\pi_0 = 0$ so that a transition in the state of Y always occurs when an individual dies. When jumps of Y occur, they are independent and always distributed according to $\{\pi_i : i = -1, 0, 1, \dots\}$. The idea now is to adjust time accordingly with the evolution of Y in such a way that these jumps are spaced out with inter-arrival times that are independent and exponentially distributed. Crucial to the following exposition is the simple and well-known fact that the minimum of $n \in \{1, 2, \dots\}$ independent and exponentially distributed random variables, with common parameter λ, is exponentially distributed with parameter λn. Further, if $\mathbf{e}_\alpha$ is exponentially distributed with parameter $\alpha > 0$, then for $\beta > 0$, $\beta \mathbf{e}_\alpha$ is equal in distribution to $\mathbf{e}_{\alpha/\beta}$.

Write, for $t \geq 0$,

$$J_t = \int_0^t Y_u \mathrm{d}u,$$

set

$$\varphi_t = \inf\{s \geq 0 : J_s > t\},$$

with the usual convention that $\inf \emptyset = \infty$, and define

$$X_t = Y_{\varphi_t}, \tag{1.21}$$

with the understanding that when $\varphi_t = \infty$, we put $X_t = 0$. Observe that, when $Y_0 = y \in \{1, 2, \ldots\}$, the first jump of Y occurs at a time, say T_1, which is the minimum of y independent exponential random variables, each with parameter $\lambda > 0$ (and hence T_1 is exponentially distributed with parameter λy). Moreover, the size of the jump is distributed according to $\{\pi_i : i = -1, 0, 1, 2, \ldots\}$. Note that $J_{T_1} = yT_1$ is the first time that the process $X = \{X_t : t \geq 0\}$ jumps. This time is exponentially distributed with parameter λ. The jump at this time is independent of the historical evolution to that point in time and distributed according to $\{\pi_i : i = -1, 0, 1, 2, \ldots\}$.

Given the information $\mathcal{G}_1 = \sigma(Y_t : t \leq T_1)$, the lack-of-memory property implies that the continuation, $\{Y_{T_1+t} : t \geq 0\}$, has the same law as Y under P_y, with $y = Y_{T_1}$. Hence, if T_2 is the time of the second jump of Y, then conditional on $\mathcal{G}_1$, we have that $T_2 - T_1$ is exponentially distributed with parameter λY_{T_1} and $J_{T_2} - J_{T_1} = Y_{T_1}(T_2 - T_1)$, which is again exponentially distributed with parameter λ and further, is independent of $\mathcal{G}_1$. Note that J_{T_2} is the time of the second jump of X and the size of the second jump is again independent and distributed according to $\{\pi_i : i = -1, 0, 1, \ldots\}$. Iterating in this way it becomes clear that X is nothing more than a compound Poisson process with arrival rate λ and jump distribution

$$F(\mathrm{d}x) = \sum_{i=-1}^{\infty} \pi_i \delta_i(\mathrm{d}x), \quad x \in \mathbb{R}, \tag{1.22}$$

stopped on first hitting the origin.

A converse to this construction is also possible. Suppose now that $X = \{X_t : t \geq 0\}$ is a compound Poisson process with arrival rate $\lambda > 0$ and jump distribution $F(\mathrm{d}x) = \sum_{i=-1}^{\infty} \pi_i \delta_i(\mathrm{d}x)$, $x \in \mathbb{R}$. Write

$$I_t = \int_0^t X_u^{-1} \mathrm{d}u$$

and set

$$\theta_t = \inf\{s \geq 0 : I_s > t\}, \tag{1.23}$$

again with the understanding that $\inf \emptyset = \infty$, Define

$$Y_t = X_{\theta_t \wedge \tau^{\{0\}}},$$

where $\tau^{\{0\}} = \inf\{t > 0 : X_t = 0\}$. By analysing the behaviour of $Y = \{Y_t : t \geq 0\}$ at the jump times of X in a similar way to above, one readily shows that the process Y is a continuous-time Markov branching process. The details are left as an exercise to the reader.

The relationship between compound Poisson processes and continuous-time Markov branching processes, as described above, turns out to hold in a much more general setting. In the work of Lamperti (1967a, 1967b), it is shown that there exists a correspondence between a class of branching processes, called continuous-state branching processes, and Lévy processes with no negative jumps ($\Pi(-\infty, 0) = 0$).

In brief, a continuous-state branching process is a $[0, \infty)$-valued Markov process having paths that are right-continuous with left limits and probabilities $\{P_x : x > 0\}$ that satisfy the branching property in Definition 1.14. Note in particular that, now, the quantities y_1 and y_2 may be chosen from the non-negative real numbers. Lamperti's characterisation of continuous-state branching processes states that they can be identified as time-changed Lévy processes with no negative jumps precisely via the transformations given in (1.21), with an inverse transformation analogous to (1.23). We explore this relationship in more detail in Chap. 12 by looking at issues such as explosion, extinction and conditioning on survival.

Exercises

1.1 Prove that, in order to check for stationary and independent increments of the process $\{X_t : t \geq 0\}$, it suffices to check that, for all $n \in \mathbb{N}$ and $0 \leq s_1 \leq t_1 \leq \cdots \leq s_n \leq t_n < \infty$ and $\theta_1, \ldots, \theta_n \in \mathbb{R}$,

$$\mathbb{E}\left[\prod_{j=1}^{n} \mathrm{e}^{\mathrm{i}\theta_j (X_{t_j} - X_{s_j})}\right] = \prod_{j=1}^{n} \mathbb{E}\left[\mathrm{e}^{\mathrm{i}\theta_j X_{t_j - s_j}}\right].$$

Show, moreover, that the sum of two (or indeed any finite number of) independent Lévy processes is again a Lévy process.

1.2 Suppose that $S = \{S_n : n \geq 0\}$ is any random walk and $\mathbf{\Gamma}_p$ is an independent random variable with a geometric distribution on $\{0, 1, 2, \ldots\}$, with parameter p.

(i) Show that $\mathbf{\Gamma}_p$ is infinitely divisible.
(ii) Show that $S_{\mathbf{\Gamma}_p}$ is infinitely divisible.

1.3 (Proof of Lemma 1.7) In this exercise, we derive the Frullani identity.

(i) Show for any function f, such that f' exists and is continuous and $f(0)$ and $f(\infty)$ are finite, that

$$\int_0^\infty \frac{f(ax) - f(bx)}{x} \mathrm{d}x = \big(f(0) - f(\infty)\big) \log\left(\frac{b}{a}\right),$$

where $b > a > 0$.
(ii) By choosing $f(x) = \mathrm{e}^{-x}$, $a = \alpha > 0$ and $b = \alpha - z$, where $z < 0$, show that

$$\frac{1}{(1 - z/\alpha)^\beta} = \mathrm{e}^{-\int_0^\infty (1 - \mathrm{e}^{zx}) \frac{\beta}{x} \mathrm{e}^{-\alpha x} \mathrm{d}x} \tag{1.24}$$

and hence, by analytic extension, show that the above identity is still valid for all $z \in \mathbb{C}$ such that $\Re z \leq 0$.

1.4 Establishing formulae (1.9) and (1.10) from the Lévy measure given in (1.11) is the result of a series of technical manipulations of special integrals. In this exercise, we work through them. In the following text, we will use the gamma function $\Gamma(z)$, defined by

$$\Gamma(z) = \int_0^\infty t^{z-1} \mathrm{e}^{-t} \mathrm{d}t,$$

for $z > 0$. Note the gamma function can also be analytically extended so that it is also defined on $\mathbb{R}\backslash\{0, -1, -2, \ldots\}$ (see Lebedev 1972). Whilst the specific definition of the gamma function for negative numbers will not play an important role in this exercise, the following two facts, which can be derived from it, will. For $z \in \mathbb{R}\backslash\{0, -1, -2, \ldots\}$ the gamma function observes the recursion $\Gamma(1+z) = z\Gamma(z)$ and $\Gamma(1/2) = \sqrt{\pi}$.

(i) Suppose that $0 < \alpha < 1$. Prove that for $u > 0$,

$$\int_0^\infty (\mathrm{e}^{-ur} - 1) r^{-\alpha-1} \mathrm{d}r = \Gamma(-\alpha) u^\alpha$$

and show that the same equality is valid when $-u$ is replaced by any complex number $w \neq 0$ with $\Re w \leq 0$. Conclude, by considering $w = i$, that

$$\int_0^\infty (1 - \mathrm{e}^{\mathrm{i}r}) r^{-\alpha-1} \mathrm{d}r = -\Gamma(-\alpha) \mathrm{e}^{-\mathrm{i}\pi\alpha/2} \tag{1.25}$$

and similarly for the complex conjugate of both sides of (1.25). Deduce (1.9) by considering the integral

$$\int_0^\infty (1 - \mathrm{e}^{\mathrm{i}\xi\theta r}) r^{-\alpha-1} \mathrm{d}r$$

for $\xi = \pm 1$ and $\theta \in \mathbb{R}$. Note that you will have to take $a = \eta - \int_{\mathbb{R}} x \mathbf{1}_{(|x|<1)} \times \Pi(\mathrm{d}x)$, which you should check is finite.

(ii) Now suppose that $\alpha = 1$. First prove that

$$\int_{|x|<1} \mathrm{e}^{\mathrm{i}\theta x} (1 - |x|) \mathrm{d}x = 2\left(\frac{1 - \cos\theta}{\theta^2}\right),$$

for $\theta \in \mathbb{R}$. Hence by, Fourier inversion, show that

$$\int_0^\infty \frac{1 - \cos r}{r^2} \mathrm{d}r = \frac{\pi}{2}.$$

Use this identity to show that for $z > 0$,

$$\int_0^\infty (1 - \mathrm{e}^{\mathrm{i}rz} + \mathrm{i}zr\mathbf{1}_{(r<1)}) \frac{1}{r^2} \mathrm{d}r = \frac{\pi}{2} z + \mathrm{i}z \log z - \mathrm{i}kz,$$

for some constant $k \in \mathbb{R}$. By considering the complex conjugate of the above integral, establish the expression in (1.10). Note that you will need a different choice of a to part (i).

(iii) Now suppose that $1 < \alpha < 2$. Integrate (1.25) by parts to get

$$\int_0^\infty (e^{ir} - 1 - ir) r^{-\alpha-1} dr = \Gamma(-\alpha) e^{-i\pi\alpha/2}.$$

Deduce the identity (1.9) in a similar manner to the proof of (i) and (ii).

1.5 For any $\theta \in \mathbb{R}$, prove that

$$\exp\{i\theta X_t + t\Psi(\theta)\}, \quad t \geq 0,$$

is a martingale where $\{X_t : t \geq 0\}$ is a Lévy process with characteristic exponent Ψ.

1.6 In this exercise, we will work out in detail some features of the inverse Gaussian process discussed earlier on in this chapter. Recall that $\tau = \{\tau_s : s \geq 0\}$ is a non-decreasing Lévy process defined by $\tau_s = \inf\{t \geq 0 : B_t + bt > s\}$, $s \geq 0$, where $B = \{B_t : t \geq 0\}$ is a standard Brownian motion and $b > 0$.

(i) Argue along the lines of Exercise 1.5 to show that, for each $\lambda > 0$,

$$e^{\lambda B_t - \frac{1}{2}\lambda^2 t}, \quad t \geq 0,$$

is a martingale. Use Doob's Optional Sampling Theorem to obtain

$$\mathbb{E}\left(e^{-(\frac{1}{2}\lambda^2 + b\lambda)\tau_s}\right) = e^{-\lambda s}.$$

Use analytic extension to deduce further that τ_s has characteristic exponent

$$\Psi_s(\theta) = s\left(\sqrt{-2i\theta + b^2} - b\right),$$

for all $\theta \in \mathbb{R}$.

(ii) Defining the measure $\Pi(dx) = (2\pi x^3)^{-1/2} e^{-xb^2/2} dx$ on $x > 0$, check, using (1.25) from Exercise 1.4, that

$$\int_0^\infty (1 - e^{i\theta x}) \Pi(dx) = \Psi(\theta),$$

for all $\theta \in \mathbb{R}$. Confirm that the triple (a, σ, Π), appearing in the Lévy–Khintchine formula, is thus $\sigma = 0$, Π as above and $a = -2sb^{-1} \int_0^b (2\pi)^{-1/2} \times e^{-y^2/2} dy$.

(iii) Taking

$$\mu_s(dx) = \frac{s}{\sqrt{2\pi x^3}} e^{sb} e^{-\frac{1}{2}(s^2 x^{-1} + b^2 x)} dx, \quad x > 0,$$

show that

$$\int_0^\infty e^{-\lambda x}\mu_s(dx) = e^{bs - s\sqrt{b^2+2\lambda}}\int_0^\infty \frac{s}{\sqrt{2\pi x^3}} e^{-\frac{1}{2}(\frac{s}{\sqrt{x}} - \sqrt{(b^2+2\lambda)x})^2} dx$$

$$= e^{bs - s\sqrt{b^2+2\lambda}}\int_0^\infty \sqrt{\frac{2\lambda+b^2}{2\pi u}} e^{-\frac{1}{2}(\frac{s}{\sqrt{u}} - \sqrt{(b^2+2\lambda)u})^2} du.$$

Hence, by adding the last two integrals together deduce that

$$\int_0^\infty e^{-\lambda x}\mu_s(dx) = e^{-s(\sqrt{b^2+2\lambda}-b)},$$

thereby confirming both that $\mu_s(dx)$ is a probability distribution on $(0,\infty)$, and that it is the probability distribution of τ_s.

1.7 Show that for a simple Brownian motion $B = \{B_t : t > 0\}$ the first passage process $\tau = \{\tau_s : s > 0\}$, where $\tau_s = \inf\{t \geq 0 : B_t \geq s\}$, is a stable process with parameters $\alpha = 1/2$ and $\beta = 1$.

1.8 (Proof of Theorem 1.9) As we shall see in this exercise, the proof of Theorem 1.9 follows from the proof of a more general result given by the conclusion of parts (i)–(iv) below for random walks.

(i) Suppose that $S = \{S_n : n \geq 0\}$ is a random walk with $S_0 = 0$ and jump distribution Q on $\mathbb{R}$. By considering the variables $S^*_k := S_n - S_{n-k}$ for $k = 0, 1, \ldots, n$ and noting that the joint distributions of $(S_0, \ldots, S_n)$ and $(S^*_0, \ldots, S^*_n)$ are identical, show that for all $y > 0$ and $n \geq 1$,

$$P(S_n \in dy \text{ and } S_n > S_j \text{ for } j = 0, \ldots, n-1)$$
$$= P(S_n \in dy \text{ and } S_j > 0 \text{ for } j = 1, \ldots, n).$$

Hint: it may be helpful to draw a diagram of the path of the first n steps of S and to rotate it by 180°.

(ii) Define

$$T_0^- = \inf\{n > 0 : S_n \leq 0\} \quad \text{and} \quad T_0^+ = \inf\{n > 0 : S_n > 0\}.$$

By summing both sides of the equality

$$P(S_1 > 0, \ldots, S_n > 0, S_{n+1} \in dx)$$
$$= \int_{(0,\infty)} P(S_1 > 0, \ldots, S_n > 0, S_n \in dy) Q(dx - y)$$

over n, show that for $x \leq 0$,

$$P(S_{T_0^-} \in dx) = \int_{[0,\infty)} V(dy) Q(dx - y),$$

where, for $y \geq 0$,

$$V(\mathrm{d}y) = \delta_0(\mathrm{d}y) + \sum_{n\geq 1} P(H_n \in \mathrm{d}y)$$

and $H = \{H_n : n \geq 0\}$ is a random walk with $H_0 = 0$ and step distribution given by $P(S_{T_0^+} \in \mathrm{d}z)$, for $z \geq 0$.

(iii) Embedded in the Cramér–Lundberg model is a random walk S whose increments are equal in distribution to that of $c\mathbf{e}_\lambda - \xi_1$, where $\mathbf{e}_\lambda$ is an independent exponential random variable with mean $1/\lambda$. Noting (using obvious notation) that $c\mathbf{e}_\lambda$ has the same distribution as $\mathbf{e}_\beta$, where $\beta = \lambda/c$, show that the step distribution of this random walk satisfies

$$Q(z,\infty) = \left(\int_0^\infty \mathrm{e}^{-\beta u} F(\mathrm{d}u)\right)\mathrm{e}^{-\beta z}, \quad z \geq 0,$$

and

$$Q(-\infty,-z) = E\big(\overline{F}(\mathbf{e}_\beta + z)\big), \quad z > 0,$$

where $\overline{F}(x) = 1 - F(x)$, for all $x \geq 0$, and E is expectation with respect to the random variable $\mathbf{e}_\beta$.

(iv) Since upward jumps are exponentially distributed in this random walk, use the lack-of-memory property to reason that

$$V(\mathrm{d}y) = \delta_0(\mathrm{d}y) + \beta \mathrm{d}y, \quad y \geq 0.$$

Hence deduce from parts (ii) and (iii) that

$$P(-S_{T_0^-} > z) = E\left(\overline{F}(\mathbf{e}_\beta) + \int_x^\infty \beta\overline{F}(\mathbf{e}_\beta + z)\mathrm{d}z\right)$$

and so, by writing out the above identity with the density of the exponential distribution, show that the conclusions of Theorem 1.9 hold.

1.9 (Proof of Theorem 1.11) Suppose that X is a compound Poisson process of the form

$$X_t = t - \sum_{i=1}^{N_t} \xi_i, \quad t \geq 0,$$

where the process $N = \{N_t : t \geq 0\}$ is a Poisson process with rate $\lambda > 0$ and $\{\xi_i : i \geq 1\}$ positive, independent and identically distributed with common distribution F having finite mean μ.

(i) Show by direct computation that, for all $\theta \geq 0$, $\mathbb{E}(\mathrm{e}^{\theta X_t}) = \mathrm{e}^{\psi(\theta)t}$, where

$$\psi(\theta) = \theta - \lambda \int_{(0,\infty)} \big(1 - \mathrm{e}^{-\theta x}\big) F(\mathrm{d}x).$$

Show that ψ is strictly convex, is equal to zero at the origin and tends to infinity at infinity. Further, show that $\psi(\theta)=0$ has one additional root in $[0,\infty)$ other than $\theta=0$ if and only if $\psi'(0+)<0$.

(ii) Show that $\{\exp\{\theta^* X_{t\wedge\tau_x^+}\}: t\geq 0\}$ is a martingale, where $\tau_x^+=\inf\{t>0: X_t>x\}$, $x>0$ and θ^* is the largest root described in the previous part of the question. Show further that

$$\mathbb{P}(\overline{X}_\infty > x) = \mathrm{e}^{-\theta^* x},$$

for all $x>0$.

(iii) Show that for all $t\geq 0$,

$$\int_0^t \mathbf{1}_{(W_s=0)}\mathrm{d}s = (\overline{X}_t - w)\vee 0,$$

where $W_t = (w\vee \overline{X}_t) - X_t$.

(iv) Deduce that $I := \int_0^\infty \mathbf{1}_{(W_s=0)}\mathrm{d}s = \infty$ if $\lambda\mu\leq 1$.

(v) Assume that $\lambda\mu>1$. Show that

$$\mathbb{P}\big(I\in\mathrm{d}x;\tau_w^+=\infty|W_0=w\big) = \big(1-\mathrm{e}^{-\theta^* w}\big)\delta_0(\mathrm{d}x), \quad x\geq 0.$$

Next use the lack-of-memory property to deduce that

$$\mathbb{P}\big(I\in\mathrm{d}x;\tau_w^+<\infty|W_0=w\big) = \theta^*\mathrm{e}^{-\theta^*(w+x)}\mathrm{d}x.$$

1.10 Here, we solve a considerably simpler optimal stopping problem than (1.20). Suppose, as in the aforementioned problem, that X is a linear Brownian motion with scaling parameter $\sigma>0$ and drift $\gamma\in\mathbb{R}$. Fix $K>0$ and let

$$v(x) = \sup_{a\in\mathbb{R}} \mathbb{E}_x\big(\mathrm{e}^{-q\tau_a^-}\big(K-\mathrm{e}^{X_{\tau_a^-}}\big)^+\big), \tag{1.26}$$

where

$$\tau_a^- = \inf\{t>0: X_t<a\}.$$

(i) Following similar arguments to those in Exercises 1.5 and 1.9, show that $\{\exp\{\theta X_t - \psi(\theta)t\}: t\geq 0\}$ is a martingale, where $\psi(\theta)=\sigma^2\theta^2/2+\gamma\theta$.

(ii) By considering the martingale in part (i) at the stopping time $t\wedge\tau_x^+$ and then letting $t\uparrow\infty$, deduce that

$$\mathbb{E}\big(\mathrm{e}^{-q\tau_x^+}\big) = \mathrm{e}^{-x(\sqrt{\gamma^2+2\sigma^2 q}-\gamma)/\sigma^2}$$

and hence deduce that for $a\geq 0$,

$$\mathbb{E}\big(\mathrm{e}^{-q\tau_{-a}^-}\big) = \mathrm{e}^{-a(\sqrt{\gamma^2+2\sigma^2 q}+\gamma)/\sigma^2}.$$

(iii) Let $v(x,a) = \mathbb{E}_x(\mathrm{e}^{-q\tau^-_{-a}}(K - \exp\{X_{\tau^-_{-a}}\}))$. For each fixed x differentiate $v(x,a)$ in the variable a and show that the solution to (1.26) is the same as the solution given in Theorem 1.13.

1.11 In this exercise, we characterise the Laplace exponent of the continuous-time Markov branching process, Y, described in Sect. 1.3.4.

(i) Show that for $\phi > 0$ and $t \geq 0$ there exists some function $u_t(\phi) > 0$ satisfying

$$E_y\left(\mathrm{e}^{-\phi Y_t}\right) = \mathrm{e}^{-y u_t(\phi)},$$

where $y \in \{0, 1, 2, \ldots\}$.

(ii) Show that for $s, t \geq 0$,

$$u_{t+s}(\phi) = u_s\left(u_t(\phi)\right).$$

(iii) Appealing to the infinitesimal behaviour of the Markov chain Y, show that

$$\frac{\partial u_t(\phi)}{\partial t} = \psi\left(u_t(\phi)\right)$$

and $u_0(\phi) = \phi$, where

$$\psi(q) = \lambda \int_{[-1,\infty)} \left(1 - \mathrm{e}^{-qx}\right) F(\mathrm{d}x)$$

and F is given in (1.22).

Chapter 2
The Lévy–Itô Decomposition and Path Structure

The main aim of this chapter is to establish a rigorous understanding of the structure of the paths of Lévy processes. The way we shall do this is to prove the assertion in Theorem 1.6 that, given any characteristic exponent, Ψ, belonging to an infinitely divisible distribution, there exists a Lévy process with the same characteristic exponent. This will be done by establishing the so-called Lévy–Itô decomposition, which describes the structure of a general Lévy process in terms of three independent auxiliary Lévy processes, each with a different type of path behaviour. In doing so it will be necessary to digress temporarily into the theory of Poisson random measures and associated square-integrable martingales. Understanding the Lévy–Itô decomposition will allow us to distinguish a number of important, but nonetheless general, subclasses of Lévy processes according to their path type. The chapter is concluded with a discussion of the interpretation of the Lévy–Itô decomposition in the context of some of the applied probability models mentioned in Chap. 1.

2.1 The Lévy–Itô Decomposition

According to Theorem 1.3, any characteristic exponent Ψ belonging to an infinitely divisible distribution can be written, after some simple reorganisation, in the form

$$\begin{aligned}\Psi(\theta) &= \left\{ \mathrm{i}a\theta + \frac{1}{2}\sigma^2\theta^2 \right\} \\ &\quad + \left\{ \Pi(\mathbb{R}\backslash(-1,1)) \int_{|x|\geq 1} \left(1 - \mathrm{e}^{\mathrm{i}\theta x}\right) \frac{\Pi(\mathrm{d}x)}{\Pi(\mathbb{R}\backslash(-1,1))} \right\} \\ &\quad + \left\{ \int_{0<|x|<1} \left(1 - \mathrm{e}^{\mathrm{i}\theta x} + \mathrm{i}\theta x\right) \Pi(\mathrm{d}x) \right\}, \end{aligned} \tag{2.1}$$

for all $\theta \in \mathbb{R}$, where $a \in \mathbb{R}$, $\sigma \in \mathbb{R}$ and Π is a measure on $\mathbb{R}\backslash\{0\}$ satisfying $\int_{\mathbb{R}} (1 \wedge x^2)\Pi(\mathrm{d}x) < \infty$. Note that this condition on Π implies that $\Pi(A) < \infty$ for

A.E. Kyprianou, *Fluctuations of Lévy Processes with Applications*, Universitext,
DOI 10.1007/978-3-642-37632-0_2, © Springer-Verlag Berlin Heidelberg 2014

all Borel A such that 0 is in the interior of A^c and, in particular, that $\Pi(\mathbb{R}\backslash(-1,1)) \in [0,\infty)$. In the case that $\Pi(\mathbb{R}\backslash(-1,1)) = 0$, one should think of the second set of curly brackets in (2.1) as absent. Call the contents of the three sets of curly brackets in (2.1) $\Psi^{(1)}(\theta)$, $\Psi^{(2)}(\theta)$ and $\Psi^{(3)}(\theta)$. The essence of the Lévy–Itô decomposition revolves around showing that $\Psi^{(1)}(\theta)$, $\Psi^{(2)}(\theta)$ and $\Psi^{(3)}(\theta)$ correspond to the characteristic exponents of three different types of Lévy processes. Therefore, Ψ may be considered as the characteristic exponent of the independent sum of these three Lévy processes, which is again a Lévy process (cf. Exercise 1.1). Indeed, as we have already seen in Chap. 1, $\Psi^{(1)}$ and $\Psi^{(2)}$ correspond, respectively, to a linear Brownian motion, say, $X^{(1)} = \{X_t^{(1)} : t \geq 0\}$, where

$$X_t^{(1)} = \sigma B_t - at, \quad t \geq 0, \tag{2.2}$$

and an independent compound Poisson process, say $X^{(2)} = \{X_t^{(2)} : t \geq 0\}$, where,

$$X_t^{(2)} = \sum_{i=1}^{N_t} \xi_i, \quad t \geq 0, \tag{2.3}$$

$\{N_t : t \geq 0\}$ is a Poisson process with rate $\Pi(\mathbb{R}\backslash(-1,1))$ and $\{\xi_i : i \geq 1\}$ are independent and identically distributed with common distribution $\Pi(\mathrm{d}x)/\Pi(\mathbb{R}\backslash(-1,1))$ concentrated on $\{x : |x| \geq 1\}$ (unless $\Pi(\mathbb{R}\backslash(-1,1)) = 0$ in which case $X^{(2)}$ is the process which is identically zero).

The proof of existence of a Lévy process with characteristic exponent given by (2.1) thus boils down to showing the existence of a Lévy process, $X^{(3)}$, whose characteristic exponent is given by $\Psi^{(3)}$. Note that

$$\begin{aligned} &\int_{0<|x|<1} \left(1 - \mathrm{e}^{\mathrm{i}\theta x} + \mathrm{i}\theta x\right) \Pi(\mathrm{d}x) \\ &\quad = \sum_{n\geq 0} \left\{ \lambda_n \int_{2^{-(n+1)} \leq |x| < 2^{-n}} \left(1 - \mathrm{e}^{\mathrm{i}\theta x}\right) F_n(\mathrm{d}x) \right. \\ &\qquad \left. + \mathrm{i}\theta\lambda_n \left(\int_{2^{-(n+1)} \leq |x| < 2^{-n}} x F_n(\mathrm{d}x) \right) \right\}, \end{aligned} \tag{2.4}$$

where $\lambda_n = \Pi(\{x : 2^{-(n+1)} \leq |x| < 2^{-n}\})$ and $F_n(\mathrm{d}x) = \Pi(\mathrm{d}x)/\lambda_n$, restricted to $\{x : 2^{-(n+1)} \leq |x| < 2^{-n}\}$ (again with the understanding that the n-th integral is absent if $\lambda_n = 0$). It would appear from (2.4) that the process $X^{(3)}$ consists of the superposition of (at most) a countable number of independent compound Poisson processes with different arrival rates and additional linear drift. To understand the mathematical sense of this superposition, we shall need to establish some facts concerning Poisson random measures and related martingales. This is done in Sects. 2.2 and 2.3. The precise construction of $X^{(3)}$ is given in Sect. 2.5.

The identification of a Lévy process, X, as the independent sum of processes $X^{(1)}$, $X^{(2)}$ and $X^{(3)}$ is attributed to Lévy (1954) and Itô (1942) and is thus known as the *Lévy–Itô decomposition*. Formally speaking, and in a little more detail, we quote the Lévy–Itô decomposition in the form of a theorem.

Theorem 2.1 (Lévy–Itô decomposition) *Given any $a \in \mathbb{R}$, $\sigma \in \mathbb{R}$ and measure Π concentrated on $\mathbb{R}\backslash\{0\}$ satisfying*

$$\int_{\mathbb{R}} (1 \wedge x^2)\Pi(\mathrm{d}x) < \infty,$$

there exists a probability space on which three independent Lévy processes exist, $X^{(1)}$, $X^{(2)}$ and $X^{(3)}$, where $X^{(1)}$ is a linear Brownian motion given by (2.2)*, $X^{(2)}$ is a compound Poisson process given by* (2.3) *and $X^{(3)}$ is a square-integrable martingale with an almost surely countable number of path discontinuities (or jumps) on each finite time interval, which are of magnitude less than unity, and with characteristic exponent given by $\Psi^{(3)}$. Moreover, by taking $X = X^{(1)} + X^{(2)} + X^{(3)}$, the conclusion of Theorem* 1.6 *holds, namely that there exists a probability space on which a Lévy process is defined with characteristic exponent*

$$\Psi(\theta) = ai\theta + \frac{1}{2}\sigma^2\theta^2 + \int_{\mathbb{R}} \big(1 - \mathrm{e}^{\mathrm{i}\theta x} + \mathrm{i}\theta x \mathbf{1}_{(|x|<1)}\big)\Pi(\mathrm{d}x), \tag{2.5}$$

for $\theta \in \mathbb{R}$.

2.2 Poisson Random Measures

Poisson random measures turn out to be the right mathematical mechanism to describe the jump structure embedded in any Lévy process. Before engaging in an abstract study of Poisson random measures, we give a rough idea of how they are related to the jump structure of Lévy processes by considering the less complicated case of a compound Poisson process.

Suppose that $X = \{X_t : t \geq 0\}$ is a compound Poisson process with a drift taking the form

$$X_t = \delta t + \sum_{i=1}^{N_t} \xi_i, \quad t \geq 0,$$

where $\delta \in \mathbb{R}$ and, as usual, $\{\xi_i : i \geq 1\}$ are independent and identically distributed random variables with common distribution function F. Further, let $\{T_i : i \geq 1\}$ be the times of arrival of the Poisson process $N = \{N_t : t \geq 0\}$ with rate $\lambda > 0$. See Fig. 2.1.

Suppose now that we pick any set in $A \in \mathcal{B}[0, \infty) \times \mathcal{B}(\mathbb{R}\backslash\{0\})$. Define

$$N(A) = \#\{i \geq 1 : (T_i, \xi_i) \in A\} = \sum_{i=1}^{\infty} \mathbf{1}_{((T_i,\xi_i)\in A)}. \tag{2.6}$$

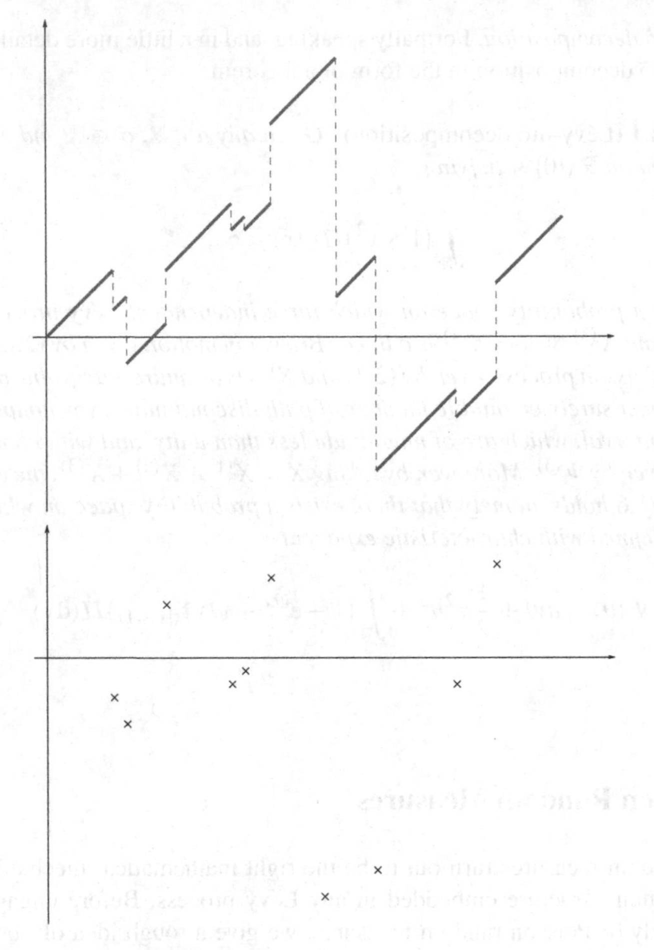

Fig. 2.1 The initial period of a sample path of a compound Poisson process with drift $\{X_t : t \geq 0\}$ and the field of points it generates.

Clearly, since X experiences an almost surely finite number of jumps over a finite period of time, it follows that $N(A) < \infty$ almost surely when $t \geq 0$ and $A \subseteq \mathcal{B}[0,t) \times \mathcal{B}(\mathbb{R}\backslash\{0\})$.

Lemma 2.2 *Choose $k \geq 1$. If $A_1, \ldots, A_k$ are disjoint sets in $\mathcal{B}[0,\infty) \times \mathcal{B}(\mathbb{R}\backslash\{0\})$, then $N(A_1), \ldots, N(A_k)$ are mutually independent and Poisson distributed*[1] *with*

[1] We understand a Poisson random variable whose parameter is infinite to be infinite valued with probability 1.

parameters $\lambda_i := \lambda \int_{A_i} \mathrm{d}t \times F(\mathrm{d}x)$, *respectively. Further, for* $\mathbb{P}$*-almost every realisation of* X, $N : \mathcal{B}[0,\infty) \times \mathcal{B}(\mathbb{R}\backslash\{0\}) \to \{0,1,2,\dots\} \cup \{\infty\}$ *is a measure.*[2]

Proof First recall a classic result concerning the Poisson process $\{N_t : t \geq 0\}$. That is, when $t > 0$, the law of $\{T_1, \dots, T_n\}$ conditional on the event $\{N_t = n\}$ is the same as the law of an ordered independent sample of size n from the uniform distribution on $[0,t]$. (See Exercise 2.2.) This, together with the fact that the variables $\{\xi_i : i = 1, \dots, n\}$ are independent and identically distributed with common law F, implies that, conditional on $\{N_t = n\}$, the joint law of the pairs $\{(T_i, \xi_i) : i = 1, \dots, n\}$ is that of n independent bivariate random variables, with common distribution $t^{-1}\mathrm{d}s \times F(\mathrm{d}x)$ on $[0,t] \times (\mathbb{R}\backslash\{0\})$, ordered in time. In particular, for any $A \in \mathcal{B}[0,t] \times \mathcal{B}(\mathbb{R}\backslash\{0\})$, the random variable $N(A)$ conditional on the event $\{N_t = n\}$ is a binomial random variable with probability of success given by $\int_A t^{-1}\mathrm{d}s \times F(\mathrm{d}x)$. A generalisation of this statement for the k-tuple $(N(A_1), \dots, N(A_k))$, where $A_1, \dots, A_k$ are mutually disjoint and chosen from $\mathcal{B}[0,t] \times \mathcal{B}(\mathbb{R}\backslash\{0\})$, is the following. Suppose that $A_0 = \{[0,t] \times \mathbb{R}\}\backslash\{A_1 \cup \cdots \cup A_k\}$, $\sum_{i=1}^k n_i \leq n$, $n_0 = n - \sum_{i=1}^k n_i$ and $\lambda_0 = \int_{A_0} \lambda \mathrm{d}s \times F(\mathrm{d}x) = \lambda t - \lambda_1 - \cdots - \lambda_k$, then $(N(A_1), \dots, N(A_k))$ has the following multinomial law,

$$\mathbb{P}\big(N(A_1) = n_1, \dots, N(A_k) = n_k | N_t = n\big)$$

$$= \frac{n!}{n_0! n_1! \cdots n_k!} \prod_{i=0}^{k} \left(\frac{\lambda_i}{\lambda t}\right)^{n_i}.$$

Summing out the conditioning on N_t, it follows that

$$\mathbb{P}\big(N(A_1) = n_1, \dots, N(A_k) = n_k\big)$$

$$= \sum_{n \geq \sum_{i=1}^k n_i} \mathrm{e}^{-\lambda t} \frac{(\lambda t)^n}{n!} \frac{n!}{n_0! n_1! \cdots n_k!} \prod_{i=0}^{k} \left(\frac{\lambda_i}{\lambda t}\right)^{n_i}$$

$$= \sum_{n \geq \sum_{i=1}^k n_i} \mathrm{e}^{-\lambda_0} \frac{\lambda_0^{(n - \sum_{i=1}^k n_i)}}{(n - \sum_{i=1}^k n_i)!} \left(\prod_{i=1}^{k} \mathrm{e}^{-\lambda_i} \frac{\lambda_i^{n_i}}{n_i!}\right)$$

$$= \prod_{i=1}^{k} \mathrm{e}^{-\lambda_i} \frac{\lambda_i^{n_i}}{n_i!},$$

[2] Specifically, $\mathbb{P}$-almost surely, $N(\emptyset) = 0$ and for disjoint $A_1, A_2, \dots$ in $\mathcal{B}[0,\infty) \times \mathcal{B}(\mathbb{R}\backslash\{0\})$, we have

$$N\Big(\bigcup_{i \geq 1} A_i\Big) = \sum_{i \geq 1} N(A_i).$$

showing that $N(A_1),\ldots,N(A_k)$ are independent and Poisson distributed, as claimed.

To complete the proof for arbitrary disjoint $A_1,\ldots,A_k$, for each $i=1,\ldots,k$, write A_i as a countable union of disjoint sets, each of which belongs to $\mathcal{B}[0,t')\times\mathcal{B}(\mathbb{R}\backslash\{0\})$ for some $t'>0$. Recall that the sum of an independent sequence of Poisson random variables is Poisson distributed with the sum of their rates. If we agree that a Poisson random variable with infinite rate is infinite with probability one (see Exercise 2.1), then the proof is complete.

Finally the fact that N is a measure $\mathbb{P}$-almost surely follows immediately from its definition. □

Lemma 2.2 shows that $N:\mathcal{B}[0,\infty)\times\mathcal{B}(\mathbb{R}\backslash\{0\})\rightarrow\{0,1,\ldots\}\cup\{\infty\}$ fulfils the following definition of a Poisson random measure.

Definition 2.3 (Poisson random measure) Let $(S,\mathcal{S},\eta)$ be an arbitrary sigma-finite measure space and $(\Omega,\mathcal{F},P)$ a probability space. Let $N:\Omega\times\mathcal{S}\rightarrow\{0,1,2,\ldots\}\cup\{\infty\}$ in such a way that the family $\{N(\cdot,A):A\in\mathcal{S}\}$ are random variables defined on $(\Omega,\mathcal{F},P)$. Henceforth, for convenience, we shall suppress the dependency of N on ω. Then N is called a Poisson random measure on $(S,\mathcal{S},\eta)$ (or sometimes a Poisson random measure on S with intensity η) if

(i) for mutually disjoint $A_1,\ldots,A_n$ in $\mathcal{S}$, the variables $N(A_1),\ldots,N(A_n)$ are independent,
(ii) for each $A\in\mathcal{S}$, $N(A)$ is Poisson distributed with parameter $\eta(A)$ (here we allow $0\le\eta(A)\le\infty$),
(iii) $N(\cdot)$ is a measure P-almost surely.

In the second condition, we note that, if $\eta(A)=0$, then it is understood that $N(A)=0$ with probability one and if $\eta(A)=\infty$ then $N(A)$ is infinite with probability one.

In the case of (2.6), we have $S=[0,\infty)\times(\mathbb{R}\backslash\{0\})$ and $\mathrm{d}\eta=\lambda\mathrm{d}t\times\mathrm{d}F$. Note also that, by construction of the compound Poisson process on the probability space $(\Omega,\mathcal{F},\mathbb{P})$, for each $A\in\mathcal{B}[0,\infty)\times\mathcal{B}(\mathbb{R}\backslash\{0\})$, the random variable $\mathbf{1}_{((T_i,\xi_i)\in A)}$ is $\mathcal{F}$-measurable, and hence so is the variable $N(A)$.

We complete this section by proving that a Poisson random measure, as defined above, exists. This is done in Theorem 2.4 below, the proof of which has many similarities to the proof of Lemma 2.2.

Theorem 2.4 *There exists a Poisson random measure $N(\cdot)$ as in Definition* 2.3.

Proof First suppose that S is such that $0<\eta(S)<\infty$. There exists a standard construction of an infinite product space, say $(\Omega,\mathcal{F},P)$, defined on which are the independent random variables

$$\mathrm{N}\quad\text{and}\quad\{\upsilon_1,\upsilon_2,\ldots\},$$

such that N has a Poisson distribution with parameter $\eta(S)$ and each of the variables υ_i has distribution $\eta(\mathrm{d}x)/\eta(S)$ on S. Define for each $A \in \mathcal{S}$,

$$N(A) = \sum_{i=1}^{\mathrm{N}} \mathbf{1}_{(\upsilon_i \in A)}, \tag{2.7}$$

so that $\mathrm{N} = N(S)$. For each $A \in \mathcal{S}$ and $i \geq 1$, the random variables $\mathbf{1}_{(\upsilon_i \in A)}$ and N are $\mathcal{F}$-measurable, hence so are the random variables $N(A)$.

When presented with mutually disjoint sets of $\mathcal{S}$, say $A_1, \ldots, A_k$, a calculation identical to the one given in the proof of Lemma 2.2 shows, again, that

$$P\big(N(A_1) = n_1, \ldots, N(A_k) = n_k\big) = \prod_{i=1}^{k} \mathrm{e}^{-\eta(A_i)} \frac{\eta(A_i)^{n_i}}{n_i!},$$

for non-negative integers $n_1, n_2, \ldots, n_k$. Returning to Definition 2.3, it is now clear from the previous calculation that conditions (i)–(iii) are met by $N(\cdot)$. In particular, similar to the case dealt with in Lemma 2.2, the third condition is automatic as $N(\cdot)$ is a counting measure by definition.

Next, we turn to the case that $(S, \mathcal{S}, \eta)$ is a sigma-finite measure space. The sigma-finite assumption means that there exists a countable disjoint exhaustive sequence of sets $B_1, B_2, \ldots$ in S such that $0 < \eta(B_i) < \infty$ for each $i \geq 1$. Define, for each $i \geq 1$, the measures $\eta_i(\cdot) = \eta(\cdot \cap B_i)$. The first part of this proof shows that, for each $i \geq 1$, there exists some probability space, say $(\Omega_i, \mathcal{F}_i, P_i)$, on which we can define a Poisson random measure, say $N_i(\cdot)$, in $(B_i, \mathcal{S} \cap B_i, \eta_i)$, where $\mathcal{S} \cap B_i = \{A \cap B_i : A \in \mathcal{S}\}$ (the reader should verify easily that $\mathcal{S} \cap B_i$ is indeed a sigma-algebra on B_i). The idea is now to show that

$$N(\cdot) = \sum_{i \geq 1} N_i(\cdot \cap B_i)$$

is a Poisson random measure on S, with intensity η, defined on the product space

$$(\Omega, \mathcal{F}, P) := \prod_{i \geq 1} (\Omega_i, \mathcal{F}_i, P_i).$$

First note, again from its definition, that $N(\cdot)$ is P-almost surely a measure. In particular with the help of Fubini's Theorem, for disjoint $A_1, A_2, \ldots$, we have

$$\begin{aligned} N\Big(\bigcup_{j \geq 1} A_j\Big) &= \sum_{i \geq 1} N_i\Big(\bigcup_{j \geq 1} A_j \cap B_i\Big) = \sum_{i \geq 1} \sum_{j \geq 1} N(A_j \cap B_i) \\ &= \sum_{j \geq 1} \sum_{i \geq 1} N(A_j \cap B_i) \\ &= \sum_{j \geq 1} N(A_j). \end{aligned}$$

Next, for each $i \geq 1$, we have that $N_i(A \cap B_i)$ is Poisson distributed with parameter $\eta_i(A)$; Exercise 2.1 tells us that under P the random variable $N(A)$ is Poisson distributed with parameter $\eta(A)$. The proof is complete once we show that, for disjoint $A_1, \ldots, A_k$ in $\mathcal{S}$, the variables $N(A_1), \ldots, N(A_k)$ are all independent under P. However this is obvious since the double array of variables,

$$\{N_i(A_j \cap B_i) : i = 1, 2, \ldots \text{ and } j = 1, \ldots, k\},$$

is also an independent sequence of variables. □

From the construction of the Poisson random measure, the following two corollaries should be clear.

Corollary 2.5 *Suppose that $N(\cdot)$ is a Poisson random measure on $(S, \mathcal{S}, \eta)$. Then for each $A \in \mathcal{S}$, $N(\cdot \cap A)$ is a Poisson random measure on $(S \cap A, \mathcal{S} \cap A, \eta(\cdot \cap A))$. Further, if $A, B \in \mathcal{S}$ and $A \cap B = \emptyset$, then $N(\cdot \cap A)$ and $N(\cdot \cap B)$ are independent.*

Corollary 2.6 *Suppose that $N(\cdot)$ is a Poisson random measure on $(S, \mathcal{S}, \eta)$, then the support of $N(\cdot)$ is P-almost surely countable. If, in addition, η is a finite measure, then the support is P-almost surely finite.*

Finally, note that, if η is a measure with an atom at, say, the singleton $s \in S$ and $\{s\} \in \mathcal{S}$, then it follows from the definition of $N(\cdot)$ in the proof of Theorem 2.4 that $P(N(\{s\}) \geq 1) > 0$. Conversely, if η has no atoms then $P(N(\{s\}) = 0) = 1$ for all singletons $s \in S$ such that $\{s\} \in \mathcal{S}$. For further discussion on this point, the reader is referred to Kingman (1993).

2.3 Functionals of Poisson Random Measures

Suppose as in Sect. 2.2 that $N(\cdot)$ is a Poisson random measure on the measure space $(S, \mathcal{S}, \eta)$. As $N(\cdot)$ is P-almost surely a measure, classical measure theory now allows us to talk of

$$\int_S f(x) N(\mathrm{d}x) \tag{2.8}$$

as a well-defined $[0, \infty]$-valued random variable, for measurable functions $f : S \to [0, \infty]$. Further, (2.8) is still well defined and $[-\infty, \infty]$ valued for signed measurable f provided at most one of the integrals of $f^+ = f \vee 0$ and $f^- = (-f) \vee 0$ is infinite. Note however, from the construction of the Poisson random measure in the proof of Theorem 2.4, the integral in (2.8) may be interpreted as equal to

$$\sum_{\upsilon \in \Upsilon} f(\upsilon) m_\upsilon,$$

where Υ is the support of $N(\cdot)$, which, from Corollary 2.6, is countable, and m_υ is the multiplicity of points at υ. Recalling the remarks following Corollary 2.6, if η has no atoms then $m_\upsilon = 1$ for all $\upsilon \in \Upsilon$.

We move to the main theorem of this section for which the reader is referred to Sect. 9.8 of Moran (1968), Kingman (1967) and the earlier work of Campbell (1909, 1910).

Theorem 2.7 *Suppose that N is a Poisson random measure on $(S, \mathcal{S}, \eta)$. Let $f : S \to \mathbb{R}$ be a measurable function.*

(i) *Then*

$$X = \int_S f(x) N(\mathrm{d}x)$$

is almost surely absolutely convergent if and only if

$$\int_S \big(1 \wedge |f(x)|\big)\eta(\mathrm{d}x) < \infty. \tag{2.9}$$

(ii) *When condition (2.9) holds, then (with E as expectation with respect to P)*

$$E\big(\mathrm{e}^{\mathrm{i}\beta X}\big) = \exp\left\{-\int_S \big(1 - \mathrm{e}^{\mathrm{i}\beta f(x)}\big)\eta(\mathrm{d}x)\right\} \tag{2.10}$$

for any $\beta \in \mathbb{R}$.

(iii) *Further*

$$E(X) = \int_S f(x)\eta(\mathrm{d}x) \quad \textit{when} \quad \int_S |f(x)|\eta(\mathrm{d}x) < \infty \tag{2.11}$$

and

$$E\big(X^2\big) = \int_S f(x)^2 \eta(\mathrm{d}x) + \left(\int_S f(x)\eta(\mathrm{d}x)\right)^2$$

when

$$\int_S f(x)^2 \eta(\mathrm{d}x) < \infty \quad \textit{and} \quad \int_S |f(x)|\eta(\mathrm{d}x) < \infty. \tag{2.12}$$

Proof (i) We begin by defining simple functions to be those of the form

$$f(x) = \sum_{i=1}^n f_i \mathbf{1}_{A_i}(x),$$

where f_i is constant and $\{A_i : i = 1, \ldots, n\}$ are disjoint sets in $\mathcal{S}$ and further $\eta(A_1 \cup \cdots \cup A_n) < \infty$.

For such functions, we have

$$X = \sum_{i=1}^{n} f_i N(A_i),$$

which is clearly finite with probability one since each $N(A_i)$ has a Poisson distribution with parameter $\eta(A_i) < \infty$. Recall the well-known fact that the moment-generating function of a Poisson distribution with parameter $\lambda > 0$ is $\exp\{-\lambda(1 - \mathrm{e}^{-\theta})\}$, for $\theta \geq 0$. For the same range of θ, we have

$$\begin{aligned} E\left(\mathrm{e}^{-\theta X}\right) &= \prod_{i=1}^{n} E\left(\mathrm{e}^{-\theta f_i N(A_i)}\right) \\ &= \prod_{i=1}^{n} \exp\{-(1-\mathrm{e}^{-\theta f_i})\eta(A_i)\} \\ &= \exp\left\{-\sum_{i=1}^{n}(1-\mathrm{e}^{-\theta f_i})\eta(A_i)\right\}. \end{aligned}$$

Since $1 - \mathrm{e}^{-\theta f(x)} = 0$ on $S\backslash(A_1 \cup \cdots \cup A_n)$, we may thus conclude that

$$E\left(\mathrm{e}^{-\theta X}\right) = \exp\left\{-\int_S (1-\mathrm{e}^{-\theta f(x)})\eta(\mathrm{d}x)\right\}.$$

Next we establish the above equality for a general positive measurable f. For this class of f, there exists a pointwise increasing sequence of positive simple functions, $\{f_n : n \geq 0\}$, such that $\lim_{n\uparrow\infty} f_n = f$, where the limit is also understood in the pointwise sense. Since N is an almost surely sigma-finite measure, we have that

$$\lim_{n\uparrow\infty} \int_S f_n(x)N(\mathrm{d}x) = \int_S f(x)N(\mathrm{d}x) = X$$

almost surely. An application of bounded convergence followed by an application of monotone convergence tells us that, for any $\theta > 0$,

$$\begin{aligned} E\left(\mathrm{e}^{-\theta X}\right) &= E\left(\exp\left\{-\theta\int f(x)N(\mathrm{d}x)\right\}\right) \\ &= \lim_{n\uparrow\infty} E\left(\exp\left\{-\theta\int f_n(x)N(\mathrm{d}x)\right\}\right) \\ &= \lim_{n\uparrow\infty} \exp\left\{-\int_S (1-\mathrm{e}^{-\theta f_n(x)})\eta(\mathrm{d}x)\right\} \\ &= \exp\left\{-\int_S (1-\mathrm{e}^{-\theta f(x)})\eta(\mathrm{d}x)\right\}. \end{aligned} \tag{2.13}$$

Note that the integral on the right-hand side of (2.13) is either infinite, for all $\theta > 0$, or finite, for all $\theta > 0$, accordingly as $X = \infty$ with probability one or $X = \infty$ with probability less than one, respectively. If $\int_S (1 - \mathrm{e}^{-\theta f(x)})\eta(\mathrm{d}x) < \infty$ for all $\theta > 0$ then as, for each $x \in S$, $(1 - \mathrm{e}^{-\theta f(x)}) \le (1 - \mathrm{e}^{-f(x)})$, for all $0 < \theta < 1$, dominated convergence implies that

$$\lim_{\theta \downarrow 0} \int_S \left(1 - \mathrm{e}^{-\theta f(x)}\right)\eta(\mathrm{d}x) = 0,$$

and hence dominated convergence, as $\theta \downarrow 0$, applied again in (2.13) tells us that $P(X = \infty) = 0$.

In conclusion, we have that $X < \infty$ almost surely if and only if $\int_S (1 - \mathrm{e}^{-\theta f(x)})\eta(\mathrm{d}x) < \infty$, for all $\theta > 0$. Moreover, it can be checked (see Exercise 2.3) that this happens if and only if

$$\int_S \left(1 \wedge f(x)\right)\eta(\mathrm{d}x) < \infty.$$

Note that both sides of (2.13) may be analytically continued by replacing θ by $\theta - \mathrm{i}\beta$ for $\beta \in \mathbb{R}$. Then taking limits on both sides as $\theta \downarrow 0$, we deduce (2.10).

Now we shall remove the restriction that f is positive. Henceforth assume, as in the statement of the theorem, that f is a measurable function. We may write $f = f^+ - f^-$ where $f^+ = f \vee 0$ and $f^- = (-f) \vee 0$ are both measurable. The sum X can be written $X_+ - X_-$ where

$$X_+ = \int_S f(x) N_+(\mathrm{d}x) \quad \textit{and} \quad X_- = \int_S f(x) N_-(\mathrm{d}x)$$

and $N_+ = N(\cdot \cap \{x \in S : f(x) \ge 0\})$ and $N_- = N(\cdot \cap \{x \in S : f(x) < 0\})$. From Corollary 2.5, we know that N_+ and N_- are both Poisson random measures with respective intensities $\eta(\cdot \cap \{f \ge 0\})$ and $\eta(\cdot \cap \{f < 0\})$. Further, they are independent and hence the same is true of X_+ and X_-. It is now clear that, almost surely, X converges absolutely if and only if X_+ and X_- are convergent. The analysis of the case when f is positive applied to the sums X_+ and X_- now tells us that absolute convergence of X occurs if and only if

$$\int_S \left(1 \wedge |f(x)|\right)\eta(\mathrm{d}x) < \infty, \tag{2.14}$$

and the proof of (i) is complete.

To complete the proof of (ii), assume that (2.14) holds. Using the independence of X_+ and X_-, as well as the conclusion of part (i), we have that, for any $\beta \in \mathbb{R}$,

$$\begin{aligned} E(\mathrm{e}^{\mathrm{i}\beta X}) &= E(\mathrm{e}^{\mathrm{i}\beta X_+})E(\mathrm{e}^{-\mathrm{i}\beta X_-}) \\ &= \exp\left\{-\int_{\{f\ge 0\}}(1-\mathrm{e}^{\mathrm{i}\beta f^+(x)})\eta(\mathrm{d}x)\right\} \\ &\quad\times \exp\left\{-\int_{\{f<0\}}(1-\mathrm{e}^{-\mathrm{i}\beta f^-(x)})\eta(\mathrm{d}x)\right\} \\ &= \exp\left\{-\int_S(1-\mathrm{e}^{\mathrm{i}\beta f(x)})\eta(\mathrm{d}x)\right\}, \end{aligned}$$

and the proof of (ii) is complete.

Part (iii) is dealt with similarly as in the above treatment. That is, first consider positive, simple f, then extend to positive measurable f and then to a general measurable f by treating its positive and negative parts separately.

Alternatively one may take the identity (2.10) and differentiate in β, once for $E(X)$ and twice for $E(X^2)$, and then set $\beta = 0$. The integrability conditions in (2.11) and (2.12) are used in applying the Dominated Convergence Theorem to differentiate through the integral on the right-hand side of (2.10). The details are left to the reader. □

2.4 Square-Integrable Martingales

We shall predominantly use the identities in Theorem 2.7 for a Poisson random measure, $N(\cdot)$, on $([0,\infty)\times\mathbb{R}, \mathcal{B}[0,\infty)\times\mathcal{B}(\mathbb{R}), \mathrm{d}t\times\Pi(\mathrm{d}x))$, where Π is a measure concentrated on $\mathbb{R}\backslash\{0\}$. We shall be interested in integrals of the form

$$\int_{[0,t]}\int_B xN(\mathrm{d}s\times\mathrm{d}x), \tag{2.15}$$

where $B\in\mathcal{B}(\mathbb{R})$. The relevant integrals appearing in (2.9)–(2.12), with $f(x)=x$, for the above Poisson random measure, can now be checked to take the form

$$t\int_B(1\wedge|x|)\Pi(\mathrm{d}x), \qquad t\int_B(1-\mathrm{e}^{\mathrm{i}\beta x})\Pi(\mathrm{d}x),$$

$$t\int_B|x|\Pi(\mathrm{d}x), \quad \text{and} \quad t\int_B x^2\Pi(\mathrm{d}x),$$

with the appearance of the factor t in front of each of the integrals being a consequence of the involvement of Lebesgue measure in the intensity of N. The following two lemmas capture the context in which we use sums of the form (2.15). The first may be considered as a converse to Lemma 2.2 and the second shows the relationship with martingales.

Lemma 2.8 *Suppose that $N(\cdot)$ is a Poisson random measure on* $([0,\infty)\times\mathbb{R}, \mathcal{B}[0,\infty)\times\mathcal{B}(\mathbb{R}), \mathrm{d}t\times\Pi(\mathrm{d}x))$, *where Π is a measure concentrated on* $\mathbb{R}\backslash\{0\}$

and $B \in \mathcal{B}(\mathbb{R})$ such that $0 < \Pi(B) < \infty$. Then

$$X_t := \int_{[0,t]} \int_B x N(\mathrm{d}u \times \mathrm{d}x), \quad t \geq 0,$$

is a compound Poisson process with arrival rate $\Pi(B)$ and jump distribution $\Pi(B)^{-1}\Pi(\mathrm{d}x)|_B$.

Proof First note that since it is assumed $\Pi(B) < \infty$, from Corollary 2.6, we know that, for each $t > 0$, X_t may be written as an almost surely finite sum. This explains why $X = \{X_t : t \geq 0\}$ is right-continuous with left limits. (One may also see finiteness of X_t from Theorem 2.7 (i).) Next note that, for all $0 \leq s < t < \infty$,

$$X_t - X_s = \int_{(s,t]} \int_B x N(\mathrm{d}u \times \mathrm{d}x),$$

which is independent of $\{X_u : u \leq s\}$ as $N(\cdot)$ has independent counts over disjoint regions. From the construction of $N(\cdot)$, see for example (2.7), and the fact that its intensity measure takes the specific form $\mathrm{d}t \times \Pi(\mathrm{d}x)$, it also follows that $X_t - X_s$ has the same distribution as X_{t-s}. Further, according to Theorem 2.7 (ii), we have that, for all $\theta \in \mathbb{R}$ and $t \geq 0$,

$$E\left(\mathrm{e}^{\mathrm{i}\theta X_t}\right) = \exp\left\{-t \int_B \left(1 - \mathrm{e}^{\mathrm{i}\theta x}\right)\Pi(\mathrm{d}x)\right\}. \tag{2.16}$$

The Lévy–Khintchine exponent in (2.16) is that of a compound Poisson process with jump distribution and arrival rate given by $\Pi(B)^{-1}\Pi(\mathrm{d}x)|_B$ and $\Pi(B)$, respectively. □

Just as in the discussion following Definition 1.1, we assume that $\mathbb{F} = \{\mathcal{F}_t : t \geq 0\}$ is the filtration generated by X satisfying the conditions of natural enlargement.

Lemma 2.9 *Suppose that N is the same as in the previous lemma and B is such that $\int_B |x|\Pi(\mathrm{d}x) < \infty$.*

(i) *The compound Poisson process with drift*

$$M_t := \int_{[0,t]} \int_B x N(\mathrm{d}s \times \mathrm{d}x) - t \int_B x\Pi(\mathrm{d}x), \quad t \geq 0,$$

is a P-martingale with respect to the filtration $\mathbb{F}$.

(ii) *If further, $\int_B x^2\Pi(\mathrm{d}x) < \infty$ then it is a square-integrable martingale.*

Proof (i) First note that the process $M = \{M_t : t \geq 0\}$ is adapted to the filtration $\mathbb{F}$. Next note that, for each $t > 0$,

$$E\left(|M_t|\right) \leq E\left(\int_{[0,t]} \int_B |x| N(\mathrm{d}s \times \mathrm{d}x) + t \int_B |x|\Pi(\mathrm{d}x)\right),$$

which, from Theorem 2.7 (iii), is finite because $\int_B |x|\Pi(\mathrm{d}x)$ is. Next use the fact that M has stationary independent increments to deduce that, for $0 \le s \le t < \infty$,

$$\begin{aligned} &E(M_t - M_s|\mathcal{F}_s) \\ &\quad = E(M_{t-s}) \\ &\quad = E\left(\int_{(s,t]}\int_B xN(\mathrm{d}u \times \mathrm{d}x)\right) - (t-s)\int_B x\Pi(\mathrm{d}x) \\ &\quad = 0, \end{aligned}$$

where in the final equality we have used Theorem 2.7 (iii) again.

(ii) To see that M is square-integrable, we may yet again appeal to Theorem 2.7 (iii), together with the assumption that $\int_B x^2\Pi(\mathrm{d}x) < \infty$, to deduce that

$$E\left(\left\{M_t + t\int_B x\Pi(\mathrm{d}x)\right\}^2\right) = t\int_B x^2\Pi(\mathrm{d}x) + t^2\left(\int_B x\Pi(\mathrm{d}x)\right)^2.$$

Recalling from the martingale property that $E(M_t) = 0$, it follows by developing the left-hand side in the previous display that

$$E(M_t^2) = t\int_B x^2\Pi(\mathrm{d}x) < \infty,$$

as required. □

The conditions in both Lemmas 2.8 and 2.9 mean that we may consider sets, for example, of the form $B_\varepsilon := (-1, -\varepsilon) \cup (\varepsilon, 1)$ for any $\varepsilon \in (0, 1)$. However, it is not necessarily the case that we may consider sets of the form $B = (-1, 0) \cup (0, 1)$. Consider for example the case that $\Pi(\mathrm{d}x) = \mathbf{1}_{(x>0)}x^{-(1+\alpha)}\mathrm{d}x + \mathbf{1}_{(x<0)}|x|^{-(1+\alpha)}\mathrm{d}x$ for $\alpha \in (1, 2)$. In this case, we have that $\int_B |x|\Pi(\mathrm{d}x) = \infty$ whereas $\int_B x^2\Pi(\mathrm{d}x) < \infty$. It will turn out to be quite important in the proof of the Lévy–Itô decomposition to understand the limit of the martingale in Lemma 2.8 for sets of the form B_ε as $\varepsilon \downarrow 0$. For this reason, let us now state and prove the following theorem.

Theorem 2.10 *Suppose that $N(\cdot)$ is as in Lemma* 2.8 *and* $\int_{(-1,1)} x^2\Pi(\mathrm{d}x) < \infty$. *For each $\varepsilon \in (0, 1)$ define the martingale*

$$M_t^\varepsilon = \int_{[0,t]}\int_{B_\varepsilon} xN(\mathrm{d}s \times \mathrm{d}x) - t\int_{B_\varepsilon} x\Pi(\mathrm{d}x), \quad t \ge 0.$$

Then there exists a martingale $M = \{M_t : t \geq 0\}$ *with the following properties*:

(i) *for each* $T > 0$, *there exists a deterministic subsequence* $\{\varepsilon_n^T : n = 1, 2, \dots\}$ *with* $\varepsilon_n^T \downarrow 0$ *along which*

$$P\Big(\lim_{n\uparrow\infty} \sup_{0\leq s\leq T} (M_s^{\varepsilon_n^T} - M_s)^2 = 0\Big) = 1,$$

(ii) *it is adapted to the filtration* $\mathbb{F}$,

(iii) *it has right-continuous paths with left limits almost surely*,

(iv) *it has, at most, a countable number of discontinuities on* $[0, T]$ *almost surely and*

(v) *it has stationary and independent increments.*

In short, there exists a Lévy process, which is also a martingale with a countable number of jumps to which, for any fixed $T > 0$, *the sequence of martingales* $\{M_t^\varepsilon : t \leq T\}$ *converges uniformly on* $[0, T]$ *with probability one along a subsequence in* ε *which may depend on* T.

Before proving Theorem 2.10, we need to remind ourselves of some general facts concerning square-integrable martingales. In our account, we shall recall a number of well-established facts coming from straightforward L^2 theory, measure theory and continuous-time martingale theory. The reader is referred to Sects. 2.4, 2.5 and 9.6 of Ash and Doléans-Dade (2000) for a clear account of the necessary background.

Fix a time horizon $T > 0$. Let us assume that $(\Omega, \mathcal{F}, \{\mathcal{F}_t : t \in [0, T]\}, P)$ is a filtered probability space in which the filtration $\{\mathcal{F}_t : t \geq 0\}$ satisfies the conditions of natural enlargement.

Definition 2.11 Fix $T > 0$. Define $\mathcal{M}_T^2 = \mathcal{M}_T^2(\Omega, \mathcal{F}, \{\mathcal{F}_t : t \in [0, T]\}, P)$ to be the space of real-valued, zero mean, almost surely right-continuous, square-integrable P-martingales with respect to the given filtration over the finite time period $[0, T]$.

One luxury that follows from the assumptions on $\{\mathcal{F}_t : t \geq 0\}$ is that any zero mean square-integrable martingale with respect to this filtration has a right-continuous modification[3] which is also a member of $\mathcal{M}_T^2$.

If we quotient out the equivalent classes of versions[4] of each martingale, it is straightforward to deduce that $\mathcal{M}_T^2$ is a vector space over the real numbers with

[3]Recall that $M' = \{M'_t : t \in [0, T]\}$ is a modification of M if, for every $t \geq 0$, we have $P(M'_t = M_t) = 1$.

[4]Recall that $M' = \{M'_t : t \in [0, T]\}$ is a version of M if it is defined on the same probability space and $\{\exists t \in [0, T] : M'_t \neq M_t\}$ is measurable with zero probability. Note that, if M' is a modification of M, then it is not necessarily a version of M. However, it is obviously the case that, if M' is a version of M, then it also fulfils the requirement of being a modification.

zero element $M_t = 0$ for all $t \in [0, T]$ and all $\omega \in \Omega$. In fact, as we shall shortly see, $\mathcal{M}_T^2$ is a Hilbert space[5] with respect to the inner product

$$\langle M, N\rangle = E(M_T N_T),$$

where $M, N \in \mathcal{M}_T^2$. It is left to the reader to verify the fact that $\langle \cdot, \cdot\rangle$ forms an inner product. The only mild technical difficulty in this verification is showing that, for $M \in \mathcal{M}_T^2$, $\langle M, M\rangle = 0$ implies that $M = 0$, the zero element. Note that, if $\langle M, M\rangle = 0$, then by Doob's Maximal Inequality, which says that for $M \in \mathcal{M}_T^2$,

$$E\Big(\sup_{0\leq s\leq T} M_s^2\Big) \leq 4E\big(M_T^2\big),$$

we have that $\sup_{0\leq t\leq T} |M_t| = 0$ almost surely. It follows necessarily that $M_t = 0$ for all $t \in [0, T]$ with probability one. This corresponds to the zero element in the quotient space.

As alluded to above, we can show without too much difficulty that $\mathcal{M}_T^2$ is a Hilbert space. To do that, we are required to show that, if $\{M^{(n)} : n = 1, 2, \ldots\}$ is a Cauchy sequence of martingales taken from $\mathcal{M}_T^2$, then there exists an $M \in \mathcal{M}_T^2$ such that

$$\|M^{(n)} - M\| \to 0,$$

as $n \uparrow \infty$, where $\|\cdot\| := \langle \cdot, \cdot\rangle^{1/2}$. To this end let us assume that the sequence of processes $\{M^{(n)} : n = 1, 2, \ldots\}$ is a Cauchy sequence, in other words,

$$E\big[\big(M_T^{(m)} - M_T^{(n)}\big)^2\big]^{1/2} \to 0 \quad \text{as } m, n \uparrow \infty.$$

Necessarily the sequence of *random variables* $\{M_T^{(k)} : k \geq 1\}$ is a Cauchy sequence in the Hilbert space of zero mean, square-integrable random variables defined on $(\Omega, \mathcal{F}_T, P)$, say $L^2(\Omega, \mathcal{F}_T, P)$, endowed with the inner product $\langle M, N\rangle = E(MN)$. Hence, there exists a limiting variable, say M_T in $L^2(\Omega, \mathcal{F}_T, P)$, satisfying

$$E\big[\big(M_T^{(n)} - M_T\big)^2\big]^{1/2} \to 0,$$

as $n \uparrow \infty$. Define the martingale M to be the right-continuous version[6] of

$$E(M_T|\mathcal{F}_t) \text{ for } t \in [0, T]$$

---

[5]Recall that $\langle \cdot, \cdot\rangle$: $L \times L \to \mathbb{R}$ is an inner product on a vector space L over the reals if it satisfies the following properties, for $f, g \in L$ and $a, b \in \mathbb{R}$; (i) $\langle af + bg, h\rangle = a\langle f, h\rangle + b\langle g, h\rangle$ for all $h \in L$, (ii) $\langle f, g\rangle = \langle g, f\rangle$, (iii) $\langle f, f\rangle \geq 0$ and (iv) $\langle f, f\rangle = 0$ if and only if $f = 0$.

For each $f \in L$, let $\|f\| = \langle f, f\rangle^{1/2}$. The pair $(L, \langle \cdot, \cdot\rangle)$ are said to form a Hilbert space if all sequences, $\{f_n : n = 1, 2, \ldots\}$ in L that satisfy $\|f_n - f_m\| \to 0$ as $m, n \to \infty$, i.e. so-called Cauchy sequences, have a limit in L.

[6]Here, we use the fact that $\{\mathcal{F}_t : t \in [0, T]\}$ satisfies the conditions of natural enlargement.

and note that, by definition,

$$\|M^{(n)} - M\| \to 0,$$

as n tends to infinity. Clearly it is an $\mathcal{F}_t$-adapted process and by Jensen's inequality

$$\begin{aligned} E(M_t^2) &= E\big(E(M_T|\mathcal{F}_t)^2\big) \\ &\le E\big(E(M_T^2|\mathcal{F}_t)\big) \\ &= E(M_T^2), \end{aligned}$$

which is finite. Hence Cauchy sequences converge in $\mathcal{M}_T^2$ and we see that $\mathcal{M}_T^2$ is indeed a Hilbert space.

We are now ready to return to Theorem 2.10.

Proof of Theorem 2.10 (i) Choose $0 < \eta < \varepsilon < 1$, fix $T > 0$ and define $M^\varepsilon = \{M_t^\varepsilon : t \in [0, T]\}$. A calculation similar to the one in Lemma 2.9 (ii) gives

$$\begin{aligned} &E\big((M_T^\varepsilon - M_T^\eta)^2\big) \\ &= E\left(\left\{\int_{[0,T]}\int_{\eta\le|x|<\varepsilon} xN(\mathrm{d}s \times \mathrm{d}x) - T\int_{\eta<|x|<\epsilon} x\Pi(\mathrm{d}x)\right\}^2\right) \\ &= T\int_{\eta\le|x|<\varepsilon} x^2\Pi(\mathrm{d}x). \end{aligned}$$

Note, however, that the left-hand side above is also equal to $\|M^\varepsilon - M^\eta\|^2$ (where as in the previous discussion, $\|\cdot\|$ is the norm induced by the inner product on $\mathcal{M}_T^2$).

Thanks to the assumption that $\int_{(-1,1)} x^2\Pi(\mathrm{d}x) < \infty$, we now have that $\lim_{\varepsilon,\eta\downarrow 0}\|M^\varepsilon - M^\eta\| = 0$ and hence that $\{M^\varepsilon : 0 < \varepsilon < 1\}$ is a Cauchy sequence in $\mathcal{M}_T^2$. As $\mathcal{M}_T^2$ is a Hilbert space, we know that there exists a right-continuous martingale $M = \{M_s : s \in [0, T]\} \in \mathcal{M}_T^2$ such that

$$\lim_{\varepsilon\downarrow 0}\|M - M^\varepsilon\| = 0.$$

An application of Doob's Maximal Inequality tells us that, in fact,

$$\lim_{\varepsilon\downarrow 0} E\Big[\sup_{0\le s\le T}(M_s - M_s^\varepsilon)^2\Big] \le 4\lim_{\varepsilon\downarrow 0}\|M - M^\varepsilon\| = 0. \tag{2.17}$$

From this, one may deduce that the limit $\{M_s : s \in [0, T]\}$ does not depend on T. Indeed, suppose it did and we adjust our notation accordingly so that $\{M_{s,T} : s \le T\}$ represents the limit. Then from (2.17), we see that, for any $0 < T' < T$,

$$\lim_{\varepsilon\downarrow 0} E\Big[\sup_{0\le s\le T'}(M_s^\varepsilon - M_{s,T'})^2\Big] = 0$$

as well as

$$\lim_{\varepsilon\downarrow 0} E\Big[\sup_{0\le s\le T'} \big(M_s^{\varepsilon} - M_{s,T}\big)^2\Big] \le \lim_{\varepsilon\downarrow 0} E\Big[\sup_{0\le s\le T} \big(M_s^{\varepsilon} - M_{s,T}\big)^2\Big] = 0,$$

where the inequality is the result of a trivial upper bound. Hence, using that, for any two sequences of real numbers $\{a_n\}$ and $\{b_n\}$, $\sup_n a_n^2 = (\sup_n |a_n|)^2$ and $\sup_n |a_n + b_n| \le \sup_n |a_n| + \sup_n |b_n|$, we have, together with an application of Minkowski's inequality, that

$$\begin{aligned} E\Big[\sup_{0\le s\le T'} (M_{s,T'} - M_{s,T})^2\Big]^{1/2} &\le \lim_{\varepsilon\downarrow 0} E\Big[\sup_{0\le s\le T'} \big(M_s^{\varepsilon} - M_{s,T'}\big)^2\Big]^{1/2} \\ &\quad + \lim_{\varepsilon\downarrow 0} E\Big[\sup_{0\le s\le T'} \big(M_s^{\varepsilon} - M_{s,T}\big)^2\Big]^{1/2} \\ &= 0, \end{aligned}$$

thus showing that the processes $M_{\cdot,T}$ and $M_{\cdot,T'}$ are almost surely uniformly equal on $[0, T']$. Since T' and T may be arbitrarily chosen, we may now speak of a well-defined limiting martingale, $M = \{M_t : t \ge 0\}$.

From the limit (2.17), we may also deduce that there exists a deterministic subsequence $\{\varepsilon_n^T : n \ge 0\}$ along which

$$\lim_{\varepsilon_n^T \downarrow 0} \sup_{0\le s\le T} \big(M_s^{\varepsilon_n^T} - M_s\big)^2 = 0$$

P-almost surely. This follows from the well-established fact that L^2 convergence of a sequence of random variables implies almost sure convergence on a deterministic subsequence.

(ii) and (iii) Since, for each $T < \infty$, $\{M_s : s \in [0, T]\} \in \mathcal{M}_T^2$, it is automatic from the definition of this space of martingales that M is $\mathbb{F}$-adapted with right-continuous paths. It remains to show that the paths of M have left limits. To this end, note that the paths of M^{ε} are right-continuous with left limits. Hence, almost sure uniform convergence (along a subsequence) on finite time intervals implies that the limiting process, M, also has paths which are right-continuous with, in particular, left limits. We are using here the fact that, if $D[0, 1]$ is the space of functions $f : [0, 1] \to \mathbb{R}$ which are right-continuous with left limits, then $D[0, 1]$ contains all its limit points under the metric $d(f, g) = \sup_{t\in[0,1]} |f(t) - g(t)|$ for $f, g \in D[0, 1]$. See Exercise 2.4.

(iv) According to Corollary 2.6, there are at most an almost surely countable number of points in the support of N. Further, recalling the discussion after Corollary 2.6, as the measure $\mathrm{d}t \times \Pi(\mathrm{d}x)$ has no atoms, the random measure $N(\cdot)$ is necessarily $\{0, 1\}$-valued on time-space singletons. Hence every discontinuity in $\{M_s : s \ge 0\}$ corresponds to a unique point in the support of $N(\cdot)$. It follows that M has at most a countable number of discontinuities. Another way to see that there are, at most, a countable number of discontinuities is simply to note that the same is true of functions in the space $D[0,1]$; see Exercise 2.4.

(v) For any $n \in \mathbb{N}$, $0 \le s_1 \le t_1 \le \cdots \le s_n \le t_n \le T < \infty$ and $\theta_1, \ldots, \theta_n \in \mathbb{R}$, dominated convergence and almost sure uniform convergence along the subsequence $\{\epsilon_n^T : t \ge 0\}$ gives

$$E\left[\prod_{j=1}^n \mathrm{e}^{\mathrm{i}\theta_j (M_{t_j} - M_{s_j})}\right] = \lim_{n\uparrow\infty} E\left[\prod_{j=1}^n \mathrm{e}^{\mathrm{i}\theta_j (M_{t_j}^{\varepsilon_n^T} - M_{s_j}^{\varepsilon_n^T})}\right]$$

$$= \lim_{n\uparrow\infty} \prod_{j=1}^n E\left[\mathrm{e}^{\mathrm{i}\theta_j M_{t_j - s_j}^{\varepsilon_n^T}}\right]$$

$$= \prod_{j=1}^n E\left[\mathrm{e}^{\mathrm{i}\theta_j M_{t_j - s_j}}\right],$$

which, thanks to Exercise 1.1, is sufficient to deduce that M has stationary and independent increments. This concludes the proof. □

2.5 Proof of the Lévy–Itô Decomposition

As previously indicated in Sect. 2.1, we will take $X^{(1)}$ to be the linear Brownian motion (2.2), now defined on some probability space $(\Omega^\#, \mathcal{F}^\#, P^\#)$.

Given Π in the statement of Theorem 2.1, we know from Theorem 2.4 that there exists a probability space, say $(\Omega^*, \mathcal{F}^*, P^*)$, on which we may construct a Poisson random measure, N, on $([0,\infty) \times \mathbb{R}, \mathcal{B}[0,\infty) \times \mathcal{B}(\mathbb{R}), \mathrm{d}t \times \Pi(\mathrm{d}x))$. We may think of the points in the support of N as having a time and space coordinate, or alternatively, as points in $\mathbb{R}\backslash\{0\}$ arriving in time.

Now define

$$X_t^{(2)} = \int_{[0,t]} \int_{|x|\ge 1} x N(\mathrm{d}s \times \mathrm{d}x), \quad t \ge 0,$$

and note from Lemma 2.8 that, since $\Pi(\mathbb{R}\backslash(-1,1)) < \infty$, it is a compound Poisson process with rate $\Pi(\mathbb{R}\backslash(-1,1))$ and jump distribution

$$\Pi\big(\mathbb{R}\backslash(-1,1)\big)^{-1} \Pi(\mathrm{d}x)|_{\mathbb{R}\backslash(-1,1)}.$$

(We can assume without loss of generality that $\Pi(\mathbb{R}\backslash(-1,1)) > 0$ as otherwise, we may take the process $X^{(2)}$ as the process which is identically zero.)

Next, we construct a Lévy process having only small jumps. For each $1 > \varepsilon > 0$, define similarly the compound Poisson process with drift,

$$X_t^{(3,\varepsilon)} = \int_{[0,t]} \int_{\varepsilon \le |x| < 1} x N(\mathrm{d}s \times \mathrm{d}x) - t \int_{\varepsilon \le |x| < 1} x \Pi(\mathrm{d}x), \quad t \ge 0. \tag{2.18}$$

(As in the definition of $X^{(2)}$, we shall assume without loss of generality $\Pi(\{x : |x| < 1\}) > 0$, otherwise the process $X^{(3)}$ may be taken as the process which is identically zero.) Using Theorem 2.7 (ii), we can compute its characteristic exponent,

$$\Psi^{(3,\varepsilon)}(\theta) := \int_{\varepsilon\leq|x|<1} (1 - \mathrm{e}^{\mathrm{i}\theta x} + \mathrm{i}\theta x)\Pi(\mathrm{d}x).$$

According to Theorem 2.10, there exists a Lévy process, which is also a square-integrable martingale, defined on $(\Omega^*, \mathcal{F}^*, P^*)$, to which $X^{(3,\varepsilon)}$ converges uniformly on $[0, T]$ along an appropriate deterministic subsequence in ε. Note that it is precisely at this point that we use the assumption that $\int_{(-1,1)} x^2\Pi(\mathrm{d}x) < \infty$. It is clear that the characteristic exponent of the aforementioned Lévy process is equal to

$$\Psi^{(3)}(\theta) = \int_{|x|<1} \big(1 - \mathrm{e}^{\mathrm{i}\theta x} + \mathrm{i}\theta x\big)\Pi(\mathrm{d}x).$$

From Corollary 2.5, we know that, for each $t > 0$, N has independent counts over the two domains $[0, t] \times \{\mathbb{R}\backslash(-1, 1)\}$ and $[0, t] \times (-1, 1)$. It follows that $X^{(2)}$ and $X^{(3)}$ are independent.

To conclude the proof of the Lévy–Itô decomposition in line with the statement of Theorem 2.1, define the process

$$X_t = X_t^{(1)} + X_t^{(2)} + X_t^{(3)}, \quad t \geq 0. \tag{2.19}$$

This process is defined on the product space

$$(\Omega, \mathcal{F}, \mathbb{P}) = \big(\Omega^\#, \mathcal{F}^\#, P^\#\big) \times \big(\Omega^*, \mathcal{F}^*, P^*\big),$$

has stationary independent increments, has paths that are right-continuous with left limits and has characteristic exponent

$$\begin{aligned}\Psi(\theta) &= \Psi^{(1)}(\theta) + \Psi^{(2)}(\theta) + \Psi^{(3)}(\theta)\\ &= \mathrm{i}a\theta + \frac{1}{2}\sigma^2\theta^2 + \int_{\mathbb{R}} \big(1 - \mathrm{e}^{\mathrm{i}\theta x} + \mathrm{i}\theta x\mathbf{1}_{(|x|<1)}\big)\Pi(\mathrm{d}x),\end{aligned}$$

as required. □

Let us conclude this section with some additional remarks on the Lévy–Itô decomposition.

Recall from (2.4) that the exponent $\Psi^{(3)}$ appears to have the form of the infinite sum of characteristic exponents belonging to compound Poisson processes with drift. This suggests that $X^{(3)}$ may be taken as the superposition of such processes. We now see from the above proof that this is exactly the case. Indeed, moving ε to zero through the sequence $\{2^{-k} : k \geq 0\}$ shows us that in the appropriate sense of L^2 convergence

$$\begin{aligned}\lim_{k\uparrow\infty} X_t^{(3,2^{-k})} &= \lim_{k\uparrow\infty} \int_{[0,t]}\int_{2^{-k}<|x|<1} xN(\mathrm{d}s\times\mathrm{d}x) - t\int_{2^{-k}<|x|<1} x\Pi(\mathrm{d}x)\\ &= \lim_{k\uparrow\infty}\sum_{i=0}^{k-1}\left\{\int_{[0,t]}\int_{2^{-(i+1)}<|x|<2^{-i}} xN(\mathrm{d}s\times\mathrm{d}x)\right.\\ &\qquad \left. - t\int_{2^{-(i+1)}<|x|<2^{-i}} x\Pi(\mathrm{d}x)\right\}.\end{aligned}$$

It is also worth remarking that the definition of $X^{(2)}$ and $X^{(3)}$ in the proof of the Lévy–Itô decomposition, corresponding to the partition of $\mathbb{R}\backslash\{0\}$ into $\mathbb{R}\backslash(-1,1)$ and $(-1,1)\backslash\{0\}$, is to some extent arbitrary. The point is that one needs to deal differently with the contributions to the path from N which come from a neighbourhood of the origin, and which come from its complement. In this respect one could have redrafted the proof replacing $(-1,1)$ by (α,β), for any $\alpha<0$ and $\beta>0$. In which case, one would need to choose a different value of a in the definition of $X^{(1)}$ in order to make terms add up precisely to the expression given in the Lévy–Khintchine exponent. To be more precise, if for example $\alpha<-1$ and $\beta>1$, then one should take $X_t^{(1)}=a't+\sigma B_t$ where

$$a' = a - \int_{\alpha<|x|\le -1} x\Pi(\mathrm{d}x) - \int_{1\le|x|<\beta} x\Pi(\mathrm{d}x).$$

This also shows that the Lévy–Khintchine formula (2.1) is not a unique representation and, indeed, the indicator $\mathbf{1}_{(|x|<1)}$ in (2.1) may be replaced by $\mathbf{1}_{(\alpha<x<\beta)}$ with an appropriate adjustment in the constant a.

Taking a much deeper view of things, the Lévy–Itô decomposition illustrates one of many examples where a Markov process can be decomposed according to an endogenous Poisson point process. This approach was pursued by K. Itô. See for example Itô (2004, 1970). Later on, in Chap. 6, we shall see another path-decomposition of Lévy processes in this spirit. In that case, the path is decomposed according to a Poisson point process of excursions of the Lévy process from its maximum.

2.6 Lévy Processes Distinguished by Their Path Type

As is clear from the proof of the Lévy–Itô decomposition, we should think of the measure Π given in the Lévy–Khintchine formula as characterising a Poisson random measure which encodes the rate at which the jumps of the associated Lévy process occur. In this section we shall re-examine elements of the proof of the Lévy–Itô decomposition and show that, with additional assumptions on Π, we may further identify special classes of Lévy processes embedded within the general class.

2.6.1 Path Variation

It is clear from the Lévy–Itô decomposition that the presence of the linear Brownian motion $X^{(1)}$ would imply that paths of the Lévy process have unbounded variation. On the other hand, should it be the case that $\sigma = 0$, then the Lévy process may or may not have unbounded variation. The term $X^{(2)}$, being a compound Poisson process, has only bounded variation. Hence, in the case $\sigma = 0$, understanding whether the Lévy process has unbounded variation is an issue determined by the limiting process $X^{(3)}$; that is to say, the process of compensated small jumps.

Reconsidering the definition of $X^{(3)}$, it is natural to ask under what circumstances

$$\lim_{\varepsilon\downarrow 0}\int_{[0,t]}\int_{\varepsilon\le|x|<1} xN(\mathrm{d}s\times\mathrm{d}x)$$

exists almost surely without the need for compensation by its mean as in (2.18). Once again, the answer is given by Theorem 2.7 (i). Here we are told that

$$\int_{[0,t]}\int_{|x|<1}|x|N(\mathrm{d}s\times\mathrm{d}x)<\infty$$

if and only if $\int_{|x|<1}|x|\Pi(\mathrm{d}x)<\infty$. In that case, we may identify $X^{(3)}$ directly via

$$X_t^{(3)} = \int_{[0,t]}\int_{|x|<1} xN(\mathrm{d}s\times\mathrm{d}x) - t\int_{|x|<1} x\Pi(\mathrm{d}x),\quad t\ge 0.$$

This also tells us that $X^{(3)}$ will be of bounded variation if and only if $\int_{|x|<1}|x|\Pi(\mathrm{d}x) < \infty$. Note that this is a stronger integrability condition than the general integrability condition $\int_{\mathbb{R}}(1\wedge x^2)\Pi(\mathrm{d}x)<\infty$. We get the following lemma.

Lemma 2.12 *A Lévy process with Lévy–Khintchine exponent corresponding to the triple (a,σ,Π) has paths of bounded variation if and only if*

$$\sigma = 0 \quad \textit{and} \quad \int_{\mathbb{R}}(1\wedge|x|)\Pi(\mathrm{d}x)<\infty. \tag{2.20}$$

Note that the finiteness of the integral in (2.20) also allows for the Lévy–Khintchine exponent of any such bounded variation process to be rewritten as

$$\Psi(\theta) = -\mathrm{i}\delta\theta + \int_{\mathbb{R}}(1-\mathrm{e}^{\mathrm{i}\theta x})\Pi(\mathrm{d}x), \tag{2.21}$$

where the constant $\delta\in\mathbb{R}$ relates to the constant a and Π via

$$\delta = -\left(a + \int_{|x|<1} x\Pi(\mathrm{d}x)\right).$$

In this case, we may write the Lévy process in the form

$$X_t = \delta t + \int_{[0,t]} \int_{\mathbb{R}} x N(\mathrm{d}s \times \mathrm{d}x), \quad t \geq 0. \tag{2.22}$$

In view of the decomposition of the Lévy–Khintchine formula for a process of bounded variation and the corresponding representation (2.22), the term δ is often referred to as the *drift*. Strictly speaking, one should not talk of drift in the case of a Lévy process whose jump part is a process of unbounded variation. If drift is to be understood in terms of a purely deterministic trend, then it is ambiguous on account of the "infinite limiting compensation" that one sees in $X^{(3)}$ coming from the second term on the right-hand side of (2.18).

From the expression given in (1.4) of Chap. 1, we see that, if X is a compound Poisson process with drift, then its characteristic exponent takes the form of (2.21) with $\Pi(\mathbb{R}) < \infty$. Conversely, if $\sigma = 0$ and Π has finite total mass, then we know from Lemma 2.8 that (2.22) is a compound Poisson process with drift δ. In conclusion, we have the following lemma.

Lemma 2.13 *A Lévy process is a compound Poisson process with drift if and only if* $\sigma = 0$ *and* $\Pi(\mathbb{R}) < \infty$.

2.6.2 One-Sided Jumps

Suppose now that $\Pi(-\infty, 0) = 0$. From the proof of the Lévy–Itô decomposition, we see that the corresponding Lévy process has no negative jumps. If further we have that $\int_{(0,\infty)} (1 \wedge x)\Pi(\mathrm{d}x) < \infty$, $\sigma = 0$ and, in the representation (2.21) of the characteristic exponent, $\delta \geq 0$, then from the representation (2.22) it becomes clear that the Lévy process has non-decreasing paths. Conversely, if a Lévy process has non-decreasing paths, then necessarily it has bounded variation. Hence $\int_{(0,\infty)} (1 \wedge x)$ $\Pi(\mathrm{d}x) < \infty$, $\sigma = 0$ and then it is easy to see that in the representation (2.21) of the characteristic exponent, we necessarily have $\delta \geq 0$. Examples of such a process were given in Chap. 1 (the gamma process and the inverse Gaussian process) and were named *subordinators*. Summarising, we have the following.

Lemma 2.14 *A Lévy process is a subordinator if and only if* $\Pi(-\infty, 0) = 0$, $\int_{(0,\infty)} (1 \wedge x)\Pi(\mathrm{d}x) < \infty$, $\sigma = 0$ *and* $\delta = -(a + \int_{(0,1)} x\Pi(\mathrm{d}x)) \geq 0$.

For the sake of clarity, we note that, when X is a subordinator, further to (2.21), its Lévy–Khintchine formula may be written as

$$\Psi(\theta) = -i\delta\theta + \int_{(0,\infty)} (1 - \mathrm{e}^{\mathrm{i}\theta x})\Pi(\mathrm{d}x). \tag{2.23}$$

If $\Pi(-\infty, 0) = 0$ and X does not have monotone paths, that is to say, it is not a subordinator and it is not a pure negative linear drift, then it is referred to in general as a *spectrally positive* Lévy process. A Lévy process, X, will then be referred to as a *spectrally negative Lévy process* if $-X$ is spectrally positive. Together, these two classes of processes are called *spectrally one-sided*. Spectrally one-sided Lévy processes may be of bounded or unbounded variation and, in the latter case, may or may not possess a Gaussian component. Note in particular that when $\sigma = 0$, it is still possible to have paths of unbounded variation. If a spectrally positive Lévy process has bounded variation, then it must take the form

$$X_t = -\delta t + S_t, \quad t \geq 0,$$

where $\{S_t : t \geq 0\}$ is a pure jump subordinator and, necessarily, $\delta > 0$. Note that if $\delta \leq 0$, then X would conform to the definition of a subordinator. Note that the above decomposition implies that if $\mathbb{E}(X_1) \leq 0$, then $\mathbb{E}(S_1) < \infty$, as opposed to the case that $\mathbb{E}(X_1) > 0$, in which case it is possible that $\mathbb{E}(S_1) = \infty$.

A special feature of spectrally positive processes is that, if $\tau_x^- = \inf\{t > 0 : X_t < x\}$, where $x < 0$, then $\mathbb{P}(\tau_x^- < \infty) > 0$. Hence, as there are no downwards jumps,

$$\mathbb{P}\big(X_{\tau_x^-} = x \,|\, \tau_x^- < \infty\big) = 1, \tag{2.24}$$

with a similar property for first passage upwards being true for spectrally negative processes. A rigorous proof of the first of the above two facts will be given in Corollary 3.13, at the end of Sect. 3.3. It turns out that (2.24) plays a very important role in the simplification of a number of theorems we shall encounter later on in this text, which concern the fluctuations of general Lévy processes.

2.7 Interpretations of the Lévy–Itô Decomposition

Let us return to some of the models considered in Chap. 1 and consider how our understanding of the Lévy–Itô decomposition helps to justify working with more general classes of Lévy processes.

2.7.1 The Structure of Insurance Claims

Recall from Sect. 1.3.1 that the Cramér–Lundberg model corresponds to a Lévy process with characteristic exponent given by

$$\Psi(\theta) = -\mathrm{i}c\theta + \lambda \int_{(-\infty,0)} \big(1 - \mathrm{e}^{\mathrm{i}\theta x}\big) F(\mathrm{d}x),$$

for $\theta \in \mathbb{R}$. In other words, a compound Poisson process with arrival rate $\lambda > 0$ and negative jumps, corresponds to claims having common distribution F, as well as a drift $c > 0$ corresponding to a steady income due to premiums. Suppose instead we work with a general spectrally negative Lévy process, that is a process for which $\Pi(0,\infty)=0$ (but without monotone paths). In this case, the Lévy–Itô decomposition offers an interpretation for large-scale insurance companies as follows. The Lévy–Khintchine exponent may be written in the form

$$\Psi(\theta)=\left\{\frac{1}{2}\sigma^2\theta^2\right\}+\left\{-\mathrm{i}\theta c+\int_{(-\infty,-1]}(1-\mathrm{e}^{\mathrm{i}\theta x})\Pi(\mathrm{d}x)\right\} \\ +\left\{\int_{(-1,0)}(1-\mathrm{e}^{\mathrm{i}\theta x}+\mathrm{i}\theta x)\Pi(\mathrm{d}x)\right\} \tag{2.25}$$

for $\theta\in\mathbb{R}$. Assume that $\Pi(-\infty,0)=\infty$, and so Ψ is genuinely different from the characteristic of a Cramér–Lundberg model. We may understand the third bracket in (2.25) as a Lévy process representing a countably infinite number of arbitrarily small claims, compensated by a deterministic positive drift (which may be infinite in the case that $\int_{(-1,0)}|x|\Pi(\mathrm{d}x)=\infty$), corresponding to the accumulation of premiums over an infinite number of contracts. Roughly speaking, the way in which claims occur is such that, in any arbitrarily small period of time $\mathrm{d}t$, a claim of size $|x|$ (for $x<0$) is made independently with probability $\Pi(\mathrm{d}x)\mathrm{d}t+o(\mathrm{d}t)$. The insurance company thus counterbalances such claims by ensuring that it collects premiums in such a way that in any $\mathrm{d}t$, $|x|\Pi(\mathrm{d}x)\mathrm{d}t$ of its income is devoted to the compensation of claims of size $|x|$. The second bracket in (2.25) can be understood as coming from large claims, which occur occasionally and are compensated for by a steady income at rate $c>0$, just as in the Cramér–Lundberg model. Here "large" is taken to mean claims of size one or more and $c=-a$, in the terminology of the Lévy–Khintchine formula given in Theorem 1.6. Finally, the first bracket in (2.25) may be seen as a stochastic perturbation of the system of claims and premium income.

Since the contents of the first and third set of curly brackets in (2.25) correspond to martingales, the company may guarantee that its surplus drifts to infinity over an infinite time horizon by assuming that such behaviour applies to the compensated process of large claims corresponding to the second bracket in (2.25).

2.7.2 General Storage Models

The workload of the $M/G/1$ queue was presented in Sect. 1.3.2 as a spectrally negative compound Poisson process with rate $\lambda>0$ and jump distribution F with positive unit drift, reflected in its supremum. In other words, the underlying Lévy process has characteristic exponent

$$\Psi(\theta)=-\mathrm{i}\theta+\lambda\int_{(-\infty,0)}(1-\mathrm{e}^{\mathrm{i}\theta x})F(\mathrm{d}x),$$

for all $\theta\in\mathbb{R}$. A general storage model, described for example in the classic books of Prabhu (1998) and Takács (1966), consists of working with a Lévy process, X, which is the difference of a positive drift and a subordinator and then reflected in its supremum. Its Lévy–Khintchine exponent thus takes the form

$$\Psi(\theta) = -\mathrm{i}\delta\theta + \int_{(-\infty,0)} \left(1 - \mathrm{e}^{\mathrm{i}\theta x}\right)\Pi(\mathrm{d}x),$$

where $\delta > 0$ and $\int_{(-\infty,0)}(1 \wedge |x|)\Pi(\mathrm{d}x) < \infty$. As with the case of the $M/G/1$ queue, the reflected process

$$W_t = (w \vee \overline{X}_t) - X_t, \quad t \geq 0,$$

may be thought of as the stored volume or workload of some system, where $\overline{X}$ is the running supremum and w is the initial volume in the system. The Lévy–Itô decomposition tells us that, during the periods of time that X is away from its supremum, there is a natural "drainage" of volume or "processing" of workload, corresponding to the downward movement of W in a linear fashion with rate δ. At the same time new "volume for storage" or equivalently new "jobs" arrive independently so that in each $\mathrm{d}t$, one arrives of size $|x|$ (where $x < 0$) with probability $\Pi(\mathrm{d}x)\mathrm{d}t + o(\mathrm{d}t)$ (thus giving similar interpretation to the occurrence of jumps in the insurance risk model described above). When $\Pi(-\infty, 0) = \infty$, the number of jumps are countably infinite over any finite time interval, thus indicating that our model is processing with "infinite frequency" in comparison to the finite activity of the workload of the $M/G/1$ process.

Of course one may also envisage working with a jump measure which has some mass on the positive half-line. This would correspond to negative jumps in the process W. This, in turn, can be interpreted as follows. Over and above the natural drainage or processing at rate δ, in each $\mathrm{d}t$ there is independent removal of a "volume" or "processing time of job" of size $y > 0$ with probability $\Pi(\mathrm{d}y)\mathrm{d}t + o(\mathrm{d}t)$. One may also consider moving to models of unbounded variation. However, in this case, the interpretation of drift is lost.

2.7.3 Financial Models

Financial mathematics has become a field of applied probability which has also embraced the use of Lévy processes, in particular, for the purpose of modelling the evolution of risky assets. We shall not attempt to give anything like a comprehensive exposure of this topic here, nor elsewhere in this book, especially since textbooks of Boyarchenko and Levendorskii (2002b), Schoutens (2003) and Cont and Tankov (2004) already offer a clear and up-to-date overview between them. It is worth mentioning briefly some of the connections between path properties of Lévy processes seen above and modern perspectives within financial modelling.

One may say that financial mathematics proper begins with the thesis of Louis Bachelier who proposed the use of linear Brownian motion to model the value of a risky asset, say the value of a stock. See Bachelier (1900, 1901). However, the *classical* model for the evolution of a risky asset, proposed by Samuelson (1965), is generally accepted to be that of an exponential linear Brownian motion with drift;

$$S_t = s\exp\{\sigma B_t + \mu t\}, \quad t \geq 0, \tag{2.26}$$

where $s > 0$ is the initial value of the asset, $B = \{B_t : t \geq 0\}$ is a standard Brownian motion, $\sigma > 0$ and $\mu \in \mathbb{R}$. This choice of model offers the feature that asset values have multiplicative stationarity and independence in the sense that for any $0 \leq u < t < \infty$,

$$S_t = S_u \times \widetilde{S}_{t-u}, \tag{2.27}$$

where $\widetilde{S}_{t-u}$ is independent of $\{S_v : v \leq u\}$ and has the same distribution as S_{t-u}. Whether or not this is a realistic assumption in terms of temporal correlations in financial markets is open to debate. Nonetheless, for the purpose of a theoretical framework in which one may examine certain economic mechanisms, such as risk-neutrality, hedging and arbitrage, as well as giving sense to the value of certain financial products such as option contracts, exponential Brownian motion has proved to be a successful model in capturing the imagination of mathematicians, economists and financial practitioners alike. Indeed, what makes (2.26) "classical" is that Black and Scholes (1973) and Merton (1973) demonstrated how one may construct rational arguments leading to the pricing of a call option on a risky asset driven by exponential Brownian motion.

Two particular points, of the many, where the above model of a risky asset can be shown to be inadequate, concern the continuity of the paths and the distribution of log-returns of the value of a risky asset. Clearly (2.26) has continuous paths and therefore cannot accommodate for jumps which arguably are present in observed historical data of certain risky assets due to shocks in the market. The feature (2.27) suggests that for a fixed period of time Δ, for each $n \geq 1$, the innovations $\log(S_{(n+1)\Delta}/S_{n\Delta})$ are independent and normally distributed with mean $\mu\Delta$ and standard deviation $\sigma\sqrt{\Delta}$. Empirical data suggests that the tails of the distribution of the log-returns are asymmetric as well as having heavier tails than those of normal distributions. Note that the tails of normal distributions are particularly light as they decay like $\exp\{-x^2\}$ for large values of $|x|$. See for example the discussion in Schoutens (2003).

Recent literature suggests that a possible remedy is to work with

$$S_t = s\mathrm{e}^{X_t}, \quad t \geq 0,$$

instead of (2.26), where again $s > 0$ is the initial value of the risky asset and $X = \{X_t : t \geq 0\}$ is a Lévy process. This preserves multiplicative stationary and independent increments, as well as allowing for jumps, distributional asymmetry and the possibility of heavier tails than the normal distribution can offer. A simple example of how this may happen is simply to take for X a compound Poisson process whose jump distribution is asymmetric and heavy tailed. A more sophisticated example, and indeed quite a popular model in the research literature, is the so-called *variance-gamma* process, introduced by Madan and Seneta (1990). This Lévy process is pure jump, that is to say $\sigma = 0$, and has Lévy measure given by

$$\Pi(\mathrm{d}x) = \mathbf{1}_{(x<0)}\frac{C}{|x|}\mathrm{e}^{Gx}\mathrm{d}x + \mathbf{1}_{(x>0)}\frac{C}{x}\mathrm{e}^{-Mx}\mathrm{d}x,$$

where $C, G, M > 0$. It is easily seen by computing explicitly the integral $\int_{\mathbb{R}}(1 \wedge |x|)\Pi(\mathrm{d}x)$ and the total mass $\Pi(\mathbb{R})$ that the variance-gamma process has paths of bounded variation and further is not a compound Poisson process. It turns out that the exponential weighting in the Lévy measure ensures that the distribution of the variance-gamma process at a fixed time t has exponentially decaying tails (as opposed to the much lighter tails of the Gaussian distribution).

Working with pure jump processes implies that there is no diffusive nature to the evolution of risky assets. Diffusive behaviour is often found attractive for modelling purposes as it has the taste of a physical interpretation in which increments in infinitesimal periods of time are explained through the Central Limit Theorem as the aggregate effect of many simultaneous conflicting external forces.[7] Geman et al. (2001) argue the case for modelling the value of risky assets with Lévy processes which have paths of bounded variation which are not compound Poisson processes. In their reasoning, such processes have a countable number of jumps over finite periods of time, which correspond to the countable, but nonetheless infinite number of purchases and sales of the asset which collectively dictate its value as a net effect. In particular, being of bounded variation means the Lévy process can be written as the difference to two independent subordinators (see Exercise 2.8). These two subordinators should be thought of the total prevailing price buy orders and total prevailing price sell orders on the logarithmic price scale.

Despite the fundamental difference between modelling with bounded variation Lévy processes and Brownian motion, Geman et al. (2001) also provide an interesting link to the classical model (2.26) via time change. The basis of their ideas lies with the following lemma.

Lemma 2.15 *Suppose that $X = \{X_t : t \geq 0\}$ is a Lévy process with characteristic exponent Ψ and $\tau = \{\tau_s : s \geq 0\}$ is an independent subordinator with characteristic exponent $\Xi(\theta)$. Then $Y = \{X_{\tau_s} : s \geq 0\}$ is again a Lévy process with characteristic exponent $\Xi(\mathrm{i}\Psi(\theta))$.*

Proof First let us make some remarks about Ξ. We already know that the formula

$$\mathbb{E}(\mathrm{e}^{\mathrm{i}\theta\tau_s}) = \mathrm{e}^{-\Xi(\theta)s}$$

[7] See for example the second volume of Lucretius (ca. 99 BC–ca. 55 BC) and the formalisation in Einstein (1905).

holds for all $\theta \in \mathbb{R}$. However, since τ is a non-negative valued process, via analytical continuation, we may claim that the previous equality is still valid for[8] $\theta \in \{z \in \mathbb{C} : \Im z \geq 0\}$. Note in particular that, since

$$\Re\Psi(u) = \frac{1}{2}\sigma^2 u^2 + \int_{\mathbb{R}} (1 - \cos(ux))\Pi(\mathrm{d}x) \geq 0,$$

for all $u \in \mathbb{R}$, the equalities

$$\mathbb{E}(\mathrm{e}^{\mathrm{i}uX_{\tau_s}}) = \mathbb{E}(\mathrm{e}^{-\Psi(u)\tau_s}) = \mathbb{E}(\mathrm{e}^{\mathrm{i}(\mathrm{i}\Psi(u))\tau_s}) = \mathrm{e}^{-\Xi(\mathrm{i}\Psi(u))s} \tag{2.28}$$

hold.

Since X and τ have right-continuous paths, then so does Y. Next consider $n \in \mathbb{N}$, $0 \leq s_1 \leq t_1 \leq \cdots \leq s_n \leq t_n < \infty$ and $\theta_1, \ldots, \theta_n \in \mathbb{R}$. Then, by first conditioning on τ and noting that $0 \leq \tau_{s_1} \leq \tau_{t_1} \leq \cdots \leq \tau_{s_n} \leq \tau_{t_n} < \infty$, we have

$$\begin{aligned}
\mathbb{E}\left(\prod_{j=1}^{n} \mathrm{e}^{\mathrm{i}\theta_j(Y_{t_j} - Y_{s_j})}\right) &= \mathbb{E}\left(\prod_{j=1}^{n} \mathrm{e}^{-\Psi(\theta_j)(\tau_{t_j} - \tau_{s_j})}\right) \\
&= \mathbb{E}\left(\prod_{j=1}^{n} \mathrm{e}^{-\Psi(\theta_j)\tau_{t_j - s_j}}\right) \\
&= \prod_{j=1}^{n} \mathrm{e}^{-\Xi(\mathrm{i}\Psi(\theta_j))(t_j - s_j)},
\end{aligned}$$

where in the final equality, we have used the fact that τ has stationary independent increments together with (2.28). Exercise 1.1 now allows us to conclude that Y has stationary and independent increments. □

Suppose in the above lemma, we take for X a linear Brownian motion with drift as in the exponent of (2.26). By sampling this continuous path process along the range of an independent subordinator, one recovers another Lévy process. Geman et al. (2001) suggest that one may consider the value of a risky asset to evolve as the process (2.26) on an abstract time scale suitable to the rate of business transactions, called *business time*. The link between business time and real time is given by the subordinator τ. That is to say, one assumes that the value of a given risky asset follows the process $Y = X \circ \tau$ because, at real time $s > 0$, τ_s units of business time have passed and hence the value of the risky asset is positioned at X_{τ_s}.

Returning to the example of the variance-gamma process given above, it turns out that one may recover it from a linear Brownian motion by applying a time change using a gamma subordinator. See Exercise 2.9 for more details on the facts mentioned here concerning the variance-gamma process as well as Exercise 2.12 for more examples of Lévy processes which may be written in terms of a subordinated Brownian motion with drift.

[8] The notation $\Im z$ refers to the imaginary part of z.

Exercises

2.1 The objective of this exercise is to give a reminder of the additive property of Poisson distributions (which is also the reason why they belong to the class of infinite divisible distributions). Suppose that $\{N_i : i = 1, 2, \ldots\}$ is an independent sequence of random variables defined on $(\Omega, \mathcal{F}, P)$ which are Poisson distributed with parameters λ_i, for $i = 1, 2, \ldots$, respectively. Let $S = \sum_{i\geq 1} N_i$. Show that

(i) if $\sum_{i\geq 1} \lambda_i < \infty$ then S is Poisson distributed with parameter $\sum_{i\geq 1} \lambda_i$ and hence in particular $P(S < \infty) = 1$,
(ii) if $\sum_{i\geq 1} \lambda_i = \infty$ then $P(S = \infty) = 1$.

2.2 Denote by $\{T_i : i \geq 1\}$ the arrival times in the Poisson process $N = \{N_t : t \geq 0\}$ with parameter λ.

(i) By recalling that inter-arrival times are independent and exponentially distributed, show that, for any $A \in \mathcal{B}([0, \infty)^n)$,

$$P\big((T_1, \ldots, T_n) \in A | N_t = n\big) = \int_A \frac{n!}{t^n} \mathbf{1}_{(0\leq t_1 \leq \cdots \leq t_n \leq t)} \mathrm{d}t_1 \times \cdots \times \mathrm{d}t_n.$$

(ii) Deduce that the distribution of $(T_1, \ldots, T_n)$, conditional on $N_t = n$, has the same law as the distribution of an ordered independent sample of size n taken from the uniform distribution on $[0, t]$.

2.3 If η is a measure on $(S, \mathcal{S})$ and $f : S \to [0, \infty)$ is measurable then show that $\int_S (1 - \mathrm{e}^{-\phi f(x)})\eta(\mathrm{d}x) < \infty$ for all $\phi > 0$ if and only if $\int_S (1 \wedge f(x))\eta(\mathrm{d}x) < \infty$.

2.4 Recall that $D[0, 1]$ is the space of functions $f : [0, 1] \to \mathbb{R}$ which are right-continuous with left limits.

(i) Define the norm $\|f\| = \sup_{x\in[0,1]} |f(x)|$. Use the triangle inequality to deduce that, if $\{f_n : n \geq 1\}$ is a sequence in $D[0, 1]$ and $f : [0, 1] \to \mathbb{R}$ such that $\lim_{n\uparrow\infty} \|f_n - f\| = 0$, then $f \in D[0, 1]$.
(ii) Suppose that $f \in D[0, 1]$ and let $\Delta = \{t \in [0, 1] : |f(t) - f(t-)| \neq 0\}$ (the set of discontinuity points). Show that Δ is countable if Δ_c is countable, for all $c > 0$, where $\Delta_c = \{t \in [0, 1] : |f(t) - f(t-)| > c\}$. Next fix $c > 0$. Suppose for contradiction that Δ_c has an accumulation point, say x. Show that the existence of either a left or right limit at x leads to the conclusion that there is no left or right limit of f at x. Deduce that Δ_c, and hence Δ, is countable.

2.5 The explicit construction of a Lévy process given in the Lévy–Itô decomposition begs the question as to whether one may construct examples of *deterministic* functions which have similar properties to those of the paths of Lévy processes. The objective of this exercise is to do precisely that. The reader is warned, however, that this is purely an analytical exercise and one should not necessarily think of the paths of Lévy processes as being entirely similar to the functions constructed below in all respects.

(i) Let us recall the definition of the Cantor function, which we shall use to construct a deterministic function that has bounded variation, that is right-continuous with left limits and whose points of discontinuity are dense in its domain. Take the interval $C_0 := [0, 1]$ and perform the following iteration. For $n \geq 0$ define C_n as the union of intervals which remain when removing the middle third of each of the intervals which make up C_{n-1}. The Cantor set C is the limiting object, $\bigcap_{n\geq 0} C_n$ and can be described by

$$C = \left\{ x \in [0, 1] : x = \sum_{k\geq 1} \frac{\alpha_k}{3^k} \text{ such that } \alpha_k \in \{0, 2\} \text{ for each } k \geq 1 \right\}.$$

One sees that the Cantor set is simply the remaining points in $[0, 1]$ after omitting numbers whose tertiary expansion contains the digit 1. To describe the Cantor function, for each $x \in [0, 1]$, let $j(x)$ be the smallest j for which $\alpha_j = 1$ in the tertiary expansion of $\sum_{k\geq 1} \alpha_k/3^k$ of x. If $x \in C$, then $j(x) = \infty$ and otherwise, if $x \in [0, 1]\backslash C$, then $1 \leq j(x) < \infty$. The Cantor function is defined as follows

$$f(x) = \frac{1}{2^{j(x)}} + \sum_{i=1}^{j(x)-1} \frac{\alpha_i}{2^{i+1}} \quad \text{for } x \in [0, 1].$$

Now consider the function $g : [0, 1] \to \mathbb{R}$, given by $g(x) = f^{-1}(x) - ax$ for $a \in \mathbb{R}$. Here, we understand $f^{-1}(x) = \inf\{\theta : f(\theta) > x\}$. Note that g is monotone if and only if $a \leq 0$. Show that g has only positive jumps and the values of x for which g jumps form a dense set in $[0, 1]$. Show further that g has bounded variation on $[0, 1]$.

(ii) Now let us construct an example of a deterministic function which has unbounded variation and that is right-continuous with left limits. Denote by $\mathbb{Q}_2$ the dyadic rationals. Consider a function $J : [0, \infty) \to \mathbb{R}$ as follows. For all $x \geq 0$ which are not in $\mathbb{Q}_2$, set $J(x) = 0$. It remains to assign a value to J for each $x = a/2^n$ where $a = 1, 3, 5, \ldots$ (even values of a cancel). Let

$$J(a/2^n) = \begin{cases} 2^{-n} & \text{if } a = 1, 5, 9, \ldots \\ -2^{-n} & \text{if } a = 3, 7, 11, \ldots \end{cases}$$

and define

$$f(x) = \sum_{s\in[0,x]\cap\mathbb{Q}_2} J(s).$$

Show that f is uniformly bounded on $[0, 1]$, is right-continuous with left limits and has unbounded variation over $[0, 1]$.

2.6 Suppose that X is a Lévy process with Lévy measure Π.

(i) For each $n \geq 2$ show that for each $t > 0$,

$$\mathbb{E}\left[\int_{[0,t]}\int_{\mathbb{R}} |x|^n N(\mathrm{d}s \times \mathrm{d}x)\right] < \infty$$

almost surely if and only if

$$\int_{|x|\geq 1} |x|^n \Pi(\mathrm{d}x) < \infty.$$

(ii) Suppose now that Π satisfies $\int_{|x|\geq 1} |x|^n \Pi(\mathrm{d}x) < \infty$ for $n \geq 2$. Show that

$$\int_{[0,t]}\int_{\mathbb{R}} x^n N(\mathrm{d}s \times \mathrm{d}x) - t\int_{\mathbb{R}} x^n \Pi(\mathrm{d}x), \quad t \geq 0,$$

is a martingale.

2.7 Let X be a Lévy process with Lévy measure Π. Denote by N the Poisson random measure associated with its jumps.

(i) Show that

$$\mathbb{P}\Big(\sup_{0<s\leq t} |X_s - X_{s-}| \geq a\Big) = 1 - \mathrm{e}^{-t\Pi(\mathbb{R}\setminus(-a,a))},$$

for $a > 0$.

(ii) Show that the paths of X are continuous if and only if $\Pi = 0$.

(iii) Show that the paths of X are piecewise linear if and only if it is a compound Poisson process with drift if and only if $\sigma = 0$ and $\Pi(\mathbb{R}) < \infty$. (Recall that a function $f : [0,\infty) \to \mathbb{R}$ is right-continuous and piecewise linear if there exist sequence of times $0 = t_0 < t_1 < \cdots < t_n < \cdots$ with $\lim_{n\uparrow\infty} t_n = \infty$ such that on $[t_{j-1}, t_j)$ the function f is linear.)

(iv) Now suppose that $\Pi(\mathbb{R}) = \infty$. Argue by contradiction that, for each positive rational $q \in \mathbb{Q}$, there exists a decreasing sequence of jump times for X, say $\{T_n(\omega) : n \geq 0\}$, such that $\lim_{n\uparrow\infty} T_n = q$. Hence deduce that the set of jump times are dense in $[0,\infty)$.

2.8 Show that any Lévy process of bounded variation may be written as the difference of two independent subordinators.

2.9 This exercise gives another explicit example of a Lévy process, the variance-gamma process, introduced by Madan and Seneta (1990) to model financial data.

(i) Suppose that $\Gamma = \{\Gamma_t : t \geq 0\}$ is a gamma subordinator with parameters α, β and that $B = \{B_t : t \geq 0\}$ is an independent standard Brownian motion. Show that, for $c \in \mathbb{R}$ and $\sigma > 0$, the variance-gamma process

$$X_t := c\Gamma_t + \sigma B_{\Gamma_t}, \quad t \geq 0,$$

is a Lévy process with characteristic exponent

$$\Psi(\theta) = \beta \log\left(1 - \mathrm{i}\frac{\theta c}{\alpha} + \frac{\sigma^2\theta^2}{2\alpha}\right), \quad \theta \in \mathbb{R}.$$

(ii) Show that the variance-gamma process is equal in law to the Lévy process

$$\Gamma^{(1)} - \Gamma^{(2)} = \{\Gamma_t^{(1)} - \Gamma_t^{(2)} : t \geq 0\},$$

where $\Gamma^{(1)}$ is a gamma subordinator with parameters

$$\alpha^{(1)} = \left(\sqrt{\frac{1}{4}\frac{c^2}{\alpha^2} + \frac{1}{2}\frac{\sigma^2}{\alpha}} + \frac{1}{2}\frac{c}{\alpha}\right)^{-1} \quad \text{and} \quad \beta^{(1)} = \beta$$

and $\Gamma^{(2)}$ is a gamma subordinator, independent of $\Gamma^{(1)}$, with parameters

$$\alpha^{(2)} = \left(\sqrt{\frac{1}{4}\frac{c^2}{\alpha^2} + \frac{1}{2}\frac{\sigma^2}{\alpha}} - \frac{1}{2}\frac{c}{\alpha}\right)^{-1} \quad \text{and} \quad \beta^{(2)} = \beta.$$

2.10 Suppose that d is an integer greater than one. Choose $\mathbf{a} \in \mathbb{R}^d$ and let Π be a measure concentrated on $\mathbb{R}^d\backslash\{0\}$ satisfying

$$\int_{\mathbb{R}^d} (1 \wedge |\mathbf{x}|^2)\Pi(\mathrm{d}\mathbf{x}) < \infty,$$

where $|\cdot|$ is the standard Euclidean norm. Show that it is possible to construct a d-dimensional process $\mathbf{X} = \{\mathbf{X}_t : t \geq 0\}$ on a probability space $(\Omega, \mathcal{F}, \mathbb{P})$ having the following properties.

(i) The paths of $\mathbf{X}$ are right-continuous with left limits $\mathbb{P}$-almost surely in the sense that, for each $t \geq 0$,

$$\mathbb{P}\left(\lim_{s \downarrow t} \mathbf{X}_s = \mathbf{X}_t\right) = 1 \quad \text{and} \quad \mathbb{P}\left(\lim_{s \uparrow t} \mathbf{X}_s \text{ exists}\right) = 1.$$

(ii) $\mathbb{P}(\mathbf{X}_0 = \mathbf{0}) = 1$, the zero vector in $\mathbb{R}^d$.
(iii) For $0 \leq s \leq t$, $\mathbf{X}_t - \mathbf{X}_s$ is independent of $\{\mathbf{X}_u : u \leq s\}$.
(iv) For $0 \leq s \leq t$, $\mathbf{X}_t - \mathbf{X}_s$ is equal in distribution to $\mathbf{X}_{t-s}$.
(v) For any $t \geq 0$ and $\theta \in \mathbb{R}^d$,

$$\mathbb{E}(\mathrm{e}^{\mathrm{i}\theta\cdot\mathbf{X}_t}) = \mathrm{e}^{-\Psi(\theta)t}$$

and

$$\Psi(\theta) = \mathrm{i}\mathbf{a}\cdot\theta + \frac{1}{2}\theta\cdot\mathbf{A}\theta + \int_{\mathbb{R}^d} \left(1 - \mathrm{e}^{\mathrm{i}\theta\cdot\mathbf{x}} + \mathrm{i}(\theta\cdot\mathbf{x})\mathbf{1}_{(|\mathbf{x}|<1)}\right)\Pi(\mathrm{d}\mathbf{x}), \quad (2.29)$$

where for any two vectors $\mathbf{x}$ and $\mathbf{y}$ in $\mathbb{R}^d$, $\mathbf{x}\cdot\mathbf{y}$ is the usual inner product and $\mathbf{A}$ is a $d \times d$ Gaussian covariance matrix.

2.11 Suppose that X is a subordinator.

(i) Show that it has a Laplace exponent given by

$$-\log \mathbb{E}(\mathrm{e}^{-qX_1}) =: \Phi(q) = \delta q + \int_{(0,\infty)} (1 - \mathrm{e}^{-qx})\Pi(\mathrm{d}x),$$

for $q \geq 0$, where $\delta \geq 0$ and $\int_{(0,\infty)} (1 \wedge x)\Pi(\mathrm{d}x) < \infty$.

(ii) Show using integration by parts that

$$\Phi(q) = \delta q + q \int_0^\infty \mathrm{e}^{-qx} \Pi(x,\infty)\mathrm{d}x$$

and hence that the drift term δ may be recovered from the limit

$$\lim_{q\uparrow\infty} \frac{\Phi(q)}{q} = \delta.$$

(iii) Show that

$$\lim_{q\downarrow 0} \frac{\Phi(q)}{q} = \mathbb{E}(X_1) = \delta + \int_{(0,\infty)} x\Pi(\mathrm{d}x) \in (0,\infty].$$

(iv) Finally, prove that $\Phi(\infty) < \infty$ if and only if X is a compound Poisson subordinator. That is to say, $\delta = 0$ and $\Pi(0,\infty) < \infty$, in which case $\Phi(\infty) = \delta + \Pi(0,\infty)$.

2.12 Here are some more examples of Lévy processes which may be written as a subordinated Brownian motion.

(i) Let $\alpha \in (0,2)$. Show that a Brownian motion subordinated by a stable process of index $\alpha/2$ is a symmetric stable process of index α.

(ii) Suppose that $X = \{X_t : t \geq 0\}$ is a compound Poisson process with Lévy measure given by

$$\Pi(\mathrm{d}x) = \left\{\mathbf{1}_{(x<0)}\mathrm{e}^{-a|x|} + \mathbf{1}_{(x>0)}\mathrm{e}^{-ax}\right\}\mathrm{d}x,$$

for $a > 0$. Now let $\tau = \{\tau_s : s \geq 0\}$ be a pure jump subordinator with Lévy measure

$$\pi(\mathrm{d}x) = \mathbf{1}_{(x>0)} 2a\mathrm{e}^{-a^2 x}\mathrm{d}x.$$

Show that $\{\sqrt{2}B_{\tau_s} : s \geq 0\}$ has the same law as X, where $B = \{B_t : t \geq 0\}$ is a standard Brownian motion independent of τ.

(iii) Suppose now that $X = \{X_t : t \geq 0\}$ is a compound Poisson process with Lévy measure given by

$$\Pi(\mathrm{d}x) = \frac{\lambda\sqrt{2}}{\sigma\sqrt{\pi}}\mathrm{e}^{-x^2/2\sigma^2}\mathrm{d}x,$$

for $x \in \mathbb{R}$. Show that $\{\sigma B_{N_t} : t \geq 0\}$ has the same law as X, where B is as in part (ii) and $\{N_s : s \geq 0\}$ is a Poisson process with rate 2λ independent of B.

The final part of this question gives a simple example of Lévy processes which may be written as a subordinated Lévy process.

(iv) Suppose that X is a symmetric stable process of index $\alpha \in (0, 2)$. Show that X can be written as a symmetric stable process of index α/β subordinated by an independent stable subordinator of index $\beta \in (0, 1)$.

Chapter 3
More Distributional and Path-Related Properties

In this chapter, we consider some more distributional and path-related properties of general Lévy processes. Specifically, we examine the strong Markov property, duality, moments and exponential change of measure.

We recall here our notation that any Lévy process, $X = \{X_t : t \geq 0\}$, is assumed to be defined on a probability space $(\Omega, \mathcal{F}, \mathbb{P})$, which is endowed with a filtration $\mathbb{F} = \{\mathcal{F}_t : t \geq 0\}$, which is the natural enlargement of the filtration generated by X.

3.1 The Strong Markov Property

The process $X = \{X_t : t \geq 0\}$ possesses the Markov property if, for each $B \in \mathcal{B}(\mathbb{R})$ and $s, t \geq 0$,

$$\mathbb{P}(X_{t+s} \in B | \mathcal{F}_t) = \mathbb{P}\big(X_{t+s} \in B | \sigma(X_t)\big). \tag{3.1}$$

It is easy to see that the Markov property is satisfied for all Lévy processes. Indeed, Lévy processes satisfy the stronger condition that the law of $X_{t+s} - X_t$ is independent of $\mathcal{F}_t$, for all $s, t \geq 0$.

A non-negative random variable, say τ, is called a *stopping time* if

$$\{\tau \leq t\} \in \mathcal{F}_t,$$

for all $t \geq 0$. It is possible that a stopping time may have the property that $\mathbb{P}(\tau = \infty) > 0$. In addition, for any stopping time τ,

$$\{\tau < t\} = \bigcup_{n \geq 1} \{\tau \leq t - 1/n\} \in \bigcup_{n \geq 1} \mathcal{F}_{t-1/n} \subseteq \mathcal{F}_t.$$

However, we also have that any random time τ which has the property that $\{\tau < t\} \in \mathcal{F}_t$ for all $t \geq 0$ must also be a stopping time. To see why, note that

$$\{\tau \leq t\} = \bigcap_{n \geq 1} \{\tau < t + 1/n\} \in \bigcap_{n \geq 1} \mathcal{F}_{t+1/n} = \mathcal{F}_{t+} = \mathcal{F}_t,$$

A.E. Kyprianou, *Fluctuations of Lévy Processes with Applications*, Universitext,
DOI 10.1007/978-3-642-37632-0_3, © Springer-Verlag Berlin Heidelberg 2014

where in the last equality, we use the right-continuity of the filtration $\mathbb{F}$. In other words, for a Lévy process whose filtration is right-continuous, we may also say that τ is a stopping time if and only if $\{\tau < t\} \in \mathcal{F}_t$ for all $t \geq 0$.

Associated with a given stopping time τ is the sigma-algebra

$$\mathcal{F}_\tau := \big\{A \in \mathcal{F} : A \cap \{\tau \leq t\} \in \mathcal{F}_t \text{ for all } t \geq 0\big\}.$$

(Note, it is a simple exercise to verify that $\mathcal{F}_\tau$ is a sigma-algebra.) The process X is said to satisfy the strong Markov property if, for each stopping time, τ,

$$\mathbb{P}(X_{\tau+s} \in B|\mathcal{F}_\tau) = \mathbb{P}\big(X_{\tau+s} \in B|\sigma(X_\tau)\big) \quad \text{on } \{\tau < \infty\}.$$

The next theorem shows, in particular, that all Lévy processes satisfy the strong Markov property.

Theorem 3.1 *Suppose that τ is a stopping time. Define on $\{\tau < \infty\}$ the process $\widetilde{X} = \{\widetilde{X}_t : t \geq 0\}$ where*

$$\widetilde{X}_t = X_{\tau+t} - X_\tau, \quad t \geq 0.$$

Then, on the event $\{\tau < \infty\}$, the process $\widetilde{X}$ is independent of $\mathcal{F}_\tau$, has the same law as X and hence in particular is a Lévy process.

Proof We need to check that $\widetilde{X}$ has stationary and independent increments which belong to the same family of infinite divisible distributions as X. (Note that $\widetilde{X}$ clearly has paths that are right-continuous with left limits, issued from the origin.) Referring to Exercise 1.1, we see that it would suffice to prove that, for any $n \in \mathbb{N}$, $0 \leq s_1 \leq t_1 \leq \cdots \leq s_n \leq t_n < \infty$, $H \in \mathcal{F}_\tau$ and $\theta_1, \ldots, \theta_n \in \mathbb{R}$,

$$\mathbb{E}\left(\prod_{i=1}^n \mathrm{e}^{\mathrm{i}\theta_i(X_{\tau+t_i} - X_{\tau+s_i})}; H \cap \{\tau < \infty\}\right) = \prod_{i=1}^n \mathrm{e}^{-\Psi(\theta_i)(t_i - s_i)} \mathbb{P}\big(H \cap \{\tau < \infty\}\big),$$

where Ψ is the characteristic exponent of X.

To this end, define a sequence of stopping times $\{\tau^{(n)} : n \geq 1\}$ by

$$\tau^{(n)} = \begin{cases} k2^{-n} & \text{if } (k-1)2^{-n} < \tau \leq k2^{-n} \text{ for } k = 1, 2, \ldots \\ 0 & \text{if } \tau = 0 \\ \infty & \text{if } \tau = \infty. \end{cases} \tag{3.2}$$

Stationary independent increments and the fact that $H \cap \{\tau^{(n)} = k2^{-n}\} \in \mathcal{F}_{k2^{-n}}$ allow us to write

$$\begin{aligned} &\mathbb{E}\left(\prod_{i=1}^n \mathrm{e}^{\mathrm{i}\theta_i(X_{\tau^{(n)}+t_i} - X_{\tau^{(n)}+s_i})}; H \cap \big\{\tau^{(n)} < \infty\big\}\right) \\ &\quad = \sum_{k \geq 0} \mathbb{E}\left(\prod_{i=1}^n \mathrm{e}^{\mathrm{i}\theta_i(X_{\tau^{(n)}+t_i} - X_{\tau^{(n)}+s_i})}; H \cap \big\{\tau^{(n)} = k2^{-n}\big\}\right) \end{aligned}$$

$$
\begin{aligned}
&= \sum_{k\geq 0} \mathbb{E}\left[\mathbb{E}\left(\prod_{i=1}^{n} \mathrm{e}^{\mathrm{i}\theta_i (X_{k2^{-n}+t_i} - X_{k2^{-n}+s_i})} | \mathcal{F}_{k2^{-n}}\right); H \cap \{\tau^{(n)} = k2^{-n}\}\right] \\
&= \sum_{k\geq 0} \prod_{i=1}^{n} \mathbb{E}(\mathrm{e}^{\mathrm{i}\theta_i X_{t_i - s_i}}) \mathbb{P}(H \cap \{\tau^{(n)} = k2^{-n}\}) \\
&= \prod_{i=1}^{n} \mathrm{e}^{-\Psi(\theta_i)(t_i - s_i)} \mathbb{P}(H \cap \{\tau^{(n)} < \infty\}).
\end{aligned}
$$

The paths of X are almost surely right-continuous and $\tau^{(n)} \downarrow \tau$ on $\{\tau < \infty\}$ as n tends to infinity. Hence, $X_{\tau^{(n)}+s} \to X_{\tau+s}$ almost surely on $\{\tau < \infty\}$, for all $s \geq 0$ as n tends to infinity. It follows by the Dominated Convergence Theorem that

$$
\begin{aligned}
&\mathbb{E}\left(\prod_{i=1}^{n} \mathrm{e}^{\mathrm{i}\theta_i (X_{\tau+t_i} - X_{\tau+s_i})}; H \cap \{\tau < \infty\}\right) \\
&\quad = \lim_{n\uparrow\infty} \mathbb{E}\left(\prod_{i=1}^{n} \mathrm{e}^{\mathrm{i}\theta_i (X_{\tau^{(n)}+t_i} - X_{\tau^{(n)}+s_i})}; H \cap \{\tau^{(n)} < \infty\}\right) \\
&\quad = \lim_{n\uparrow\infty} \prod_{i=1}^{n} \mathrm{e}^{-\Psi(\theta_i)(t_i - s_i)} \mathbb{P}(H \cap \{\tau^{(n)} < \infty\}) \\
&\quad = \prod_{i=1}^{n} \mathrm{e}^{-\Psi(\theta_i)(t_i - s_i)} \mathbb{P}(H \cap \{\tau < \infty\}).
\end{aligned}
$$

This shows that $\widetilde{X}$ is independent of $\mathcal{F}_\tau$ on $\{\tau < \infty\}$ and has the same law as X. □

Examples of $\mathbb{F}$-stopping times which will repeatedly occur in the remaining text are those of the *first-entrance time* and *first-hitting time* of a given open or closed set $B \subseteq \mathbb{R}$. They are defined as

$$
T^B = \inf\{t \geq 0 : X_t \in B\} \quad \text{and} \quad \tau^B = \inf\{t > 0 : X_t \in B\},
$$

respectively. We take the usual definition $\inf \emptyset = \infty$ here. At many places throughout this book, we shall work with the special cases that B is equal to (x, ∞), $[x, \infty)$, $(-\infty, x)$, $(-\infty, x]$ and $\{x\}$ where $x \in \mathbb{R}$. The two times T^B and τ^B are very closely related. They are equal when $X_0 \notin \overline{B}$; but they may possibly differ in value when $X_0 \in \overline{B}$. Consider for example the case that $B = [0, \infty)$ and X is a compound Poisson process with strictly negative drift. When $X_0 = 0$, we have $\mathbb{P}(T^B = 0) = 1$ whereas $\mathbb{P}(\tau^B > 0) = 1$.[1]

---

[1] As we shall see later, this is a phenomenon which is not exclusive to compound Poisson processes with strictly negative drift. The same behaviour is experienced by, for example, Lévy processes of bounded variation with strictly negative drift.

To some extent, it is intuitively obvious why T^B and τ^B are stopping times. Nonetheless, we complete this section by justifying this claim. The justification comes in the form of a supporting lemma and a theorem establishing the claim. The lemma illustrates that there exists a sense of left-continuity of Lévy processes when appropriately sampling the path with an increasing sequence of stopping times; that is, *quasi-left-continuity*. The proofs of the forthcoming lemma and theorem are quite technical and it will do no harm if the reader chooses to bypass their proofs and continue reading on to the next section. The arguments given are rooted in the works of Dellacherie and Meyer (1975–1993) and Blumenthal and Getoor (1968) who give a comprehensive and highly detailed account of the theory of Markov processes in general.

Lemma 3.2 (Quasi-Left-Continuity) *If T is an $\mathbb{F}$-stopping time and $\{T_n: n \geq 1\}$ is an increasing sequence of $\mathbb{F}$-stopping times such that $\lim_{n\uparrow\infty} T_n = T$ almost surely, then $\lim_{n\uparrow\infty} X_{T_n} = X_T$ on $\{T < \infty\}$. Hence, if $T_n < T$ almost surely for each $n \geq 1$, then X is left-continuous at T on $\{T < \infty\}$.*

Note that for any fixed $t > 0$, $\mathbb{P}(N(\{t\} \times \mathbb{R}) = 0) = 1$, where N is the Poisson random measure describing the jumps of X, and hence t is a jump time with probability zero. If $\{t_n: n = 1, 2, \ldots\}$ is a sequence of deterministic times satisfying $t_n \to t$ as $n \uparrow \infty$, then with probability one $X_{t_n} \to X_t$. In other words, t is a point of continuity of X.[2] The statement in the above lemma thus asserts that this property extends to the case of increasing stopping times.

Proof of Lemma 3.2 First suppose that $\mathbb{P}(T < \infty) = 1$. As the sequence $\{T_n : n \geq 1\}$ is almost surely increasing, we can identify the limit of $\{X_{T_n} : n \geq 0\}$ by

$$Z = \mathbf{1}_A X_T + \mathbf{1}_{A^c} X_{T-},$$

where $A = \bigcup_{n\geq 1} \bigcap_{k\geq n} \{T_k = T\} = \{T_k = T \text{ eventually}\}$. Suppose that f and g are two continuous functions, each with compact support. Appealing to bounded convergence (twice), together with the right-continuity and left limits of paths, we have

$$\begin{aligned}\lim_{t\downarrow 0} \lim_{n\uparrow\infty} \mathbb{E}\big(f(X_{T_n})g(X_{T_n+t})\big) &= \lim_{t\downarrow 0} \mathbb{E}\big(f(Z)g(X_{(T+t)-})\big) \\ &= \mathbb{E}\big(f(Z)g(X_T)\big).\end{aligned} \tag{3.3}$$

Now write, for short, $P_t g(x) = \mathbb{E}(g(x + X_t)) = \mathbb{E}_x(g(X_t))$, which is uniformly bounded in x and t and, by bounded convergence, continuous in x. Note that right-continuity of X, together with bounded convergence, also implies that

[2] It is worth reminding oneself, for the sake of clarity, that $X_{t_n} \to X_t$ $\mathbb{P}$-a.s. as $n \uparrow \infty$ means that, for all $\varepsilon > 0$, there exists an almost surely finite $N > 0$ such that $|X_{t_n} - X_t| < \varepsilon$ for all $n > N$. This does not contradict the fact that there might be an infinite number of discontinuities in the path of X in an arbitrary small neighbourhood of t.

$\lim_{t\downarrow 0} P_t g(x) = g(x)$ for each $x \in \mathbb{R}$. These facts, together with the Markov property applied at time T_n and bounded convergence, imply that

$$\begin{aligned}\lim_{t\downarrow 0}\lim_{n\uparrow\infty}\mathbb{E}\big(f(X_{T_n})g(X_{T_n+t})\big) &= \lim_{t\downarrow 0}\lim_{n\uparrow\infty}\mathbb{E}\big(f(X_{T_n})P_t g(X_{T_n})\big)\\ &= \lim_{t\downarrow 0}\mathbb{E}\big(f(Z)P_t g(Z)\big)\\ &= \mathbb{E}\big(f(Z)g(Z)\big).\end{aligned} \tag{3.4}$$

Equating (3.3) and (3.4), we see that, for all uniformly bounded continuous functions f and g,

$$\mathbb{E}\big(f(Z)g(X_T)\big) = \mathbb{E}\big(f(Z)g(Z)\big).$$

From this equality, we may deduce (by splitting into real and imaginary parts) that

$$\mathbb{E}\big(e^{i\theta_1 Z + i\theta_2 X_T}\big) = \mathbb{E}\big(e^{i\theta_1 Z + i\theta_2 Z}\big),$$

where $\theta_1, \theta_1 \in \mathbb{R}$, and hence $Z = X_T$ almost surely.

When $T_n < T$ almost surely for all $n \geq 1$, it is clear that $Z = X_{T-}$ and the concluding sentence in the statement of the lemma follows for the case that $\mathbb{P}(T < \infty) = 1$.

To remove the requirement that $\mathbb{P}(T < \infty) = 1$, recall that for each $t > 0$, $T \wedge t$ is a finite stopping time. We have that $T_n \wedge t \uparrow T \wedge t$ as $n \uparrow \infty$ and hence, from the previous part of the proof, $\lim_{n\uparrow\infty} X_{T_n\wedge t} = X_{T\wedge t}$ almost surely. In other words, $\lim_{n\uparrow\infty} X_{T_n} = X_T$ on $\{T \leq t\}$. Since we may take t arbitrarily large the result follows. □

Theorem 3.3 *Suppose that B is open or closed. Then,*
(i) T^B is a stopping time and $X_{T^B} \in \overline{B}$ on $\{T^B < \infty\}$ and
(ii) τ^B is a stopping time and $X_{\tau^B} \in \overline{B}$ on $\{\tau^B < \infty\}$.

(Note that $\overline{B} = B$ when B is closed.)

Proof (i) First, we deal with the case that B is open. Since any Lévy process $X = \{X_t : t \geq 0\}$ has right-continuous paths and B is open, we may describe the event $\{T^B < t\}$ in terms of the path of X at rational times. That is to say,

$$\{T^B < t\} = \bigcup_{s\in\mathbb{Q}\cap[0,t)} \{X_s \in B\}. \tag{3.5}$$

Since each of the sets in the union is $\mathcal{F}_t$-measurable and sigma-algebras are closed under countable set operations, we have that $\{T^B < t\}$ is also $\mathcal{F}_t$-measurable. Recalling that $\mathbb{F}$ is right-continuous, we have that $\{T^B < t\}$ is $\mathcal{F}_t$-measurable if and only if $\{T^B \leq t\}$ is $\mathcal{F}_t$-measurable and hence T^B fulfils the definition of an $\mathbb{F}$-stopping time. Now note that, on $\{T^B < \infty\}$, we have that either $X_{T^B} \in B$ or that at the time T^B, X is at the boundary of B and at the next instant moves into B.

That is to say, on $\{T^B < \infty\}$, there exists a sequence of (random) times $\{\sigma_n : n \geq 1\}$, such that $\sigma_n \downarrow T^B$ with $X_{\sigma_n} \in B$ for all $n \geq 1$, in which case, right-continuity of paths implies that $X_{T^B} \in \overline{B}$. For illustrative purposes, consider the example where $B = (x, \infty)$ for some $x > 0$ and X is any compound Poisson process with strictly positive drift and negative jumps. It is clear that $\mathbb{P}(X_{T^{(x,\infty)}} = x) > 0$ as the process may drift up to the boundary point $\{x\}$ and then continue into (x, ∞) before, for example, the first jump occurs.

For the case of closed B, the argument given above is not subtle enough for the proof. The reason why lies with the possibility that X may enter B simply by touching its boundary, which is now included in B. Further, this may occur in a way that cannot be described in terms of a countable sequence of events.

We thus employ another technique for the proof of (i) when B is closed. Suppose that $\{B_n : n \geq 1\}$ is a sequence of open sets given by

$$B_n = \{x \in \mathbb{R} : |x - y| < 1/n \text{ for some } y \in B\}.$$

Note that $B \subset B_n$ for all $n \geq 1$ and $\bigcap_{n\geq 1} \overline{B}_n = B$. From the previous paragraph, we have that T^{B_n} are $\mathbb{F}$-stopping times and, clearly, they are increasing. Denote their limit by T. Since, for all $t \geq 0$,

$$\{T \leq t\} = \left\{\sup_{n\geq 1} T^{B_n} \leq t\right\} = \bigcap_{n\geq 1} \{T^{B_n} \leq t\} \in \mathcal{F}_t,$$

we see that T is an $\mathbb{F}$-stopping time. Obviously $T^{B_n} \leq T^B$ for all n and hence $T \leq T^B$. On the other hand, according to quasi-left-continuity described in the previous lemma, $\lim_{n\uparrow\infty} X_{T^{B_n}} = X_T$ on the event $\{T < \infty\}$, showing that $X_T \in \overline{B} = B$ and hence that $T \geq T^B$ on $\{T < \infty\}$. In conclusion, we have that $T = T^B$ and $X_{T^B} \in B$ on $\{T^B < \infty\}$.

(ii) Suppose now that B is open. Let $T^B_\varepsilon = \inf\{t \geq \varepsilon : X_t \in B\}$. Note that $\{T^B_\varepsilon < t\} = \emptyset \in \mathcal{F}_t$ for all $t < \varepsilon$ and for $t \geq \varepsilon$,

$$\{T^B_\varepsilon < t\} = \bigcup_{s\in\mathbb{Q}\cap[\varepsilon,t)} \{X_s \in B\},$$

which is $\mathcal{F}_t$. Hence by right-continuity of $\mathbb{F}$, T^B_ε is an $\mathbb{F}$-stopping time. Now suppose that B is closed. Following the arguments in part (i) but with $T^{B_n}_\varepsilon := \inf\{t \geq \varepsilon : X_t \in B_n\}$, we conclude for closed B that T^B_ε is again an $\mathbb{F}$-stopping time. In both cases, when B is open or closed, we also see, as in part (i), that $X_{T^B_\varepsilon} \in \overline{B}$ on $\{T^B_\varepsilon < \infty\}$.

Now suppose that B is open or closed. The sequence of stopping times $\{T^B_\varepsilon : \varepsilon > 0\}$ forms a decreasing sequence as $\varepsilon \downarrow 0$ and hence has an almost sure limit, which is equal to τ^B by definition. Note also that $\{T^B_\varepsilon < \infty\}$ increases to $\{\tau^B < \infty\}$ as $\varepsilon \downarrow 0$. Since for all $t \geq 0$ and decreasing sequences $\varepsilon_n \downarrow 0$,

$$\{\tau^B \leq t\}^c = \left\{\inf_{n\geq 1} T^B_{\varepsilon_n} > t\right\} = \bigcap_{n\geq 1} \{T^B_{\varepsilon_n} > t\} \in \mathcal{F}_t,$$

we see that τ^B is an $\mathbb{F}$-stopping time. Right-continuity of the paths of X tell us that $\lim_{\varepsilon\downarrow 0} X_{T^B_\varepsilon} = X_{\tau^B}$ on $\{\tau^B < \infty\}$. Hence $X^{\tau^B} \in \overline{B}$ whenever $\{\tau^B < \infty\}$. □

3.2 Duality

In this section, we discuss a simple feature of all Lévy processes, which follows as a direct consequence of stationary independent increments. That is, when the path of a Lévy process, taken over a finite time horizon, is time reversed (in an appropriate sense), the new path is equal in law to the negative of the original process. This property will prove to be of crucial importance in a number of fluctuation calculations later on.

Lemma 3.4 (Duality Lemma) *For each fixed $t > 0$, define the reversed process*

$$\{X_{(t-s)-} - X_t : 0 \leq s \leq t\}$$

and the dual process,

$$\{-X_s : 0 \leq s \leq t\}.$$

Then the two processes have the same law under $\mathbb{P}$.

Proof Define the time reversed process $Y_s = X_{(t-s)-} - X_t$ for $0 \leq s \leq t$ and note that, under $\mathbb{P}$, we have $Y_0 = 0$ almost surely since t is a jump time with probability zero. As can be seen from Fig. 3.1, the paths of Y are obtained from those of X by a reflection about the vertical axis, with an adjustment of the continuity at the jump times so that its paths are almost surely right-continuous with left limits. The stationary independent increments of X imply directly that the same is true for Y. Moreover, for each $0 \leq s \leq t$, the distribution of $X_{(t-s)-} - X_t$ is identical to that of $-X_s$ and hence, since the finite time distributions of Y determine its law, the proof is complete. □

The Duality Lemma is also well known for (and in fact originates from the theory of) random walks, the discrete-time analogue of Lévy processes, and is justified using an identical proof. See for example Sect. 2 of Chap. XII in Feller (1971).

One interesting feature that follows as a consequence of the Duality Lemma, is the relationship between the running supremum, the running infimum, the process reflected in its supremum and the process reflected in its infimum. The last four objects are, respectively,

$$\overline{X}_t := \sup_{0\leq s\leq t} X_s, \qquad \underline{X}_t := \inf_{0\leq s\leq t} X_s$$

$$\{\overline{X}_t - X_t : t \geq 0\} \quad \text{and} \quad \{X_t - \underline{X}_t : t \geq 0\}.$$

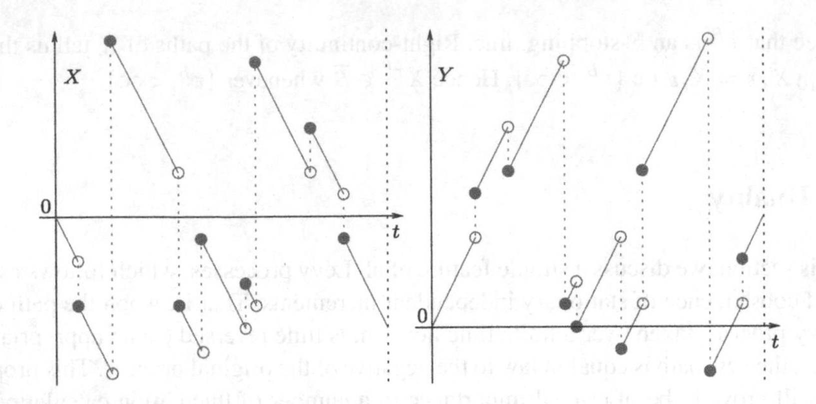

Fig. 3.1 Duality of the processes $X=\{X_s : s\le t\}$ and $Y=\{X_{(t-s)-}-X_t : s\le t\}$. The path of Y is a reflection of the path of X with an adjustment of continuity at jump times.

Lemma 3.5 *For each fixed $t>0$, the pairs $(\overline{X}_t, \overline{X}_t - X_t)$ and $(X_t - \underline{X}_t, -\underline{X}_t)$ have the same distribution under $\mathbb{P}$.*

Proof For $0\le s\le t$, define $\widetilde{X}_s = X_t - X_{(t-s)-}$ and write $\underline{\widetilde{X}}_t = \inf_{0\le s\le t}\widetilde{X}_s$. Using right-continuity and left limits of paths, we may deduce that

$$(\overline{X}_t, \overline{X}_t - X_t) = (\widetilde{X}_t - \underline{\widetilde{X}}_t, -\underline{\widetilde{X}}_t)$$

almost surely. One may visualise this in Fig. 3.2. By rotating the picture by 180° one sees the almost sure equality of the pairs $(\overline{X}_t, \overline{X}_t - X_t)$ and $(\widetilde{X}_t - \underline{\widetilde{X}}_t, -\underline{\widetilde{X}}_t)$. Now appealing to the Duality Lemma, we have that $\{\widetilde{X}_s : 0\le s\le t\}$ is equal in law to $\{X_s : 0\le s\le t\}$ under $\mathbb{P}$. The result now follows. □

3.3 Exponential Moments and Martingales

It is well known that the distribution of the position of a Brownian motion at a fixed time has moments of all orders. It is natural therefore to cast an eye on similar issues for Lévy processes. In general, the picture is not so straightforward. One needs only to consider compound Poisson processes to see how things can differ. Suppose we write the aforementioned process in the form

$$X_t = \sum_{i=1}^{N_t}\xi_i, \quad t\ge 0,$$

where $N=\{N_t\colon t\ge 0\}$ is a Poisson process and $\{\xi_i\colon i\ge 0\}$ are independent and identically distributed. By choosing the jump distribution of each ξ_i in such a way

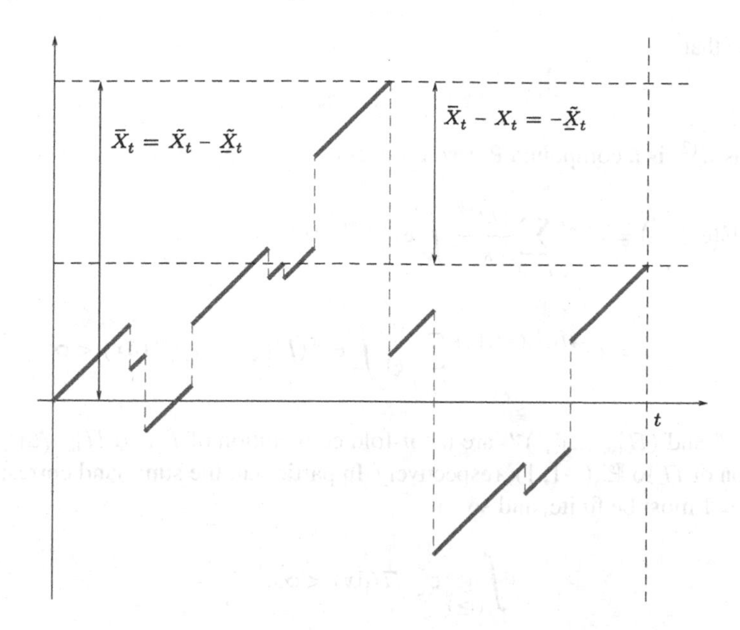

Fig. 3.2 Duality of the pairs $(\overline{X}_t, \overline{X}_t - X_t)$ and $(X_t - \underline{X}_t, -\underline{X}_t)$.

that it has infinite first moment (for example any stable distribution on $(0, \infty)$ with index $\alpha \in (0, 1)$), it is clear that

$$\mathbb{E}(X_t) = \lambda t \mathbb{E}(\xi_1) = \infty,$$

for all $t > 0$.

As one might suspect, there is an intimate relationship between the moments of the Lévy measure and the moments of the distribution of the associated Lévy process at any fixed time. This is indeed the case and we have the following theorem.

Theorem 3.6 *Let $\beta \in \mathbb{R}$, then*

$$\mathbb{E}\big(\mathrm{e}^{\beta X_t}\big) < \infty, \quad \textit{for all } t \geq 0, \quad \textit{if and only if } \int_{|x|\geq 1} \mathrm{e}^{\beta x} \Pi(\mathrm{d}x) < \infty.$$

Proof The statement of the theorem is obvious when $\beta = 0$. Therefore, we shall always assume that $\beta \neq 0$. First suppose that $\mathbb{E}(\mathrm{e}^{\beta X_t}) < \infty$ for some $t > 0$. Recall $X^{(1)}$, $X^{(2)}$ and $X^{(3)}$ given in the Lévy–Itô decomposition. Note, in particular, that $X^{(2)}$ is a compound Poisson process with arrival rate $\lambda := \Pi(\mathbb{R}\backslash(-1, 1))$ and jump distribution $F(\mathrm{d}x) := \mathbf{1}_{(|x|\geq 1)}\Pi(\mathrm{d}x)/\Pi(\mathbb{R}\backslash(-1, 1))$, and $X^{(1)} + X^{(3)}$ is a Lévy process with Lévy measure $\mathbf{1}_{(|x|<1)}\Pi(\mathrm{d}x)$. Since

$$\mathbb{E}(\mathrm{e}^{\beta X_t}) = \mathbb{E}(\mathrm{e}^{\beta X_t^{(2)}})\mathbb{E}(\mathrm{e}^{\beta(X_t^{(1)}+X_t^{(3)})}),$$

it follows that

$$\mathbb{E}\big(\mathrm{e}^{\beta X_t^{(2)}}\big) < \infty. \tag{3.6}$$

Hence, as $X^{(2)}$ is a compound Poisson process,

$$\begin{aligned}\mathbb{E}\big(\mathrm{e}^{\beta X_t^{(2)}}\big) &= \mathrm{e}^{-\lambda t} \sum_{k\geq 0} \frac{(\lambda t)^k}{k!} \int_{\mathbb{R}} \mathrm{e}^{\beta x} F^{*k}(\mathrm{d}x) \\ &= \mathrm{e}^{-\Pi(\mathbb{R}\backslash(-1,1))t} \sum_{k\geq 0} \frac{t^k}{k!} \int_{\mathbb{R}} \mathrm{e}^{\beta x} (\Pi|_{\mathbb{R}\backslash(-1,1)})^{*k}(\mathrm{d}x) < \infty, \end{aligned} \tag{3.7}$$

where F^{*n} and $(\Pi|_{\mathbb{R}\backslash(-1,1)})^{*n}$ are the n-fold convolution of F and $\Pi|_{\mathbb{R}\backslash(-1,1)}$, the restriction of Π to $\mathbb{R}\backslash(-1,1)$, respectively. In particular, the summand corresponding to $k=1$ must be finite, and so

$$\int_{|x|\geq 1} \mathrm{e}^{\beta x}\, \Pi(\mathrm{d}x) < \infty.$$

Now suppose that $\int_{\mathbb{R}} \mathrm{e}^{\beta x} \mathbf{1}_{(|x|\geq 1)} \Pi(\mathrm{d}x) < \infty$ for some $\beta \neq 0$. Without loss of generality, it suffices to take $\beta > 0$. (The case that $\beta < 0$ can be dealt with by considering the forthcoming proof, but for the process $-X$.) Since, for all $n \in \mathbb{N}$,

$$\int_{\mathbb{R}} \mathrm{e}^{\beta x} (\Pi|_{\mathbb{R}\backslash(-1,1)})^{*n}(\mathrm{d}x) = \left(\int_{|x|\geq 1} \mathrm{e}^{\beta x}\, \Pi(\mathrm{d}x)\right)^n < \infty,$$

one easily argues that (3.6) holds, for all $t \geq 0$, through (3.7). The proof is thus complete once we show that, for all $t \geq 0$,

$$\mathbb{E}\big(\mathrm{e}^{\beta(X_t^{(1)}+X_t^{(3)})}\big) < \infty. \tag{3.8}$$

However, since $X^{(1)} + X^{(3)}$ is a Lévy process whose Lévy measure has bounded support, it follows that its characteristic exponent,

$$\begin{aligned} &-\frac{1}{t}\log \mathbb{E}\big(\mathrm{e}^{\mathrm{i}\theta(X_t^{(1)}+X_t^{(3)})}\big) \\ &\quad = \mathrm{i}a\theta + \frac{1}{2}\sigma^2\theta^2 + \int_{(-1,1)} \big(1 - \mathrm{e}^{\mathrm{i}\theta x} + \mathrm{i}\theta x\big)\Pi(\mathrm{d}x), \quad \theta \in \mathbb{R}, \end{aligned} \tag{3.9}$$

can be extended to an entire function (analytic on the whole of $\mathbb{C}$). To see why, note that

$$\int_{(-1,1)} \big(1 - \mathrm{e}^{\mathrm{i}\theta x} + \mathrm{i}\theta x\big)\Pi(\mathrm{d}x) = -\int_{(-1,1)} \sum_{k\geq 0} \frac{(\mathrm{i}\theta x)^{k+2}}{(k+2)!}\Pi(\mathrm{d}x).$$

The sum and the integral may be exchanged using Fubini's Theorem and the estimate

$$\sum_{k\geq 0}\int_{(-1,1)}\frac{|\theta x|^{k+2}}{(k+2)!}\Pi(\mathrm{d}x)\leq\sum_{k\geq 0}\frac{|\theta|^{k+2}}{(k+2)!}\int_{(-1,1)}x^2\Pi(\mathrm{d}x)<\infty.$$

Hence, the right-hand side of (3.9) can be written as a power series for all $\theta\in\mathbb{C}$ and is thus entire. In turn this guarantees that $\widehat{\mu}_t(\theta):=\exp\{-\Psi^{(1)}(\theta)t-\Psi^{(3)}(\theta)t\}$ is also an entire function. Note that $\widehat{\mu}_t(\theta)$ is nothing more than the Fourier transform of the measure $\mu_t(\mathrm{d}x)=\mathbb{P}(X_t^{(1)}+X_t^{(3)}\in\mathrm{d}x)$, $x\in\mathbb{R}$. Since $\widehat{\mu}_t(\theta)$ is an entire function, it follows that all the moments of μ_t exist with $\mathrm{d}^n\widehat{\mu}_t(\theta)/\mathrm{d}\theta^n|_{\theta=0}=\mathrm{i}^n m_n(t)$, where $m_n(t)=\int_{\mathbb{R}}x^n\mu_t(\mathrm{d}x)$ for $n\in\mathbb{N}$. Expanding $\widehat{\mu}_t$ as a power series about 0, we have

$$\widehat{\mu}_t(\theta)=\sum_{n\geq 0}\frac{1}{n!}\mathrm{i}^n m_n(t)\theta^n,\quad\theta\in\mathbb{C}.\tag{3.10}$$

The entire nature of the previous sum implies, in particular, that it is absolutely convergent for all $\theta\in\mathbb{C}$.

Now define $a_n(t)=\int_{\mathbb{R}}|x|^n\mu_t(\mathrm{d}x)$ for $n\in\mathbb{N}$. It is straightforward to note that, for $k\in\mathbb{N}$, $a_{2k}(t)=m_{2k}(t)$ and $a_{2k+1}(t)\leq(m_{2k+2}(t)+m_{2k}(t))$ where the latter follows on account of the fact that

$$|x|^{2k+1}\leq|x|^{2k+2}+|x|^{2k}=x^{2k+2}+x^{2k},\quad x\in\mathbb{R}.$$

We thus have that

$$\mathbb{E}(\mathrm{e}^{\beta(X_t^{(1)}+X_t^{(3)})})\leq\mathbb{E}(\mathrm{e}^{\beta|X_t^{(1)}+X_t^{(3)}|})=\int_{\mathbb{R}}\mathrm{e}^{\beta|x|}\mu_t(\mathrm{d}x)=\sum_{n\geq 0}\frac{1}{n!}a_n(t)\beta^n<\infty,$$

where the final equality is justified by writing $\mathrm{e}^{\beta|x|}$ as a power series and then interchanging the operation of integration with summation using Fubini's Theorem, the estimates for $a_n(t)$ and the absolute convergence of the series (3.10). This also justifies the final inequality. □

The conclusion of the previous theorem can be extended to a larger class of functions over and above the exponential functions.

Definition 3.7 A measurable function, $g:\mathbb{R}\to[0,\infty)$, is called submultiplicative if there exists a constant $c>0$ such that $g(x+y)\leq cg(x)g(y)$ for all $x,y\in\mathbb{R}$.

It follows easily from this definition that, for example, the product of two submultiplicative functions is submultiplicative. Again, working directly with the definition, it is also easy to show that, if $g(x)$ is submultiplicative, then so is $g(cx+\gamma)^\alpha$, where $c\in\mathbb{R}$, $\gamma\in\mathbb{R}$ and $\alpha>0$. An easy way to see this is first to prove the statement for $g(cx)$, then for $g(x+\gamma)$ and finally for $g(x)^\alpha$, and then to combine the conclusions.

Theorem 3.8 *Suppose that g is submultiplicative and bounded on compacts. Then*

$$\int_{|x|\geq 1} g(x)\Pi(\mathrm{d}x) < \infty \quad \textit{if and only if } \mathbb{E}\big(g(X_t)\big) < \infty \quad \textit{for all } t > 0.$$

The proof is essentially the same once one has established that for each submultiplicative function, g, which is bounded on compacts, there exist constants $a_g > 0$ and $b_g > 0$ such that $g(x) \leq a_g \exp\{b_g|x|\}$, $x \in \mathbb{R}$. See Exercise 3.3 where examples of submultiplicative functions, other than exponential functions, can be found.

Theorem 3.6 gives us a criterion under which we can perform an exponential change of measure. Define the Laplace exponent

$$\psi(\beta) = \frac{1}{t}\log \mathbb{E}\big(\mathrm{e}^{\beta X_t}\big) = -\Psi(-\mathrm{i}\beta) \tag{3.11}$$

whenever it exits. We now know that the Laplace exponent is finite if and only if $\int_{|x|\geq 1} \mathrm{e}^{\beta x}\Pi(\mathrm{d}x) < \infty$. Following Exercise 1.5, it is easy to deduce that, under this assumption, $\mathcal{E}(\beta) = \{\mathcal{E}_t(\beta) \colon t \geq 0\}$ is a $\mathbb{P}$-martingale with respect to $\mathbb{F}$, where

$$\mathcal{E}_t(\beta) = \mathrm{e}^{\beta X_t - \psi(\beta)t}, \quad t \geq 0. \tag{3.12}$$

Since this martingale has unit mean, it may be used to perform a change of measure via

$$\left.\frac{\mathrm{d}\mathbb{P}^\beta}{\mathrm{d}\mathbb{P}}\right|_{\mathcal{F}_t} = \mathcal{E}_t(\beta), \quad t \geq 0.$$

The change of measure above is known as the *Esscher transform.* As the next theorem shows, it has the important property that the process X under $\mathbb{P}^\beta$ is still a Lévy process. This fact will play a crucial role in the analysis of spectrally negative Lévy processes later on in this text.

Theorem 3.9 *Suppose that X is a Lévy process with characteristic triple* (a, σ, Π), *and that* $\beta \in \mathbb{R}$ *is such that*

$$\int_{|x|\geq 1} \mathrm{e}^{\beta x}\Pi(\mathrm{d}x) < \infty.$$

Under the change of measure $\mathbb{P}^\beta$, *the process X is still a Lévy process with characteristic triple* (a^*, σ^*, Π^*), *where*

$$a^* = a - \beta\sigma^2 + \int_{|x|<1} \big(1 - \mathrm{e}^{\beta x}\big)x\Pi(\mathrm{d}x), \qquad \sigma^* = \sigma \quad \textit{and} \quad \Pi^*(\mathrm{d}x) = \mathrm{e}^{\beta x}\Pi(\mathrm{d}x).$$

Proof Suppose, without loss of generality, that $\beta > 0$.[3] Begin by noting from Hölder's inequality that, for any $\theta \in [0, \beta]$ and all $t \geq 0$,

$$\mathbb{E}(\mathrm{e}^{\theta X_t}) \leq \mathbb{E}(\mathrm{e}^{\beta X_t})^{\theta/\beta} < \infty.$$

Hence, $\psi(\theta) < \infty$ for all $\theta \in [0, \beta]$. (In fact, the above computation shows that ψ is convex in this interval.) In turn, this implies that $|\mathbb{E}(\mathrm{e}^{\mathrm{i}\theta X_t})| < \infty$, for all θ such that $-\Im\theta \in [0, \beta]$ and $t \geq 0$. By analytic extension, the characteristic exponent Ψ of X is thus finite on the same region of the complex plane.

Fix a time horizon, $t > 0$, and note that the density $\exp\{\beta X_t - \psi(\beta)t\}$ is almost surely positive. Hence $\mathbb{P}$ and $\mathbb{P}^\beta$ are equivalent measures on $\mathcal{F}_t$. For each $t > 0$, let

$$A_t = \left\{\forall s \in (0, t],\ \exists \lim_{u \uparrow s} X_u \text{ and } \forall s \in [0, t),\ \lim_{u \downarrow s} X_u = X_s\right\}.$$

Then, since $\mathbb{P}(A_t) = 1$ for all $t > 0$, it follows that $\mathbb{P}^\beta(A_t) = 1$ for all $t > 0$. That is to say, under $\mathbb{P}^\beta$, the process X still has paths which are almost surely continuous from the right with left limits.

Next, let $0 \leq u \leq s \leq t < \infty$ and $\theta \in \mathbb{R}$. Write $\mathbb{E}^\beta$ for expectation with respect to $\mathbb{P}^\beta$. We have, for all $A \in \mathcal{F}_u$,

$$\begin{aligned}
&\mathbb{E}^\beta\left(\mathbf{1}_A \mathrm{e}^{\mathrm{i}\theta_1(X_t - X_s)}\right) \\
&\quad = \mathbb{E}\left(\mathbf{1}_A \mathrm{e}^{\beta X_s - \psi(\beta)s} \mathrm{e}^{(\mathrm{i}\theta_1 + \beta)(X_t - X_s) - \psi(\beta)(t-s)}\right).
\end{aligned}$$

Using the martingale property of the change of measure and stationary independent increments of X under $\mathbb{P}$, by first conditioning on $\mathcal{F}_s$, and then on $\mathcal{F}_u$, we find from the previous equality that

$$\mathbb{E}^\beta\left(\mathbf{1}_A \mathrm{e}^{\mathrm{i}\theta_1(X_t - X_s)}\right) = \mathrm{e}^{(\Psi(-\mathrm{i}\beta) - \Psi(\theta_1 - \mathrm{i}\beta))(t-s)} \mathbb{P}^\beta(A).$$

Hence, under $\mathbb{P}^\beta$, we deduce that X has stationary independent increments, with characteristic exponent given by

$$\Psi_\beta(\theta) := \Psi(\theta - \mathrm{i}\beta) - \Psi(-\mathrm{i}\beta), \quad \theta \in \mathbb{R}.$$

By writing out the exponent in terms of the triple (a, σ, Π) associated with X under $\mathbb{P}$, it is a straightforward exercise to deduce that

$$\begin{aligned}
\Psi_\beta(\theta) &= \mathrm{i}\theta\left(a - \beta\sigma^2 + \int_{|x|<1} (1 - \mathrm{e}^{\beta x}) x \Pi(\mathrm{d}x)\right) + \frac{1}{2}\theta^2\sigma^2 \\
&\quad + \int_{\mathbb{R}} (1 - \mathrm{e}^{\mathrm{i}\theta x} + \mathrm{i}\theta x \mathbf{1}_{(|x|<1)}) \mathrm{e}^{\beta x} \Pi(\mathrm{d}x), \quad \theta \in \mathbb{R}.
\end{aligned} \tag{3.13}$$

We thus identify the triple (a^*, σ^*, Π^*) as given in the statement of the theorem. □

[3] In the case that $\beta < 0$, simply consider the forthcoming argument for $-X$. For $\beta = 0$ the statement of the theorem is trivial.

The effect of the Esscher transform is to exponentially tilt the Lévy measure, to introduce an additional linear drift and to leave the Gaussian contribution untouched.

Note that, in the case of a spectrally negative Lévy process, the Laplace exponent satisfies $|\psi(\theta)| < \infty$ for $\theta \geq 0$. This follows as a consequence of Theorem 3.6 together with the fact that $\Pi(0,\infty) = 0$.

Corollary 3.10 *The Esscher transform may be applied for all $\beta \geq 0$ when X is a spectrally negative Lévy process. Further, under $\mathbb{P}^\beta$, X remains within the class of spectrally negative Lévy processes. The Laplace exponent, ψ_β, of X under $\mathbb{P}^\beta$ satisfies*

$$\psi_\beta(\theta) = \psi(\theta + \beta) - \psi(\beta),$$

for all $\theta \geq -\beta$.

Proof The Esscher transform has the effect of exponentially tilting the original Lévy measure and therefore does not have any influence on the support of the Lévy measure. We have, as previously, that, for $\theta \geq -\beta$,

$$\mathrm{e}^{\psi_\beta(\theta)} = \mathbb{E}^\beta\left(\mathrm{e}^{\theta X_1}\right) = \mathbb{E}\left(\mathrm{e}^{(\theta+\beta)X_1 - \psi(\beta)}\right) = \mathrm{e}^{\psi(\theta+\beta)-\psi(\beta)},$$

which establishes the final statement of the corollary. □

Corollary 3.11 *Under the conditions of Theorem 3.9, if τ is an $\mathbb{F}$-stopping time, then*

$$\left.\frac{\mathrm{d}\mathbb{P}^\beta}{\mathrm{d}\mathbb{P}}\right|_{\mathcal{F}_\tau} = \mathcal{E}_\tau(\beta) \quad \text{on } \{\tau < \infty\}.$$

Proof By definition if $A \in \mathcal{F}_\tau$, then $A \cap \{\tau \leq t\} \in \mathcal{F}_t$. Hence,

$$\begin{aligned}\mathbb{P}^\beta\left(A \cap \{\tau \leq t\}\right) &= \mathbb{E}\left(\mathcal{E}_t(\beta)\mathbf{1}_{(A,\tau\leq t)}\right)\\ &= \mathbb{E}\left(\mathbb{E}\left(\mathcal{E}_t(\beta)\mathbf{1}_{(A,\tau\leq t)}|\mathcal{F}_\tau\right)\right)\\ &= \mathbb{E}\left(\mathcal{E}_\tau(\beta)\mathbf{1}_{(A,\tau\leq t)}\right),\end{aligned}$$

where in the third equality, we have used the strong Markov property as well as the martingale property for $\mathcal{E}(\beta)$. Now taking limits, as $t \uparrow \infty$, the result follows with the help of the Monotone Convergence Theorem. □

Remaining with spectrally negative Lévy processes, we conclude this section by giving another application of the exponential martingale $\mathcal{E}(\beta)$. Recall the stopping times

$$\tau_x^+ = \inf\{t > 0 : X_t > x\}, \tag{3.14}$$

for $x \geq 0$; also called *first-passage times.*

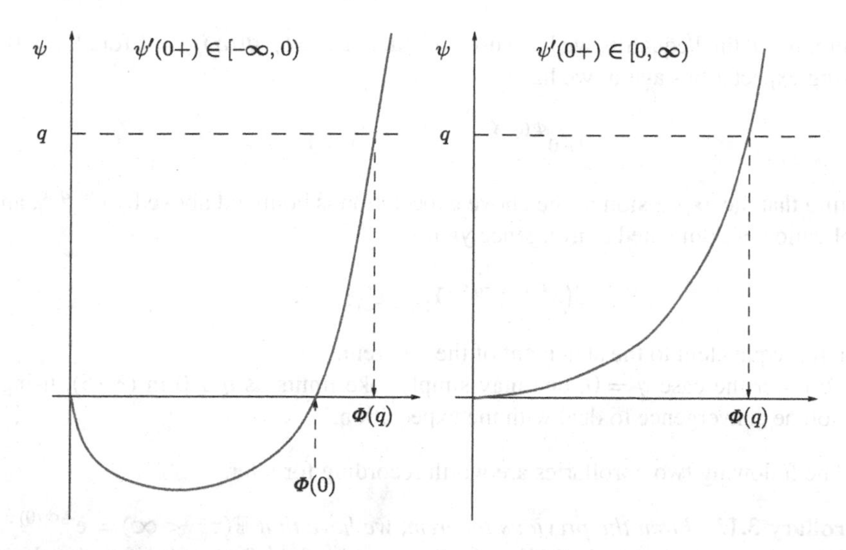

Fig. 3.3 Two examples of ψ, the Laplace exponent of a spectrally negative Lévy process, and the relation to Φ.

Theorem 3.12 *For any spectrally negative Lévy process,*

$$\mathbb{E}\big(\mathrm{e}^{-q\tau_x^+}\mathbf{1}_{(\tau_x^+<\infty)}\big) = \mathrm{e}^{-\Phi(q)x}, \tag{3.15}$$

where $q \geq 0$ and $\Phi(q)$ is the largest root of the equation $\psi(\theta) = q$.

Before proceeding to the proof, let us make some remarks about the function

$$\Phi(q) = \sup\{\theta \geq 0 : \psi(\theta) = q\}, \tag{3.16}$$

defined for all $q \geq 0$, also known as the *right inverse* of ψ. Exercise 3.5 shows that, on $[0,\infty)$, ψ is infinitely differentiable, strictly convex and that $\psi(0) = 0$, whilst $\psi(\infty) = \infty$. As a particular consequence of these facts, it follows that $\mathbb{E}(X_1) = \psi'(0+) \in [-\infty,\infty)$. In the case that $\mathbb{E}(X_1) \geq 0$, $\Phi(q)$ is the unique solution to $\psi(\theta) = q$ in $[0,\infty)$. When $\mathbb{E}(X_1) < 0$ the previous statement is true only when $q > 0$. If $\mathbb{E}(X_1) < 0$ and $q = 0$, then there are two roots to the equation $\psi(\theta) = 0$, one of them being $\theta = 0$ and the other being $\Phi(0) > 0$. See Fig. 3.3 for further clarification.

Proof of Theorem 3.12 Fix $q > 0$. Using spectral negativity to write $x = X_{\tau_x^+}$ on $\{\tau_x^+ < \infty\}$, the strong Markov property gives us

$$\begin{aligned}
&\mathbb{E}\big(\mathrm{e}^{\Phi(q)X_t - qt}\,|\,\mathcal{F}_{\tau_x^+}\big)\\
&= \mathbf{1}_{(\tau_x^+\geq t)}\mathrm{e}^{\Phi(q)X_t - qt} + \mathbf{1}_{(\tau_x^+<t)}\mathrm{e}^{\Phi(q)x - q\tau_x^+}\mathbb{E}\big(\mathrm{e}^{\Phi(q)(X_t - X_{\tau_x^+}) - q(t-\tau_x^+)}\,|\,\mathcal{F}_{\tau_x^+}\big)\\
&= \mathrm{e}^{\Phi(q)X_{t\wedge\tau_x^+} - q(t\wedge\tau_x^+)},
\end{aligned}$$

where, in the final equality, we have used the fact that $\mathbb{E}(\mathcal{E}_t(\Phi(q))) = 1$ for all $t \geq 0$. Taking expectations again, we have

$$\mathbb{E}\big(\mathrm{e}^{\Phi(q)X_{t\wedge\tau_x^+} - q(t\wedge\tau_x^+)}\big) = 1.$$

Noting that the expression in the above expectation is bounded above by $\mathrm{e}^{\Phi(q)x}$, an application of dominated convergence yields

$$\mathbb{E}\big(\mathrm{e}^{\Phi(q)x - q\tau_x^+}\mathbf{1}_{(\tau_x^+<\infty)}\big) = 1,$$

which is equivalent to the statement of the theorem.

To cover the case $q = 0$, one may simply take limits as $q \downarrow 0$ in (3.15), using monotone convergence to deal with the expectation. □

The following two corollaries are worth recording for later.

Corollary 3.13 *From the previous theorem, we have that* $\mathbb{P}(\tau_x^+ < \infty) = \mathrm{e}^{-\Phi(0)x}$, *which is one if and only if* $\Phi(0) = 0$, *if and only if* $\psi'(0+) \geq 0$, *if and only if* $\mathbb{E}(X_1) \geq 0$.

For the next corollary, we define a killed subordinator to be a subordinator which is sent to an additional "cemetery" state at an independent and exponentially distributed time.

Corollary 3.14 *If* $\mathbb{E}(X_1) \geq 0$, *then the process* $\{\tau_x^+ : x \geq 0\}$ *is a subordinator, and otherwise it is equal in law to a subordinator killed at an independent exponential time with parameter* $\Phi(0)$.

Proof First, we claim that $\Phi(q) - \Phi(0)$ is the Laplace exponent of a non-negative infinitely divisible random variable. To see this, note that, for all $x \geq 0$,

$$\mathbb{E}\big(\mathrm{e}^{-q\tau_x^+}|\tau_x^+ < \infty\big) = \mathrm{e}^{-(\Phi(q)-\Phi(0))x} = \mathbb{E}\big(\mathrm{e}^{-q\tau_1^+}|\tau_1^+ < \infty\big)^x,$$

and hence, in particular,

$$\mathbb{E}\big(\mathrm{e}^{-q\tau_1^+}|\tau_1^+ < \infty\big) = \mathbb{E}\big(\mathrm{e}^{-q\tau_{1/n}^+}|\tau_{1/n}^+ < \infty\big)^n,$$

showing that, for $z \geq 0$, $\mathbb{P}(\tau_1^+ \in \mathrm{d}z|\tau_1^+ < \infty)$ is the law of an infinitely divisible random variable. Next, using the strong Markov property, spatial homogeneity and, again, the special feature of spectral negativity that $\{X_{\tau_x^+} = x\}$ on the event $\{\tau_x^+ < \infty\}$, we have, for $x, y \geq 0$ and $q \geq 0$, that

$$\begin{aligned}
&\mathbb{E}\big(\mathrm{e}^{-q(\tau_{x+y}^+ - \tau_x^+)}\mathbf{1}_{(\tau_{x+y}^+<\infty)}|\mathcal{F}_{\tau_x^+}\big)\mathbf{1}_{(\tau_x^+<\infty)} \\
&\quad = \mathbb{E}\big(\mathrm{e}^{-q\tau_y^+}\mathbf{1}_{(\tau_y^+<\infty)}\big)\mathbf{1}_{(\tau_x^+<\infty)} \\
&\quad = \mathrm{e}^{-(\Phi(q)-\Phi(0))y}\mathrm{e}^{-\Phi(0)y}\mathbf{1}_{(\tau_x^+<\infty)}.
\end{aligned}$$

We see that the increment $\tau^+_{x+y} - \tau^+_x$ is independent of $\mathcal{F}_{\tau^+_x}$ on $\{\tau^+_x < \infty\}$ and has the same law as the subordinator with Laplace exponent $\Phi(q) - \Phi(0)$, but killed at an independent and exponentially distributed time with parameter $\Phi(0)$.

When $\mathbb{E}(X_1) \geq 0$, we have that $\Phi(0) = 0$ and hence the concluding statement of the previous paragraph indicates that $\{\tau^+_x : x \geq 0\}$ is a subordinator (without killing). On the other hand, if $\mathbb{E}(X_1) < 0$, or equivalently $\Phi(0) > 0$, then the second statement of the corollary follows. □

Note that, embedded in the previous corollary is the same reasoning which lies behind the justification that an inverse Gaussian process is a Lévy process. See Sect. 1.2.5 and Exercise 1.6.

Exercises

3.1 For a general stochastic process on a filtered probability space, the operations of completing the filtration and taking its right-continuous version must be treated separately. However, for a Lévy process it turns out that completing the filtration is already enough to make it right-continuous.

Suppose that X is a Lévy process defined on $(\Omega, \mathcal{F}, \mathbb{P})$ and that $\mathcal{F}_t$ is the sigma-algebra obtained by completing $\sigma(X_s : s \leq t)$ by the null sets of $\mathbb{P}$. We want to show that, for all $t \geq 0$,

$$\mathcal{F}_t = \bigcap_{s>t} \mathcal{F}_s.$$

(i) Fix $t_2 > t_1 \geq 0$ and show that, for any $t \geq 0$,

$$\lim_{u \downarrow t} \mathbb{E}(\mathrm{e}^{\mathrm{i}\theta_1 X_{t_1} + \mathrm{i}\theta_2 X_{t_2}} | \mathcal{F}_u) = \mathbb{E}(\mathrm{e}^{\mathrm{i}\theta_1 X_{t_1} + \mathrm{i}\theta_2 X_{t_2}} | \mathcal{F}_t)$$

almost surely, where $\theta_1, \theta_2 \in \mathbb{R}$.

(ii) Deduce that, for any sequence of times $t_1, \ldots, t_n \geq 0$,

$$\mathbb{E}\big(g(X_{t_1}, \ldots, X_{t_n}) | \mathcal{F}_t\big) = \mathbb{E}\big(g(X_{t_1}, \ldots, X_{t_n}) | \mathcal{F}_{t+}\big)$$

almost surely, for all functions g satisfying $\mathbb{E}(|g(X_{t_1}, \ldots, X_{t_n})|) < \infty$.

(iii) Conclude that for each $A \in \mathcal{F}_{t+}$, $\mathbb{E}(\mathbf{1}_A | \mathcal{F}_t) = \mathbf{1}_A$ almost surely, and hence that $\mathcal{F}_t = \mathcal{F}_{t+}$.

3.2 Show that, for any $x \geq 0$,

$$Y^{(x)}_t := (x \vee \overline{X}_t) - X_t, \quad t \geq 0, \quad \text{and} \quad Z^{(x)}_t := X_t - \big(\underline{X}_t \wedge (-x)\big), \quad t \geq 0,$$

are $[0, \infty)$-valued strong Markov processes.

Hint: following the original proof in Bingham (1975), it will be useful to show that, for $s,t \geq 0$, $Y^x_{t+s} = \widetilde{Y}^{(Y^x_t)}_s$, where, for $x \geq 0$, $\{\widetilde{Y}^{(x)}_s : s \geq 0\}$ is an independent copy of $\{Y^{(x)}_s : s \geq 0\}$.

3.3 (Proof of Theorem 3.8 and examples)

(i) Use the comments following Theorem 3.8 to prove it.
(ii) Prove that the following functions are submultiplicative: $x \vee 1$, $x^\alpha \vee 1$ $|x| \vee 1$, $|x|^\alpha \vee 1$, $\exp(|x|^\beta)$, $\log(|x| \vee \mathrm{e})$, $\log\log(|x| \vee \mathrm{e^e})$, where $\alpha > 0$ and $\beta \in (0,1]$.
(iii) Suppose that X is a stable process of index $\alpha \in (0,2)$. Show that $\mathbb{E}(|X_t|^\eta) < \infty$ for all $t \geq 0$ if and only if $\eta \in [0,\alpha)$.

3.4 A generalised tempered stable process is a Lévy process with no Gaussian component and Lévy measure given by

$$\Pi(\mathrm{d}x) = \mathbf{1}_{(x>0)} \frac{c^+}{x^{1+\alpha^+}} \mathrm{e}^{-\gamma^+ x}\,\mathrm{d}x + \mathbf{1}_{(x<0)} \frac{c^-}{|x|^{1+\alpha^-}} \mathrm{e}^{\gamma^- x}\,\mathrm{d}x,$$

where $c^\pm > 0$, $\alpha^\pm \in (-\infty, 2)$ and $\gamma^\pm > 0$. Show that if X is a generalised tempered stable process, then X may always be written in the form $X = X^+ - X^-$ where $X^+ = \{X^+_t : t \geq 0\}$ and $X^- = \{X^-_t : t \geq 0\}$ satisfy the following:

(i) If $\alpha^\pm < 0$ then $X^\pm$ is a compound Poisson process with drift.
(ii) If $\alpha^\pm = 0$ then $X^\pm$ is a gamma process with drift.
(iii) If $\alpha^\pm \in (0,2)$, then up to the addition of a linear drift, $X^\pm$ has the same law as a spectrally positive stable process with index $\alpha^\pm$, but considered under the change of measure $\mathbb{P}^{-\gamma^\pm}$.

3.5 Suppose that ψ is the Laplace exponent of a spectrally negative Lévy process. By considering the formula

$$\psi(\beta) = -a\beta + \frac{1}{2}\sigma^2\beta^2 + \int_{(-\infty,0)} \left(\mathrm{e}^{\beta x} - 1 - \beta x \mathbf{1}_{(x>-1)}\right)\Pi(\mathrm{d}x),$$

show that, on $[0,\infty)$, ψ is infinitely differentiable, strictly convex and that $\psi(0) = 0$ whilst $\psi(\infty) = \infty$.

3.6 Suppose that X is a spectrally negative Lévy process with Lévy–Khintchine exponent Ψ. Here, we give another proof of the existence of a finite Laplace exponent for all spectrally negative Lévy processes.

(i) Use spectral negativity, together with the lack-of-memory property to show that, for $x, y > 0$,

$$\mathbb{P}(\overline{X}_{\mathbf{e}_q} > x + y) = \mathbb{P}(\overline{X}_{\mathbf{e}_q} > x)\mathbb{P}(\overline{X}_{\mathbf{e}_q} > y),$$

where $\mathbf{e}_q$ is an exponentially distributed random variable[4] with parameter q, independent of X.

(ii) Deduce that $\overline{X}_{\mathbf{e}_q}$ is exponentially distributed and hence the Laplace exponent $\psi(\beta) = -\Psi(-\mathrm{i}\beta)$ exists and is finite for all $\beta \geq 0$.

(iii) By considering the Laplace transform of the first-passage time τ_x^+ as in Sect. 3.3, show that one may also deduce via a different route that $\overline{X}_{\mathbf{e}_q}$ is exponentially distributed with parameter $\Phi(q)$. In particular show that $\overline{X}_\infty$ is either infinite with probability one or is exponentially distributed accordingly as $\mathbb{E}(X_1) \geq 0$ or $\mathbb{E}(X_1) < 0$.
Hint: reconsider Exercise 3.5.

3.7 For this exercise, it will be useful to refer to Sect. 1.2.6. Suppose that X is a stable Lévy process with index $\beta = 1$; in particular, there are no negative jumps.

(i) Show that, if $\alpha \in (0, 1)$, then X is a *driftless* subordinator with Laplace exponent satisfying

$$-\log \mathbb{E}(\mathrm{e}^{-\theta X_1}) = c\theta^\alpha, \quad \theta \geq 0,$$

for some $c > 0$.

(ii) Show that, if $\alpha \in (1, 2)$, then X has a Laplace exponent satisfying

$$-\log \mathbb{E}(\mathrm{e}^{-\theta X_1}) = -C\theta^\alpha, \quad \theta \geq 0,$$

for some $C > 0$. Confirm that X has no integer moments of order 2 and above. Show, moreover, that X is a process with unbounded variation paths.

---

[4]As noted just after Definition 1.1, we are making an abuse of notation in the use of the measure $\mathbb{P}$ here. Strictly speaking, we should work with the measure $\mathbb{P} \times \mathcal{P}$, where $\mathcal{P}$ is the probability measure on the space in which the random variable $\mathbf{e}_q$ is defined. This abuse of notation will be repeated at various points throughout this text.

Chapter 4
General Storage Models and Paths of Bounded Variation

In this chapter, we return to the queueing and general storage models discussed in Sects. 1.3.2 and 2.7.2. Predominantly, we shall concentrate on the asymptotic behaviour of the two quantities that correspond to the workload process and the idle time in the $M/G/1$ queue, but now in the general setting described in Sect. 2.7.2. Along the way, we will introduce some new tools, which will be of help both in this chapter and in later chapters. Specifically, we shall spend some additional time looking at the change of variable and compensation formulae. We also spend some time discussing similarities between the mathematical description of the limiting distribution of the workload process (when it is non-trivial) and the Pollaczek–Khintchine formula. This requires a study of the small-scale behaviour of Lévy processes of bounded variation. We start, however, by briefly recalling, and expanding a little on, the mathematical background of general storage models.

4.1 General Storage Models

A general storage model consists of two processes: $\{A_t : t \geq 0\}$, the volume of incoming work, and $\{B_t : t \geq 0\}$, the total amount of work that can potentially exit from the system as a result of processing work continuously. In the case of the $M/G/1$ queue, we have $A_t = \sum_{i=1}^{N_t} \xi_i$, $t \geq 0$, where $\{N_t : t \geq 0\}$ is a Poisson process and $\{\xi_i : i = 1, 2, \ldots\}$ are the independent service times of the ordered customers. Further, as the server processes at a constant unit rate, we have simply that $B_t = t$. For all $t \geq 0$, let $D_t = A_t - B_t$. The process $D = \{D_t : t \geq 0\}$ is clearly related to the workload of the system, although it is itself *not* a suitable candidate to model the workload. Indeed, D may become negative and the workload is clearly a non-negative quantity. The work stored in the system, $W = \{W_t : t \geq 0\}$, is defined instead by

$$W_t = D_t + L_t, \quad t \geq 0,$$

where $L = \{L_t : t \geq 0\}$ is increasing with paths that are right continuous (and left limits are of course automatic by monotonicity), and is added to the process D to

A.E. Kyprianou, *Fluctuations of Lévy Processes with Applications*, Universitext,
DOI 10.1007/978-3-642-37632-0_4, © Springer-Verlag Berlin Heidelberg 2014

ensure that $W_t \geq 0$ for all $t \geq 0$. The process L must only increase when $W = 0$, so in particular

$$\int_0^\infty \mathbf{1}_{(W_t>0)} \mathrm{d}L_t = 0.$$

It is easy to check that we may take $L_t = -(\inf_{s\leq t} D_s \wedge 0)$, $t \geq 0$. Indeed with this choice of L, we have that $\{W_t = 0\}$ if and only if $D_t = \inf_{s\leq t} D_s \wedge 0$ if and only if t is in the support of the measure $\mathrm{d}L$. It can also be proved that there is no other choice of L fulfilling these requirements (see for example Kella and Whitt, 1996).

We are concerned with the case that the process A is a pure jump subordinator and B is a linear trend. Specifically, $D_t = w - X_t$ where $w \geq 0$ is the workload already in the system at time $t = 0$ and X is a spectrally negative Lévy process of bounded variation. A little algebra with the given expressions for D and L shows that

$$W_t = (w \vee \overline{X}_t) - X_t, \quad t \geq 0,$$

where $\overline{X}_t = \sup_{s\leq t} X_s$.

We know from the discussion in Sect. 3.3 (see also Exercise 3.6) that the process X has Laplace exponent $\psi(\theta) = \log \mathbb{E}(\mathrm{e}^{\theta X_1})$, $\theta \geq 0$. Writing X in the form $\delta t - S_t$, $t \geq 0$, where $\delta > 0$ and $S = \{S_t : t \geq 0\}$ is a pure jump subordinator, it is convenient to write the Laplace exponent of X in the form

$$\psi(\theta) = \delta\theta - \int_{(0,\infty)} (1 - \mathrm{e}^{-\theta x})\nu(\mathrm{d}x), \quad \theta \geq 0,$$

where ν is the Lévy measure of the subordinator S, which satisfies $\int_{(0,\infty)} (1 \wedge x)\nu(\mathrm{d}x) < \infty$.

4.2 Idle Times

We start by introducing the parameter

$$\rho := \frac{\delta - \psi'(0+)}{\delta}.$$

Note that regimes $0 < \rho < 1$, $\rho = 1$ and $\rho > 1$ correspond precisely to the regimes $\psi'(0+) > 0$, $\psi'(0+) = 0$ and $\psi'(0+) < 0$, respectively. For the first two of these cases, we also have that $\Phi(0) = 0$ and, in the third case, we have $\Phi(0) > 0$, where Φ is the right inverse of ψ, defined in (3.16). When $\delta = 1$ and $\nu = \lambda F$, where F is a distribution function and $\lambda > 0$ is the arrival rate, the process W is the workload of an $M/G/1$ queue. In that case $\rho = \lambda \mathbb{E}(\xi)$, where ξ is a random variable with distribution F, and this constant is called the *traffic intensity*.

The main purpose of this section is to prove the following result, which includes Theorem 1.11 as a corollary.

Theorem 4.1 *Suppose that $\rho > 1$. The total time that the storage process spends idle,*

$$I := \int_0^\infty \mathbf{1}_{(W_t=0)} \mathrm{d}t,$$

has the distribution

$$\mathbb{P}(I \in \mathrm{d}x \,|\, W_0 = w) = \left(1 - \mathrm{e}^{-\Phi(0)w}\right)\delta_0(\mathrm{d}x) + \delta\Phi(0)\mathrm{e}^{-\Phi(0)(w+x\delta)}\mathrm{d}x, \quad x \geq 0.$$

Otherwise, if $0 < \rho \leq 1$, then I is infinite with probability one.

Proof Essentially the proof mimics the steps of Exercise 1.9. As one sees for the case of the $M/G/1$ queue, a key ingredient to the proof is that one may identify the processes $\{\delta \int_0^t \mathbf{1}_{(W_s=0)}\mathrm{d}s : t \geq 0\}$ and $\{\overline{X}_t : t \geq 0\}$ as one and the same. To see why this is true in the general storage model, recall from the Lévy–Itô decomposition that X has a countable number of jumps over finite intervals of time, hence the same is true of W. Further, since X has negative jumps, $W_s = 0$ only if there is no jump at time s. Hence, given that X is the difference of a linear drift with rate δ and a subordinator S, it follows that, for each $t \geq 0$,

$$\begin{aligned} \overline{X}_t &= \int_0^t \mathbf{1}_{(\overline{X}_s = X_s)} \mathrm{d}X_s \\ &= \delta \int_0^t \mathbf{1}_{(\overline{X}_s = X_s)} \mathrm{d}s - \int_0^t \mathbf{1}_{(\overline{X}_s = X_s)} \mathrm{d}S_s \\ &= \delta \int_0^t \mathbf{1}_{(\overline{X}_s = X_s)} \mathrm{d}s \end{aligned}$$

almost surely, where the final equality follows as a consequence of the fact that

$$\int_0^t \mathbf{1}_{(\overline{X}_s = X_s)} \mathrm{d}S_s \leq \int_0^t \mathbf{1}_{(\Delta S_s = 0)} \mathrm{d}S_s = 0, \quad t \geq 0.$$

It is important to note that this calculation only works for spectrally negative Lévy processes of bounded variation on account of the particular form of the Lévy–Itô decomposition.

Now, following Exercise 3.6 (iii), we can use the equivalence of the events $\{\overline{X}_\infty > x\}$ and $\{\tau_x^+ < \infty\}$, where τ_x^+ is the first-hitting time of (x, ∞) defined in (3.14), to deduce that $\overline{X}_\infty$ is exponentially distributed with parameter $\Phi(0)$. When $\Phi(0) = 0$, the previous statement is understood to mean that $\mathbb{P}(\overline{X}_\infty = \infty) = 1$. When $w = 0$, we have that

$$\frac{\overline{X}_\infty}{\delta} = \int_0^\infty \mathbf{1}_{(\overline{X}_s = X_s)} \mathrm{d}s = \int_0^\infty \mathbf{1}_{(W_s=0)} \mathrm{d}s. \tag{4.1}$$

Hence, we see that I is exponentially distributed with parameter $\delta\Phi(0)$. Recalling the values of ρ which imply that $\Phi(0) > 0$, we see that the statement of the theorem follows for the case $w = 0$.

In general, when $w > 0$, the equality (4.1) is not valid. Instead, we have that

$$\begin{aligned}\int_0^\infty \mathbf{1}_{(W_s=0)}\mathrm{d}s &= \int_0^{\tau_w^+} \mathbf{1}_{(W_s=0)}\mathrm{d}s + \int_{\tau_w^+}^\infty \mathbf{1}_{(W_s=0)}\mathrm{d}s \\ &= \mathbf{1}_{(\tau_w^+<\infty)} \int_{\tau_w^+}^\infty \mathbf{1}_{(W_s=0)}\mathrm{d}s \\ &= \mathbf{1}_{(\overline{X}_\infty \geq w)} I^*, \end{aligned} \tag{4.2}$$

where I^* is independent of $\mathcal{F}_{\tau_w^+}$ on $\{\tau_w^+ < \infty\}$ and equal in distribution to $\int_0^\infty \mathbf{1}_{(W_s=0)}\mathrm{d}s$ with $w = 0$. Note that the first integral in the right-hand side of the first equality disappears on account of the fact that $W_s > 0$ for all $s < \tau_w^+$. The statement of the theorem now follows for $0 < \rho \leq 1$ by once again recalling that, in this regime, $\Phi(0) = 0$ and hence, from (4.2), $\overline{X}_\infty = \infty$ with probability one, which, in turn, implies that $I = I^*$. This quantity has previously been shown to be infinite with probability one. On the other hand, when $\rho > 1$, we see from (4.2) that there is an atom at zero, corresponding to the event $\{\overline{X}_\infty < w\}$, with probability $1 - \mathrm{e}^{-\Phi(0)w}$. Otherwise, with probability $\mathrm{e}^{-\Phi(0)w}$, the integral I has the same distribution as I^*. Again, from previous calculations for the case $w = 0$, we have seen that this is exponential with parameter $\delta\Phi(0)$, and the proof is complete. □

4.3 Change of Variable and Compensation Formulae

Next, we spend a little time introducing the change of variable formula and the compensation formula. Both formulae pertain to a form of stochastic calculus. The theory of stochastic calculus is an avenue which we choose not to pursue in full generality over and above making some brief remarks. Our exposition will suffice to study in more detail the storage processes discussed in Chap. 1, as well as a number of other applications in later chapters.

4.3.1 The Change of Variable Formula

We assume that $X = \{X_t : t \geq 0\}$ is a Lévy process of bounded variation. Referring back to Chap. 2, (2.21) and (2.22), we recall that we may always write its Lévy–Khintchine exponent as

$$\Psi(\theta) = -\mathrm{i}\delta\theta + \int_{\mathbb{R}} (1 - \mathrm{e}^{\mathrm{i}\theta x})\Pi(\mathrm{d}x),$$

where $\delta \in \mathbb{R}$ and $\int_{\mathbb{R}} (1 \wedge |x|)\Pi(\mathrm{d}x) < \infty$. Accordingly, we identify X pathwise in the form

$$X_t = \delta t + \int_{[0,t]} \int_{\mathbb{R}} xN(\mathrm{d}s \times \mathrm{d}x), \quad t \geq 0,$$

where, as usual, N is the Poisson random measure associated with the jumps of X. Our goal in this section is to prove the following change of variable formula.

Theorem 4.2 *Let $C^{1,1}([0,\infty)\times\mathbb{R})$ be the space of functions $f:[0,\infty)\times\mathbb{R}\to\mathbb{R}$ which are continuously differentiable in each variable (in the case of the derivative in the first variable at the origin, a right derivative is understood). If $f\in C^{1,1}([0,\infty)\times\mathbb{R})$ then, for $t\geq 0$,*

$$f(t,X_t)=f(0,X_0)+\int_0^t\frac{\partial f}{\partial t}(s,X_s)\mathrm{d}s+\delta\int_0^t\frac{\partial f}{\partial x}(s,X_s)\mathrm{d}s$$
$$+\int_{[0,t]}\int_{\mathbb{R}}\big(f(s,X_{s-}+x)-f(s,X_{s-})\big)N(\mathrm{d}s\times\mathrm{d}x).$$

It will become apparent from the proof of this theorem that the final integral with respect to N is well defined.

It is worth mentioning that the change of variable formula exists in a much more general form. For example, it is known (cf. Sect. 7 of Chap. II of Protter 2004) that if $V=\{V_t:t\geq 0\}$ is any right-continuous mapping from $[0,\infty)$ to $\mathbb{R}$ (random or deterministic) of bounded variation and $f(s,x)\in C^{1,1}([0,\infty)\times\mathbb{R})$, then $\{f(t,V_t):t\geq 0\}$ is a mapping from $[0,\infty)$ to $\mathbb{R}$ of bounded variation which satisfies, for $t\geq 0$,

$$f(t,V_t)=f(0,V_0)+\int_0^t\frac{\partial f}{\partial t}(s,V_s)\mathrm{d}s+\int_{(0,t]}\frac{\partial f}{\partial x}(s,V_{s-})\mathrm{d}V_s$$
$$+\sum_{0<s\leq t}\left\{f(s,V_s)-f(s,V_{s-})-\Delta V_s\frac{\partial f}{\partial x}(s,V_{s-})\right\},\qquad(4.3)$$

where $\Delta V_s=V_s-V_{s-}$. Note also that since V is of bounded variation, it has a decomposition as the difference of two increasing functions mapping $[0,\infty)$ to $[0,\infty)$. Hence, the existence of left-limits in the paths of V is automatically guaranteed. This means that V has a countable number of discontinuities (see Exercise 2.4). One may, therefore, understand the final term on the right-hand side of (4.3) as a convergent sum over the discontinuities of V. In the case that V is a Lévy process of bounded variation, it is a straightforward exercise to deduce that when one represents the discontinuities of V via a Poisson random measure, Eq. (4.3) and the conclusion of Theorem 4.2 agree.

Proof of Theorem 4.2 Define, for all $\varepsilon>0$,

$$X_t^\varepsilon=\delta t+\int_{[0,t]}\int_{\{|x|\geq\varepsilon\}}xN(\mathrm{d}s\times\mathrm{d}x),\quad t\geq 0.$$

As $\Pi(\mathbb{R}\backslash(-\varepsilon,\varepsilon))<\infty$, it follows that N counts an almost surely finite number of jumps over $[0,t]\times\{\mathbb{R}\backslash(-\varepsilon,\varepsilon)\}$. Moreover, $X^\varepsilon=\{X_t^\varepsilon:t\geq 0\}$ is a compound Poisson process with drift. Suppose the collection of jumps of X^ε up to time $t \geq 0$ are described by the time-space points $\{(T_i, \xi_i) : i = 1, \ldots, \mathrm{N}\}$, where $\mathrm{N} = N([0,t] \times \{\mathbb{R}\backslash(-\varepsilon, \varepsilon)\})$. Let $T_0 = 0$. Then a telescopic sum gives

$$f(t, X_t^\varepsilon) = f(0, X_0^\varepsilon) + \sum_{i=1}^{\mathrm{N}} \left(f(T_i, X_{T_i}^\varepsilon) - f(T_{i-1}, X_{T_{i-1}}^\varepsilon)\right) + \left(f(t, X_t^\varepsilon) - f(T_\mathrm{N}, X_{T_\mathrm{N}}^\varepsilon)\right).$$

Now noting that X^ε is piecewise linear, we have

$$\begin{aligned} f(t, X_t^\varepsilon) &= f(0, X_0^\varepsilon) \\ &\quad + \sum_{i=1}^{\mathrm{N}} \left(\int_{T_{i-1}}^{T_i} \frac{\partial f}{\partial t}(s, X_s^\varepsilon) + \delta \frac{\partial f}{\partial x}(s, X_s^\varepsilon) \mathrm{d}s + \left(f(T_i, X_{T_i-}^\varepsilon + \xi_i) - f(T_i, X_{T_i-}^\varepsilon)\right) \right) \\ &\quad + \int_{T_\mathrm{N}}^t \frac{\partial f}{\partial t}(s, X_s^\varepsilon) + \delta \frac{\partial f}{\partial x}(s, X_s^\varepsilon) \mathrm{d}s \\ &= f(0, X_0^\varepsilon) + \int_0^t \frac{\partial f}{\partial t}(s, X_s^\varepsilon) + \delta \frac{\partial f}{\partial x}(s, X_s^\varepsilon) \mathrm{d}s \\ &\quad + \int_{[0,t]} \int_{\mathbb{R}\backslash\{0\}} \left(f(s, X_{s-}^\varepsilon + x) - f(s, X_{s-}^\varepsilon)\right) \mathbf{1}_{(|x| \geq \varepsilon)} N(\mathrm{d}s \times \mathrm{d}x). \end{aligned} \tag{4.4}$$

(Note that the smoothness of f has been used here.)

From Exercise 2.8, we know that any Lévy process of bounded variation may be written as the difference of two independent subordinators. In this spirit, write $X_t = X_t^{(+)} - X_t^{(-)}$, where

$$X_t^{(+)} = (\delta \vee 0)t + \int_{[0,t]} \int_{(0,\infty)} x N(\mathrm{d}s \times \mathrm{d}x), \quad t \geq 0,$$

and

$$X_t^{(-)} = |\delta \wedge 0|t - \int_{[0,t]} \int_{(-\infty,0)} x N(\mathrm{d}s \times \mathrm{d}x), \quad t \geq 0.$$

Now let

$$X_t^{(+,\varepsilon)} = (\delta \vee 0)t + \int_{[0,t]} \int_{[\varepsilon,\infty)} x N(\mathrm{d}s \times \mathrm{d}x) \quad t \geq 0,$$

and

$$X_t^{(-,\varepsilon)} = |\delta \wedge 0|t - \int_{[0,t]} \int_{(-\infty,-\varepsilon]} x N(\mathrm{d}s \times \mathrm{d}x), \quad t \geq 0,$$

and note, by almost sure monotone convergence, that as $\varepsilon \downarrow 0$, for each fixed $t \geq 0$, $X_t^{(\pm,\varepsilon)} \uparrow X_t^{(\pm)}$, for $i = 1, 2$. Since $X_t^\varepsilon = X_t^{(+,\varepsilon)} - X_t^{(-,\varepsilon)}$, we see that, for each fixed $t > 0$, we have $\lim_{\varepsilon\downarrow 0} X^\varepsilon_t = X_t$ almost surely. By replacing $[0,t]$ by $[0,t)$ in the delimiters of the definitions above it is also clear that, for each fixed $t > 0$, $\lim_{\varepsilon\downarrow 0} X^\varepsilon_{t-} = X_{t-}$ almost surely.

Now define the random region $B = \{0 \le x \le |X^\varepsilon_s| : s \le t \text{ and } \varepsilon > 0\}$. Note that B is almost surely bounded in $\mathbb{R}$ since it is contained in

$$\{0 \le x \le X^{(+)}_s : s \le t\} \cup \{0 \ge x \ge -X^{(-)}_s : s \le t\},$$

which is the union of two almost surely bounded sets, thanks to the right-continuity of paths. Due to the assumed smoothness of f, both derivatives of f are uniformly bounded (by a random value) on $[0,t] \times \overline{B}$, where $\overline{B}$ is the closure of the set B. Using the limiting behaviour of X^ε in ε and boundedness of the derivatives of f on $[0,t] \times \overline{B}$ together with almost sure dominated convergence, we see that

$$\lim_{\varepsilon\downarrow 0} \int_0^t \frac{\partial f}{\partial t}(s, X^\varepsilon_s) + \delta \frac{\partial f}{\partial x}(s, X^\varepsilon_s)\mathrm{d}s = \int_0^t \frac{\partial f}{\partial t}(s, X_s) + \delta \frac{\partial f}{\partial x}(s, X_s)\mathrm{d}s.$$

Again, using uniform boundedness of $\partial f/\partial x$, but this time on $[0,t] \times \{x + \overline{B} : |x| \le 1\}$, we note, with the help of the Mean Value Theorem, that, for all $\varepsilon > 0$ and $s \in [0,t]$,

$$\left|\left(f(s, X^\varepsilon_{s-} + x) - f(s, X^\varepsilon_{s-})\right)\mathbf{1}_{(\varepsilon \le |x| < 1)}\right| \le C|x|\mathbf{1}_{(|x|<1)},$$

where $C > 0$ is some random variable, independent of s, ε and x. The function $|x|$ integrates against N on $[0,t] \times (-1,1)$, thanks to the assumption that X has bounded variation. Now appealing to almost sure dominated convergence again, we have that

$$\begin{aligned}\lim_{\varepsilon\downarrow 0} &\int_{[0,t]} \int_{(-1,1)} \left(f(s, X^\varepsilon_{s-} + x) - f(s, X^\varepsilon_{s-})\right)\mathbf{1}_{(|x|\ge\varepsilon)} N(\mathrm{d}s \times \mathrm{d}x) \\ &= \int_{[0,t]} \int_{(-1,1)} f(s, X_{s-} + x) - f(s, X_{s-}) N(\mathrm{d}s \times \mathrm{d}x).\end{aligned}$$

A similar limit holds when the delimiters in the double integrals above are replaced by $[0,t] \times \{\mathbb{R}\backslash(-1,1)\}$ as there are, at most, a finite number of atoms in the support of N in this domain. Now taking limits on both sides of (4.4), the statement of the theorem follows. □

It is clear from the above proof that one could not expect such a formula to be valid for a general Lévy process. In order to write down a change of variable formula for a general Lévy process, X, one must first have an understanding of stochastic integrals with respect to X. At the very least, we need to have a definition for integrals of the form

$$\int_0^t g(s, X_{s-})\mathrm{d}X_s, \tag{4.5}$$

for continuous functions g. Roughly speaking, this integral may be understood as the limit

$$\lim_{\|\mathcal{P}\|\downarrow 0}\sum_{i\geq 1} g(t_{i-1}, X_{t_{i-1}})(X_{t\wedge t_i} - X_{t\wedge t_{i-1}}),$$

where $\mathcal{P} = \{0 = t_0 \leq t_1 \leq t_2 \leq \cdots\}$ is a partition of $[0,\infty)$, $\|\mathcal{P}\| = \sup_{i\geq 1}(t_i - t_{i-1})$ and the limit is taken in probability, uniformly in t on $[0, T]$, where $T > 0$ is some finite time horizon. This is not the only way to make sense of (4.5), although all definitions must be equivalent; see for example Exercise 4.4. In the case that X has bounded variation, the integral (4.5) takes the recognisable form

$$\int_0^t g(s, X_{s-})\mathrm{d}X_s = \delta\int_0^t g(s, X_s)\mathrm{d}s + \int_{[0,t]}\int_{\mathbb{R}} g(s, X_{s-})xN(\mathrm{d}s\times\mathrm{d}x). \quad (4.6)$$

Establishing these facts is of course non-trivial, and, taking account of the main theme of this book (fluctuation theory), we shy away from their proofs. The reader is otherwise directed to Applebaum (2004) for a focused account of the necessary calculations. Protter (2004) also gives the much broader picture for integration with respect to a general semi-martingale. A Lévy process is an example of a broader family of stochastic processes, called semi-martingales, which form a natural class from which to construct a theory of stochastic integration. We finish this section by simply stating Itô's formula for a general Lévy process,[1] which serves as a change of variable for the cases not covered by Theorem 4.2.

Theorem 4.3 *Let $C^{1,2}([0,\infty)\times\mathbb{R})$ be the space of functions $f : [0,\infty)\times\mathbb{R}$ which are continuously differentiable in the first variable (understood as the right-derivative at the origin) and twice continuously differentiable in the second variable. Then, for a general Lévy process, X, with Gaussian coefficient $\sigma\in\mathbb{R}$ and $f\in C^{1,2}([0,\infty)\times\mathbb{R})$, we have, for $t\geq 0$,*

$$\begin{aligned}
f(t, X_t) = f(0, X_0) &+ \int_0^t \frac{\partial f}{\partial t}(s, X_s)\mathrm{d}s + \int_0^t \frac{\partial f}{\partial x}(s, X_{s-})\mathrm{d}X_s \\
&+ \int_0^t \frac{1}{2}\sigma^2\frac{\partial^2 f}{\partial x^2}(s, X_s)\mathrm{d}s \\
&+ \int_{[0,t]}\int_{\mathbb{R}}\left(f(s, X_{s-}+x) - f(s, X_{s-}) - x\frac{\partial f}{\partial x}(s, X_{s-})\right)N(\mathrm{d}s\times\mathrm{d}x).
\end{aligned}$$

[1] As with the change of variable formula, a more general form of Itô's formula exists which includes the statement of Theorem 4.3. The natural setting as indicated above is the case that X is a semi-martingale.

4.3.2 The Compensation Formula

Although it was indicated that this chapter principally concerns processes of bounded variation, the compensation formula, which we will shortly discuss, is applicable to all Lévy processes. Suppose that X is a general Lévy process with Lévy measure Π. Recall our running assumption that X is defined on the filtered probability space $(\Omega, \mathcal{F}, \mathbb{F}, \mathbb{P})$, where $\mathbb{F} = \{\mathcal{F}_t : t \geq 0\}$ is assumed to satisfy *les conditions habituelles*. As usual, N will denote Poisson random measure with intensity $\mathrm{d}t \times \Pi(\mathrm{d}x)$ describing the jumps of X. The main result of this section may be considered as a generalisation of the results in Theorem 2.7.

Theorem 4.4 *Suppose $\phi : [0, \infty) \times \mathbb{R} \times \Omega \to [0, \infty)$ is a random time-space function such that*

(i) *as a trivariate function $\phi = \phi(t, x)[\omega]$ is measurable,*
(ii) *for each $t \geq 0$, $\phi(t, x)[\omega]$ is $\mathcal{B}(\mathbb{R}) \times \mathcal{F}_t$-measurable and*
(iii) *for each $x \in \mathbb{R}$, with probability one, $\{\phi(t, x) : t \geq 0\}$ is a left continuous process.*

Then, for all $t \geq 0$,

$$\mathbb{E}\left(\int_{[0,t]} \int_{\mathbb{R}} \phi(s, x) N(\mathrm{d}s \times \mathrm{d}x)\right) = \mathbb{E}\left(\int_0^t \int_{\mathbb{R}} \phi(s, x) \Pi(\mathrm{d}x) \mathrm{d}s\right) \tag{4.7}$$

with the understanding that the right-hand side is infinite if and only if the left-hand side is.

Note that, for each $t, \varepsilon > 0$,

$$\int_{[0,t]} \int_{\mathbb{R}\setminus(-\varepsilon,\varepsilon)} \phi(s, x) N(\mathrm{d}s \times \mathrm{d}x)$$

is nothing but the sum over a finite number of terms of positive random objects and hence, under the first assumption on ϕ, is measurable in ω. By (almost sure) monotone convergence, the integral $\int_{[0,t]} \int_{\mathbb{R}} \phi(s, x) N(\mathrm{d}s \times \mathrm{d}x)$ is well defined as

$$\lim_{\varepsilon \downarrow 0} \int_{[0,t]} \int_{\mathbb{R}\setminus(-\varepsilon,\varepsilon)} \phi(s, x) N(\mathrm{d}s \times \mathrm{d}x),$$

and is measurable in ω (recall when the limit of a sequence of measurable functions exists it is also measurable). Hence the left-hand side of (4.7) is well defined, even if infinite in value.

On the other hand, under the first assumption on ϕ, Fubini's Theorem implies that,

$$\int_0^t \int_{\mathbb{R}} \phi(s, x)[\omega] \Pi(\mathrm{d}x) \mathrm{d}s$$

is measurable in ω. Hence, the expression on the right-hand side of (4.7) is also well defined, even when infinite in value.

Proof of Theorem 4.4 Suppose initially that, in addition to the assumptions of the theorem, ϕ is uniformly bounded by $C(1 \wedge x^2)$, for some $C > 0$. This ensures the finiteness of the expressions on the left-hand and right-hand sides of (4.7). Write, for $t \geq 0$ and $x \in \mathbb{R}$,

$$\phi^n(t,x) = \phi(0,x)\mathbf{1}_{(t=0)} + \sum_{k\geq 0} \phi(k/2^n, x)\mathbf{1}_{(t\in(k/2^n,(k+1)/2^n])}, \tag{4.8}$$

noting that ϕ^n also satisfies the assumptions (i)–(iii) of the theorem. Hence, as remarked above, for each $\varepsilon > 0$,

$$\int_{[0,t]} \int_{\mathbb{R}\setminus(-\varepsilon,\varepsilon)} \phi^n(s,x) N(\mathrm{d}s \times \mathrm{d}x)$$

is well defined and measurable in ω. We have, for $t, \varepsilon > 0$,

$$\begin{aligned}
&\mathbb{E}\left(\int_{[0,t]} \int_{\mathbb{R}\setminus(-\varepsilon,\varepsilon)} \phi^n(s,x) N(\mathrm{d}s \times \mathrm{d}x)\right) \\
&= \mathbb{E}\left(\sum_{k\geq 0} \int_{(\frac{k}{2^n}\wedge t, \frac{k+1}{2^n}\wedge t]} \int_{\mathbb{R}\setminus(-\varepsilon,\varepsilon)} \phi(k/2^n, x) N(\mathrm{d}s \times \mathrm{d}x)\right) \\
&= \mathbb{E}\left(\sum_{k\geq 0} \mathbb{E}\left(\int_{(\frac{k}{2^n}\wedge t, \frac{k+1}{2^n}\wedge t]} \int_{\mathbb{R}\setminus(-\varepsilon,\varepsilon)} \phi(k/2^n, x) N(\mathrm{d}s \times \mathrm{d}x) \middle| \mathcal{F}_{\frac{k}{2^n}\wedge t}\right)\right) \\
&= \mathbb{E}\left(\sum_{k\geq 0} \int_{(\frac{k}{2^n}\wedge t, \frac{k+1}{2^n}\wedge t]} \int_{\mathbb{R}\setminus(-\varepsilon,\varepsilon)} \phi(k/2^n, x) \Pi(\mathrm{d}x)\mathrm{d}s\right) \\
&= \mathbb{E}\left(\int_{[0,t]} \int_{\mathbb{R}\setminus(-\varepsilon,\varepsilon)} \phi^n(s,x) \Pi(\mathrm{d}x)\mathrm{d}s\right),
\end{aligned} \tag{4.9}$$

where, in the third equality, we have used the fact that N has independent counts on disjoint domains, the measurability of $\phi^n(k/2^n, x)$ and an application of Theorem 2.7 (iii). Since it is assumed that ϕ is uniformly bounded by $C(1 \wedge x^2)$, we may apply dominated convergence on both sides of (4.9) as $n \uparrow \infty$, together with the fact that $\lim_{n\uparrow\infty} \phi^n(t,x) = \phi(t-,x) = \phi(t,x)$ almost surely (by the assumed left continuity of ϕ), to conclude that

$$\mathbb{E}\left(\int_{[0,t]} \int_{\mathbb{R}\setminus(-\varepsilon,\varepsilon)} \phi(s,x) N(\mathrm{d}s \times \mathrm{d}x)\right) = \mathbb{E}\left(\int_0^t \int_{\mathbb{R}\setminus(-\varepsilon,\varepsilon)} \phi(s,x) \Pi(\mathrm{d}x)\mathrm{d}s\right),$$

for all $t, \varepsilon > 0$. Now take limits as $\varepsilon \downarrow 0$ and apply the Monotone Convergence Theorem on each side of the above equality to deduce (4.7), for the case that ϕ is uniformly bounded by $C(1 \wedge x^2)$.

To remove the aforementioned condition, note that it has been established that (4.7) holds for $\phi \wedge C(1 \wedge x^2)$, where ϕ is given in the statement of the theorem. By taking limits as $C \uparrow \infty$ in the aforementioned equality, again with the help of the Monotone Convergence Theorem, the required result follows. □

Reviewing the proof of this result, there is a rather obvious corollary which follows. We leave its proof to the reader as an exercise.

Corollary 4.5 *Under the same conditions as Theorem* 4.4, *we have for all* $0 \leq u \leq t < \infty$,

$$\mathbb{E}\left(\int_{(u,t]}\int_{\mathbb{R}} \phi(s,x) N(\mathrm{d}s \times \mathrm{d}x) \middle| \mathcal{F}_u\right) = \mathbb{E}\left(\int_u^t \int_{\mathbb{R}} \phi(s,x) \Pi(\mathrm{d}x)\mathrm{d}s \middle| \mathcal{F}_u\right).$$

The last corollary also implies the martingale result below.

Corollary 4.6 *Assuming the same conditions as Theorem* 4.4 *and that, for all* $t \geq 0$,

$$\mathbb{E}\left(\int_{[0,t]}\int_{\mathbb{R}} \phi(s,x) \mathrm{d}s \Pi(\mathrm{d}x)\right) < \infty,$$

we have that

$$M_t := \int_{[0,t]}\int_{\mathbb{R}} \phi(s,x) N(\mathrm{d}s \times \mathrm{d}x) - \int_{[0,t]}\int_{\mathbb{R}} \phi(s,x) \Pi(\mathrm{d}x)\mathrm{d}s, \quad t \geq 0,$$

is a martingale.

Proof The additional integrability condition on ϕ and Theorem 4.4 implies that, for each $t \geq 0$,

$$\mathbb{E}|M_t| \leq 2\mathbb{E}\left(\int_{[0,t]}\int_{\mathbb{R}} \phi(s,x) \mathrm{d}s \Pi(\mathrm{d}x)\right) < \infty.$$

For $0 \leq u \leq t$, we see that

$$\begin{aligned} \mathbb{E}(M_t|\mathcal{F}_u) &= M_u + \mathbb{E}\left(\int_{(u,t]}\int_{\mathbb{R}} \phi(s,x) N(\mathrm{d}s \times \mathrm{d}x) \middle| \mathcal{F}_u\right) \\ &\quad - \mathbb{E}\left(\int_u^t \int_{\mathbb{R}} \phi(s,x) \Pi(\mathrm{d}x)\mathrm{d}s \middle| \mathcal{F}_u\right) \\ &= M_u, \end{aligned}$$

where the last equality is a consequence of Corollary 4.5. □

4.4 The Kella–Whitt Martingale

In this section, we introduce a martingale, the Kella–Whitt martingale, which will prove to be useful for the analysis concerning the existence of a stationary distribution of the workload process W. The martingale itself is of implicit interest as far as fluctuation theory of general spectrally negative Lévy processes is concerned, since one may derive a number of important identities from it. These identities also appear later in this text as a consequence of other techniques, centred around the Wiener–Hopf factorisation. See in particular Exercise 4.7.

The Kella–Whitt martingale takes its name from Kella and Whitt (1992) and is presented in the theorem below.

Theorem 4.7 *Suppose that X is a spectrally negative Lévy process of bounded variation. For each $\alpha \geq 0$, the process*

$$\psi(\alpha)\int_0^t \mathrm{e}^{-\alpha(\overline{X}_s - X_s)}\mathrm{d}s + 1 - \mathrm{e}^{-\alpha(\overline{X}_t - X_t)} - \alpha\overline{X}_t, \quad t \geq 0,$$

is a zero-mean $\mathbb{P}$-martingale with respect to $\mathbb{F}$.

Proof The proof of this theorem will rely on the change of variable and compensation formulae. To be more precise, we will make use of the slightly more general version of the change of variable formula, given in Exercise 4.2, which takes the form:

$$\begin{aligned} f(\overline{X}_t, X_t) &= f(\overline{X}_0, X_0) + \delta\int_0^t \frac{\partial f}{\partial x}(\overline{X}_s, X_s)\mathrm{d}s + \int_0^t \frac{\partial f}{\partial y}(\overline{X}_s, X_s)\mathrm{d}\overline{X}_s \\ &\quad + \int_{[0,t]}\int_{(-\infty,0)} \big(f(\overline{X}_s, X_{s-} + x) - f(\overline{X}_s, X_{s-})\big)N(\mathrm{d}s \times \mathrm{d}x) \end{aligned}$$

for $f(y, x) \in C^{1,1}([0,\infty) \times \mathbb{R})$ and $t \geq 0$. From this, we have that

$$\begin{aligned} \mathrm{e}^{-\alpha(\overline{X}_t - X_t)} &= 1 + \alpha\delta\int_0^t \mathrm{e}^{-\alpha(\overline{X}_s - X_s)}\mathrm{d}s - \alpha\int_0^t \mathrm{e}^{-\alpha(\overline{X}_s - X_s)}\mathrm{d}\overline{X}_s \\ &\quad + \int_{[0,t]}\int_{(-\infty,0)} \big(\mathrm{e}^{-\alpha(\overline{X}_s - X_{s-} - x)} - \mathrm{e}^{-\alpha(\overline{X}_s - X_{s-})}\big)N(\mathrm{d}s \times \mathrm{d}x) \\ &= 1 + \alpha\delta\int_0^t \mathrm{e}^{-\alpha(\overline{X}_s - X_s)}\mathrm{d}s - \alpha\int_0^t \mathrm{e}^{-\alpha(\overline{X}_s - X_s)}\mathrm{d}\overline{X}_s \\ &\quad + \int_0^t\int_{(0,\infty)} \big(\mathrm{e}^{-\alpha(\overline{X}_s - X_{s-} + x)} - \mathrm{e}^{-\alpha(\overline{X}_s - X_{s-})}\big)\nu(\mathrm{d}x)\mathrm{d}s \\ &\quad + M_t, \end{aligned} \tag{4.10}$$

(recall that ν is the Lévy measure of $-X$, defined at the end of Sect. 4.1), where, for each $t \geq 0$,

$$M_t = \int_{[0,t]} \int_{(-\infty,0)} \left(\mathrm{e}^{-\alpha(\overline{X}_s - X_{s-} - x)} - \mathrm{e}^{-\alpha(\overline{X}_s - X_{s-})}\right) N(\mathrm{d}s \times \mathrm{d}x)$$
$$- \int_0^t \int_{(0,\infty)} \left(\mathrm{e}^{-\alpha(\overline{X}_s - X_{s-} + x)} - \mathrm{e}^{-\alpha(\overline{X}_s - X_{s-})}\right) \nu(\mathrm{d}x)\mathrm{d}s. \quad (4.11)$$

Note that the second integral on the right-hand side of (4.10) can be replaced by $\overline{X}_t$ since the process $\overline{X}$ increases if and only if the integrand is equal to one. Note also that the final double integral on the right-hand side of (4.10) combines with the first integral to give

$$\alpha\delta \int_0^t \mathrm{e}^{-\alpha(\overline{X}_s - X_s)} \mathrm{d}s + \int_0^t \mathrm{e}^{-\alpha(\overline{X}_s - X_s)} \mathrm{d}s \int_{(0,\infty)} \left(\mathrm{e}^{-\alpha x} - 1\right) \nu(\mathrm{d}x)$$
$$= \psi(\alpha) \int_0^t \mathrm{e}^{-\alpha(\overline{X}_s - X_s)} \mathrm{d}s.$$

The theorem is thus proved once we show that $M = \{M_t : t \geq 0\}$ is a martingale. However, this is a consequence of Corollary 4.6. □

For the reader who is more familiar with stochastic calculus and Itô's formula for a general Lévy process, the conclusion of the previous theorem is still valid when we replace X by a general spectrally negative Lévy process. See Exercise 4.6. The interested reader is also encouraged to consult Kella and Whitt (1992), where general complex-valued martingales of this type are derived, as well as Kennedy (1976), Jacod and Shiryaev (1987) and Nguyen-Ngoc and Yor (2005).

The theorem below, taken from Kyprianou and Palmowski (2005), is an example of how one may use the Kella–Whitt martingale to study the distribution of the running infimum $\underline{X} = \{\underline{X}_t : t \geq 0\}$ where $\underline{X}_t := \inf_{s \leq t} X_s$.

Theorem 4.8 *Suppose that X is a general spectrally negative Lévy process with Laplace exponent ψ and that $\mathbf{e}_q$ is a random variable which is exponentially distributed with parameter q and independent of X. Then, for all $\beta \geq 0$ and $q > 0$,*

$$\mathbb{E}\left(\mathrm{e}^{-\beta(\overline{X}_{\mathbf{e}_q} - X_{\mathbf{e}_q})}\right) = \frac{q}{\Phi(q)} \frac{\beta - \Phi(q)}{\psi(\beta) - q}. \quad (4.12)$$

Proof As indicated in the remarks following its proof, Theorem 4.7 is still valid when X is a general spectrally negative Lévy process. We will assume this fact without proof here (otherwise refer to Exercise 4.6).

Since M, defined in (4.11), is a martingale, it follows that $\mathbb{E}(M_{\mathbf{e}_q}) = 0$. That is to say, for all $\alpha \geq 0$,

$$\psi(\alpha)\mathbb{E}\left(\int_0^{\mathbf{e}_q} \mathrm{e}^{-\alpha(\overline{X}_s - X_s)} \mathrm{d}s\right) + 1 - \mathbb{E}\left(\mathrm{e}^{-\alpha(\overline{X}_{\mathbf{e}_q} - X_{\mathbf{e}_q})}\right) - \alpha\mathbb{E}(\overline{X}_{\mathbf{e}_q}) = 0. \quad (4.13)$$

Taking the first of the three expectations, note that

$$\begin{aligned}\mathbb{E}\left(\int_0^{\mathbf{e}_q} \mathrm{e}^{-\alpha(\overline{X}_s - X_s)} \mathrm{d}s\right) &= \mathbb{E}\left(\int_0^\infty \mathrm{d}u \cdot q\mathrm{e}^{-qu} \int_0^\infty \mathbf{1}_{(s\le u)} \mathrm{e}^{-\alpha(\overline{X}_s - X_s)} \mathrm{d}s\right) \\ &= \frac{1}{q}\mathbb{E}\left(\int_0^\infty q\mathrm{e}^{-qs}\mathrm{e}^{-\alpha(\overline{X}_s - X_s)}\mathrm{d}s\right) \\ &= \frac{1}{q}\mathbb{E}\left(\mathrm{e}^{-\alpha(\overline{X}_{\mathbf{e}_q} - X_{\mathbf{e}_q})}\right).\end{aligned}$$

To compute the third expectation of (4.13), we recall from Exercise 3.6 that $\overline{X}_{\mathbf{e}_q}$ is exponentially distributed with parameter $\Phi(q)$. Hence, the aforesaid expectation is equal to $1/\Phi(q)$. Now returning to (4.13), we may rewrite it as (4.12). □

Corollary 4.9 *For all* $\beta \ge 0$,

$$\mathbb{E}(\mathrm{e}^{\beta \underline{X}_\infty}) = (0 \vee \psi'(0+))\frac{\beta}{\psi(\beta)}. \tag{4.14}$$

In particular, this shows that $-\underline{X}_t$*, and then by duality* $\overline{X}_t - X_t$*, has a non-defective limiting distribution if and only if* $\psi'(0+) > 0$.

Proof By monotonicity, $-\underline{X}_t$ has an almost sure limit as $t \uparrow \infty$. Recalling that $-\underline{X}_t$ is equal in distribution to $\overline{X}_t - X_t$, its limiting distribution is characterised by taking limits in (4.12) as $q \downarrow 0$. To this end, note that when $\Phi(0) = 0$, equivalently $\psi'(0+) \ge 0$,

$$\psi'(0+) = \lim_{\theta\downarrow 0}\frac{\psi(\theta)}{\theta} = \lim_{q\downarrow 0}\frac{q}{\Phi(q)}.$$

On the other hand, when $\Phi(0) > 0$, equivalently $\psi'(0+) < 0$,

$$\lim_{q\downarrow 0}\frac{q}{\Phi(q)} = 0.$$

Using these limits, the Final Value Theorem for Laplace transforms gives us (4.14). The limiting distribution is clearly defective when $\psi'(0+) \le 0$. When $\psi'(0+) > 0$, non-defectiveness can be seen by taking $\beta \downarrow 0$. □

4.5 Stationary Distribution of the Workload

In this section, we turn to the stationary distribution of the workload process W, making use of the conclusion in Corollary 4.9, which itself is drawn from the Kella–Whitt martingale. The setting is as in the introduction to this chapter.

Theorem 4.10 *Suppose that $0 < \rho < 1$. Then, for all $w \geq 0$, the workload has a stationary distribution,*

$$\lim_{t\uparrow\infty} \mathbb{P}(W_t \in \mathrm{d}x \,|\, W_0 = w) = (1-\rho)\sum_{k=0}^{\infty} \rho^k \eta^{*k}(\mathrm{d}x), \tag{4.15}$$

where

$$\eta(\mathrm{d}x) = \frac{1}{\delta\rho}\nu(x,\infty)\mathrm{d}x. \tag{4.16}$$

Here, we understand $\eta^{*0}(\mathrm{d}x) = \delta_0(\mathrm{d}x)$, so that the limiting distribution has an atom at zero. Otherwise, when $\rho \geq 1$, there is no stationary distribution.

Proof First suppose that $\rho \geq 1$. In this case, we know that $\psi'(0+) \leq 0$. Since $W_t = (w \vee \overline{X}_t) - X_t \geq \overline{X}_t - X_t$, it follows that, for all $M > 0$,

$$\lim_{t\uparrow\infty} \mathbb{P}(W_t > M) \geq \lim_{t\uparrow\infty} \mathbb{P}(\overline{X}_t - X_t > M) = 1,$$

where the final equality follows from Corollary 4.9. This shows that W_t does not converge in distribution.

Now suppose that $0 < \rho < 1$. In this case $\psi'(0+) > 0$ and hence, from Corollary 3.13, we know that $\mathbb{P}(\tau_w^+ < \infty) = 1$. It follows that, for all $t \geq \tau_w^+$, $W_t = \overline{X}_t - X_t$ and so, from Corollary 4.9, we see that, for all $\beta > 0$,

$$\lim_{t\uparrow\infty} \mathbb{E}(\mathrm{e}^{-\beta W_t}) = \psi'(0+)\frac{\beta}{\psi(\beta)}. \tag{4.17}$$

The remainder of the proof thus requires us to show that the right-hand side of (4.15) has Laplace–Stieltjes transform equal to the right-hand side of (4.17).

To this end, using integration by parts in the definition of ψ, note that

$$\frac{\psi(\beta)}{\beta} = \delta - \int_0^\infty \mathrm{e}^{-\beta x}\nu(x,\infty)\mathrm{d}x. \tag{4.18}$$

As $\psi'(0+) > 0$, we have that $\delta^{-1}\int_0^\infty \nu(x,\infty)\mathrm{d}x < 1$; indeed, for all $\beta \geq 0$, we have that $\delta^{-1}\int_0^\infty \mathrm{e}^{-\beta x}\nu(x,\infty)\mathrm{d}x < 1$. We may thus develop the right-hand side of (4.17) as follows:

$$\psi'(0+)\frac{\beta}{\psi(\beta)} = \frac{\psi'(0+)}{\delta}\sum_{k\geq 0}\left(\frac{1}{\delta}\int_0^\infty \mathrm{e}^{-\beta x}\nu(x,\infty)\mathrm{d}x\right)^k, \quad \beta \geq 0.$$

Now define the measure $\eta(\mathrm{d}x) = (\delta\rho)^{-1}\nu(x,\infty)\mathrm{d}x$. We have

$$\psi'(0+)\frac{\beta}{\psi(\beta)} = \frac{\psi'(0+)}{\delta}\sum_{k\geq 0}\rho^k\int_0^\infty \mathrm{e}^{-\beta x}\eta^{*k}(\mathrm{d}x), \quad \beta \geq 0, \tag{4.19}$$

with the understanding that $\eta^{*0}(\mathrm{d}x) = \delta_0(\mathrm{d}x)$. Note that $\psi'(0+)/\delta = 1 - \rho$. The result now follows by comparing (4.19) against (4.17). Note in particular that the stationary distribution, as one would expect, is independent of the initial value of the workload. □

Theorem 4.10 contains Theorem 1.12. To see this, simply set $\delta = 1$, $\nu = \lambda F$, where F is the distribution with mean μ.

As noted earlier in Sect. 1.3.2, for the case of the $M/G/1$ queue with $0 < \rho < 1$, the expression for the stationary distribution, given in statement of Theorem 4.10, is remarkably similar to the expression for the Pollaczek–Khintchine formula, given in Theorem 1.8. The similarity of these two can be explained in a simple way using the Duality Lemma 3.4. Duality implies that, for each fixed $t \geq 0$, $\overline{X}_t - X_t$ is equal in distribution to $-\underline{X}_t$. As was noted in the proof of Theorem 4.10, when $0 < \rho < 1$, the limit in distribution of W is independent of w and equal to the distributional limit of $\overline{X}_t - X_t$ and hence by the previous remarks, is also equal to the distribution of $-\underline{X}_\infty$. Noting further that

$$\mathbb{P}(-\underline{X}_\infty \leq x) = \mathbb{P}_x(\tau_0^- = \infty),$$

where $\tau_0^- = \inf\{t > 0 : X_t < 0\}$, we see that Theorem 4.10 also reads: For all $x > 0$,

$$\mathbb{P}_x(\tau_0^- = \infty) = (1 - \rho)\sum_{k=0}^{\infty} \rho^k \eta^{*k}(x), \tag{4.20}$$

where, now, $\eta^{*0}(x) = 1$. However, this is precisely the combined statements of Theorems 1.8 and 1.9, but now for a general spectrally negative Lévy process of bounded variation.

4.6 Small-Time Behaviour and the Pollaczek–Khintchine Formula

Within the context of either the stationary distribution of the workload process or the ruin problem, the reason for the appearance of a geometric-type sum in both cases is related to how spectrally negative Lévy processes of bounded variation behave at arbitrarily small times, and consequently how the entire path of the process X decomposes into objects called *excursions*. This section is dedicated to explaining this phenomenon.

We start the discussion with a lemma, essentially due to Shtatland (1965); see also Chap. IV of Gikhman and Skorokhod (1975).

Lemma 4.11 *Suppose that X is a spectrally negative Lévy process of bounded variation. Then*

$$\lim_{t \downarrow 0} \frac{X_t}{t} = \delta$$

almost surely.

Proof Recall from the Lévy–Itô decomposition that jumps of Lévy processes are described by a Poisson random measure with intensity $dt \times \nu(dx)$. From this, it follows that the first jump of X of magnitude greater than ϵ appears after a length of time which is exponentially distributed with parameter $\nu(\epsilon, \infty)$. Since we are interested in small-time behaviour, it therefore is of no consequence if we assume that ν is concentrated on $(0, \epsilon)$. That is to say, there are no negative jumps of magnitude greater than ϵ.

Recall that X is written in the form $X_t = \delta t - S_t$, for $t \geq 0$, where $S = \{S_t : t \geq 0\}$ is a pure jump subordinator with Lévy measure ν. The proof is then completed by showing that

$$\lim_{t \downarrow 0} \frac{S_t}{t} = 0.$$

To this end, set $M_n = S_{2^{-n}}/2^{-n}$ and note that, on the one hand,

$$\mathbb{E}(M_{n+1}|M_1, \ldots, M_n) = 2M_n - 2^{n+1}\mathbb{E}(S_{2^{-n}} - S_{2^{-(n+1)}}|M_1, \ldots, M_n). \tag{4.21}$$

On the other hand, time reversing the path $\{S_t : t \leq 2^{-n}\}$ and using the stationarity and independence of increments, we have that the law of $S_{2^{-(n+1)}} - S_0$ given $\{S_{2^{-n}}, S_{2^{-(n-1)}}, \ldots, S_{1/2}\}$ is equal to the law of $S_{2^{-n}} - S_{2^{-(n+1)}}$ given $\{S_{2^{-n}}, S_{2^{-(n-1)}}, \ldots, S_{1/2}\}$. Hence,

$$\mathbb{E}(S_{2^{-n}} - S_{2^{-(n+1)}}|M_1, \ldots, M_n) = \mathbb{E}(S_{2^{-(n+1)}}|M_1, \ldots, M_n).$$

Substituting back into (4.21), we see that $\mathbb{E}(M_{n+1}|M_1, \ldots, M_n) = M_n$ and hence the sequence $M = \{M_n : n \geq 1\}$ is a positive $\mathbb{P}$-martingale. The Martingale Convergence Theorem implies that $M_\infty := \lim_{n\uparrow\infty} M_n$ exists and Fatou's Lemma implies that

$$\mathbb{E}(M_\infty) \leq \mathbb{E}(M_1) = \int_{(0,\epsilon)} x\nu(dx).$$

Note that for the last equality, we have appealed to Exercise 2.11. Since for $t \in [2^{-(n+1)}, 2^{-n})$,

$$\frac{S_t}{t} \leq \frac{S_{2^{-n}}}{2^{-(n+1)}} = 2M_n,$$

we thus have that

$$\mathbb{E}\left(\limsup_{t\downarrow 0} \frac{S_t}{t}\right) \leq 2\mathbb{E}\left(\limsup_{n\uparrow\infty} M_n\right) = 2\mathbb{E}(M_\infty) \leq 2\int_{(0,\epsilon)} x\nu(dx). \tag{4.22}$$

Since $\int_{(0,1)} x\nu(dx) < \infty$, the right-hand side above can be made arbitrarily small by letting $\epsilon \downarrow 0$. This shows that the expectation on the left-hand side of (4.22) is equal to zero, and hence so is the limsup in the expectation in the almost sure sense. □

The lemma shows that, for all sufficiently small times, $X_t > 0$ and hence $\mathbb{P}(\tau_0^- > 0) = 1$. That is to say, when starting from zero, it takes a strictly positive amount of time before X visits $(-\infty, 0)$. Compare this with, for example, the situation for Brownian motion. It is intuitively clear that it will visit both sides of the origin immediately. To be rigorous about this, recall from Exercise 1.7 that the first-passage process of a Brownian motion is a stable-$\frac{1}{2}$ subordinator. Since this subordinator is not a compound Poisson process, and hence does not remain at the origin for an almost surely strictly positive period of time, first passage strictly above level zero of B occurs immediately. By symmetry, the same can be said about first passage strictly below the level zero.

In order to complete our explanation of the geometric-type sum appearing in (4.20), let us proceed by showing that $\mathbb{P}(\sigma_x^+ = \infty)$ takes the form given in the right-hand side of (4.15), where, now, we take $Y = -X$ and, for each $x \geq 0$, $\sigma_x^+ = \inf\{t > 0 : Y_t > x\}$. Lemma 4.11 shows that $\mathbb{P}(\sigma_0^+ > 0) = 1$. This information allows us to make the following path decomposition.

Define $T_0 = 0$ and $H_0 = 0$. Let $T_1 := \sigma_0^+$ and

$$H_1 = \begin{cases} Y_{T_1} & \text{if } T_1 < \infty \\ \infty & \text{if } T_1 = \infty. \end{cases}$$

Next, we construct iteratively the variables $T_1, T_2, \ldots$ and $H_1, H_2, \ldots$ in such a way that

$$T_n := \begin{cases} \inf\{t > T_{n-1} : Y_t > H_{n-1}\} & \text{if } T_{n-1} < \infty \\ \infty & \text{if } T_{n-1} = \infty \end{cases}$$

and

$$H_n := \begin{cases} Y_{T_n} & \text{if } T_n < \infty \\ \infty & \text{if } T_n = \infty. \end{cases}$$

Note in particular, $T_1 = \sigma_0^+$ is a stopping time and, for each $n \geq 1$, $T_{n+1} - T_n$ is equal in distribution to T_1. The strong Markov property and stationary independent increments imply that, on $\{T_{n-1} < \infty\}$, the path

$$\epsilon_n = \{Y_t - Y_{T_{n-1}} : T_{n-1} < t \leq T_n\}, \tag{4.23}$$

also known as an *excursion of Y from its maximum* (equiv. an excursion of X from its minimum), is independent of $\mathcal{F}_{T_{n-1}}$ and has the same law as

$$\{Y_t : 0 < t \leq \sigma_0^+\}.$$

In particular, on the event $\{T_{n-1} < \infty\}$, the pair $(T_n - T_{n-1}, H_n - H_{n-1})$ is independent of $\mathcal{F}_{T_{n-1}}$ and has the same distribution as $(\sigma_0^+, Y_{\sigma_0^+})$ under $\mathbb{P}$.

The sequence of pairs $\{(T_n, H_n) : n \geq 1\}$ are nothing more than the jump times and the consecutive heights of the new maxima of Y, so long as they are finite. The assumption that X drifts to infinity (equivalently Y drifts to $-\infty$) implies that the distribution of σ_0^+ under $\mathbb{P}$ is defective. To see this, recall that, by duality, the limiting distribution of $\overline{X}_t - X_t$ is equal to that of the limiting distribution of $-\underline{X}_t$, which, in turn, is equal to the limiting distribution of $\overline{Y}_t$. Note that $\overline{Y}_t = -\underline{X}_t$ has an

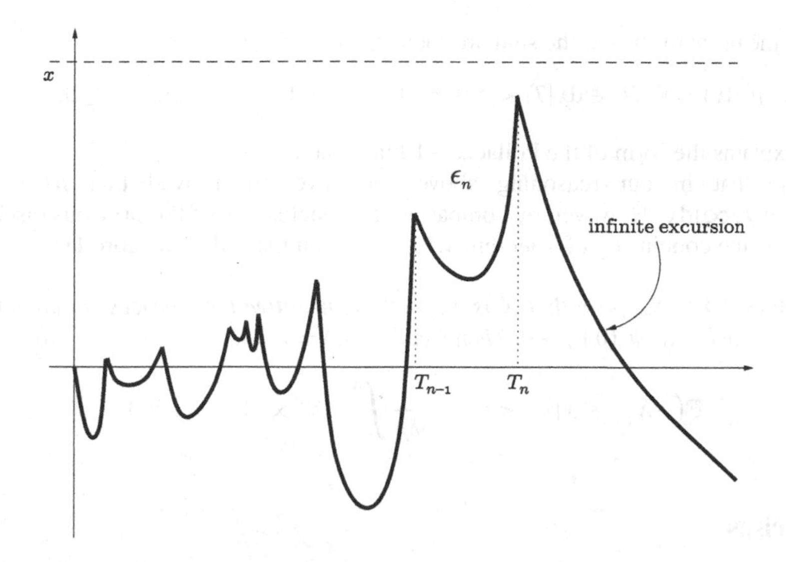

Fig. 4.1 A symbolic sketch of the decomposition of the path of Y when it fails to cross the level x.

almost sure limiting distribution on account of it being monotone in t. From (4.18), we see that $\lim_{\beta\uparrow\infty}\psi(\beta)/\beta=\delta$. Hence, when it is assumed that $0<\rho<1$, or equivalently that $\psi'(0+)>0$, we see from Corollary 4.9 that

$$1-\rho=\frac{\psi'(0+)}{\delta}=\lim_{\beta\uparrow\infty}\mathbb{E}(\mathrm{e}^{-\beta\overline{Y}_\infty})=\mathbb{P}(\overline{Y}_\infty=0)=\mathbb{P}(\sigma_0^+=\infty).$$

It follows that there exists an almost surely finite $N\in\{0,1,2,\ldots\}$ such that each member of the pair (T_n,H_n) is finite for all $n\le N$, and infinite for all $n>N$. We say that the excursion ϵ_n is *finite* if $T_n-T_{n-1}<\infty$ and otherwise, at the first index, n, for which $T_n-T_{n-1}=\infty$, we say that the n-th excursion is *infinite*. The total number of excursions, $N+1$, is the first time to success in a sequence of Bernoulli trials, where "success" means the occurrence of an infinite excursion and, as noted above, "success" has probability $1-\rho$. That is to say, $N+1$ is geometrically distributed with parameter $1-\rho$. As the process Y is assumed to drift to ∞, the structure of the path of Y must correspond to the juxtaposition of N i.i.d. excursions conditioned to be finite, followed by a final infinite excursion. Figure 4.1 gives a symbolic impression of this decomposition, leaving out details of the path within excursions.

Using the above decomposition, it is now clear that the event $\{\sigma_x^+=\infty\}$ corresponds to the event that there are N i.i.d. finite excursions of Y which, when pasted end to end, have a right end point which is no higher than x, followed by an infinite excursion. As $N+1$ is geometrically distributed with parameter $1-\rho$, it follows that

$$\mathbb{P}(\sigma_x^+=\infty)=\sum_{n\ge0}(1-\rho)\rho^n\mathbb{P}(H_n\le x|\epsilon_1,\ldots,\epsilon_n\text{ are finite}),$$

where the probabilities in the sum are each equal to $\mu^{*n}(x)$, with

$$\mu(\mathrm{d}x) = \mathbb{P}(H_1 \in \mathrm{d}x | T_1 < \infty) = \mathbb{P}\big(-X_{\tau_0^-} \in \mathrm{d}x | \tau_0^- < \infty\big), \quad x \geq 0.$$

This explains the form of the Pollaczek–Khintchine formula.

Note that in our reasoning above, we have not proved that $\mu(\mathrm{d}x) = (\delta\rho)^{-1}\nu(x,\infty)\mathrm{d}x$. However, by comparing the conclusions of the previous discussion with the conclusion of Theorem 4.10, we obtain the following corollary.

Corollary 4.12 *Suppose that X is a spectrally negative Lévy process of bounded variation such that $\psi'(0+) > 0$. Then $\mathbb{P}(\tau_0^- < \infty) = \rho$ and*

$$\mathbb{P}\big(-X_{\tau_0^-} \leq x | \tau_0^- < \infty\big) = \frac{1}{\delta\rho}\int_0^x \nu(y,\infty)\mathrm{d}y, \quad x \geq 0.$$

Exercises

4.1 Suppose that $X = \{X_t : t \geq 0\}$ is a spectrally negative process of bounded variation with drift δ (see the discussion following Lemma 2.14). Define, for each $t \geq 0$,

$$L_t^0 = \#\{0 < s \leq t : X_s = 0\}.$$

(i) Show that the process $\{L_t^0 : t \geq 0\}$ is almost surely integer-valued with paths that are right-continuous with left limits.

(ii) Suppose now that f is a function which is equal to a $C^1(\mathbb{R})$ function[2] on $(-\infty, 0)$ and equal to another $C^1(\mathbb{R})$ function on $(0, \infty)$ but may have a discontinuity at 0. Its derivative at 0 may also be undefined. Show that for each $t \geq 0$,

$$\begin{aligned} f(X_t) = {} & f(X_0) + \delta\int_0^t f'(X_s)\mathrm{d}s \\ & + \int_{(0,t]}\int_{(-\infty,0)} \big(f(X_{s-}+x) - f(X_{s-})\big)N(\mathrm{d}s \times \mathrm{d}x) \\ & + \int_{(0,t]} \big(f(X_s) - f(X_{s-})\big)\mathrm{d}L_s^0. \end{aligned}$$

4.2 Suppose that $X = \{X_t : t \geq 0\}$ is a spectrally negative Lévy process of bounded variation with drift δ. Show that, for $f(y,x) \in C^{1,1}([0,\infty) \times \mathbb{R})$ and $t > 0$,

$$f(\overline{X}_t, X_t) = f(\overline{X}_0, X_0) + \delta\int_0^t \frac{\partial f}{\partial x}(\overline{X}_s, X_s)\mathrm{d}s + \int_0^t \frac{\partial f}{\partial y}(\overline{X}_s, X_s)\mathrm{d}\overline{X}_s$$

[2] A $C^1(\mathbb{R})$ function is a continuously differentiable mapping from $\mathbb{R}$ to $\mathbb{R}$.

$$+\int_{[0,t]}\int_{(-\infty,0)} f(\overline{X}_s, X_{s-}+x) - f(\overline{X}_s, X_{s-})N(\mathrm{d}s\times \mathrm{d}x).$$

4.3 Suppose that ϕ fulfils the conditions of Theorem 4.4 and that, for each $t>0$, $\mathbb{E}(\int_{[0,t]}\int_{\mathbb{R}}\phi(s,x)\Pi(\mathrm{d}x)\mathrm{d}s)<\infty$. If $M=\{M_t: t\geq 0\}$ is the martingale given in Corollary 4.6 and, further, it is assumed that, for all $t\geq 0$, $\mathbb{E}(\int_{[0,t]}\int_{\mathbb{R}}\phi(s,x)^2\Pi(\mathrm{d}x)\mathrm{d}s)<\infty$ show that

$$\mathbb{E}(M_t^2)=\mathbb{E}\left(\int_{[0,t]}\int_{\mathbb{R}}\phi(s,x)^2\mathrm{d}s\Pi(\mathrm{d}x)\right),\quad t\geq 0.$$

4.4 In this exercise, we use ideas coming from the proof of the Lévy–Itô decomposition to prove Itô's formula in Theorem 4.3 for the case that $\sigma=0$. Henceforth, we will assume that X is a Lévy process with no Gaussian component and $f(t,x)\in C^{1,2}([0,\infty)\times\mathbb{R})$ is uniformly bounded, along with its first derivative in s and first two derivatives in x.

(i) Suppose that X has characteristic exponent

$$\Psi(\theta)=\mathrm{i}\theta a+\int_{\mathbb{R}}\left(1-\mathrm{e}^{\mathrm{i}\theta x}+\mathrm{i}\theta x\mathbf{1}_{(|x|<1)}\right)\Pi(\mathrm{d}x),\quad \theta\in\mathbb{R}.$$

For each $1>\varepsilon>0$, let $X^{(\varepsilon)}=\{X_t^{(\varepsilon)}: t\geq 0\}$ be the Lévy process with characteristic exponent

$$\Psi^{(\varepsilon)}(\theta)=\mathrm{i}\theta a+\int_{\mathbb{R}\setminus(-\varepsilon,\varepsilon)}(1-\mathrm{e}^{\mathrm{i}\theta x}+\mathrm{i}\theta x\mathbf{1}_{(|x|<1)})\Pi(\mathrm{d}x).$$

Show that

$$\begin{aligned} f(t,X_t^{(\varepsilon)}) &= f(0,X_0)+\int_0^t\frac{\partial f}{\partial t}(s,X_s^{(\varepsilon)})\mathrm{d}s\\ &+\int_{[0,t]}\int_{|x|\geq\varepsilon}\left(f(s,X_{s-}^{(\varepsilon)}+x)-f(s,X_{s-}^{(\varepsilon)})-x\frac{\partial f}{\partial x}(s,X_{s-}^{(\varepsilon)})\right)N(\mathrm{d}s\times\mathrm{d}x)\\ &+\int_0^t\frac{\partial f}{\partial x}(s,X_{s-}^{(\varepsilon)})\mathrm{d}X_s^*+M_t^{(\varepsilon)}, \end{aligned}\tag{4.24}$$

where X^* is a Lévy process with characteristic exponent $a\mathrm{i}\theta+\int_{|x|\geq 1}(1-\mathrm{e}^{\mathrm{i}\theta x})\Pi(\mathrm{d}x)$ and $M^{(\varepsilon)}=\{M_t^{(\varepsilon)}: t\geq 0\}$ is a right-continuous, square-integrable martingale.

(ii) Fix $T>0$. Show that $\{M^{(\varepsilon)}: 0<\varepsilon<1\}$ is a Cauchy family in the martingale space $\mathcal{M}_T^2$ (see Definition 2.11).

(iii) Denote the limiting martingale in part (ii) by M. By taking limits as $\varepsilon \downarrow 0$ along a suitable subsequence, show that the Itô formula holds, where

$$\int_0^t \frac{\partial f}{\partial x}(s, X_{s-})\mathrm{d}X_s := \int_0^t \frac{\partial f}{\partial x}(s, X_{s-})\mathrm{d}X^* + M_t.$$

Explain why the left-hand side above is a suitable choice of notation.

(iv) Show that if the restrictions of uniform boundedness of f and its derivatives are removed, then the same conclusion may be drawn as in (iii), except now there exists an increasing sequence of stopping times tending to infinity, say $\{T_n : n \geq 1\}$, such that, for each $n \geq 1$, the process M is a martingale when stopped at time T_n. In other words, M is a local martingale and not necessarily a martingale.

4.5 Consider the workload process W of an $M/G/1$ queue as described in Sect. 1.3.2. Suppose that $W_0 = w = 0$ and the service distribution F has Laplace transform $\hat{F}(\beta) = \int_{(0,\infty)} \mathrm{e}^{-\beta x} F(\mathrm{d}x)$, $\beta \geq 0$.

(i) Show that the first busy period (the time from the moment of first service to the first moment thereafter that the queue is again empty), denoted B, fulfils

$$\mathbb{E}(\mathrm{e}^{-\beta B}) = \hat{F}(\Phi(\beta))$$

where $\Phi(\beta)$ is the largest solution to the equation

$$\theta - \int_{(0,\infty)} (1 - \mathrm{e}^{-\theta x})\lambda F(\mathrm{d}x) = \beta.$$

(ii) When $\rho > 1$, show that there are a geometrically distributed number of busy periods. Hence, give another proof of the first part of Theorem 4.1 when $w = 0$ by using this fact.

(iii) Suppose further that the service distribution F is that of an exponential random variable with parameter $\mu > \lambda$. This is the case of an $M/M/1$ queue. Show that the workload process has limiting distribution given by

$$\left(1 - \frac{\lambda}{\mu}\right)\left(\delta_0(\mathrm{d}x) + \mathbf{1}_{(x>0)}\lambda \mathrm{e}^{-(\mu-\lambda)x}\mathrm{d}x\right).$$

4.6 This exercise is only for the reader familiar with the general theory of stochastic calculus with respect to semi-martingales. Suppose that X is a general spectrally negative Lévy process. Recall the notation $\mathcal{E}_t(\alpha) = \exp\{\alpha X_t - \psi(\alpha)t\}$, for $t \geq 0$.

(i) If M is the Kella–Whitt martingale, show that

$$\mathrm{d}M_t = -\mathrm{e}^{-\overline{X}_t + \psi(\alpha)t}\mathrm{d}\mathcal{E}_t(\alpha), \quad t \geq 0,$$

and hence deduce that M is a local martingale.

(ii) Show that $\mathbb{E}(\overline{X}_t) < \infty$ for all $t > 0$.

(iii) Deduce that $\mathbb{E}(\sup_{s\le t}|M_s|) < \infty$, and hence that M is a martingale.

4.7 Suppose that X is a spectrally negative Lévy process of bounded variation with characteristic exponent Ψ.

(i) Show that, for each $\alpha, \beta \in \mathbb{R}$,

$$M_t = -\Psi(\alpha)\int_0^t \mathrm{e}^{\mathrm{i}\alpha(X_s-\overline{X}_s)+\mathrm{i}\beta\overline{X}_s}\,\mathrm{d}s + 1 - \mathrm{e}^{\mathrm{i}\alpha(X_t-\overline{X}_t)+\mathrm{i}\beta\overline{X}_t} - \mathrm{i}(\alpha-\beta)\int_0^t \mathrm{e}^{\mathrm{i}\alpha(X_s-\overline{X}_s)+\mathrm{i}\beta\overline{X}_s}\,\mathrm{d}\overline{X}_s, \quad t\ge 0$$

is a martingale. Note, for the reader familiar with general stochastic calculus for semi-martingales, one may equally prove that $\{M_t : t\ge 0\}$ is a martingale for a general spectrally negative Lévy process.

(ii) Use the fact that $\mathbb{E}(M_{\mathbf{e}_q}) = 0$, where $\mathbf{e}_q$ is an independent exponentially distributed random variable with parameter q, to show that, for $\alpha, \beta \ge 0$,

$$\mathbb{E}\big(\mathrm{e}^{\mathrm{i}\alpha(X_{\mathbf{e}_q}-\overline{X}_{\mathbf{e}_q})+\mathrm{i}\beta\overline{X}_{\mathbf{e}_q}}\big) = \frac{q(\Phi(q)-\mathrm{i}\alpha)}{(\Psi(\alpha)+q)(\mathrm{i}\beta-\Phi(q))}, \tag{4.25}$$

where Φ is the right inverse of the Laplace exponent $\psi(\beta) = -\Psi(-\mathrm{i}\beta)$.

(iii) Deduce that $\overline{X}_{\mathbf{e}_q} - X_{\mathbf{e}_q}$ and $\overline{X}_{\mathbf{e}_q}$ are independent.

4.8 Suppose that X is *any* Lévy process of bounded variation with drift $\delta > 0$ (excluding the case of a subordinator or the negative of a subordinator).

(i) Show that

$$\lim_{t\downarrow 0}\frac{X_t}{t} = \delta$$

almost surely.

(ii) Define $\tau_0^- = \inf\{t>0 : X_t < 0\}$. By reasoning along similar lines for the case of a spectrally negative process, show that $\mathbb{P}(\tau_0^- > 0) > 0$.

(iii) Suppose now that $\lim_{t\uparrow\infty} X_t = \infty$. Let $\eta(\mathrm{d}x) = \mathbb{P}(-X_{\tau_0^-}\in \mathrm{d}x\,|\,\tau_0^- < \infty)$, $x\ge 0$. Conclude that the Pollaczek–Khintchine formula,

$$\mathbb{P}_x\big(\tau_0^- = \infty\big) = (1-\rho)\sum_{k=0}^{\infty}\rho^k\eta^{*k}(x), \quad x\ge 0,$$

is still valid under these circumstances.

Chapter 5
Subordinators at First Passage and Renewal Measures

In this chapter, we look at subordinators. Recall that these are Lévy processes which have paths that are non-decreasing. In addition, we consider *killed subordinators*, that is, subordinators which are sent to a "cemetery state" (in other words an additional point that is not in $[0, \infty)$) at an independent time that is exponentially distributed. Principally, we are interested in first passage over a fixed level, and some asymptotic features thereof, as the level tends to infinity. In particular, we will study the (asymptotic) law of the overshoot and undershoot, as well as the phenomenon of crossing a level by hitting it. These three points of interest turn out to be very closely related to renewal measures. The results obtained in this chapter will be of significance later on when we consider first passage over a fixed level of a general Lévy process. As part of the presentation on asymptotic first passage, we will review some basic facts about regular variation. Regular variation will also be of use in later chapters. We conclude with a brief introduction to the theory of special subordinators which, amongst other things, permits the construction of a number of concrete examples of some of the theory discussed earlier in the chapter.

5.1 Killed Subordinators and Renewal Measures

In the setting of Sect. 2.6.1 a subordinator is a Lévy process of bounded variation, drift $\delta \geq 0$ and jump measure concentrated on $(0, \infty)$. In this section we shall consider a slightly more general class of processes, *killed subordinators*. Let Y be a subordinator and $\mathbf{e}_\eta$ an independent exponentially distributed random variable with rate $\eta > 0$. Then a killed subordinator is the process

$$X_t = \begin{cases} Y_t & \text{if } t < \mathbf{e}_\eta \\ \partial & \text{if } t \geq \mathbf{e}_\eta, \end{cases}$$

where ∂ is a "cemetery state". We shall also refer to X as "Y killed at rate η". If we agree that $\mathbf{e}_\eta = \infty$ when $\eta = 0$, then the definition of a killed subordinator

A.E. Kyprianou, *Fluctuations of Lévy Processes with Applications*, Universitext,
DOI 10.1007/978-3-642-37632-0_5, © Springer-Verlag Berlin Heidelberg 2014

includes the class of regular subordinators.[1] This will prove to be useful for making general statements. The Laplace exponent of a killed subordinator X is defined, for all $\theta \geq 0$, by the formula

$$\Phi(\theta) = -\log\mathbb{E}(\mathrm{e}^{-\theta X_1}) = -\log\mathbb{E}(\mathrm{e}^{-\theta Y_1}\mathbf{1}_{(1<\mathbf{e}_\eta)}) = \eta - \log\mathbb{E}(\mathrm{e}^{-\theta Y_1}) = \eta + \Psi(i\theta),$$

where Ψ is the Lévy–Khintchine exponent of Y. From the Lévy–Khintchine formula given in the form (2.21), we easily deduce that

$$\Phi(\theta) = \eta + \delta\theta + \int_{(0,\infty)} (1 - \mathrm{e}^{-\theta x})\Pi(\mathrm{d}x), \tag{5.1}$$

where $\delta \geq 0$ and $\int_{(0,\infty)}(1 \wedge x)\Pi(\mathrm{d}x) < \infty$; recall Exercise 2.11.

With each killed subordinator, we associate a family of potential measures. Define for each $q \geq 0$ the *q-potential measure* on $[0, \infty)$ by

$$U^{(q)}(\mathrm{d}x) = \mathbb{E}\left(\int_0^\infty \mathrm{e}^{-qt}\mathbf{1}_{(X_t\in\mathrm{d}x)}\mathrm{d}t\right) = \int_0^\infty \mathrm{e}^{-qt}\mathbb{P}(X_t \in \mathrm{d}x)\mathrm{d}t. \tag{5.2}$$

For notational ease, we shall simply write $U^{(0)} = U$ and call it the *potential measure*. Note that the q-potential measure of a killed subordinator with killing at rate $\eta > 0$ is equal to the $(q + \eta)$-potential measure of the same subordinator without killing. Note also that, for each $q > 0$, $(q + \eta)U^{(q)}$ is a probability measure on $[0, \infty)$ and also that, for each $q \geq 0$, $U^{(q)}(x) := U^{(q)}[0, x]$ is right-continuous. Roughly speaking, a q-potential measure is a discounted measure of how long the process X occupies different regions of space on average.[2]

These potential measures will play an important role in the study of how subordinators cross fixed levels. For this reason, we will devote the remainder of this section to studying some of their analytical properties. One of the most important facts about q-potential measures is that they are closely related to renewal measures.

We recall briefly that a renewal process, $N = \{N_x : x \geq 0\}$, counts the number of points in $[0, x]$, for $x \geq 0$, of an arrival process on $[0, \infty)$ in which points are laid down as follows. Let F be a distribution function on $(0, \infty)$ and suppose that $\{\xi_i : i = 1, 2, \ldots\}$ is a sequence of independent random variables with common distribution F. Points are positioned at $\{T_1, T_2, \ldots\}$, where, for each $k \geq 1$, $T_k = \sum_{i=1}^k \xi_i$. In other words, the underlying arrival process is nothing more than the range of a random walk with jump distribution F. For each $x \geq 0$, we may now identify $N_x = \sup\{i \geq 1 : T_i \leq x\}$, where we use the notational convention $\sup\emptyset = 0$. Note that if F is an exponential distribution, then N is nothing more than a Poisson process.

[1] A killed subordinator is only a Lévy process when $\eta = 0$, but it is still a Markov process even when $\eta > 0$.

[2] From the general theory of Markov processes, $U^{(q)}$ also comes under the name of resolvent measure or Green's measure.

The associated renewal measure is defined by

$$V(\mathrm{d}x) = \sum_{k\geq 0} F^{*k}(\mathrm{d}x), \quad x \geq 0,$$

where we understand $F^{*0}(\mathrm{d}x) := \delta_0(\mathrm{d}x)$. As with potential measures, we work with the notation $V(x) := V[0,x]$, $x \geq 0$. For future reference, let us recall some of the classical renewal theorems.

Theorem 5.1 (Renewal Theorem) *Suppose that V is the renewal function given above and let $\mu := \int_{(0,\infty)} xF(\mathrm{d}x) \in (0,\infty]$.*

(i) [3]*If F does not have lattice support, then, for all $y > 0$,*

$$\lim_{x\uparrow\infty} \{V(x+y) - V(x)\} = \frac{y}{\mu}.$$

(ii) [4]*If F does not have lattice support and $h : [0,\infty) \to \mathbb{R}$ is directly Riemann integrable, then*

$$\lim_{x\uparrow\infty} \int_0^x h(x-y)V(\mathrm{d}y) = \frac{1}{\mu}\int_0^\infty h(y)\mathrm{d}y.$$

(iii) *Without restriction on the support of F,*

$$\lim_{x\uparrow\infty} \frac{V(x)}{x} = \frac{1}{\mu}.$$

Here, we understand $\mu^{-1} = 0$ if $\mu = \infty$.

The reader may find more on the different aspects of the Renewal Theorem in Chap. XI of Feller (1971). See also Chap. 4 of Durrett (2004).

The precise relationship between q-potential measures of subordinators and renewal measures is given in the following lemma.

Lemma 5.2 *Suppose that X is a subordinator (no killing). Let $F = U^{(1)}$ and let V be the renewal measure associated with the distribution F. Then $V(\mathrm{d}x)$ is equal to the measure $\delta_0(\mathrm{d}x) + U(\mathrm{d}x)$ on $[0,\infty)$.*

Proof First note that, for all $\theta > 0$,

$$\int_{[0,\infty)} \mathrm{e}^{-\theta x} U^{(1)}(\mathrm{d}x) = \int_0^\infty \mathrm{d}t \cdot \mathrm{e}^{-t} \int_{[0,\infty)} \mathrm{e}^{-\theta x} \mathbb{P}(X_t \in \mathrm{d}x)$$

[3]This part of the theorem is known as Blackwell's Renewal Theorem.

[4]This part of the theorem is also known on its own as the Key Renewal Theorem.

$$= \int_0^\infty \mathrm{d}t \cdot \mathrm{e}^{-(1+\Phi(\theta))t}$$

$$= \frac{1}{1+\Phi(\theta)},$$

where Φ is the Laplace exponent of the underlying subordinator. In the final equality, we have used the fact that $\Phi(\theta) > 0$.

Next compute the Laplace transform of V for all $\theta > 0$ as follows:

$$\int_{[0,\infty)} \mathrm{e}^{-\theta x} V(\mathrm{d}x) = \sum_{k\geq 0} \left(\int_{[0,\infty)} \mathrm{e}^{-\theta x} U^{(1)}(\mathrm{d}x) \right)^k$$

$$= \sum_{k\geq 0} \left(\frac{1}{1+\Phi(\theta)} \right)^k$$

$$= \frac{1}{1-(1+\Phi(\theta))^{-1}}$$

$$= 1 + \frac{1}{\Phi(\theta)}. \tag{5.3}$$

In the third equality, we have used the fact that $|1/(1+\Phi(\theta))| < 1$.

On the other hand, a similar computation to the one in the first paragraph of this proof shows us that the Laplace transform of $\delta_0(\mathrm{d}x) + U(\mathrm{d}x)$ equals the right-hand side of (5.3). Since distinct measures have distinct Laplace transforms, the proof is complete. □

The conclusion of the previous lemma means that the Renewal Theorem can be employed to understand the asymptotic behaviour of U. Specifically, we have the following two asymptotics.

Corollary 5.3 *Suppose that X is a subordinator (no killing) such that $\mu := \mathbb{E}(X_1)$.*

(i) *If U does not have lattice support, then for all $y > 0$,*

$$\lim_{x\uparrow\infty} \{U(x+y) - U(x)\} = \frac{y}{\mu}.$$

(ii) *Without restriction on the support of U,*

$$\lim_{x\uparrow\infty} \frac{U(x)}{x} = \frac{1}{\mu}.$$

As before, we understand $\mu^{-1} = 0$ when $\mu = \infty$.

Proof The proof is a direct consequence of Theorem 5.1, as soon as one notes that

$$\mu = \int_{[0,\infty)} x U^{(1)}(\mathrm{d}x) = \int_0^\infty \mathrm{e}^{-t}\mathbb{E}(X_t)\mathrm{d}t = \int_0^\infty t\mathrm{e}^{-t}\mathbb{E}(X_1)\mathrm{d}t = \mathbb{E}(X_1)$$

and that $U^{(1)}$ has the same support as U. □

The requirement that U does not have a lattice support is not a serious restriction as there are analogues to Corollary 5.3 (i); see for example Chap. XI of Feller (1971). The following theorem, for (killed) subordinators, shows that the only examples of potential measures with lattice support occur when X is a (killed) compound Poisson subordinator whose jump distribution has lattice support.

Theorem 5.4 *Suppose that X is a (killed) subordinator with Lévy measure Π and drift $\delta \geq 0$.*

(i) *If $\delta > 0$ or $\Pi(0,\infty) = \infty$, then for any $q \geq 0$, $U^{(q)}$ has no atoms.*
(ii) *If $\delta = 0$, $\Pi(0,\infty) < \infty$ and Π has a non-lattice support, then for all $q \geq 0$, $U^{(q)}$ does not have a lattice support.*
(iii) *If $\delta = 0$, $\Pi(0,\infty) < \infty$ and Π has a lattice support, then for all $q \geq 0$, $U^{(q)}$ has the same lattice support in $(0,\infty)$.*

Proof (i) Recall the definition

$$U^{(q)}(\mathrm{d}x) = \mathbb{E}\left(\int_0^\infty \mathrm{e}^{-qt}\mathbf{1}_{(X_t\in \mathrm{d}x)}\mathrm{d}t\right), \quad x \geq 0$$

and note that, on account of monotonicity of the paths of X, an atom at $x > 0$ occurs only if, with positive probability, the path of X remains at level x over some interval of time (a,b), where $0 \leq a < b < \infty$. However, since $\Pi(0,\infty) = \infty$, we know that this behaviour is impossible; see Exercise 2.7. This is also the case when $\delta > 0$. In that case, all increments of X are almost surely strictly positive and hence X is almost surely strictly increasing.

(ii)–(iii) Now suppose that X is equal in law to a compound Poisson subordinator, with jump distribution F and arrival rate $\lambda > 0$, which is killed at rate $\eta \geq 0$. (Note $\lambda F = \Pi$.) By conditioning on the number of jumps up to time $t > 0$, we have

$$\mathbb{P}(X_t \in \mathrm{d}x) = \mathrm{e}^{-\eta t}\sum_{k\geq 0} \mathrm{e}^{-\lambda t}\frac{(\lambda t)^k}{k!}F^{*k}(\mathrm{d}x), \quad x \geq 0,$$

where, as usual, we understand $F^{*0} = \delta_0(\mathrm{d}x)$. Using this representation of the transition measure, we compute, for $x \geq 0$,

$$\begin{aligned} U^{(q)}(\mathrm{d}x) &= \sum_{k\geq 0}\frac{1}{k!}F^{*k}(\mathrm{d}x)\int_0^\infty \mathrm{e}^{-(\lambda+q+\eta)t}(\lambda t)^k\mathrm{d}t \\ &= \frac{\rho}{\lambda}\sum_{k\geq 0}\rho^k F^{*k}(\mathrm{d}x), \end{aligned} \tag{5.4}$$

where $\rho = \lambda/(\lambda + \eta + q)$. The second and third statements of the theorem now follow from the last equality. If F does not have a lattice support in $(0, \infty)$, then neither does F^{*k} for any $k \geq 1$, and hence neither does $U^{(q)}$. On the other hand, if F has a lattice support in $(0, \infty)$, then so does F^{*k} for any $k \geq 1$ (the sum of k independent and identically distributed lattice valued random variables is also lattice valued). □

Note that the above theorem also shows that rescaling the Lévy measure of a subordinator by a constant (i.e. $\Pi \mapsto c\Pi$ for some $c > 0$) has no effect on the presence of atoms in the potential measure.

In addition to the close association of the potential measure, U, with classical renewal measures, the connection of a subordinator with renewal processes can be seen in a pathwise sense when X is a compound Poisson subordinator with arrival rate $\lambda > 0$ and non-negative jumps with distribution F. In this case, it is clear that the range of the process X, i.e. the projection of the graph of $\{X_t : t \geq 0\}$ onto the spatial axis, is nothing more than a renewal process. Note that, in this renewal process, the spatial domain of X plays the role of time and the inter-arrival times are distributed according to F. See Fig. 5.1.

As in Sect. 5.1, denote this renewal process by $N = \{N_x : x \geq 0\}$ and let $\{T_i : i \geq 0\}$ be the renewal epochs, starting with $T_0 = 0$. Then the excess lifetime of N at time $x > 0$ is defined by $T_{N_x+1} - x$, and the current lifetime is defined by $x - T_{N_x}$. On the other hand, recall the stopping time (first-passage time)

$$\tau_x^+ = \inf\{t > 0 : X_t > x\}.$$

Then the overshoot and undershoot at first passage of level x are given by $X_{\tau_x^+} - x$ and $x - X_{\tau_x^+-}$, respectively. Excess and current lifetimes and overshoots and undershoots are thus related by

$$X_{\tau_x^+} - x = T_{N_x+1} - x \quad \text{and} \quad x - X_{\tau_x^+-} = x - T_{N_x}. \tag{5.5}$$

See Fig. 5.1.

Classical renewal theory presents the following result for the excess and current lifetime; see for example Chap. XI of Feller (1971) or Dynkin (1961). We give the proof for the sake of later reference.

Lemma 5.5 *Suppose that N is a renewal process with F as the distribution for the spacings. Then the following hold.*

(i) *For $x, u > 0$ and $y \in (0, x]$,*

$$\mathbb{P}(T_{N_x+1} - x \in \mathrm{d}u, x - T_{N_x} \in \mathrm{d}y) = V(x - \mathrm{d}y)F(\mathrm{d}u + y), \tag{5.6}$$

where V is the renewal measure constructed from F.

(ii) *Suppose that F has mean $\mu < \infty$ and is non-lattice, then, for $u > 0$ and $y > 0$,*

$$\lim_{x\uparrow\infty} \mathbb{P}(T_{N_x+1} - x > u, x - T_{N_x} > y) = \frac{1}{\mu}\int_{u+y}^{\infty} \overline{F}(z)\mathrm{d}z,$$

where $\overline{F}(x) = 1 - F(x)$.

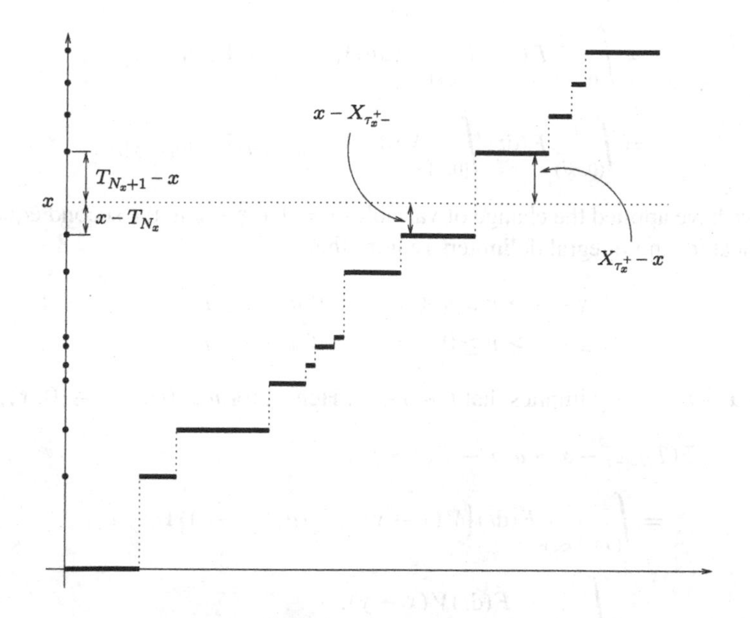

Fig. 5.1 A realisation of a compound Poisson subordinator. The range of the process, projected onto the *vertical axis*, forms a renewal process thus relating overshoot and undershoot to excess and current lifetimes.

Proof (i) The key to the proof of the first part is to partition the event of interest by the number of renewal epochs at time x. We have, for $k \geq 0$,

$$\mathbb{P}(T_{N_x+1} - x > u, x - T_{N_x} > y, N_x = k) = \int_{[0,x-y)} F^{*k}(\mathrm{d}v)\overline{F}(x - v + u).$$

The event in the probability on the left-hand side requires that the k-th renewal epoch occurs sometime before $x - y$. Further, this epoch occurs in $\mathrm{d}v$ with probability $F^{*k}(\mathrm{d}v)$ and, hence, the probability that the excess exceeds u requires that the next inter-arrival time exceeds $x - v + u$. This occurs with probability $\overline{F}(x - v + u)$. Summing over k and changing variable in the integral via $z = x - v$, we have

$$\mathbb{P}(T_{N_x+1} - x > u, x - T_{N_x} > y) = \int_{(y,x]} V(x - \mathrm{d}z)\overline{F}(z + u).$$

In differential form, this gives the distribution given in the statement of part (i).

(ii) From part (i), we may write, for $u > 0$ and $y \in [0, x)$,

$$\mathbb{P}(T_{N_x+1} - x > u, x - T_{N_x} \geq y)$$

$$= \int_{(u,\infty)} \int_{[0,x-y]} V(\mathrm{d}v)F(x - v + \mathrm{d}\theta)$$

$$= \int_{(0,\infty)} F(\mathrm{d}t) \int_{[0,x]} V(\mathrm{d}v) \mathbf{1}_{(t>u+x-v)} \mathbf{1}_{(v\in[0,x-y])}$$

$$= \int_{(0,\infty)} F(\mathrm{d}t) \int_{[0,x]} V(\mathrm{d}v) \mathbf{1}_{(v>u+x-t)} \mathbf{1}_{(v\in[0,x-y])},$$

where we have applied the change of variables $t = \theta + x - v$ in the second equality. The indicators and integral delimiters require that

$$\begin{aligned} x - y \geq v > u + x - t \quad & \text{if } u + x \geq t, \\ x - y \geq v \geq 0 \quad & \text{if } u + x < t, \end{aligned}$$

and $u + x - t < x - y$ implies that $t > u + y$. Hence, for $u > 0$ and $y \in [0, x)$,

$$\begin{aligned} \mathbb{P}(T_{N_x+1} - x > u, x - T_{N_x} > y) &= \int_{(u+y,\infty)} F(\mathrm{d}t)\{V(x-y) - V(u+x-t)\}\mathbf{1}_{(t\leq u+x)} \\ &\quad + \int_{(u+x,\infty)} F(\mathrm{d}t)V(x-y). \end{aligned} \tag{5.7}$$

To deal with the second term on the right-hand side of (5.7), we may use the Renewal Theorem 5.1 (iii) to show that, for some $\varepsilon > 0$ and x sufficiently large,

$$\int_{(u+x,\infty)} F(\mathrm{d}t)V(x-y) \leq \frac{1+\varepsilon}{\mu} \int_{(u+x,\infty)} tF(\mathrm{d}t).$$

The right-hand side above tends to zero as x tends to infinity, since $\mu = \int_{(0,\infty)} tF(\mathrm{d}t) < \infty$.

For the first term on the right-hand side of (5.7), suppose that X is a compound Poisson subordinator whose jump distribution is F and arrival rate is 1. For this subordinator,

$$\mathbb{E}(\tau_x^+) = \int_0^\infty \mathbb{P}(\tau_x^+ > t)\mathrm{d}t = \int_0^\infty \mathbb{P}(X_t \leq x)\mathrm{d}t = V(x),$$

where the final equality follows from (5.4), with $q = \eta = 0$ and $\lambda = 1$. Now applying the strong Markov property, we can establish that

$$\begin{aligned} V(x+y) &= \mathbb{E}(\tau_{x+y}^+) \\ &= \mathbb{E}(\tau_x^+ + \mathbb{E}_{X_{\tau_x^+}}(\tau_{x+y}^+)) \\ &\leq \mathbb{E}(\tau_x^+) + \mathbb{E}(\tau_y^+) \\ &= V(x) + V(y). \end{aligned}$$

Using the bound $V(x-y) - V(u+x-t) \leq V(t-u-y) \leq V(t)$, the right-continuity of V and the Renewal Theorem 5.1 (iii), we know that the integrand in the first term on the right-hand side of (5.7) is bounded by a multiple of t. Hence, as $\int_{(0,\infty)} tF(\mathrm{d}t) < \infty$, dominated convergence, together with Theorem 5.1 (i), gives us

$$
\begin{aligned}
&\lim_{x\uparrow\infty} \int_{(u+y,\infty)} F(\mathrm{d}t)\big\{V(x-y) - V(u+x-t)\big\}\mathbf{1}_{(t<u+x)} \\
&= \frac{1}{\mu}\int_{(u+y,\infty)} (t-u-y)F(\mathrm{d}t) \\
&= \frac{1}{\mu}\int_{u+y}^{\infty} \overline{F}(t)\mathrm{d}t,
\end{aligned}
$$

where the final equality follows after an integration by parts. □

In light of (5.5), we see that Lemma 5.5 gives the exact and asymptotic distribution of the overshoot and undershoot at first passage of a compound Poisson subordinator with jump distribution F (with finite mean and non-lattice support in the case of the asymptotic behaviour). In this spirit, we shall proceed to study the exact and asymptotic joint distributions of the overshoot and undershoot of a killed subordinator at first passage.

There are a number of differences concerning the range of a killed subordinator when compared to the range of a compound Poisson subordinator. Firstly, in the case of a killed subordinator, the process may be killed before reaching a specified fixed level. Hence one should expect an atom in the distribution of the overshoot at ∞. Secondly, the number of jumps over a finite time horizon may be infinite, in which case the analysis in the proof of Lemma 5.5 (i) is no longer valid. Finally, in the case of a compound Poisson subordinator, when F has no atoms, it is clear that the probability that there is first passage over a given level by hitting the level is zero. However, for a killed subordinator, for which either $\Pi(0,\infty) = \infty$ or there is a drift present, one should not exclude the possibility that first passage over a fixed level occurs by hitting the level with positive probability. This behaviour is called *creeping over a fixed level* and is equivalent to there being an atom at zero in the distribution of the overshoot at that level. As one might intuitively expect, creeping over a specified fixed level turns out to occur only in the presence of a drift, in which case, by spatial homogeneity, it is possible to creep over all fixed levels. These points will be dealt with in more detail in Sect. 5.3.

5.2 Overshoots and Undershoots

We begin with the following theorem, which gives a generalisation of Lemma 5.5 (i), in the sense that it contains it as a corollary. Weaker versions of this theorem can be found in Kesten (1969) and Horowitz (1972). The format we give is from Bertoin (1996a).

Theorem 5.6 *Suppose that X is a killed subordinator. Then for $u > 0$ and $y \in [0, x]$,*

$$\mathbb{P}(X_{\tau_x^+} - x \in \mathrm{d}u, x - X_{\tau_x^+-} \in \mathrm{d}y) = U(x - \mathrm{d}y)\Pi(y + \mathrm{d}u). \tag{5.8}$$

Proof The proof makes use of the compensation formula. Suppose that f and g are two strictly positive, bounded, continuous functions satisfying $f(0) = f(\infty) = 0$. This last requirement ensures that the product $f(X_{\tau_x^+} - x)g(x - X_{\tau_x^+-})$ is non-zero only if X jumps strictly above x when first crossing x before killing occurs. We may write its expectation in terms of the Poisson random measure associated with the jumps of X whilst avoiding the issue of creeping. To this end, let us assume that X is equal in law to a subordinator Y killed at rate $\eta \geq 0$. Then

$$\mathbb{E}\big(f(X_{\tau_x^+} - x)g(x - X_{\tau_x^+-})\big) = \mathbb{E}\left(\int_{[0,\infty)}\int_{(0,\infty)} \mathrm{e}^{-\eta t}\phi(t,\theta)N(\mathrm{d}t \times \mathrm{d}\theta)\right),$$

where

$$\phi(t,\theta) = \mathbf{1}_{(Y_{t-} \leq x)}\mathbf{1}_{(Y_{t-}+\theta > x)} f(Y_{t-} + \theta - x)g(x - Y_{t-}),$$

and N is the Poisson random measure associated with the jumps of Y. It is straightforward to see that ϕ satisfies the conditions of Theorem 4.4; in particular, it is left-continuous in t. Then, with the help of the aforementioned theorem,

$$\begin{aligned}
&\int_{[0,x]} g(y)\int_{(0,\infty)} f(u)\mathbb{P}(X_{\tau_x^+} - x \in \mathrm{d}u, x - X_{\tau_x^+-} \in \mathrm{d}y) \\
&= \mathbb{E}\left(\int_0^\infty \mathrm{d}t \cdot \mathrm{e}^{-\eta t}\mathbf{1}_{(Y_{t-} \leq x)}g(x - Y_{t-})\int_{(x-Y_{t-},\infty)} f(Y_{t-} + \theta - x)\Pi(\mathrm{d}\theta)\right) \\
&= \mathbb{E}\left(\int_0^\infty \mathrm{d}t \cdot \mathrm{e}^{-\eta t}\mathbf{1}_{(Y_t \leq x)}g(x - Y_t)\int_{(x-Y_t,\infty)} f(Y_t + \theta - x)\Pi(\mathrm{d}\theta)\right) \\
&= \int_{[0,x]} g(x-z)\int_{(x-z,\infty)} f(z+\theta - x)\Pi(\mathrm{d}\theta)\int_0^\infty \mathrm{d}t \cdot \mathrm{e}^{-\eta t}\mathbb{P}(Y_t \in \mathrm{d}z) \\
&= \int_{[0,x]} g(x-z)\int_{(x-z,\infty)} f(z+\theta - x)\Pi(\mathrm{d}\theta)U(\mathrm{d}z) \\
&= \int_{[0,x]} g(y)\int_{(0,\infty)} f(u)\Pi(\mathrm{d}u + y)U(x - \mathrm{d}y),
\end{aligned} \tag{5.9}$$

where the final equality follows by changing variables, first with $y = x - z$ and then with $u = \theta - y$. (Note also that U is the potential measure of X and not Y.) As f and g are arbitrary within their prescribed classes (which themselves are sufficient to characterise the desired law), we read off from the left- and right-hand sides of (5.9) the required distributional identity. □

Intuitively speaking, the proof of Theorem 5.6 follows the logic of the proof of Lemma 5.5 (i). The compensation formula serves as a way of "decomposing" the event of first passage by a jump over level x according to the position of X prior to its first passage, even when there are an unbounded number of jumps over finite time horizons.

To make the connection with the expression given for renewal processes in Lemma 5.5 (i), recall from (5.4) that $U(\mathrm{d}x) = \lambda^{-1}V(\mathrm{d}x)$ on $(0,\infty)$, where U is the potential measure associated with a compound Poisson subordinator with jump distribution F and arrival rate $\lambda > 0$, and V is the renewal measure associated with the distribution F. For this compound Poisson subordinator, we also know that $\Pi(\mathrm{d}x) = \lambda F(\mathrm{d}x)$, so that $U(x-\mathrm{d}y)\Pi(\mathrm{d}u+y) = V(x-\mathrm{d}y) F(\mathrm{d}u+y)$.

As (5.8) is the analogue of the statement in Lemma 5.5 (i), it is now natural to reconsider the proof of part (ii) of the same lemma in the current, more general context. The following result is lifted from Bertoin et al. (1999).

Theorem 5.7 *Suppose that X is a subordinator (no killing) with finite mean $\mu := \mathbb{E}(X_1)$, such that U does not have lattice support (cf. Theorem 5.4). Then for $u > 0$ and $y \geq 0$, in the sense of weak convergence*

$$\lim_{x\uparrow\infty} \mathbb{P}(X_{\tau_x^+} - x \in \mathrm{d}u, x - X_{\tau_x^+-} \in \mathrm{d}y) = \frac{1}{\mu}\mathrm{d}y\Pi(y+\mathrm{d}u).$$

In particular, by integrating out u and y in the above limit, it follows that the asymptotic probability of creeping satisfies

$$\lim_{x\uparrow\infty} \mathbb{P}(X_{\tau_x^+} = x) = \frac{\delta}{\mu}.$$

The proof of this result is a straightforward adaptation of the proof of Lemma 5.5 (ii) taking advantage of Corollary 5.3 and is left to the reader to verify in Exercise 5.3.

5.3 Creeping

Now let us turn to the issue of creeping. Although τ_x^+ is the first time that X strictly exceeds the level $x > 0$, it is possible that $\mathbb{P}(X_{\tau_x^+} = x) > 0$; recall the statement and proof of Theorem 3.3. The following conclusion, found for example in Horowitz (1972), shows that, in the case where the jump measure is infinite or that X has a drift, crossing the level $x > 0$ by hitting it cannot occur by jumping onto it from a position strictly below x. In other words, if our killed subordinator makes first passage above x with a jump, then it must do so by jumping it clear, so $\{X_{\tau_x^+} = x\} = \{X_{\tau_x^+} - x = 0, x - X_{\tau_x^+-} = 0\}$. This is of implicit relevance when computing the atom at zero in the overshoot distribution.

Lemma 5.8 *Let X be any killed subordinator with $\Pi(0,\infty)=\infty$ or $\delta>0$. For all $x>0$, we have*

$$\mathbb{P}(X_{\tau_x^+}-x=0, x-X_{\tau_x^+-}>0)=0. \tag{5.10}$$

Proof Suppose, for a given $x>0$, that

$$\mathbb{P}(X_{\tau_x^+}-x=0, x-X_{\tau_x^+-}>0)>0.$$

Then this implies that there exists a $y<x$ such that

$$\mathbb{P}(X_{\tau_y^+}=x)>0.$$

However, this cannot happen because of the combined conclusions of Theorem 5.6 and Theorem 5.4 (i). (It is also useful to note that Π can have at most a countable number of atoms.) Hence, by contradiction (5.10) holds. □

Although one may write, with the help of Theorem 5.6 and Lemma 5.8,

$$\mathbb{P}(X_{\tau_x^+}=x)=1-\mathbb{P}(X_{\tau_x^+}>x)=1-\int_{(0,x]}U(x-\mathrm{d}y)\Pi(y,\infty),$$

this does not necessarily bring one closer to understanding when the probability on the left-hand side above is strictly positive. In fact, although the answer to this question is intuitively obvious, namely that a drift term must be present, it turns out to be difficult to prove. It was resolved by Kesten (1969); see also Bretagnolle (1971). The result is given below.

Theorem 5.9 *For any killed subordinator such that $\Pi(0,\infty)=\infty$ or $\delta>0$, we have the following*:

(i) *If $\delta=0$, then $\mathbb{P}(X_{\tau_x^+}=x)=0$ for all $x>0$.*
(ii) *If $\delta>0$, then U has a strictly positive and continuous density on $(0,\infty)$, say u, satisfying*

$$\mathbb{P}(X_{\tau_x^+}=x)=\delta u(x).$$

The version of the proof we give here follows the reasoning in Andrew (2006) (see also Sect. III.2 of Bertoin (1996a)) and first requires two auxiliary lemmas, given below. In the proof of both, we shall make use of the following two key estimates for the probabilities $p_x:=\mathbb{P}(X_{\tau_x^+}=x)$, $x\geq 0$. For all $0<y<x$,

$$p_x\leq p_y p_{x-y}+(1-p_{x-y}) \tag{5.11}$$

and

$$p_x\geq p_y p_{x-y}. \tag{5.12}$$

The upper bound is a direct consequence of the fact that

$$\begin{aligned}\mathbb{P}(X_{\tau_x^+}=x) &= \mathbb{P}(X_{\tau_{x-y}^+}=x-y,\ X_{\tau_x^+}=x)\\ &\quad +\mathbb{P}(X_{\tau_{x-y}^+}>x-y,\ X_{\tau_x^+}=x)\\ &\leq \mathbb{P}(X_{\tau_{x-y}^+}=x-y)\mathbb{P}(X_{\tau_x^+}=x|X_0=x-y)\\ &\quad +\mathbb{P}(X_{\tau_{x-y}^+}>x-y),\end{aligned}$$

where in the last line the strong Markov property has been used. In a similar way, the lower bound is a consequence of the fact that

$$\mathbb{P}(X_{\tau_x^+}=x) \geq \mathbb{P}(X_{\tau_{x-y}^+}=x-y)\mathbb{P}(X_{\tau_x^+}=x|X_0=x-y).$$

Lemma 5.10 *Assume the setting of Theorem* 5.9.

(i) *If, for some* $x>0$, *we have* $p_x>0$, *then* $\lim_{\varepsilon\downarrow 0}\sup_{\eta\in(0,\varepsilon)} p_\eta = 1$.
(ii) *If, for some* $x>0$, *we have* $p_x>3/4$, *then*

$$p_y \geq 1/2+\sqrt{p_x-3/4}$$

for all $y\in(0,x)$.

Proof (i) From Lemma 5.8, we know that X cannot jump onto x. In other words, we have

$$\mathbb{P}(X_{\tau_x^+}=x>X_{\tau_x^+-})=0.$$

This implies that

$$\{X_{\tau_x^+}=x\}\subseteq\bigcap_{n\geq 1}\{X \text{ visits } (x-1/n,x)\}$$

almost surely. On the other hand, on the event $\bigcap_{n\geq 1}\{X \text{ visits } (x-1/n,x)\}$, we also have by quasi-left-continuity (cf. Lemma 3.2) that $X_\sigma=x$, where $\sigma=\lim_{n\uparrow\infty}\tau_{x-1/n}^+$ (the limit exists because of monotonicity). Note that, on the one hand, by its definition, $\sigma\leq\tau_x^+$. Since

$$\{\sigma\leq t\}=\bigcap_{n\geq 1}\{\tau_{x-1/n}^+\leq t\}$$

almost surely, it follows that σ is a stopping time with respect to $\mathbb{F}$. Since $X_\sigma=x$ on $\bigcap_{n\geq 1}\{X \text{ visits } (x-1/n,x)\}$ and X is not a compound Poisson subordinator, applying the strong Markov property at time σ, we have that $X_t>x$ for all $t>\sigma$. This shows that, on $\bigcap_{n\geq 1}\{X \text{ visits } (x-1/n,x)\}$, we have $\sigma=\tau_x^+$. In conclusion, $\sigma=\tau_x^+$ on $\bigcap_{n\geq 1}\{X \text{ visits } (x-1/n,x)\}$ and hence

$$\{X_{\tau_x^+} = x\} = \bigcap_{n\geq 1}\{X \text{ visits } (x-1/n, x)\}$$

almost surely.

We may now write

$$p_x = \lim_{n\uparrow\infty} \mathbb{P}\big(X \text{ visits } (x-1/n, x)\big). \tag{5.13}$$

Also, we have the upper estimate

$$p_x \leq \mathbb{P}\big(X \text{ visits } (x-1/n, x)\big) \sup_{z\in(0,1/n)} p_z.$$

Letting $n \uparrow \infty$ in the above inequality and taking (5.13) into account, we see that $\lim_{\varepsilon\downarrow 0}\sup_{\eta\in(0,\varepsilon)} p_\eta = 1$.

(ii) Suppose that $0 < y < x$. We may assume without loss of generality that $p_y < p_x$, otherwise it is clear that $p_y \geq p_x \geq 1/2 + \sqrt{p_x - 3/4}$, when $p_x > 3/4$.

From (5.11) it is a simple algebraic manipulation, replacing y by $x - y$, to show that

$$p_y \leq \frac{1-p_x}{1-p_{x-y}}.$$

Again, replacing y by $x - y$ in the above inequality, we obtain that

$$1 - p_{x-y} \geq \frac{p_x - p_y}{1-p_y}.$$

Combining the last two inequalities, we therefore have

$$p_y \leq \frac{(1-p_x)(1-p_y)}{p_x - p_y}$$

and hence the quadratic inequality $p_y^2 - p_y + 1 - p_x \geq 0$. This in turn implies that

$$p_y \in \big[0, 1/2 - \sqrt{p_x - 3/4}\big] \cup \big[1/2 + \sqrt{p_x - 3/4}, 1\big]. \tag{5.14}$$

The remainder of the proof is thus dedicated to showing that the inclusion of p_y in the first of the two intervals in (5.14) cannot happen.

Suppose, for contradiction, that (5.14) holds for all $y \in (0, x)$ and, moreover, that there exists a $y \in (0, x)$ such that $p_y \leq 1/2 - \sqrt{p_x - 3/4}$. Now define

$$g = \sup\big\{z \in [0, y);\, p_z \geq 1/2 + \sqrt{p_x - 3/4}\big\},$$

which is well defined since $p_0 = 1$. From this definition, it could be the case that $g = y$. Reconsidering the definition of g and (5.14), we see that either there exists an $\varepsilon > 0$ such that $p_z \leq 1/2 - \sqrt{p_x - 3/4}$ for all $z \in (g - \varepsilon, g)$ or, for all $\varepsilon > 0$, there exists a sequence of $z \in (g - \varepsilon, g)$ such that $p_z \geq 1/2 + \sqrt{p_x - 3/4}$. In the former case, it is clear by the definition of g that $p_g \geq 1/2 + \sqrt{p_x - 3/4}$. In the latter case, we have with the help of (5.13) that

$$p_g = \lim_{z \uparrow g} \mathbb{P}\big(X \text{ visits } (z, g)\big) \geq \lim_{\varepsilon \downarrow 0} \sup_{\eta \in (0,\varepsilon)} p_{g-\eta},$$

and hence $p_g \geq 1/2 + \sqrt{p_x - 3/4}$. For both cases, we have the implication that $g < y$. On the other hand, using (5.12) and the conclusion of part (i), we see that

$$\lim_{\varepsilon \downarrow 0} \sup_{\eta \in (0,\varepsilon)} p_{g+\eta} \geq p_g \times \lim_{\varepsilon \downarrow 0} \sup_{\eta \in (0,\varepsilon)} p_\eta = p_g \geq 1/2 + \sqrt{p_x - 3/4}.$$

Since (5.14) is in force for all $y < x$ and $g < y$, this implies that there exists a $g' > g$ such that $p_{g'} \geq 1/2 + \sqrt{p_x - 3/4}$, which contradicts the definition of g. The consequence of this contradiction is that there does not exist a $y \in (0, x)$ for which $p_y < 1/2 + \sqrt{p_x - 3/4}$, and hence, from (5.14), it necessarily follows that $p_y \geq 1/2 + \sqrt{p_x - 3/4}$, for all $y \in (0, x)$. □

Lemma 5.11 *Assume the setting of Theorem* 5.9. *If there exists an* $x > 0$ *such that* $p_x > 0$, *then*

(i) $\lim_{\varepsilon \downarrow 0} p_\varepsilon = 1$ *and*
(ii) $x \mapsto p_x$ *is strictly positive and continuous on* $[0, \infty)$.

Proof (i) The first part is a direct consequence of parts (i) and (ii) of Lemma 5.10.

(ii) Positivity follows from a repeated use of the lower estimate in (5.12) to obtain $p_x \geq (p_{x/n})^n$ and, hence, the conclusion of part (i).

To show continuity, note, with the help of (5.11), that

$$\limsup_{\varepsilon \downarrow 0} p_{x+\varepsilon} \leq \limsup_{\varepsilon \downarrow 0} \{p_\varepsilon p_x + 1 - p_\varepsilon\} = p_x,$$

and from (5.12) and part (i),

$$\liminf_{\varepsilon \downarrow 0} p_{x+\varepsilon} \geq \liminf_{\varepsilon \downarrow 0} p_x p_\varepsilon = p_x.$$

Further, arguing in a similar manner,

$$\limsup_{\varepsilon \downarrow 0} p_{x-\varepsilon} \leq \limsup_{\varepsilon \downarrow 0} \frac{p_x}{p_\varepsilon} = p_x,$$

and

$$\liminf_{\varepsilon \downarrow 0} p_{x-\varepsilon} \geq \liminf_{\varepsilon \downarrow 0} \frac{p_x + p_\varepsilon - 1}{p_\varepsilon} = p_x.$$

Thus, continuity is confirmed. □

Finally, we return to the proof of Theorem 5.9.

Proof of Theorem 5.9 Consider the function

$$M(a) := \mathbb{E}\left(\int_0^a \mathbf{1}_{(X_{\tau_x^+}=x)}\mathrm{d}x\right) = \int_0^a p_x \mathrm{d}x$$

for all $a \geq 0$.

For convenience, suppose further that X is equal in law to a subordinator Y, killed at rate η. Let N be the Poisson random measure associated with the jumps of X (or equivalently Y). Then we may write, with the help of the Lévy–Itô decomposition for subordinators,

$$M(a) = \mathbb{E}\left(Y_{(\tau_a^+ \wedge \mathbf{e}_\eta)-} - \int_{[0,\tau_a^+\wedge\mathbf{e}_\eta)}\int_{(0,\infty)} xN(\mathrm{d}s \times \mathrm{d}x)\right) = \delta\mathbb{E}\big(\tau_a^+ \wedge \mathbf{e}_\eta\big).$$

(i) If $\delta = 0$, then $p_x = 0$ for Lebesgue almost every x. Lemma 5.11 now implies that $p_x = 0$ for all $x > 0$.

(ii) If $\delta > 0$, then there exists an $x > 0$ such that $p_x > 0$. Hence, from Lemma 5.11, $x \mapsto p_x$ is strictly positive and continuous. Further, we see that

$$M(a) = \delta\int_0^\infty \mathbb{P}\big(\tau_a^+ \wedge \mathbf{e}_\eta > t\big)\mathrm{d}t = \delta\int_0^\infty \mathbb{P}(X_t \leq a)\mathrm{d}t = \delta U(a).$$

The above implies that U has a density, which may be taken as equal to $\delta^{-1}p_x$ for all $x \geq 0$. □

Theorem 5.9 excludes the possibility that $\Pi(0,\infty) < \infty$ when $\delta = 0$. Here, one may easily envisage a scenario where a given $x_0 > 0$ is in the range of the subordinator with positive probability. Indeed, it suffices to consider the case that Π has an atom at x_0. Note, however, that, because of the strict inequality in the definition of $\tau_{x_0}^+$, it is not the case that $X_{\tau_{x_0}^+} = x_0$. In general, creeping for compound Poisson processes cannot occur.

5.4 Regular Variation and Tauberian Theorems

The inclusion of the forthcoming discussion on regular variation and Tauberian theorems is a prerequisite to Sect. 5.5, which gives the Dynkin–Lamperti asymptotics for the joint law of the overshoot and undershoot of a subordinator at a threshold. However, the necessary facts concerning regular variation will also appear in later sections and chapters.

Suppose that $U : [0,\infty) \to [0,\infty)$ is a non-decreasing, right-continuous function. Denote by $U(\mathrm{d}x)$, $x \geq 0$, its associated measure, with the convention that there is an atom of size $U(0)$ at $x = 0$. For its Laplace–transform, write

$$\Lambda(\theta) = \int_{[0,\infty)} \mathrm{e}^{-\theta x}U(\mathrm{d}x), \quad \theta \geq 0.$$

If there exists a θ_0 such that $\Lambda(\theta_0) < \infty$, then $\Lambda(\theta) < \infty$ for all $\theta \geq \theta_0$. The point of this section is to present some classic results which equivalently relate certain types of tail behaviour of the measure U to a similar type of behaviour of Λ. Our presentation will only offer the bare essentials based on Karamata's theory of regularly varying functions. Aside from their intrinsic analytic curiosity, regularly varying functions have proved to be of great practical value within probability theory, not least in the current context. The highly readable account given in Chap. VIII.8 of Feller (1971) is an important bridging text, embedding into probability theory the classic work of Karamata and his collaborators, which dates back to the period between 1930 and the 1960s. For a complete account, the reader is referred to Bingham et al. (1987) or Embrechts et al. (1997). The presentation here is principally based on these books.

Definition 5.12 A measurable function $f : [0, \infty) \to (0, \infty)$ is said to be regularly varying at zero with index $\rho \in \mathbb{R}$ (written $f \in \mathcal{R}_0(\rho)$) if, for all $\lambda > 0$,

$$\lim_{x \downarrow 0} \frac{f(\lambda x)}{f(x)} = \lambda^\rho.$$

If the above limit holds as x tends to infinity, then f is said to be regularly varying at infinity with index ρ (written $f \in \mathcal{R}_\infty(\rho)$). The case that $\rho = 0$ is referred to as slow variation (written for short as just $\mathcal{R}_0$ and $\mathcal{R}_\infty$, respectively).

Note that any regularly varying function, f, may always be written in the form

$$f(x) = x^\rho L(x),$$

where L is a slowly varying function. Any function which has a strictly positive and finite limit at infinity (resp. zero) belongs to $\mathcal{R}_\infty$ (resp. $\mathcal{R}_0$), so the class of slowly (and hence regularly) varying functions is clearly non-empty due to this trivial example. There are, however, many non-trivial examples of slowly varying functions. Examples in $\mathcal{R}_\infty$ include (for x sufficiently large) $L(x) = \log x$, $L(x) = \log_k x$ (the k-th iterate of $\log x$) and $L(x) = \exp\{(\log x)/\log\log x\}$. All of these examples have the property that they are functions which tend to infinity at infinity. The function

$$L(x) = \exp\{(\log x)^{\frac{1}{3}} \cos[(\log x)^{\frac{1}{3}}]\}$$

is an example in $\mathcal{R}_\infty$ which oscillates, that is to say, $\liminf_{x \uparrow \infty} L(x) = 0$ and $\limsup_{x \uparrow \infty} L(x) = \infty$.

The main concerns of this section are the following remarkable results.

Theorem 5.13 *Suppose that $L \in \mathcal{R}_\infty$, $\rho \in [0, \infty)$. Then the following two statements are equivalent:*

(i) $\Lambda(\theta) \sim \theta^{-\rho} L(1/\theta)$ *as* $\theta \to 0$,
(ii) $U(x) \sim x^\rho L(x)/\Gamma(1 + \rho)$ *as* $x \to \infty$.

In the above theorem, we are using the notation $f \sim g$ for functions f and g to mean that $\lim f(x)/g(x) = 1$.

Theorem 5.14 *Suppose that $L \in \mathcal{R}_\infty$, $\rho \in (0, \infty)$ and $U(\mathrm{d}x) = u(x)\mathrm{d}x$, $x \geq 0$, where the density, u, is ultimately monotone. Then the following two statements are equivalent:*

(i) $\Lambda(\theta) \sim \theta^{-\rho} L(1/\theta)$ *as* $\theta \to 0$,
(ii) $u(x) \sim x^{\rho-1} L(x)/\Gamma(\rho)$ *as* $x \to \infty$.

Recalling that $\Gamma(1+\rho) = \rho\Gamma(\rho)$, Theorem 5.14 is a natural statement next to Theorem 5.13. It says that, up to a slowly varying function, the asymptotic behaviour of the derivative of $U(x)$ behaves like the derivative of the polynomial that $U(x)$ asymptotically mimics; providing, of course, the density u exists and is ultimately monotone. The methods used to prove these results also produce the following corollary with virtually no change at all.

Corollary 5.15 *The statements of Theorems* 5.13 *and* 5.14 *are still valid when, instead, $\mathcal{R}_\infty$ is replaced by $\mathcal{R}_0$ and the limits in parts* (i) *and* (ii) *are simultaneously changed to $\theta \to \infty$ and $x \to 0$.*

We now give the proof of Theorem 5.13, which, in addition to the assumed regular variation, uses little more than the Continuity Theorem for Laplace transforms of positive random variables.

Proof of Theorem 5.13 It will be helpful for this proof to record the well-known fact that, for any $\infty > \rho \geq 0$ and $\lambda > 0$,

$$\int_0^\infty x^\rho \mathrm{e}^{-\lambda x} \mathrm{d}x = \frac{\Gamma(1+\rho)}{\lambda^{1+\rho}}. \tag{5.15}$$

In addition, we will also use the fact that, for all $\lambda > 0$ and $\theta > 0$,

$$\int_{[0,\infty)} \mathrm{e}^{-\lambda x} U(\mathrm{d}x/\theta) = \Lambda(\lambda\theta). \tag{5.16}$$

First, we prove that (i) implies (ii). Fix $\lambda_0 > 0$. From (5.16), we have, for $\theta > 0$, that $\mathrm{e}^{-\lambda_0 x} U(\mathrm{d}x/\theta)/\Lambda(\lambda_0\theta)$ is a probability distribution. Again, from (5.16), we can compute its Laplace transform as $\Lambda((\lambda + \lambda_0)\theta)/\Lambda(\lambda_0\theta)$. The regular variation assumed in (i), together with (5.15), implies that

$$\lim_{\theta \downarrow 0} \int_{[0,\infty)} \mathrm{e}^{-(\lambda+\lambda_0)x} \frac{U(\mathrm{d}x/\theta)}{\Lambda(\lambda_0\theta)} = \frac{\lambda_0^\rho}{(\lambda_0+\lambda)^\rho} = \frac{\lambda_0^\rho}{\Gamma(\rho)} \int_0^\infty x^{\rho-1} \mathrm{e}^{-(\lambda+\lambda_0)x} \mathrm{d}x,$$

where the right-hand side is the Laplace transform of a gamma distribution. It follows from the Continuity Theorem for Laplace transforms (cf. Theorem XIII.1.2a of Feller (1971)) that the measure $\mathrm{e}^{-\lambda_0 x} U(\mathrm{d}x/\theta)/\Lambda(\lambda_0\theta)$ converges vaguely to

$\mathrm{e}^{-\lambda_0 x}\lambda_0^\rho x^{\rho-1}/\Gamma(\rho)\mathrm{d}x$, as θ tends to zero. Using the regular variation of Λ again, this implies that, for all $y > 0$,

$$\lim_{\theta\downarrow 0}\frac{U(y/\theta)}{L(1/\theta)}\lambda_0^\rho\theta^\rho = \frac{\lambda_0^\rho y^\rho}{\rho\Gamma(\rho)}.$$

Now setting $y = 1$, rewriting $x = 1/\theta$ and recalling that $\Gamma(1+\rho) = \rho\Gamma(\rho)$, statement (ii) follows.

Now we prove that (ii) implies (i). The assumption in (ii) expressed in terms of vague convergence implies that, on bounded intervals of $[0,\infty)$,

$$\lim_{x\uparrow\infty}\frac{U(x\,\mathrm{d}y)}{U(x)} = \rho y^{\rho-1}\mathrm{d}y.$$

In particular, for any $t > 0$ and $\lambda > 0$,

$$\lim_{x\uparrow\infty}\int_0^t \mathrm{e}^{-\lambda y}\frac{U(x\,\mathrm{d}y)}{U(x)} = \rho\int_0^t y^{(\rho-1)}\mathrm{e}^{-\lambda y}\mathrm{d}y. \tag{5.17}$$

In view of (5.16), the Laplace transform of the measure $U(x\,\mathrm{d}y)/U(x)$ is given by $\Lambda(\lambda/x)/U(x)$, for $\lambda > 0$. Now suppose that, for some $0 < \lambda_0 < 1$ and $x_0 > 0$, the sequence $\{\Lambda(\lambda_0/x)/U(x) : x > x_0\}$ is uniformly bounded by some $C > 0$. With this additional assumption in place, we may pick a sufficiently large $t > 0$ such that

$$\int_t^\infty \mathrm{e}^{-y}\frac{U(x\,\mathrm{d}y)}{U(x)} < \mathrm{e}^{-(1-\lambda_0)t}\int_t^\infty \mathrm{e}^{-\lambda_0 y}\frac{U(x\,\mathrm{d}y)}{U(x)} < C\mathrm{e}^{-(1-\lambda_0)t}.$$

Together with (5.17), the above estimate is sufficient to deduce that

$$\lim_{x\uparrow\infty}\frac{\Lambda(1/x)}{U(x)} = \lim_{x\uparrow\infty}\int_0^\infty \mathrm{e}^{-y}\frac{U(x\,\mathrm{d}y)}{U(x)} = \rho\int_0^\infty y^{(\rho-1)}\mathrm{e}^{-y}\mathrm{d}y = \Gamma(1+\rho).$$

Choosing $\lambda = 1$ and writing $\theta = 1/x$, the statement in (i) follows.

It remains then to show that, for some $0 < \lambda_0 < 1$ and $x_0 > 0$, the sequence $\{\Lambda(\lambda_0/x)/U(x) : x > x_0\}$ is uniformly bounded. This is done by partitioning the domain of integration in (5.16) over the lattice $\{2^k x : k \geq 0\}$, for some $x > 0$. The assumed regular variation of U implies that, for all x sufficiently large, $U(2x) < 2^{\rho+1}U(x)$. This can be iterated to deduce that, for x sufficiently large, $U(2^n x) < 2^{n(1+\rho)}U(x)$ for each $n \geq 1$. With this inequality in hand, we may quite coarsely estimate, for all sufficiently large x,

$$\frac{\Lambda(\lambda_0/x)}{U(x)} \leq \sum_{n\geq 1}\mathrm{e}^{-\lambda_0 2^{n-1}}\frac{U(2^n x)}{U(x)} < \sum_{n\geq 1}2^{n(1+\rho)}\mathrm{e}^{-\lambda_0 2^{n-1}} < \infty,$$

and the proof is complete. □

Next, we turn to the proof of Theorem 5.14, which implicitly uses the statement of Theorem 5.13.

Proof of Theorem 5.14 First, we prove that (ii) implies (i). It suffices to prove that (ii) implies Theorem 5.13 (ii). However this is a simple issue of weak convergence and regular variation, since, for any $y > 0$,

$$\frac{\rho U(x\,\mathrm{d}y)}{xu(x)} = \frac{\rho u(xy)x}{xu(x)}\mathrm{d}y \to \rho y^{\rho-1}\mathrm{d}y$$

as x tends to infinity, in the sense of weak convergence. This implies that

$$\frac{\rho U(xy)}{xu(x)} \sim y^{\rho}.$$

Now choose $y = 1$. Taking account of the fact that $xu(x)/\rho \sim x^{\rho}L(x)/\Gamma(1+\rho)$ (here we use that $\Gamma(1+\rho) = \rho\Gamma(\rho)$), the result thus follows.

Next, we prove that (i) implies (ii). From Theorem 5.13, we see that $U(x) \sim x^{\rho}L(x)/\Gamma(1+\rho)$ for some $L \in \mathcal{R}_{\infty}$. Let us temporarily assume that u is eventually non-decreasing. For any $0 < a < b < \infty$, we have

$$U(bx) - U(ax) = \int_{ax}^{bx} u(y)\mathrm{d}y,$$

and hence, for x large enough,

$$\frac{(b-a)xu(ax)}{x^{\rho}L(x)/\Gamma(1+\rho)} \le \frac{U(bx) - U(ax)}{x^{\rho}L(x)/\Gamma(1+\rho)} \le \frac{(b-a)xu(bx)}{x^{\rho}L(x)/\Gamma(1+\rho)}. \tag{5.18}$$

Using the regular variation of U, we also have that

$$\lim_{x\uparrow\infty} \frac{U(bx) - U(ax)}{x^{\rho}L(x)/\Gamma(1+\rho)} = \left(b^{\rho} - a^{\rho}\right).$$

Hence, from the left inequality of (5.18), we have

$$\limsup_{x\uparrow\infty} \frac{u(ax)}{x^{\rho-1}L(x)/\Gamma(1+\rho)} \le \frac{(b^{\rho} - a^{\rho})}{(b-a)}.$$

Now taking $a = 1$ and letting $b \downarrow 1$, it follows that

$$\limsup_{x\uparrow\infty} \frac{u(x)}{x^{\rho-1}L(x)} \le \frac{\rho}{\Gamma(1+\rho)}.$$

A similar treatment for the right inequality in (5.18), taking $b = 1$ and letting $a \uparrow 1$, shows that

$$\liminf_{x\uparrow\infty} \frac{u(x)}{x^{\rho-1}L(x)} \ge \frac{\rho}{\Gamma(1+\rho)}.$$

Recalling that $\Gamma(1+\rho) = \rho\Gamma(\rho)$, the statement of the theorem follows.

The proof when u is eventually non-increasing is essentially the same with minor adjustments. □

5.5 Dynkin–Lamperti Asymptotics

Let us return to the issue of the asymptotic behaviour of overshoots and undershoots of subordinators. The following theorem is due to Dynkin (1961) and Lamperti (1962). It shows that obtaining an asymptotic bivariate limit distribution of the overshoot and undershoot, when rescaling by the level of the barrier, is equivalent to an assumption of regular variation on the Laplace exponent of the subordinator.

Theorem 5.16 *Suppose that X is any subordinator with Laplace exponent Φ, which belongs to $\mathcal{R}_0(\alpha)$ (resp. $\mathcal{R}_\infty(\alpha)$), where $\alpha \in (0,1)$. Then, in the sense of weak convergence, for $u > 0$ and $y \in [0,1)$,*

$$\mathbb{P}\left(\frac{X_{\tau_x^+} - x}{x} \in \mathrm{d}u, \frac{x - X_{\tau_x^+ -}}{x} \in \mathrm{d}y\right) \to \frac{\alpha \sin \pi\alpha}{\pi}(1-y)^{\alpha-1}(y+u)^{-\alpha-1}\mathrm{d}y\,\mathrm{d}u, \tag{5.19}$$

as x tends to infinity (resp. zero).

The statement of the theorem is not empty as any stable subordinator fulfils the assumptions. Recall from Exercise 3.7 that a stable subordinator necessarily has Laplace exponent on $[0,\infty)$ given by $\Phi(\theta) = c\theta^\alpha$, for some $c > 0$ and $\alpha \in (0,1)$.

We may take the result in the above theorem a little further. For example, one may conversely prove that the pair

$$\left(\frac{X_{\tau_x^+} - x}{x}, \frac{x - X_{\tau_x^+ -}}{x}\right)$$

has a non-degenerate limit in distribution, as $x \uparrow \infty$, only if $\Phi \in \mathcal{R}_0(\alpha)$ for $\alpha \in (0,1)$. See Exercise 5.9 or Bingham (1973a).

It is also possible to calculate the marginal laws of (5.19) as follows:

$$\mathbb{P}\left(\frac{X_{\tau_x^+} - x}{x} \in \mathrm{d}u\right) \to \frac{\sin \pi\alpha}{\pi} u^{-\alpha}(1+u)^{-1}\mathrm{d}u, \quad u \geq 0,$$

and

$$\mathbb{P}\left(\frac{x - X_{\tau_x^+ -}}{x} \in \mathrm{d}y\right) \to \frac{\sin \pi\alpha}{\pi} y^{-\alpha}(1-y)^{\alpha-1}\mathrm{d}y, \quad y \geq 0,$$

in the sense of weak convergence, as $x \uparrow \infty$ or $x \downarrow 0$. These limits are known as the generalised arcsine laws. The classical arcsine law is a special case when $\alpha = 1/2$.[5]

[5] In that case, the density $(\pi\sqrt{y(1-y)})^{-1}$ is related (via a linear transform) to the derivative of the arcsine function.

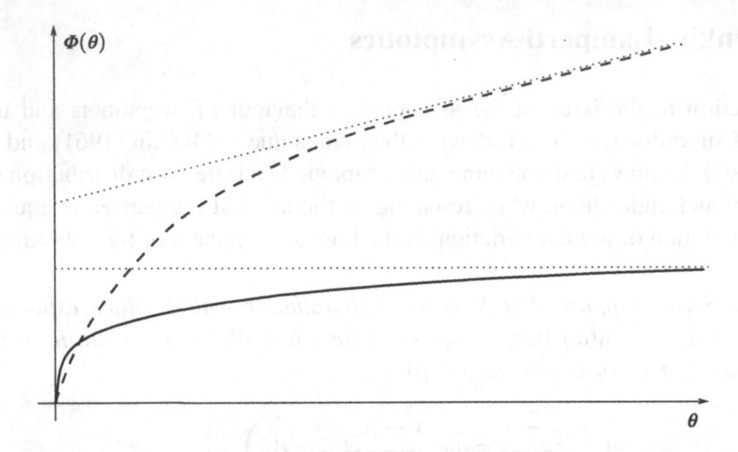

Fig. 5.2 Examples of the shape of the Laplace exponent $\Phi(\theta)$. The *solid concave curve* corresponds to the case of a compound Poisson process with infinite mean ($\Phi'(0+) = \infty$ and $\Phi(\infty) < \infty$). The *dashed concave curve* corresponds to the case of a finite mean subordinator with strictly positive linear drift ($\Phi'(0+) < \infty$ and $\lim_{\theta\uparrow\infty} \Phi(\theta)/\theta = \delta$).

Before moving to the proof of Theorem 5.16, let us make some remarks about regular variation of the Laplace exponent Φ of a subordinator. It is easy to deduce, with the help of dominated convergence, that Φ is infinitely differentiable and strictly concave. In addition, $\Phi'(0+) = \mathbb{E}(X_1) \in (0,\infty]$, $\Phi(0) = 0$ and $\Phi(\infty) = -\log \mathbb{P}(X_1 = 0)$ (which is only finite in the case that X is a compound Poisson subordinator). Finally, recall again from Exercise 2.11 that $\lim_{\theta\uparrow\infty} \Phi(\theta)/\theta = \delta$. See Fig. 5.2 for a visualisation of these facts.

Suppose now that $\Phi \in \mathcal{R}_0(\alpha)$ with $\alpha \in \mathbb{R}$, so that $\Phi(\theta) \sim \theta^\alpha L(\theta)$ as $\theta \downarrow 0$, for some slowly varying function L. As $\Phi(0) = 0$, we necessarily have that $\alpha \geq 0$. If $\mathbb{E}(X_1) < \infty$ then clearly $\Phi(\theta)/\theta \sim \mathbb{E}(X_1)$ as $\theta \downarrow 0$ forcing $\alpha = 1$. On the other hand, if $\mathbb{E}(X_1) = \infty$, then $\Phi(\theta)/\theta$ explodes as $\theta \downarrow 0$, forcing $\alpha \leq 1$. In conclusion, $\Phi \in \mathcal{R}_0(\alpha)$ implies that $\alpha \in [0,1]$.

Now suppose that $\Phi \in \mathcal{R}_\infty(\alpha)$, with $\alpha \in \mathbb{R}$. Since $\Phi(\infty) > 0$ (actually $\Phi(\infty) = \infty$ in the case that X is not a compound Poisson subordinator), we have that $\alpha \geq 0$. On the other hand, the fact that $\Phi(\theta)/\theta$ tends to the constant δ at infinity also dictates that $\alpha \leq 1$. Hence, $\Phi \in \mathcal{R}_\infty(\alpha)$, again, implies that $\alpha \in [0,1]$.

We now turn to the proof of Theorem 5.16, beginning with the following preparatory lemma. Recall that U is the potential measure of the given subordinator.

Lemma 5.17 *Suppose that the Laplace exponent of a subordinator, Φ, belongs to $\mathcal{R}_0(\alpha)$ (resp. $\mathcal{R}_\infty(\alpha)$), where $\alpha \in [0,1]$. Then, for all $\lambda > 0$,*

(i) $U(\lambda x)\Phi(1/x) \to \lambda^\alpha/\Gamma(1+\alpha)$ *as* $x \uparrow \infty$ *(resp.* $x \downarrow 0$*) and*
(ii) *when α is further restricted to* $[0,1)$, $\Pi(\lambda x,\infty)/\Phi(1/x) \to \lambda^{-\alpha}/\Gamma(1-\alpha)$ *as* $x \uparrow \infty$ *(resp.* $x \downarrow 0$*).*

Proof (i) Recall that

$$\int_{[0,\infty)} \mathrm{e}^{-qx} U(\mathrm{d}x) = \frac{1}{\Phi(q)}.$$

The assumption on Φ means that $\Phi(\theta) \sim \theta^{\alpha} L(1/\theta)$ as θ tends to zero, where $L \in \mathcal{R}_{\infty}$. That is to say, $1/\Phi(1/x) \sim x^{\alpha}/L(x)$ as x tends to infinity. Noting that $1/L \in \mathcal{R}_{\infty}$, Theorem 5.13 implies that $U(x) \sim x^{\alpha}/L(x)\Gamma(1+\alpha)$, as $x \uparrow \infty$. Regular variation now implies the statement in part (i). The same argument works when Φ is regularly varying at infinity, rather than at zero.

(ii) Now recall from Exercise 2.11 that

$$\frac{\Phi(\theta)}{\theta} = \delta + \int_0^{\infty} \mathrm{e}^{-\theta x} \Pi(x,\infty)\mathrm{d}x,$$

showing that $\Phi(\theta)/\theta$ is a Laplace transform. The assumed regular variation on Φ implies that $\Phi(\theta)/\theta \sim \theta^{-(1-\alpha)} L(1/\theta)$, as $\theta \downarrow 0$, for some $L \in \mathcal{R}_{\infty}$. Theorem 5.14 now dictates that $\Pi(x,\infty) \sim x^{-\alpha} L(x)/\Gamma(1-\alpha)$ and regular variation gives the statement in part (ii). As usual, the same argument works when, instead, it is assumed that Φ is regularly varying at infinity. Note also, in this case, the assumption that $\alpha \in [0,1)$ implies that $\delta = 0$ as, otherwise, if $\delta > 0$, then necessarily $\alpha = 1$. □

Finally, we are ready for the proof of the Dynkin–Lamperti Theorem.

Proof of Theorem 5.16 We give the proof for the case that $x \uparrow \infty$. The proof for $x \downarrow 0$ requires minor modification.

Starting from the conclusion of Theorem 5.6, we have, for $\theta \in [0,1)$ and $\phi > 0$, that

$$\mathbb{P}\left(\frac{X_{\tau_x^+} - x}{x} \in \mathrm{d}\phi, \frac{x - X_{\tau_x^+-}}{x} \in \mathrm{d}\theta\right) = U\big(x(1-\mathrm{d}\theta)\big)\Pi\big(x(\theta+\mathrm{d}\phi)\big)$$

and hence, for $0 < a < b < 1$ and $c > 0$,

$$\begin{aligned}
\mathbb{P}&\left(\frac{X_{\tau_x^+} - x}{x} > c, \frac{x - X_{\tau_x^+-}}{x} \in (a,b)\right) \\
&= \int_{(a,b)} \Pi\big(x(\theta+c),\infty\big) U\big(x(1-\mathrm{d}\theta)\big) \\
&= \int_{(1-b,1-a)} \frac{\Pi(x(1-\eta+c),\infty)}{\Phi(1/x)} U(x\,\mathrm{d}\eta)\Phi(1/x),
\end{aligned} \tag{5.20}$$

where in the last equality, we have changed variables. From Lemma 5.17 (i), we see, on the one hand, that $U(x\,\mathrm{d}\eta)\Phi(1/x)$ converges weakly to $\eta^{\alpha-1}\mathrm{d}\eta/\Gamma(\alpha)$ (we have used that $\Gamma(1+\alpha) = \alpha\Gamma(\alpha)$). On the other hand, we have seen from part (ii) of the same lemma that

$$\lim_{x\downarrow 0} \frac{\Pi(x(1-\eta+c),\infty)}{\Phi(1/x)} = \frac{(1-\eta+c)^{-\alpha}}{\Gamma(1-\alpha)}.$$

Thanks to a general technical result for regularly varying functions, it turns out that this convergence is uniform in η on compacts. We refrain from giving the details here, referring instead to Theorem 1.5.2 of Bingham et al. (1987). The right-hand side of (5.20) thus converges to

$$\int_{(1-b,1-a)} \frac{(1-\eta+c)^{-\alpha}}{\Gamma(1-\alpha)} \frac{\eta^{\alpha-1}}{\Gamma(\alpha)} \mathrm{d}\eta$$
$$= \frac{1}{\Gamma(\alpha)\Gamma(1-\alpha)} \int_{(a,b)} (\theta+c)^{-\alpha}(1-\theta)^{\alpha-1} \mathrm{d}\theta,$$

as $x \uparrow \infty$, which is tantamount to saying

$$\lim_{x\uparrow\infty} \mathbb{P}\left(\frac{X_{\tau_x^+} - x}{x} \in \mathrm{d}u, \frac{x - X_{\tau_x^+ -}}{x} \in \mathrm{d}y\right)$$
$$= \frac{\alpha}{\Gamma(\alpha)\Gamma(1-\alpha)} (y+u)^{-\alpha-1}(1-y)^{\alpha-1} \mathrm{d}y\, \mathrm{d}u,$$

for $u > 0$ and $y \in [0, 1)$, in the sense of weak convergence. Finally, Euler's reflection formula for gamma functions gives us that $1/(\Gamma(\alpha)\Gamma(1-\alpha)) = (\sin \pi\alpha)/\pi$ and hence the proof is complete. □

5.6 Special and Complete Subordinators

We close this chapter by looking at subclasses of killed subordinators which offer a greater degree of mathematical tractability with regard to the analysis of their potential measures. In particular, this will allow us to construct some concrete examples of subordinators with explicit potential measures. Moreover, later on in Chap. 9, we shall see how this plays an important role in the theory of so-called *scale functions*.

Recall that any killed subordinator, $X = \{X_t : t \geq 0\}$, can be uniquely identified by its Laplace exponent

$$\Phi(\theta) = -\frac{1}{t} \log \mathbb{E}\left(\mathrm{e}^{-\theta X_t}\right) = \eta + \delta\theta + \int_{(0,\infty)} \left(1 - \mathrm{e}^{-\theta x}\right) \Pi(\mathrm{d}x), \quad \theta \geq 0, \quad (5.21)$$

where $\eta \geq 0$ is the killing rate, $\delta \geq 0$ is the drift coefficient and the Lévy measure Π is concentrated on $(0, \infty)$ and satisfies $\int_{(0,\infty)} (1 \wedge x)\Pi(\mathrm{d}x) < \infty$. Such functions are also known as *Bernstein functions*.

Definition 5.18 (Special and conjugate subordinators) A killed subordinator, X, is said to be *special* if, for $\theta \geq 0$, the *conjugate* $\Phi^*(\theta) := \theta/\Phi(\theta)$ is also the Laplace exponent of a killed subordinator (also referred to as the conjugate killed subordinator).

Let us make some additional notational remarks to accompany this definition. For convenience, we shall often say that "Φ is a *special Bernstein function*" rather than "the killed subordinator corresponding to Φ is special". If Φ is special, then so is its conjugate. Moreover, we shall write its conjugate in the form

$$\Phi^*(\theta) = \eta^* + \delta^*\theta + \int_{(0,\infty)} (1 - \mathrm{e}^{-\theta x}) \Pi^*(\mathrm{d}x), \quad \theta \geq 0,$$

where $\eta^* \geq 0$, $\delta^* \geq 0$ and Π^* is a measure concentrated on $(0, \infty)$, satisfying $\int_{(0,\infty)} (1 \wedge x) \Pi^*(\mathrm{d}x) < \infty$.

The family of special subordinators was introduced by Song and Vondraček (2006), although the same notion can also be found in the earlier work of Bertoin (1997c) and Steutel and van Harn (1977). The theory we shall discuss here is largely based on the first of these three references. The reader is directed to the recent monograph of Schilling et al. (2010) for a global perspective on the theory of special subordinators. One of the most important consequences of this definition is that special subordinators can be equivalently characterised through their potential measures.

Theorem 5.19 *Suppose that X is a killed subordinator with potential measure U. Then X is a special subordinator if and only if*

$$U(\mathrm{d}x) = c\delta_0(\mathrm{d}x) + u(x)\mathrm{d}x, \quad x \geq 0,$$

where $c \geq 0$ and $u : (0, \infty) \mapsto (0, \infty)$ is a non-increasing function, satisfying $\int_0^1 u(x)\mathrm{d}x < \infty$. Moreover, $c = \delta^*$ and $u(x) = \eta^* + \Pi^*(x, \infty)$.

Proof Let us first suppose that Φ is a special Bernstein function. Appealing to Exercise 2.11 (ii), we may now write, for $\theta \geq 0$,

$$\begin{aligned} \frac{1}{\Phi(\theta)} &= \frac{\Phi^*(\theta)}{\theta} \\ &= \frac{\eta^*}{\theta} + \delta^* + \int_0^\infty \mathrm{e}^{-\theta x} \Pi^*(x, \infty)\mathrm{d}x \\ &= \delta^* + \int_0^\infty \mathrm{e}^{-\theta x} \big(\eta^* + \Pi^*(x, \infty)\big)\mathrm{d}x. \end{aligned} \tag{5.22}$$

Recalling that the potential measure associated with Φ satisfies

$$\int_{[0,\infty)} \mathrm{e}^{-\theta x} U(\mathrm{d}x) = \frac{1}{\Phi(\theta)}, \tag{5.23}$$

it follows directly from (5.22) that, for $x \geq 0$,

$$U(\mathrm{d}x) = c\delta_0(\mathrm{d}x) + u(x)\mathrm{d}x, \tag{5.24}$$

where $c = \delta^*$ and $u(x) = \eta^* + \Pi^*(x, \infty)$.

Now suppose that (5.24) holds such that the pair c and u satisfy the conditions in the statement of the theorem. Again, making use of (5.23), we have[6]

$$\begin{aligned}\frac{\theta}{\Phi(\theta)} &= c\theta + \int_0^\infty \theta \mathrm{e}^{-\theta x} u(x)\mathrm{d}x \\ &= c\theta + u(x)\big(1-\mathrm{e}^{-\theta x}\big)\big|_0^\infty - \int_{(0,\infty)} \big(1-\mathrm{e}^{-\theta x}\big)u(\mathrm{d}x) \\ &= c\theta + u(\infty) + \int_{(0,\infty)} \big(1-\mathrm{e}^{-\theta x}\big)\big[-u(\mathrm{d}x)\big]. \end{aligned} \tag{5.25}$$

Note that the assumption $\int_0^1 u(x)\mathrm{d}x < \infty$ implies that $\lim_{x\downarrow 0} u(x)(1-\mathrm{e}^{-\theta x}) = 0$ and, after a straightforward integration by parts, it also implies that the integral on the right-hand side of (5.25) is finite. In particular, $\int_{(0,\infty)}(1 \wedge x)[-u(\mathrm{d}x)] < \infty$. Note also that $-u(\mathrm{d}x)$ has positive increments on account of the fact that u is decreasing, in other words, it is a positive measure. Hence, writing $\eta^* = u(\infty)$, $\delta^* = c$ and $\Pi^*(\mathrm{d}x) = -u(\mathrm{d}x)$, it follows that Φ is special with conjugate triple given by $(\eta^*, \delta^*, \Pi^*)$. □

We can also identify the constants η^* and δ^* in terms of the original triple (η, δ, Π). Indeed, recall that $\eta^* = \Phi^*(0+) = \lim_{\theta\downarrow 0} \theta\Phi(\theta)^{-1}$ and, from Exercise 2.11 (ii), $\delta^* = \lim_{\theta\uparrow\infty} \Phi^*(\theta)/\theta = \lim_{\theta\uparrow\infty} \Phi(\theta)^{-1}$. Hence, appealing to Exercise 2.11 (iii), we have that

$$\eta^* = \begin{cases} 0 & \text{if } \eta > 0 \\ (\delta + \int_{(0,\infty)} x\Pi(\mathrm{d}x))^{-1} & \text{if } \eta = 0, \end{cases} \tag{5.26}$$

where we interpret the right-hand side to be zero when $\int_{(0,\infty)} x\Pi(\mathrm{d}x) = \infty$, and

$$\delta^* = \begin{cases} 0 & \text{if } \delta > 0 \text{ or } \Pi(0,\infty) = \infty \\ (\eta + \Pi(0,\infty))^{-1} & \text{if } \delta = 0 \text{ and } \Pi(0,\infty) < \infty. \end{cases} \tag{5.27}$$

It is now straightforward to deduce, from (5.26) and (5.27), that $\eta\eta^* = \delta\delta^* = 0$. Moreover, $\delta = 0$ *and* $\Pi(0,\infty) = \infty$ if and only if $\delta^* = 0$ *and* $\Pi^*(0,\infty) = \infty$.

We shall shortly give some concrete examples of special subordinators. However, before doing so, it is natural to ask where one should look to find such examples. Otherwise said, we are interested in sufficient conditions to ensure that a given subordinator is special. The next result, taken from Song and Vondraček (2010), requires a rather technical proof and hence, we omit it in favour of illustrating the result by example.

[6] Note also that u is right-continuous and non-increasing so that we may make sense of the measure $-u(\mathrm{d}x)$.

Theorem 5.20 *Suppose that X is a killed subordinator with Lévy measure Π, which has the property that $x \mapsto \log \Pi(x,\infty)$ is a convex function on $(0,\infty)$. Then X is a special subordinator.*

As a first example, consider the function

$$u(x) = \begin{cases} x^{-\alpha} & \text{for } 0 < x < 1 \\ 1 & \text{for } x \geq 1, \end{cases}$$

where $0 < \alpha < 1$. Note that u is decreasing and log-convex. Suppose we define a measure Π^* on $(0,\infty)$ by its tail, so that $\Pi^*(x,\infty) = u(x) - 1$, for $x > 0$. It is also straightforward to check that $\int_{(0,\infty)} (1 \wedge x)\Pi^*(\mathrm{d}x) < \infty$.

Now define, for $\theta \geq 0$,

$$\Phi^*(\theta) = 1 + \int_{(0,\infty)} (1 - \mathrm{e}^{-\theta x})\Pi^*(\mathrm{d}x).$$

In terms of our earlier notation, $\eta^* = 1$ and $\delta^* = 0$. According to Theorem 5.20, Φ^* is a special Bernstein function, in which case it has a conjugate, which we shall denote by Φ. As Φ is also a special Bernstein function, it follows from Theorem 5.19 that its potential measure, U, can be identified in the form $U(\mathrm{d}x) = u(x)\mathrm{d}x$, $x \geq 0$.

From the discussion preceding the proof of Theorem 5.19, since $\eta^* > 0$, we may conclude that $\eta = 0$. Since $\Pi^*(0,\infty) = \infty$, it follows from (5.27), applied to Φ^* instead of Φ, that $\delta = 0$. Finally, by Theorem 5.19, to compute Π, it suffices to compute the potential measure U^* associated with Φ^*. This is tantamount to performing a Laplace inverse of $1/\Phi^*(\theta)$, but this is not analytically tractable.

It turns out that there is a subclass of special subordinators, known as *complete subordinators*, amongst which it is much easier to find tractable examples of conjugate subordinators. By "tractable", we mean here that it is possible to compute both triples (η, δ, Π) and $(\eta^*, \delta^*, \Pi^*)$. The vast majority of known tractable examples of special subordinators fall into the class of complete subordinators.

In order to formally state the definition of a complete subordinator, let us recall the definition of *complete monotonicity*. The reader is referred to Schilling et al. (2010) and Widder (2010) for a detailed modern and classical account, respectively. A function $f : [0,\infty) \to [0,\infty)$ is called *completely monotone* if

$$(-1)^n f^{(n)}(x) \geq 0 \quad \text{for all } x \in (0,\infty) \text{ and } n = 0, 1, 2, \ldots, \tag{5.28}$$

where $f^{(n)}$ denotes the n-th derivative of f. Perhaps the most straightforward examples of completely monotone functions are the exponential functions $f(x) = \mathrm{e}^{-\theta x}$, where $\theta \geq 0$. As a simple generalisation of this family, it is also true that, for any finite Borel measure μ on $[0,\infty)$,

$$\int_{[0,\infty)} \mathrm{e}^{-\theta x} \mu(\mathrm{d}\theta) \tag{5.29}$$

is also a completely monotone function. Indeed, this follows by dominated convergence (which justifies differentiating through the integral) and the complete monotonicity of the exponential functions. It turns out that every completely monotone function can be represented in the form (5.29). The equivalence of completely monotone functions with the representation (5.29) is known as *Bernstein's Theorem.*[7]

Remarkably, there is also a deep connection between Bernstein functions and completely monotone functions.

Theorem 5.21 *The class of Bernstein functions agrees with the class of nonnegative functions whose first derivative is completely monotone.*

We can now give the definition of a complete subordinator.

Definition 5.22 (Complete subordinators) A subordinator is said to be *complete* if it has a Lévy measure which is absolutely continuous with respect to Lebesgue measure and has a completely monotone density.

In a similar spirit to previously, we shall refer to the Laplace exponent of a complete subordinator as a *complete Bernstein function*. The following theorem reiterates what we have already alluded to above, namely that complete subordinators are a subclass of special subordinators. However, it also exposes some interesting symmetric properties with regard to conjugate pairs. Again, we omit the proof for the same reasons as above.

Theorem 5.23 *A complete subordinator is also a special subordinator. Moreover, its conjugate is also a complete subordinator.*

The following corollary is a direct consequence of Theorems 5.19 and 5.23.

Corollary 5.24 *Every complete subordinator has a potential measure whose density on* $(0, \infty)$ *is completely monotone.*

The class of complete subordinators is strictly contained in the class of special subordinators. That is to say, they are not identical classes. One only needs to consider the example of a special subordinator following Theorem 5.20 to verify this fact. Nonetheless, almost all known examples of special subordinators turn out to be complete subordinators.

We conclude this section, and this chapter, with two tractable examples of complete subordinators. More examples can be found in the exercises and even more can be found in Schilling et al. (2010), together with the proof of Theorem 5.23. In addition, many complete subordinators can be found by considering inverse local times of diffusions. See for example Borodin and Salminen (2002).

[7] This result is also attributed to Hausdorff and Widder.

Example 5.25 For the first example, let $0 < \alpha \le \beta \le 1$, $a, b > 0$ and let Φ be the Bernstein function defined by

$$\Phi(\theta) = a\theta^{\beta-\alpha} + b\theta^{\beta}, \quad \theta \ge 0.$$

Hence, when $0 < \alpha < \beta < 1$, Φ is the Laplace exponent of the sum of two independent stable subordinators, one of parameter $\beta - \alpha$ and the other of parameter β, respectively. In terms of the notation in (5.21), $\eta = \delta = 0$, and

$$\Pi(\mathrm{d}x) = \left(\frac{a(\beta-\alpha)}{\Gamma(1-\beta+\alpha)} x^{-(1+\beta-\alpha)} + \frac{b\beta}{\Gamma(1-\beta)} x^{-(1+\beta)} \right) \mathrm{d}x, \quad x > 0,$$

see for example Exercise 5.8. If $\alpha = \beta < 1$, then Φ is the Laplace exponent of a stable subordinator killed at rate a. When $\alpha < \beta = 1$, Φ is the Laplace exponent of a stable subordinator with positive drift b. Finally, in the case where $\alpha = 1 = \beta$, Φ is simply the Laplace exponent of a pure drift subordinator killed at rate a. This last case will be excluded and left for the reader to explore in Exercise 5.12. In all cases, the underlying Lévy measure has a density which is completely monotone, and thus its potential density is completely monotone.

In the remainder of this example, as well as subsequent examples, we shall make heavy use of the two-parameter Mittag–Leffler function, defined by

$$\mathrm{E}_{\alpha,\beta}(x) = \sum_{n \ge 0} \frac{x^n}{\Gamma(n\alpha + \beta)}, \quad x \in \mathbb{R}, \tag{5.30}$$

where $\alpha, \beta > 0$. This function is characterised via a Laplace transform. Namely, for $\lambda \in \mathbb{R}$, $\theta \in \mathbb{C}$ and $|\theta^\alpha/\lambda| > 1$, we have

$$\int_0^\infty \mathrm{e}^{-\theta x} x^{\beta-1} \mathrm{E}_{\alpha,\beta}(\lambda x^\alpha) \mathrm{d}x = \frac{\theta^{\alpha-\beta}}{\theta^\alpha - \lambda}. \tag{5.31}$$

Recall from (5.27) that, since $\Pi(0,\infty) = \infty$, we have $\delta^* = 0$ and, hence, Theorem 5.19 predicts that, on $[0,\infty)$, $U(\mathrm{d}x) = u(x)\mathrm{d}x$ for some density u. However, using the above pseudo-Laplace transform it is not difficult to confirm using (5.23) that

$$u(x) = \frac{1}{b} x^{\beta-1} \mathrm{E}_{\alpha,\beta}\big(-ax^\alpha/b\big), \quad x > 0. \tag{5.32}$$

Taking account of Exercise 5.11, we may note that this is a completely monotone function because it is the product of the completely monotone functions $x^{\beta-1}$ and $\mathrm{E}_{\alpha,\beta}(-x^\alpha)$. Moreover the latter of these two is completely monotone because it is the composition of the completely monotone function $t \mapsto \mathrm{E}_{\alpha,\beta}(-t)$ for $t \ge 0$, with the Bernstein function x^α; see Schneider (1996).

With the potential of Φ in hand, we can now apply Theorem 5.19, again to the conjugate, to recover the complete picture for its triple and its potential measure. The conjugate Bernstein function is given by

$$\Phi^*(\theta) = \frac{\theta}{a\theta^{\beta-\alpha} + b\theta^{\beta}}, \quad \theta \ge 0.$$

From this, it is easy to check that $\eta^* = \Phi^*(0) = 0$, which can also be recovered from (5.26) by noting that $\delta = 0$ and $\int_{(0,\infty)} x\Pi(\mathrm{d}x) = \infty$. In that case, we have $U^*(\mathrm{d}x) = \Pi(x,\infty)\mathrm{d}x$, $x \geq 0$, and hence

$$U^*(x) = \frac{a}{\Gamma(2-\beta+\alpha)} x^{1-\beta+\alpha} + \frac{b}{\Gamma(2-\beta)} x^{1-\beta}, \quad x \geq 0.$$

Finally, to recover the Lévy measure of the conjugate subordinator, recall again from Theorem 5.19, that $u(x) = \Pi^*(x,\infty)$, $x > 0$.

Example 5.26 For the second example, let $c > 0$, $\nu \geq 0$ and $\lambda \in (0,1)$. We claim that

$$\Phi(\theta) = \frac{c\theta\Gamma(\nu+\theta)}{\Gamma(\nu+\theta+\lambda)}, \quad \theta \geq 0,$$

is a Bernstein function, where $\Gamma(u)$ denotes the usual gamma function with parameter $u > 0$. In order to determine its triple (η,δ,Π), let us recall that the beta function is related to the gamma function by Euler's beta integral formula: For $a, b > 0$,

$$B(a,b) := \int_0^1 x^{b-1}(1-x)^{a-1}\mathrm{d}x = \frac{\Gamma(a)\Gamma(b)}{\Gamma(a+b)}.$$

We thus have that

$$\Phi(\theta) = \frac{c\theta}{\Gamma(\lambda)} B(\theta+\nu,\lambda), \quad \theta \geq 0.$$

Making a change of variable in the expression for the beta function, we reach the identity

$$\frac{\Phi(\theta)}{\theta} = \frac{c}{\Gamma(\lambda)} \int_0^\infty \mathrm{e}^{-\theta z}\mathrm{e}^{-z\nu}\left(1-\mathrm{e}^{-z}\right)^{\lambda-1}\mathrm{d}z, \quad \theta \geq 0. \tag{5.33}$$

Recalling the computations in Exercise 2.11, this shows that $\eta = \delta = 0$ and

$$\Pi(x,\infty) = c\frac{\mathrm{e}^{-x(\nu+\lambda-1)}}{\Gamma(\lambda)}\left(\mathrm{e}^x - 1\right)^{\lambda-1}, \quad x > 0.$$

It is a straightforward computation, with the help of Exercise 5.11, to show that Π has a completely monotone density. Hence, Φ is a complete Bernstein function.

In order to determine the potential measure associated with this subordinator, observe the following elementary identity:

$$\frac{\theta}{\Phi(\theta)} = \frac{\Gamma(\nu+\theta+\lambda)\Gamma(1-\lambda)}{c\Gamma(\nu+\theta+1)} \frac{\nu+\theta}{\Gamma(1-\lambda)}, \quad \theta \geq 0. \tag{5.34}$$

Therefore, we have that

$$
\begin{aligned}
\frac{\theta}{\Phi(\theta)} &= \frac{\nu+\theta}{c\Gamma(1-\lambda)}\int_0^1 x^{\nu+\theta-1}x^{\lambda}(1-x)^{-\lambda}\mathrm{d}x \\
&= \frac{\lambda}{c\Gamma(1-\lambda)}\int_0^1 \frac{1}{x^2}\left(1-x^{\nu+\theta}\right)\left(\frac{1}{x}-1\right)^{-\lambda-1}\mathrm{d}x \\
&= \frac{\lambda}{c\Gamma(1-\lambda)}\int_0^\infty \left(1-\mathrm{e}^{-(\nu+\theta)z}\right)\left(\mathrm{e}^z-1\right)^{-\lambda-1}\mathrm{e}^z\mathrm{d}z \\
&= \frac{\lambda}{c\Gamma(1-\lambda)}\int_0^\infty \left(1-\mathrm{e}^{-\nu z}\right)\frac{\mathrm{e}^z}{(\mathrm{e}^z-1)^{\lambda+1}}\mathrm{d}z \\
&\quad + \frac{\lambda}{c\Gamma(1-\lambda)}\int_0^\infty \left(1-\mathrm{e}^{-\theta z}\right)\frac{\mathrm{e}^{z(1-\nu)}}{(\mathrm{e}^z-1)^{\lambda+1}}\mathrm{d}z \\
&= \frac{\Gamma(\nu+\lambda)}{c\Gamma(\nu)} + \frac{\lambda}{c\Gamma(1-\lambda)}\int_0^\infty \left(1-\mathrm{e}^{-\theta z}\right)\frac{\mathrm{e}^{z(1-\nu)}}{(\mathrm{e}^z-1)^{\lambda+1}}\mathrm{d}z. \qquad (5.35)
\end{aligned}
$$

Note the second equality above is obtained by using integration by parts, splitting the integrand into the product of the functions $1-x^{\nu+\theta}$ and $(x^{-1}-1)^{-\lambda}$, for $x\in(0,1)$. Moreover, the fifth equality follows also by an integration by parts and the change of variables $u=\mathrm{e}^{-z}$ as follows,

$$
\begin{aligned}
\lambda\int_0^\infty \left(1-\mathrm{e}^{-\nu z}\right)\frac{\mathrm{e}^z}{(\mathrm{e}^z-1)^{\lambda+1}}\mathrm{d}z &= -\left.\frac{(1-\mathrm{e}^{-\nu z})}{(\mathrm{e}^z-1)^{\lambda}}\right|_0^\infty + \int_0^\infty \frac{\nu\mathrm{e}^{-(\nu+1)z}\mathrm{e}^z}{(\mathrm{e}^z-1)^{\lambda}}\mathrm{d}z \\
&= \nu\int_0^\infty \frac{\mathrm{e}^{-(\nu+\lambda-1)z}\mathrm{e}^{-z}}{(1-\mathrm{e}^{-z})^{\lambda}}\mathrm{d}z \\
&= \nu\int_0^1 u^{\lambda+\nu-1}(1-u)^{1-\lambda-1}\mathrm{d}u \\
&= \frac{\Gamma(\lambda+\nu)\Gamma(1-\lambda)}{\Gamma(\nu)}.
\end{aligned}
$$

The right-hand side of (5.35) shows that Φ is a special Bernstein function whose conjugate, Φ^*, has triple $(\eta^*, \delta^*, \Pi^*)$, where $\eta^*=\Gamma(\nu+\lambda)/c\Gamma(\nu)$, $\delta^*=0$ and

$$
\Pi^*(\mathrm{d}x) = \frac{\lambda}{c\Gamma(1-\lambda)}\frac{\mathrm{e}^{x(1-\nu)}}{(\mathrm{e}^x-1)^{\lambda+1}}\mathrm{d}x.
$$

Theorem 5.19 again allows us to identify the potential measures. The potential measure of Φ is given by

$$
U(\mathrm{d}x) = \frac{\Gamma(\nu+\lambda)}{c\Gamma(\nu)}\delta_0(\mathrm{d}x) + \left\{\int_x^\infty \frac{\lambda}{c\Gamma(1-\lambda)}\frac{\mathrm{e}^{z(1-\nu)}}{(\mathrm{e}^z-1)^{\lambda+1}}\mathrm{d}z\right\}\mathrm{d}x,
$$

for $x \geq 0$. The potential measure of Φ^* is given by

$$U^*(\mathrm{d}x) = c\frac{\mathrm{e}^{-x(\upsilon+\lambda-1)}}{\Gamma(\lambda)}(\mathrm{e}^x - 1)^{\lambda-1}\mathrm{d}x, \tag{5.36}$$

for $x \geq 0$.

Exercises

5.1 In this exercise, we shall derive the form of the Laplace exponent of a killed subordinator given in (5.1), without appealing to the Lévy–Itô decomposition. To this end, suppose that $X = \{X_t : t \geq 0\}$ is a $[0, \infty]$-valued stochastic process which has non-decreasing, right-continuous paths with left limits. Here, $+\infty$ serves as a cemetery state. Denote its lifetime by

$$\zeta = \inf\{t > 0 : X_t = \infty\}.$$

Suppose that under measure $\mathbb{P}$, X has the property that, for all $s, t \geq 0$, on the event $\{t < \zeta\}$, the increment $X_{t+s} - X_t$ is independent of $\{X_u : u \leq t\}$ and equal in distribution to X_s.

(i) By agreeing to write $\mathrm{e}^{-\infty} = 0$, show that

$$\mathbb{E}(\mathrm{e}^{-\theta X_{t+s}}) = \mathbb{E}(\mathrm{e}^{-\theta X_t})\mathbb{E}(\mathrm{e}^{-\theta X_s}),$$

where $\theta, s, t \geq 0$ and $\mathbb{E}$ denotes expectation with respect to $\mathbb{P}$. Hence, deduce that, for $\theta, t \geq 0$,

$$\mathbb{E}(\mathrm{e}^{-\theta X_t}) = \mathrm{e}^{-\Phi(\theta)t},$$

where $\Phi(\theta) = -\log \mathbb{E}(\mathrm{e}^{-\theta X_1})$.

(ii) Prove that, for $\theta \geq 0$,

$$\frac{\Phi(\theta)}{\theta} = \lim_{n\uparrow\infty} \int_0^\infty \mathrm{e}^{-\theta x} n\mathbb{P}(X_{1/n} \geq x)\mathrm{d}x.$$

(iii) Hence, deduce that, for $\theta \geq 0$,

$$\Phi(\theta) = \eta + \delta\theta + \int_{(0,\infty)} (1 - \mathrm{e}^{-\theta x})\Pi(\mathrm{d}x),$$

where $\Pi(\mathrm{d}x) = -\overline{\Pi}(\mathrm{d}x)$, $\delta \geq 0$ and $\eta \geq 0$ are uniquely identified.

(iv) Explain why $\int_{(0,\infty)}(1 \wedge x)\Pi(\mathrm{d}x) < \infty$.

5.2 Suppose that, under $\mathbb{P}$, $X = \{X_t : t \geq 0\}$ is a (killed) subordinator with Laplace exponent Φ, just as in part (iii) of Exercise 5.1. Define for $q \geq 0$,

$$\left.\frac{\mathrm{d}\mathbb{P}^q}{\mathrm{d}\mathbb{P}}\right|_{\mathcal{F}_t} := \exp\{-qX_t + \Phi(q)t\}, \quad t \geq 0,$$

where, as usual, $\{\mathcal{F}_t : t \geq 0\}$ is the natural filtration generated by X. Show that $(X, \mathbb{P}^q)$ is a subordinator without killing, the same drift coefficient as $(X, \mathbb{P})$ and Lévy measure given by $e^{-qx}\Pi(dx)$, $x > 0$, where Π is the Lévy measure of $(X, \mathbb{P})$.

5.3 Prove Theorem 5.7.

5.4 Suppose that Y is a spectrally positive Lévy process of bounded variation drifting to $-\infty$, with Laplace exponent written in the usual form

$$\log \mathbb{E}(e^{-\theta Y_1}) = \psi(\theta) = \delta\theta - \int_{(0,\infty)} (1 - e^{-\theta x})\nu(dx),$$

where necessarily $\delta > 0$, $\int_{(0,\infty)}(1 \wedge x)\nu(dx) < \infty$ and $\psi'(0+) > 0$. Define $\sigma_x^+ = \inf\{t > 0 : Y_t > x\}$ and $\overline{Y}_t = \sup_{s \leq t} Y_s$.

(i) Suppose that $X = \{X_t : t \geq 0\}$ is a compound Poisson subordinator with jump distribution $(\delta - \psi'(0+))^{-1}\nu(x, \infty)dx$. By following similar reasoning to the explanation of the Pollaczek–Khintchine formula in Chap. 4, show that

$$\begin{aligned}&\mathbb{P}(Y_{\sigma_x^+} - x \in du, x - \overline{Y}_{\sigma_x^+-} \in dy \mid \sigma_x^+ < \infty)\\&\quad= \mathbb{P}(X_{\tau_x^+} - x \in du, x - X_{\tau_x^+-} \in dy).\end{aligned}$$

(ii) Deduce that if $\int_0^\infty x\nu(x, \infty)dx < \infty$, then, for $u, y > 0$, in the sense of weak convergence,

$$\begin{aligned}&\lim_{x\uparrow\infty} \mathbb{P}(Y_{\sigma_x^+} - x \in du, x - \overline{Y}_{\sigma_x^+-} \in dy | \sigma_x^+ < \infty)\\&\quad= \frac{1}{\int_0^\infty x\nu(x, \infty)dx}\nu(u + y, \infty)du\,dy.\end{aligned}$$

(iii) Give an interpretation of the result in (ii) in the context of modelling insurance claims.

5.5 Suppose that X is a finite mean subordinator and that its associated potential measure U does not have lattice support. Suppose that Z is a random variable whose distribution is equal to that of the limiting distribution of $X_{\tau_x^+} - X_{\tau_x^+-}$ as $x \uparrow \infty$. Suppose further that (V, W) is a bivariate random variable whose distribution is equal to the limiting distribution of $(X_{\tau_x^+} - x, x - X_{\tau_x^+-})$ as $x \uparrow \infty$, and that U is independent of V, W, Z and uniformly distributed on $[0, 1]$. Show that (V, W) is equal in distribution to $((1 - U)Z, UZ)$.

5.6 Let X and Y be two (possibly correlated) subordinators killed independently at the rate $\eta \geq 0$. Denote their bivariate jump measure by $\mathbf{\Pi}(\cdot, \cdot)$. Define their bivariate renewal function

$$\mathcal{U}(dx, dy) = \int_0^\infty dt \cdot \mathbb{P}(X_t \in dx, Y_t \in dy), \quad x, y \geq 0,$$

and suppose, as usual, that

$$\tau_x^+ = \inf\{t > 0 : X_t > x\}, \quad x \geq 0.$$

Use a generalised version of the compensation formula to establish the following quadruple law

$$P(\Delta X_{\tau_x^+} \in \mathrm{d}t, X_{\tau_x^+-} \in \mathrm{d}s, x - Y_{\tau_x^+-} \in \mathrm{d}y, Y_{\tau_x^+} - x \in \mathrm{d}u)$$
$$= \mathcal{U}(\mathrm{d}s, x - \mathrm{d}y)\mathbf{\Pi}(\mathrm{d}t, \mathrm{d}u + y),$$

for $u > 0$, $y \in [0, x]$ and $s, t \geq 0$. This formula will be of use later on when considering the first passage of a general Lévy process over a fixed level.

5.7 Let X be any subordinator with Laplace exponent Φ, drift coefficient $\delta \geq 0$ and recall that $\tau_x^+ = \inf\{t > 0 : X_t > x\}$. Let $\mathbf{e}_\alpha$ be an exponentially distributed random variable with rate α, which is independent of X.

(i) By applying the strong Markov property at time τ_x^+ in the expectation $\mathbb{E}(\mathrm{e}^{-\beta X_{\mathbf{e}_\alpha}} \mathbf{1}_{(X_{\mathbf{e}_\alpha} > x)})$, show that, for all $\alpha, \beta, x \geq 0$, we have

$$\mathbb{E}\big(\mathrm{e}^{-\alpha\tau_x^+ - \beta X_{\tau_x^+}}\big) = \big(\alpha + \Phi(\beta)\big) \int_{(x,\infty)} \mathrm{e}^{-\beta z} U^{(\alpha)}(\mathrm{d}z), \qquad (5.37)$$

for all $x > 0$.

(ii) Show further, with the help of the identity in (i), that, when $q > 0$ and $\beta \geq 0$,

$$\int_0^\infty \mathrm{e}^{-qx} \mathbb{E}\big(\mathrm{e}^{-\alpha\tau_x^+ - \beta(X_{\tau_x^+} - x)}\big)\mathrm{d}x = \frac{1}{q - \beta}\left(1 - \frac{\alpha + \Phi(\beta)}{\alpha + \Phi(q)}\right).$$

(iii) Deduce, with the help of Theorem 5.9, that

$$\mathbb{E}\big(\mathrm{e}^{-\alpha\tau_x^+} \mathbf{1}_{(X_{\tau_x^+} = x)}\big) = \delta u^{(\alpha)}(x),$$

where, if $\delta = 0$, the term $u^{(\alpha)}(x)$ may be taken as equal to zero and, otherwise, the potential measure $U^{(\alpha)}$ has a density such that $u^{(\alpha)}$ is a continuous and strictly positive version thereof.

(iv) Show that for this version of the density, $u^{(\alpha)}(0+) = 1/\delta$, where δ is the drift of X.

5.8 Suppose that X is a stable subordinator with parameter $\alpha \in (0, 1)$, thus having Laplace exponent $\Phi(\theta) = c\theta^\alpha$, for $\theta \geq 0$ and some $c > 0$. In this exercise, we will take $c = 1$.

(i) Show from Exercise 1.4 that the precise expression for the jump measure is

$$\Pi(\mathrm{d}x) = \frac{x^{-(1+\alpha)}}{-\Gamma(-\alpha)}\mathrm{d}x, \quad x > 0.$$

(ii) By considering the Laplace transform of the potential measure U, show that

$$U(\mathrm{d}x) = \frac{x^{\alpha-1}}{\Gamma(\alpha)}\mathrm{d}x, \quad x \geq 0.$$

(iii) Hence deduce that

$$\mathbb{P}(X_{\tau_x^+} - x \in \mathrm{d}u, x - X_{\tau_x^+-} \in \mathrm{d}y)$$
$$= \frac{\alpha \sin \alpha\pi}{\pi}(x-y)^{\alpha-1}(y+u)^{-(\alpha+1)}\mathrm{d}u\,\mathrm{d}y,$$

for $u > 0$ and $y \in [0, x]$. Note further that the distribution of the pair

$$\left(\frac{x - X_{\tau_x^+-}}{x}, \frac{X_{\tau_x^+} - x}{x}\right) \tag{5.38}$$

is independent of x.

(iv) Show directly that stable subordinators do not creep.

5.9 Suppose that X is any subordinator.

(i) Use the joint law of the overshoot and undershoot to deduce that, for $\beta, \gamma \geq 0$ and $q > 0$,

$$\int_0^\infty \mathrm{d}x \cdot \mathrm{e}^{-qx}\mathbb{E}\left(\mathrm{e}^{-\beta X_{\tau_x^+-} - \gamma(X_{\tau_x^+} - x)}\mathbf{1}_{(X_{\tau_x^+} > x)}\right)$$
$$= \frac{1}{q-\gamma}\left(\frac{\Phi(q) - \Phi(\gamma)}{\Phi(q+\beta)}\right) - \frac{\delta}{\Phi(q+\beta)}.$$

(ii) Taking account of creeping, use part (i) to deduce that

$$\int_0^\infty \mathrm{d}x \cdot \mathrm{e}^{-qx}\mathbb{E}\left(\mathrm{e}^{-\beta(X_{\tau_{tx}^+-}/t) - \gamma(X_{\tau_{tx}^+} - tx)/t}\right) = \frac{1}{(q-\gamma)}\frac{\Phi(q/t) - \Phi(\gamma/t)}{\Phi((q+\beta)/t)},$$

for $\beta, \gamma \geq 0$ and $q > 0$.

(iii) Show that if $\Phi \in \mathcal{R}_0(\alpha)$ (resp. $\Phi \in \mathcal{R}_\infty(\alpha)$) with α equal to 0 or 1, then the limiting distribution of the pair in (5.38) is trivial as x tends to infinity (resp. zero).

(iv) It is possible to show that, if for a given measurable function $f : [0, \infty) \to (0, \infty)$, there exists a $g : (0, \lambda) \to (0, \infty)$ such that

$$\lim \frac{f(\lambda t)}{f(t)} = g(\lambda),$$

for all $\lambda > 0$, as t tends to zero (resp. infinity), then f must be regularly varying. Roughly speaking, the reason for this is that, for $\lambda, \mu > 0$,

$$g(\lambda\mu) = \lim \frac{f(\mu\lambda t)}{f(\lambda t)}\frac{f(\lambda t)}{f(t)} = g(\mu)g(\lambda)$$

showing that g is a multiplicative function. With a little measure theory, one can show that $g(\lambda) = \lambda^\rho$, for some $\rho \in \mathbb{R}$. See Theorem 1.4.1 of Bingham et al. (1987) for the full details.

Use the above remarks to deduce that, if (5.38) has a limiting distribution as x tends to infinity (resp. zero), then necessarily $\Phi \in \mathcal{R}_0(\alpha)$ (resp. $\Phi \in \mathcal{R}_\infty(\alpha)$) with $\alpha \in [0, 1]$. Hence conclude that (5.38) has a non-trivial limiting distribution if and only if $\alpha \in (0, 1)$.

5.10 Suppose that F is a probability distribution function. Write $\overline{F}(x) = 1 - F(x)$. Then F belongs to $\mathcal{L}^{(\alpha)}$, where $\alpha \geq 0$, if the support of F is non-lattice in $[0, \infty)$, $\overline{F}(x) > 0$ for all $x \geq 0$ and, for all $y > 0$,

$$\lim_{x\uparrow\infty} \frac{\overline{F}(x+y)}{\overline{F}(x)} = \mathrm{e}^{-\alpha y}.$$

Note that the requirement that F is a probability measure can be weakened to a finite measure, as one may always normalise by its total mass to fulfil the conditions given earlier.

We are interested in establishing an asymptotic conditional distribution for the overshoot of a killed subordinator. To this end, we assume that X is a killed subordinator with killing rate $\eta > 0$, Laplace exponent Φ, jump measure Π, drift $\delta \geq 0$ and potential measure U which is assumed to belong to class $\mathcal{L}^{(\alpha)}$, for some $\alpha \geq 0$ such that $\Phi(-\alpha) < \infty$.

(i) Show that, for $x > 0$,

$$\mathbb{P}(\tau_x^+ < \infty) = \eta U(x, \infty),$$

where $\tau_x^+ = \inf\{t > 0 : X_t > x\}$.

(ii) Show that, for all $\beta \geq 0$,

$$\mathbb{E}\big(\mathrm{e}^{-\beta(X_{\tau_x^+} - x)} | \tau_x^+ < \infty\big) = \frac{\Phi(\beta)}{\eta U(x, \infty)} \int_{(x,\infty)} \mathrm{e}^{-\beta(y-x)} U(\mathrm{d}y).$$

(iii) Applying integration by parts, deduce that

$$\lim_{x\uparrow\infty} \mathbb{E}\big(\mathrm{e}^{-\beta(X_{\tau_x^+} - x)} | \tau_x^+ < \infty\big) = \frac{\Phi(\beta)}{\eta}\left(\frac{\alpha}{\alpha+\beta}\right).$$

(iv) Now take the distribution G on $[0, \infty)$, defined by its tail

$$G(x, \infty) = \frac{\mathrm{e}^{-\alpha x}}{\eta}\left\{\Phi(-\alpha) + \int_{(x,\infty)} (\mathrm{e}^{\alpha y} - \mathrm{e}^{\alpha x}) \Pi(\mathrm{d}y)\right\}.$$

Show that G has an atom at zero and

$$\int_{(0,\infty)} \mathrm{e}^{-\beta y} G(\mathrm{d}y) = \frac{\Phi(\beta)}{\eta}\left(\frac{\alpha}{\alpha+\beta}\right) - \frac{\delta\alpha}{\eta}.$$

(v) Deduce that, for all $u > 0$,

$$\lim_{x\uparrow\infty} \mathbb{P}\big(X_{\tau_x^+} - x > u \,|\, \tau_x^+ < \infty\big) = G(u, \infty)$$

and

$$\lim_{x\uparrow\infty} \mathbb{P}\big(X_{\tau_x^+} = x \,|\, \tau_x^+ < \infty\big) = \frac{\delta\alpha}{\eta}.$$

5.11 Suppose that f is a completely monotone function.

(i) If f is another completely monotone function show that $\alpha f + \beta g$ is completely monotone for all $\alpha, \beta \geq 0$ as well as fg.
(ii) Suppose that Φ is a Bernstein function. Show that $f \circ \Phi$ is completely monotone.

5.12 This exercise gives two more examples of complete subordinators for which the analysis in Sect. 5.6 is completely tractable.

(i) Consider the, apparently trivial, Bernstein function

$$\Phi(\theta) = \eta + \delta\theta, \quad \theta \geq 0,$$

where $\delta, \eta > 0$. This corresponds to the subordinator which is a deterministic linear drift killed at rate η. Show that

$$U(x) = \frac{1}{\eta}\big(1 - \mathrm{e}^{-x\eta/\delta}\big), \quad x \geq 0,$$

and hence deduce that $\delta^* = \eta^* = 0$, $\Pi^*(x, \infty) = \delta^{-1}\mathrm{e}^{-x\eta/\delta}$ and $U^*(x) = \delta + \eta x$ for $x \geq 0$.
(ii) Now consider, for $\nu \in (0, 1)$ and $\gamma > 0$,

$$\Phi(\theta) := \eta + \lambda\left(1 - \left(\frac{\gamma}{\gamma + \theta}\right)^{\nu}\right), \quad \theta \geq 0,$$

where $\eta, \lambda > 0$. Show that Φ is a complete Bernstein with components $\delta = 0$, the killing rate is η,

$$\Pi(\mathrm{d}x) = \frac{\lambda\gamma^{\nu}}{\Gamma(\nu)} x^{\nu-1}\mathrm{e}^{-\gamma x}\mathrm{d}x, \quad x > 0,$$

and

$$U(x) = \frac{1}{\lambda + \eta} + \frac{\rho\gamma^{\nu}}{\lambda + \eta}\int_0^x \mathrm{e}^{-\gamma y} y^{\nu-1} \mathrm{E}_{\nu,\nu}\big(\rho\gamma^{\nu} y^{\nu}\big)\mathrm{d}y,$$

where $\rho = \lambda/(\lambda + \eta)$.

Hence deduce that $\eta^* = 0$, $\delta^* = 1/(\eta+\lambda)$,

$$\Pi^*(x,\infty) = \frac{\rho\gamma^\nu}{\lambda+\eta} \mathrm{e}^{-\gamma x} x^{\nu-1} \mathrm{E}_{\nu,\nu}(\rho\gamma^\nu x^\nu)$$

and

$$U^*(x) = \eta x + \frac{\lambda\gamma^\nu}{\Gamma(\nu)} \int_0^x \left\{ \int_y^\infty z^{\nu-1} \mathrm{e}^{-\gamma z} \mathrm{d}z \right\} \mathrm{d}y.$$

5.13 This exercise concerns another transformation for Bernstein functions which produces again a Bernstein function. The origins and more details of this transformation can be found in Urbanik (2005), Gnedin (2010) and Chazal et al. (2012).

(i) Suppose that Φ is the Laplace exponent of a subordinator with triple (η, δ, Π). Show that, for $\theta, \beta \geq 0$,

$$\phi(\theta) := \frac{\theta}{\theta+\beta} \Phi(\theta+\beta), \quad \theta \geq 0,$$

is the Laplace exponent of a subordinator.

(ii) Show moreover that the triple of ϕ is equal to $(0, \delta, \nu)$, where

$$\nu(\mathrm{d}x) = \beta \mathrm{e}^{-\beta x} \Pi(x,\infty)\mathrm{d}x + \mathrm{e}^{-\beta x} \Pi(\mathrm{d}x) + \eta\beta \mathrm{e}^{-\beta x} \mathrm{d}x,$$

for $x > 0$.

5.14 In this exercise, we show that the proof of Theorem 5.9 is relatively straightforward for special subordinators. To this end, suppose that Φ is a special Bernstein function with representation (5.1). Assume that $\Pi(0,\infty) = \infty$. Recall from Theorem 5.19 that its conjugate has a potential density on $(0,\infty)$, denoted by $u^*(x)$, which satisfies $u^*(x) = \eta + \Pi(x,\infty)$.

(i) By considering the factorisation $\theta^{-1} = \Phi(\theta)^{-1} \times \Phi(\theta)/\theta$ for $\theta > 0$, show that, for all $x > 0$,

$$1 = \delta u(x) + \int_0^x u(x-y) u^*(y) \mathrm{d}y.$$

(ii) Deduce with the help of Theorem 5.6 that, for the killed subordinator X with Laplace exponent Φ,

$$\mathbb{P}(X_{\tau_x^+} = x) = \delta u(x), \quad x > 0.$$

Chapter 6
The Wiener–Hopf Factorisation

This chapter gives an account of the theory of excursions of a Lévy process from its maximum and the *Wiener–Hopf factorisation* that follows as a consequence.

In Sect. 4.6, the analytical form of the Pollaczek–Khintchine formula was explained through a decomposition of the path of the underlying Lévy process into independent and identically distributed sections of path, called excursions from the supremum. The decomposition made heavy use of the fact that, for the particular class of Lévy processes considered, namely spectrally positive processes of bounded variation, the times of new maxima form a discrete set.

For a general Lévy process, it is still possible to decompose its path into "excursions from the running maximum". Conceptually, this decomposition is *a priori* somewhat more tricky as, in principle, a general Lévy process may exhibit an infinite number of excursions from its maximum over any finite period of time. Nonetheless, when considered in the right mathematical framework, excursions from the maximum can be given a sensible definition in terms of a Poisson random measure. The theory of excursions presents one of the more mathematically challenging aspects of the theory of Lévy processes. This means that in order to keep to the level outlined in the preface of this text, there will be a number of proofs in the forthcoming sections which are excluded or discussed only at an intuitive level.

Within a very broad spectrum of probabilistic literature, *the Wiener–Hopf factorisation* may be found as a common reference to a multitude of statements concerning the distributional decomposition of the path of any Lévy process, when sampled at an independent and exponentially distributed time, in terms of its excursions from the maximum. (We devote a little time later in this text to explain the origin of the name "Wiener–Hopf factorisation".) The collection of conclusions which fall under the umbrella of the Wiener–Hopf factorisation turns out to provide a robust tool with which one may analyse a number of problems concerning the fluctuations of Lévy processes, in particular, problems which have relevance to the applications we shall consider in later chapters. This chapter concludes with some special classes of Lévy processes for which the Wiener–Hopf factorisation may be exemplified in more detail.

A.E. Kyprianou, *Fluctuations of Lévy Processes with Applications*, Universitext,
DOI 10.1007/978-3-642-37632-0_6,

6.1 Local Time at the Maximum

Unlike the Lévy processes presented in Sect. 4.6, a general Lévy process may have an infinite number of new maxima over any given finite period of time. As one of our goals is to show how to decompose events according to the behaviour of the path in individual excursions, we need a way of indexing them. To this end we introduce the notion of *local time at the maximum.*

To avoid trivialities, we shall assume throughout this section that neither X nor $-X$ is a subordinator. Recall, moreover, the definition $\overline{X}_t = \sup_{s\le t} X_s$. We shall repeatedly refer to the process $\overline{X} - X = \{\overline{X}_t - X_t : t \ge 0\}$, which we also recall, from Exercise 3.2, can be shown to be a strong Markov process.

Definition 6.1 (Local time at the maximum) A continuous, non-decreasing, $[0,\infty)$-valued, $\mathbb{F}$-adapted process, $L = \{L_t : t \ge 0\}$, is called a local time at the maximum (or just local time for short) if the following hold.

(i) The support of the Stieltjes measure $\mathrm{d}L$ is the closure of the (random) set of times $\{t \ge 0 : \overline{X}_t = X_t\}$.
(ii) For every $\mathbb{F}$-stopping time T such that $\overline{X}_T = X_T$ on $\{T < \infty\}$ almost surely, the shifted process

$$\{L_{T+t} - L_T : t \ge 0\}$$

is independent of $\mathcal{F}_T$ on $\{T < \infty\}$ and has the same law as L under $\mathbb{P}$.

Let us make some remarks about the above definition. Firstly, note that since X and $\overline{X} - X$ are strong Markov processes, it also follows, from the requirement in part (ii) of the above definition, that the shifted trivariate process

$$\{(X_{T+t} - X_T, \overline{X}_{T+t} - X_{T+t}, L_{T+t} - L_T) : t \ge 0\}$$

is independent of $\mathcal{F}_T$ on $\{T < \infty\}$ and has the same law as $(X, \overline{X} - X, L)$ under $\mathbb{P}$. Next, note that if L is a local time, then so is kL for any constant $k > 0$. Hence, local times can at best be defined uniquely up to a multiplicative constant. On occasion, we shall need to talk about both local time and the time scale on which the Lévy process itself is defined. In such cases, we shall refer to the latter as *real time*. Finally, by applying this definition of local time to $-X$, it is clear that one may talk of a local time at the minimum. This will always be referred to as $\widehat{L}$.

Local times, as defined above, do not always exist on account of the requirement of continuity. Nonetheless, in such cases, it turns out that one may construct right-continuous processes which satisfy conditions (i) and (ii) of Definition 6.1, and which serve their purpose equally well in the forthcoming analysis of the Wiener–Hopf factorisation. We provide more details shortly. We first give some examples for which a continuous local time can be identified explicitly.

Example 6.2 (Spectrally negative processes) Recall that a spectrally negative process has the properties that $\Pi(0,\infty) = 0$ and that its paths are not monotone. As there are no positive jumps, the process $\overline{X}$ must therefore be continuous. It is easy to check that $L := \overline{X}$ fulfils Definition 6.1.

Example 6.3 (Compound Poisson processes with drift $\delta \geq 0$) By considering the piecewise linearity of the paths of these processes, one has obviously that, over any finite time horizon, the time spent at the maximum has strictly positive Lebesgue measure with probability one. Hence, the quantity

$$L_t := \int_0^t \mathbf{1}_{(\overline{X}_s = X_s)} \mathrm{d}s, \quad t \geq 0, \tag{6.1}$$

is almost surely positive and may be taken as a candidate for local time. Indeed it increases on $\{t : \overline{X}_t = X_t\}$, is continuous, non-decreasing and is an $\mathbb{F}$-adapted process. Taking T as in part (ii) of Definition 6.1, we also see that on $\{T < \infty\}$,

$$L_{T+t} - L_T = \int_T^{T+t} \mathbf{1}_{(\overline{X}_s - X_s = 0)} \mathrm{d}s, \tag{6.2}$$

which is independent of $\mathcal{F}_T$ (because $\{\overline{X}_{T+t} - X_{T+t} : t \geq 0\}$ is) and has the same law as L (by the strong Markov property applied to the process $\overline{X} - X$ and the fact that $\overline{X}_T - X_T = 0$).

If we allow only negative jumps and $\delta > 0$, then, according to the previous example, $\overline{X}$ also fulfils the definition of local time. However, as we have seen in the proof of Theorem 4.1,

$$\overline{X}_t = \delta \int_0^t \mathbf{1}_{(\overline{X}_s = X_s)} \mathrm{d}s,$$

for all $t \geq 0$.

Next, we would like to identify the class of Lévy processes for which a continuous local time cannot be constructed, and for which a right-continuous alternative can be used instead. In a nutshell, the aforementioned class consists of those Lévy processes whose times of new maxima form a discrete set. The qualifying criterion for this turns out to be related to the behaviour of the Lévy process at arbitrarily small times. A sense of this has already been given in the discussion of Sect. 4.6. We spend a little time developing the relevant notions, namely regularity of points, in order to complete the discussion on local time.

Definition 6.4 For a Lévy process X, the point $x \in \mathbb{R}$ is said to be regular (resp. irregular) for an open or closed set B if

$$\mathbb{P}_x\big(\tau^B = 0\big) = 1 \quad \text{(resp. 0)},$$

where $\tau^B = \inf\{t > 0 : X_t \in B\}$. Intuitively speaking, x is regular for B if, when starting from x, the Lévy process hits B immediately.

Note that, as τ^B is a stopping time, it follows that

$$\mathbf{1}_{(\tau^B=0)} = \mathbb{P}_x(\tau^B = 0|\mathcal{F}_0).$$

On the other hand, since $\mathcal{F}_0$ is generated by null sets, Kolmogorov's definition of conditional expectation implies

$$\mathbb{P}_x(\tau^B = 0|\mathcal{F}_0) = \mathbb{P}_x(\tau^B = 0),$$

and hence $\mathbb{P}_x(\tau^B = 0)$ is either zero or one. In fact, one may replace $\{\tau^B = 0\}$ by any event $A \in \mathcal{F}_0$ and reach the same conclusion about $\mathbb{P}(A)$. This is nothing but *Blumenthal's zero-one law*. See, for example, Proposition 40.4 in Sato (1999).

We know from the Lévy–Itô decomposition that the range of a Lévy process over any finite time horizon is almost surely bounded and, thanks to right-continuity, $\lim_{t\downarrow 0} \max\{-\underline{X}_t, \overline{X}_t\} = 0$. Hence, for any given open or closed B, the points $x \in \mathbb{R}$ for which $\mathbb{P}_x(\tau^B = 0) = 1$ necessarily belong to $B \cup \partial B$. However, $x \in \partial B$ is not a sufficient condition for $\mathbb{P}_x(\tau^B = 0) = 1$. To see why, consider the case that $B = (0, \infty)$ and X is any compound Poisson process. Another example for the same B is the case when X is the difference of a driftless subordinator and a pure linear drift; cf. Sect. 4.6.

Finally note that the notion of regularity can be asserted for any Markov process, with an analogous definition to the one given earlier. However, for the special case of a Lévy process, stationary independent increments allow us to reduce the discussion of regularity of x for open and closed sets to simply the regularity of 0 for open and closed sets. Indeed, for any Lévy process, x is regular for B if and only if 0 is regular for $B - x$.

As we shall shortly see, it is regularity of 0 for $[0, \infty)$ which dictates whether one may find a continuous local time. The following result, collectively due to Rogozin (1968), Shtatland (1965) and Bertoin (1997a), gives precise conditions for the slightly different issue of regularity of 0 for $(0, \infty)$.

Theorem 6.5 *For any Lévy process, X, excluding the case of a compound Poisson process, the point 0 is regular for $(0, \infty)$ if and only if*

$$\int_0^1 \frac{1}{t}\mathbb{P}(X_t > 0)\mathrm{d}t = \infty, \tag{6.3}$$

and this holds if and only if one of the following three conditions holds:

(i) *X is a process of unbounded variation*,
(ii) *X is a process of bounded variation and $\delta > 0$*,
(iii) *X is a process of bounded variation, $\delta = 0$ and*

$$\int_{(0,1)} \frac{x\Pi(\mathrm{d}x)}{\int_0^x \Pi(-\infty, -y)\mathrm{d}y} = \infty.$$

Here, δ is the drift coefficient in the representation (2.21) *of a Lévy process of bounded variation.*

Recall that if N is the Poisson random measure associated with the jumps of X, then the time of arrival of a jump of size $\varepsilon > 0$ or greater, say $T(\varepsilon)$, is exponentially distributed since

$$\mathbb{P}(T(\varepsilon) > t) = \mathbb{P}(N([0,t] \times \{\mathbb{R}\backslash(-\varepsilon,\varepsilon)\}) = 0) = \exp\{-t\Pi(\mathbb{R}\backslash(-\varepsilon,\varepsilon))\}.$$

This tells us that jumps of size greater than ε become less and less probable as $t \downarrow 0$. Hence, the jumps that have any influence over the initial behaviour of the path of X, if at all, will necessarily be arbitrarily small. With this in mind, one may intuitively see the conditions (i)–(iii) in Theorem 6.5 in the following way.

In case (i), when $\sigma^2 > 0$, regularity follows as a consequence of the presence of Brownian motion, whose behaviour on the small time scale always dominates the path of the Lévy process. If on the other hand $\sigma = 0$, then the high intensity of small jumps causes behaviour on the small time scale to be similar to the case when a Brownian component is present. (We use the words "high intensity" here in the sense that $\int_{(-1,1)} |x|\Pi(\mathrm{d}x) = \infty$.) Case (ii) says that when the Poisson random measure describing jumps fulfils the condition $\int_{(-1,1)} |x|\Pi(\mathrm{d}x) < \infty$, over small time scales, the sum of the jumps grows sub-linearly in time almost surely. Therefore if a drift is present, this dominates the initial motion of the path. In case (iii) when there is no dominant drift, the integral test may be thought of as a statement about what the "relative weight" of the small positive jumps needs to be, when compared to the small negative jumps, in order for regularity to occur.

In the case of bounded variation, the integral $\int_0^x \Pi(-\infty,-y)\mathrm{d}y$ is finite for all $x > 0$. This can be deduced by taking the (necessarily) finite integral $\int_{(-1,0)} |x|\Pi(\mathrm{d}x)$ and then integrating by parts.

Theorem 6.5 also implies that processes of unbounded variation are such that 0 is regular for both $(0,\infty)$ and $(-\infty,0)$ and that processes of bounded variation with $\delta > 0$ have the property that 0 is irregular for $(-\infty,0)$. For processes of bounded variation with $\delta = 0$ it is possible to find examples where 0 is regular for both $(0,\infty)$ and $(-\infty,0)$. See Exercise 6.1.

We offer no proof of Theorem 6.5 here. However, it is worth recalling that, from Lemma 4.11 and the follow-up Exercise 4.8 (i), we know that, for any Lévy process, X, of bounded variation,

$$\lim_{t\downarrow 0} \frac{X_t}{t} = \delta$$

almost surely, where δ is the drift coefficient. This shows that if $\delta > 0$ (resp. $\delta < 0$), then for all $t > 0$ sufficiently small, X_t must be strictly positive (resp. negative). That is to say, 0 is regular for $(0,\infty)$ and irregular for $(-\infty,0]$ if $\delta > 0$ (resp. regular for $(-\infty,0)$ and irregular for $[0,\infty)$ if $\delta < 0$). For the case of a spectrally negative Lévy process of unbounded variation, Exercise 6.2 deduces regularity properties in agreement with Theorem 6.5. In addition, Exercise 6.8 shows how to establish criterion (6.3).

There is a slight difference between regularity of 0 for $(0,\infty)$ and regularity of 0 for $[0,\infty)$. Consider for example the case of a compound Poisson process. This process is such that 0 is regular for $[0,\infty)$ but not for $(0,\infty)$ due to the initial exponentially distributed period of time during which the process remains at the origin. It turns out that these are the only processes for which 0 is regular for $[0,\infty)$ but not $(0,\infty)$.

By definition, when 0 is irregular for $[0,\infty)$, the Lévy process takes a strictly positive period of time to reach a new maximum when starting at the origin. Hence, applying the strong Markov property at the time of first entry into $[0,\infty)$, we see that, in a finite interval of time, there are almost surely a finite number of new maxima. In other words, $\{0 < s \leq t : \overline{X}_s = X_s\}$ is a finite set. (Recall that this type of behaviour has been observed for spectrally positive Lévy process of bounded variation in Chap. 4.) In this case, we may then define the counting process $\{n_t : t \geq 0\}$ by

$$n_t = \#\{0 < s \leq t : \overline{X}_s = X_s\}. \tag{6.4}$$

We are now ready to make the distinction between those processes which admit continuous local times in Definition 6.1 and those that do not.

Theorem 6.6 *Let X be any Lévy process.*

(i) *There exists a continuous version of L if and only if 0 is regular for* $[0,\infty)$. *When it exists, it is unique up to a multiplicative constant.*
(ii) *If 0 is irregular for* $[0,\infty)$, *then we can take as our definition of local time*

$$L_t = \sum_{i=0}^{n_t} \mathbf{e}_\lambda^{(i)}, \quad t \geq 0, \tag{6.5}$$

satisfying (i) *and* (ii) *of Definition* 6.1, *where* $\{\mathbf{e}_\lambda^{(i)} : i \geq 0\}$ *are independent and exponentially distributed random variables with parameter* $\lambda > 0$ *(chosen arbitrarily).*

We offer no proof for case (i). It is a particular example of a classic result from potential theory of stochastic processes, a general account of which can be found in Blumenthal and Getoor (1968). See also Greenwood and Pitman (1980c). The proof of part (ii) is quite accessible and we leave it as an exercise. Note that one slight problem occurring in the definition of L in (6.5), aside from the fact that it is no longer a continuous process, is that it is not adapted to the filtration of X. However, this is easily resolved by simply enlarging the filtration, before completing it with null sets, to include $\sigma(\mathbf{e}_\lambda^{(i)} : i \geq 0)$. Also, the unspecified value of the parameter λ, used in (6.5), is of no effective consequence. In principle, one could always work with the definition

$$L'_t = \sum_{i=0}^{n_t} \mathbf{e}_1^{(i)}.$$

Scaling properties of exponential distributions would then allow us to construct the $\mathbf{e}_\lambda^{(i)}$ and $\mathbf{e}_1^{(i)}$ on the same probability space via the relation $\lambda \mathbf{e}_\lambda^{(i)} := \mathbf{e}_1^{(i)}$ for each

$i = 0, 1, 2, \ldots$, and this would imply that $L' = \lambda L$ where L is local time constructed using exponential distributions with parameter λ. Hence, within the specified class of local times in part (ii) of the above theorem, the only effective difference is a multiplicative constant. The reason why we do not define $L_t = n_t$ has to do with the fact that we shall require some special properties of the inverse L^{-1}. This will be discussed in Sect. 6.2.

In accordance with the conclusion of Theorem 6.6, *in the case that* 0 *is regular for* $[0, \infty)$, *we shall henceforth work with a continuous version of L and in the case that* 0 *is irregular for* $[0, \infty)$, *we shall work with the definition* (6.5) *for L, assuming that the filtration* $\mathbb{F}$ *is sufficiently enlarged so that L is adapted.*

In Example 6.3, we saw that we may use a multiple of the Lebesgue measure of the real time spent at the maximum to give a continuous version of local time. The fact that the aforesaid is non-zero is a clear consequence of piecewise linearity of the process. Although compound Poisson processes (with drift) are the only Lévy processes which are piecewise linear, it is nonetheless natural to investigate to what extent one may work with the Lebesgue measure of the time spent at the maximum for local time in the case that 0 is regular for $[0, \infty)$. Rubinovitch (1971) supplies us with the following characterisation of such processes.

Theorem 6.7 *Suppose that X is a Lévy process for which* 0 *is regular for* $[0, \infty)$. *Let L be some continuous version of local time. Then there exists a constant* $\mathtt{a} \geq 0$, *such that*

$$\int_0^t \mathbf{1}_{(\overline{X}_s = X_s)} \mathrm{d}s = \mathtt{a} L_t, \quad t \geq 0.$$

This constant is strictly positive if and only if X is a Lévy process of bounded variation and 0 *is irregular for* $(-\infty, 0)$.

Proof Note that $\int_0^\infty \mathbf{1}_{(\overline{X}_t = X_t)} \mathrm{d}t > 0$ with positive probability if and only if

$$\mathbb{E}\left(\int_0^\infty \mathbf{1}_{(\overline{X}_t - X_t = 0)} \mathrm{d}t \right) > 0.$$

By Fubini's Theorem and Lemma 3.5, this occurs if and only if $\int_0^\infty \mathbb{P}(\underline{X}_t = 0)\mathrm{d}t = \mathbb{E}(\int_0^\infty \mathbf{1}_{(\underline{X}_t = 0)} \mathrm{d}t) > 0$ (recall that $\underline{X}_t := \inf_{s \leq t} X_s$). Due to the fact that $\underline{X}$ has paths that are right-continuous and non-increasing, the strict positivity of the last expectation happens if and only if it takes an almost surely strictly positive time for X to visit $(-\infty, 0)$. In short, we have that $\int_0^\infty \mathbf{1}_{(\overline{X}_t = X_t)} \mathrm{d}t > 0$ with positive probability if and only if 0 is irregular for $(-\infty, 0)$. By Theorem 6.5, this can only occur when X has bounded variation.

Following the same reasoning as used in Example 6.3, it is straightforward to deduce that

$$\int_0^t \mathbf{1}_{(\overline{X}_s = X_s)} \mathrm{d}s, \quad t \geq 0,$$

may be used as a local time. Theorem 6.6 (i) now gives us the existence of a constant $\mathtt{a} > 0$ so that for a given local time L,

$$\mathtt{a}L_t = \int_0^t \mathbf{1}_{(\overline{X}_s = X_s)} \mathrm{d}s.$$

When 0 is regular for $(-\infty, 0)$ the reasoning above also shows that, for all $t \geq 0$, $\int_0^t \mathbf{1}_{(\overline{X}_s = X_s)} \mathrm{d}s = 0$ almost surely and hence it is clear that the constant $\mathtt{a} = 0$. □

We can now summarise the discussion on local times as follows. There are three types of Lévy processes which are associated with three types of local times.

1. *Processes of bounded variation for which* 0 *is irregular for* $[0, \infty)$. The set of maxima forms a discrete set and we take a right-continuous version of local time in the form
$$L_t = \sum_{i=0}^{n_t} \mathbf{e}_1^{(i)}, \quad t \geq 0,$$
where n_t is the count of the number of maxima up to time t and $\{\mathbf{e}_1^{(i)} : i = 0, 1, \ldots\}$ are independent and exponentially distributed random variables with parameter 1. To make the process L adapted, we assume that the filtration $\mathbb{F}$ is sufficiently enlarged.
2. *Processes of bounded variation for which* 0 *is irregular for* $(-\infty, 0)$. There exists a continuous version of local time given by
$$L_t = \mathtt{a}^{-1} \int_0^t \mathbf{1}_{(\overline{X}_s = X_s)} \mathrm{d}s, \quad t \geq 0,$$
for some arbitrary $0 < \mathtt{a} < \infty$. In the case that X is spectrally negative, we have that L is equal to a multiplicative constant times $\overline{X}$.
3. *Processes of unbounded variation.* For all such processes, 0 is regular for $[0, \infty)$. A continuous version of local time exists but cannot be identified explicitly as a functional of the path of X in general. However, if X is a spectrally negative Lévy process, then this local time may be taken as $\overline{X}$.

6.2 The Ladder Process

Define the inverse local time process, $L^{-1} := \{L_t^{-1} : t \geq 0\}$, by

$$L_t^{-1} := \begin{cases} \inf\{s > 0 : L_s > t\} & \text{if } t < L_\infty \\ \infty & \text{otherwise.} \end{cases}$$

Next, define the process $H = \{H_t : t \geq 0\}$ where

$$H_t := \begin{cases} X_{L_t^{-1}} & \text{if } t < L_\infty \\ \infty & \text{otherwise.} \end{cases}$$

The range of the inverse local time, L^{-1}, corresponds to the set of real times at which new maxima occur. The elements of this set are called the *ascending ladder times*. The range of the process H corresponds to the set of new maxima. Similarly, the elements of this set are called the *ascending ladder heights*. The bivariate process $(L^{-1}, H) := \{(L_t^{-1}, H_t) : t \geq 0\}$, called the *ascending ladder process*, is the main object of study of this section. It is implicit from their definition that L^{-1} and H are processes with paths that are right-continuous with left limits.

The word "ascending" distinguishes the process (L^{-1}, H) from the analogous object $(\widehat{L}^{-1}, \widehat{H})$, which is constructed from $-X$ and is called the *descending ladder process* (note that local time at the maximum of $-X$ is the local time at the minimum of X and was previously referred to as $\widehat{L}$). When the context is obvious, we shall drop the use of the words "ascending" or "descending".

The ladder processes we have defined here are the continuous-time analogue of the processes with the same name for random walks. In the case of random walks, one defines L_n to be the number of times a maxima is reached during the first n steps, $T_n = \min\{k \geq 1 : L_k = n\}$ as the number of steps required to achieve n new maxima (if $L_\infty \geq n$) and H_n as the n-th new maximum (if it exists).

An additional subtlety for random walks is that the count L_n may be taken to include visits to previous maxima (consider for example a simple random walk which may visit an existing maximum several times before generating a strictly greater maximum). In that case, the associated ascending ladder process is called *weak*. When $\{L_n : n \geq 0\}$ only counts the number of new maxima which exceed all previous maxima, the associated ascending ladder process is called *strict*.

The same subtlety appears in the definition of L for Lévy processes when 0 is irregular for $[0, \infty)$, and our definition of the process $\{n_t : t \geq 0\}$ is then analogous to a count of weak ascending ladder heights. This is of no consequence in the forthcoming discussion since, as we shall see, with probability one, no two maxima can be equal. (This will be discussed in greater detail just before Theorem 6.15 ahead.) When 0 is regular for $[0, \infty)$ but X is not a compound Poisson process, we shall again see in due course that ladder heights at different ladder times are distinct. Finally, when X is a compound Poisson process, the distinction between weak and strict maxima will become an issue at some point in later discussion. Indeed, the choice of local time

$$L_t = \mathrm{a}^{-1} \int_0^t \mathbf{1}_{(\overline{X}_s = X_s)} \mathrm{d}s, \quad t \geq 0,$$

is analogous to the count of weak ascending ladder heights in a random walk. Consider, for example, the continuous-time version of a simple random walk; that is a compound Poisson process with jump distribution supported on $\{-1, 1\}$.

Our task in this section will be to characterise the ladder process (L^{-1}, H). We start with the following lemma, which will be used in several places later on.

Lemma 6.8 *For each $t \geq 0$, both L_t^{-1} and L_{t-}^{-1} are $\mathbb{F}$-stopping times.*

Proof From Sect. 3.1, thanks to the assumed right-continuity of $\mathbb{F}$, it suffices to prove that, for each $s > 0$, $\{L_t^{-1} < s\} \in \mathcal{F}_s$ and that a similar notion holds for L_{t-}^{-1}. For all $s, t \geq 0$, $\{L_t^{-1} < s\} = \{L_{s-} > t\}$. Moreover, this event belongs to $\mathcal{F}_s$ as the process L is $\mathbb{F}$-adapted. To prove that L_{t-}^{-1} is a stopping time, note that, for $t \geq 0$,

$$\{L_{t-}^{-1} < s\} = \bigcap_{n\geq 1} \{L_{t-1/n}^{-1} < s\} \in \mathcal{F}_s.$$

□

In the next theorem, we shall link the process (L^{-1}, H) to a bivariate subordinator. With Exercise 2.10 in mind, recall that a bivariate subordinator is a two-dimensional $[0,\infty)^2$-valued stochastic processes, $\mathbf{X} = \{\mathbf{X}_t : t \geq 0\}$, with paths that are right-continuous with left limits, as well as having stationary independent increments and, further, each component is non-decreasing. It is important to note that, in general, it is not correct to think of a bivariate subordinator simply as a vector process composed of two independent subordinators. Correlation between the subordinators in each of the co-ordinates may be represented pathwise in the form

$$\mathbf{X}_t = \mathbf{d}t + \int_{[0,t]} \int_{(0,\infty)^2} \mathbf{x} N(\mathrm{d}s \times \mathrm{d}\mathbf{x}), \quad t \geq 0,$$

where $\mathbf{d} \in [0,\infty)^2$ and N is a Poisson random measure describing the jumps of $\mathbf{X}$. Moreover, the intensity measure of N is given by $\mathrm{d}t \times \Lambda(\mathrm{d}x, \mathrm{d}y)$, for some bivariate measure Λ concentrated on $(0,\infty)^2$ satisfying

$$\int_{(0,\infty)^2} \left(1 \wedge \sqrt{x^2 + y^2}\right) \Lambda(\mathrm{d}x, \mathrm{d}y) < \infty.$$

Independence of the two individual co-ordinate processes corresponds to the case that Λ takes the form $\Lambda(\mathrm{d}x, \mathrm{d}y) = \Lambda_1(\mathrm{d}x)\delta_0(\mathrm{d}y) + \Lambda_2(\mathrm{d}y)\delta_0(\mathrm{d}x)$, $x, y \geq 0$.

For a general bivariate subordinator, positivity allows us to talk about its Laplace exponent $\phi(\alpha, \beta)$, $\alpha, \beta \geq 0$, where

$$\mathbb{E}\left(\exp\left\{-\binom{\alpha}{\beta} \cdot \mathbf{X}_t\right\}\right) = \exp\{-\phi(\alpha,\beta)t\}, \quad t \geq 0.$$

Referring back to Chap. 2, it is a straightforward exercise to deduce that

$$\phi(\alpha,\beta) = \mathbf{d} \cdot \binom{\alpha}{\beta} + \int_{(0,\infty)^2} \left(1 - \mathrm{e}^{-\binom{\alpha}{\beta}\cdot\binom{x}{y}}\right) \Lambda(\mathrm{d}x, \mathrm{d}y).$$

Theorem 6.9 *Let X be a Lévy process and $\mathbf{e}_q$ an independent and exponentially distributed random variable with parameter $q \geq 0$. Then*

$$\mathbb{P}\left(\limsup_{t\uparrow\infty} X_t < \infty\right) = 0 \quad \text{or} \quad 1$$

and the ladder process (L^{-1}, H) satisfies the following properties:

(i) *If* $\mathbb{P}(\limsup_{t\uparrow\infty} X_t = \infty) = 1$, *then* (L^{-1}, H) *has the law of a bivariate subordinator.*

(ii) *If* $\mathbb{P}(\limsup_{t\uparrow\infty} X_t < \infty) = 1$, *then, for some* $q > 0$, $L_\infty \stackrel{d}{=} \mathbf{e}_q$ *and* $\{(L_t^{-1}, H_t) : t < L_\infty\}$ *has the same law as* $(\mathcal{L}^{-1}, \mathcal{H}) := \{(\mathcal{L}_t^{-1}, \mathcal{H}_t) : t < \mathbf{e}_q\}$, *where* $(\mathcal{L}^{-1}, \mathcal{H})$ *is a bivariate subordinator independent of* $\mathbf{e}_q$.

Proof Since

$$\left\{\limsup_{t\uparrow\infty} X_t < \infty\right\} = \left\{\limsup_{\mathbb{Q}\cap[0,\infty)\ni t\uparrow\infty} X_t < \infty\right\}$$

and this event is in the tail sigma-algebra $\bigcap_{t\in\mathbb{Q}\cap[0,\infty)} \sigma(X_s : s \geq t)$, Kolmogorov's zero-one law for tail events tells us that $\mathbb{P}(\limsup_{t\uparrow\infty} X_t < \infty) = 0$ or 1.

To deal with (i) and (ii) in the case that 0 is irregular for $[0,\infty)$, the analysis proceeds in the spirit of the discussion around the Pollaczek–Khintchine formula in Chap. 4. We give a brief outline of the arguments again.

If we agree that a geometric distribution with parameter 1 is infinite with probability one, then the total number of excursions from the maximum, n_∞, defined in (6.4), is geometrically distributed with parameter $1 - \rho = \mathbb{P}(\tau_0^+ = \infty)$, where $\tau_0^+ = \inf\{t > 0 : X_t > 0\}$. Now define the sequence of times $T_0 = 0$,

$$\begin{aligned} T_{n+1} &= \inf\{t > T_n : X_t > X_{T_n}\} \\ &= \inf\{t > T_n : \Delta L_t > 0\} \\ &= \inf\{t > T_n : \Delta n_t = 1\}, \end{aligned}$$

for $n = 0, 1, \ldots, n_\infty$, where $\Delta L_t = L_t - L_{t-}$, $\Delta n_t = n_t - n_{t-}$ and $\inf\emptyset = \infty$. It is easy to verify that these times form an increasing sequence of almost surely finite stopping times. Further, by the strong Markov property for Lévy processes, if $n_\infty < \infty$, then the successive excursions of X from its maximum,

$$\epsilon_n := \left\{X_t - X_{T_{n-1}} : t \in (T_{n-1}, T_n]\right\},$$

for $n = 1, \ldots, n_\infty$, are equal in law to an independent sample of $n_\infty - 1$ copies of the first excursion from the maximum conditioned to be finite, followed by a final independent copy conditioned to be infinite in length. If $n_\infty = \infty$, then the sequence $\{\epsilon_n : n = 1, 2, \ldots\}$ is equal in law to an independent sample of the first excursion from the maximum.

By considering Fig. 6.1, we see that L^{-1} (the reflection of L about the diagonal) is a step function and its successive jumps (the flat sections of L) correspond precisely to the sequence $\{T_{n+1} - T_n : n = 0, \ldots, n_\infty\}$. From the previous paragraph, it follows that L^{-1} has independent and identically distributed jumps and is independently sent to infinity (which we may consider as a "cemetery" state) on the n_∞-th jump, in accordance with the arrival of the first infinite excursion. As the jumps of L are independent and exponentially distributed, it also follows that the periods between jumps of L^{-1} are independent and exponentially distributed. According to

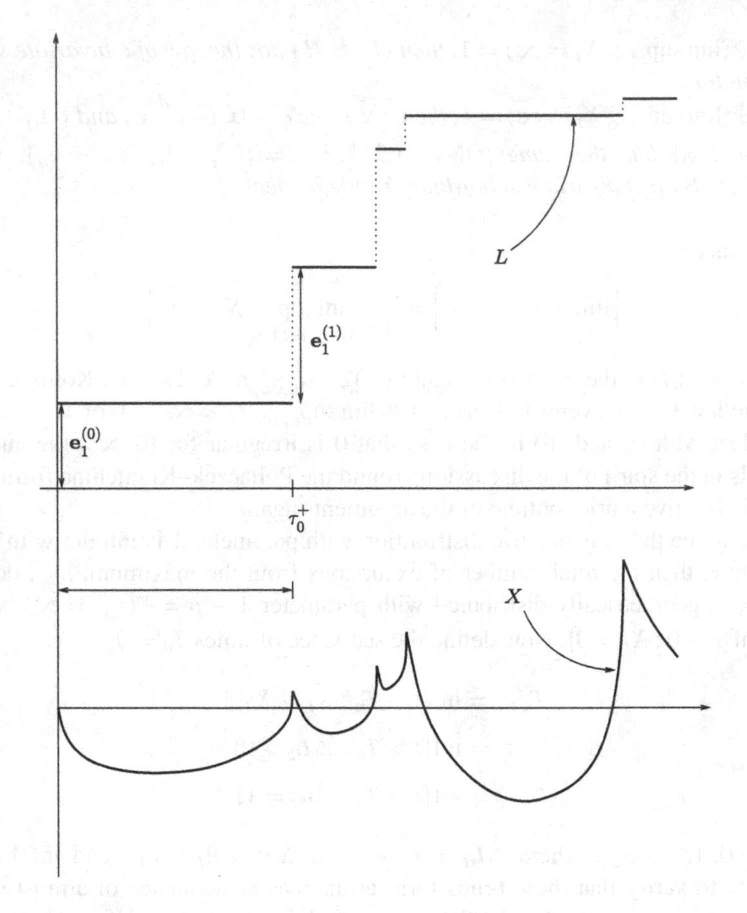

Fig. 6.1 A realisation of local time and inverse local time for a Lévy process for which 0 is irregular for $[0, \infty)$. The *upper graph* plots the paths of L and the lower graph symbolically plots the path of X in terms of the excursions from the maximum.

Exercise 6.3, the process L^{-1} is now equal in law to a compound Poisson subordinator killed independently after an exponentially distributed time with parameter $\lambda(1-\rho)$. (Again, we work with the notion that an exponential distribution with parameter 0 is infinite with probability one.) It follows by construction that H is also a compound Poisson subordinator killed at the same rate.

Next, we prove (i) and (ii) for the case that 0 is regular for $[0, \infty)$, so that the version of local time we work with has continuous paths. From Lemma 6.8, we know that L_t^{-1} is a stopping time. Hence, according to Definition 6.1, on the event $\{L_t^{-1} < \infty\}$, or equivalently on the event $\{t < L_\infty\}$, the process $\widetilde{L} := \{\widetilde{L}_s : s \geq 0\}$, where

$$\widetilde{L}_s := L_{L_t^{-1}+s} - t, \quad s \geq 0,$$

is the local time at the maximum of $\widetilde{X} := \{\widetilde{X}_s : s \geq 0\}$, where

$$\widetilde{X}_s = X_{L_t^{-1}+s} - X_{L_t^{-1}}, \quad s \geq 0.$$

From Theorem 3.1 and Definition 6.1, we have that $\widetilde{X}$ and $\widetilde{L}$ are independent of $\mathcal{F}_{L_t^{-1}}$. It is clear that, on $\{t < L_\infty\}$,

$$\widetilde{L}_s^{-1} = L_{t+s}^{-1} - L_t^{-1} \tag{6.6}$$

and

$$\widetilde{H}_s := \widetilde{X}_{\widetilde{L}_s^{-1}} = X_{L_{t+s}^{-1}} - X_{L_t^{-1}} = H_{t+s} - H_t. \tag{6.7}$$

In conclusion, we have established that, on $t < L_\infty$,

$$\left\{\left(L_{t+s}^{-1} - L_t^{-1}, H_{t+s} - H_t\right) : s \geq 0\right\}$$

is independent of $\mathcal{F}_{L_t^{-1}}$ and equal in law to (L^{-1}, H). With this in hand, note that, for any $\alpha, \beta \geq 0$,

$$\begin{aligned}
&\mathbb{E}\left(\mathrm{e}^{-\alpha L_{t+s}^{-1} - \beta H_{t+s}} \mathbf{1}_{(t+s<L_\infty)}\right) \\
&\quad = \mathbb{E}\left(\mathrm{e}^{-\alpha L_t^{-1} - \beta H_t} \mathbf{1}_{(t<L_\infty)} \mathbb{E}\left(\mathrm{e}^{-\alpha \widetilde{L}_s^{-1} - \beta \widetilde{H}_s} \mathbf{1}_{(s<\widetilde{L}_\infty)} | \mathcal{F}_{L_t^{-1}}\right)\right) \\
&\quad = \mathbb{E}\left(\mathrm{e}^{-\alpha L_t^{-1} - \beta H_t} \mathbf{1}_{(t<L_\infty)}\right) \mathbb{E}\left(\mathrm{e}^{-\alpha L_s^{-1} - \beta H_s} \mathbf{1}_{(s<L_\infty)}\right).
\end{aligned}$$

As the expectation on the left-hand side above is also right-continuous in t (on account of the same being true of L^{-1} and H), a standard argument shows that this multiplicative decomposition implies that

$$\mathbb{E}\left(\mathrm{e}^{-\alpha L_t^{-1} - \beta H_t} \mathbf{1}_{(t<L_\infty)}\right) = \mathrm{e}^{-\kappa(\alpha,\beta)t}, \quad t \geq 0, \tag{6.8}$$

where $\kappa(\alpha, \beta) = -\log \mathbb{E}(\mathrm{e}^{-\alpha L_1^{-1} - \beta H_1} \mathbf{1}_{(1<L_\infty)}) \geq 0$. In particular, we see that L_∞ must follow an exponential distribution with parameter $\kappa(0, 0)$ if $\kappa(0, 0) > 0$, and $\mathbb{P}(L_\infty = \infty) = 1$ otherwise. For each α, β, write

$$\kappa(\alpha, \beta) = \kappa(0, 0) + \phi(\alpha, \beta). \tag{6.9}$$

Formula (6.8) shows that, for all $t \geq 0$,

$$\begin{aligned}
\mathrm{e}^{-\phi(\alpha,\beta)} &= \mathbb{E}\left(\mathrm{e}^{-\alpha L_1^{-1} - \beta H_1} | 1 < L_\infty\right) \\
&= \left\{\mathbb{E}\left(\mathrm{e}^{-\alpha L_t^{-1} - \beta H_t} | t < L_\infty\right)\right\}^{1/t},
\end{aligned} \tag{6.10}$$

thus illustrating that $\phi(\alpha, \beta)$ is the Laplace exponent of the bivariate, infinitely divisible distribution

$$\eta(\mathrm{d}x, \mathrm{d}y) = \mathbb{P}\left(L_1^{-1} \in \mathrm{d}x, H_1 \in \mathrm{d}y | 1 < L_\infty\right), \quad x, y > 0.$$

(Consider (6.10) for $t = 1/n$ where n is a positive integer.) In the spirit of the Lévy–Itô decomposition, there exists a bivariate subordinator, say $(\mathcal{L}^{-1}, \mathcal{H})$, whose Laplace exponent is $\phi(\alpha, \beta)$. We now see from (6.8) and (6.9) that (L^{-1}, H) is equal in law to $(\mathcal{L}^{-1}, \mathcal{H})$ killed independently (with "cemetery" state (∞, ∞)) after an exponentially distributed time with parameter $q = \kappa(0, 0)$. In particular, $\overline{X}_\infty = \infty$ almost surely if and only if $q = 0$, and otherwise $\overline{X}_\infty$ is equal in distribution to $\mathcal{H}_{\mathbf{e}_q}$ which itself is almost surely finite. □

Corollary 6.10 *In the previous theorem, the subordinator associated with L^{-1} has drift* $\mathtt{a}$, *where* $\mathtt{a}$ *is the constant appearing in Theorem* 6.7.

Proof Let $\Delta L_t^{-1} = L_t^{-1} - L_{t-}^{-1}, t \geq 0$. Note that, for any $\varepsilon > 0$, $\Delta L_t^{-1} > \varepsilon$ whenever the path of X moves away from its maximum for a period of real time exceeding ε. That is to say, individual jumps of L^{-1} correspond to individual *excursions lengths* of X from $\overline{X}$. Let us denote by $N_{L^{-1}}$ the Poisson random measure associated with the jumps of L^{-1}. Then the time it takes to accumulate $t < L_\infty$ units of local time is the sum of the periods of time that X has spent away from its maximum plus the real time that X has spent at its maximum (if any). The last qualification is only of significance when X is of bounded variation with 0 irregular for $(-\infty, 0)$, in which case the constant $\mathtt{a}$ in Theorem 6.7 is strictly positive; then the local time is taken as the Lebesgue measure of the time spent at the maximum. We have, on $\{t < L_\infty\}$,

$$L_t^{-1} = \int_0^{L_t^{-1}} \mathbf{1}_{(\overline{X}_s = X_s)} \mathrm{d}s + \int_{[0,t]} \int_{(0,\infty)} x N_{L^{-1}}(\mathrm{d}s \times \mathrm{d}x).$$

From Theorem 6.7, we know that the integral is equal to $\mathtt{a} L_{L_t^{-1}} = \mathtt{a}t$ and hence $\mathtt{a}$ is the drift of the subordinator L^{-1}. □

Finally, we look at compound Poisson processes. For some processes in this class, with positive probability, the same maximum may be visited over two intervals of time separated by at least one excursion. Hence, it is possible that $\Delta H_t = 0$ when $\Delta L_t^{-1} > 0$; in other words, the jump measure of H may have an atom at zero. This would be the case for the earlier given example of a compound Poisson process with jumps in $\{-1, 1\}$. Strictly speaking this violates our definition of a subordinator. However, this does not present a serious problem since H is necessarily a compound Poisson subordinator and, hence, its paths are well defined with the presence of this atom. Further, this does not affect the forthcoming analysis, unless otherwise mentioned.

Example 6.11 (Spectrally negative processes) Suppose that X is a spectrally negative Lévy process with Laplace exponent ψ having right inverse Φ; see Sect. 3.3 for a reminder of what this means. As noted earlier, we may work with local time given by $L = \overline{X}$. It follows that L_x^{-1} is nothing more than the first-passage time above $x > 0$. (Note that, in general, it is *not* true that L_x^{-1} is the first-passage time above x.) As X is spectrally negative, we have, in particular, that

$H_x = X_{L_x^{-1}} = x$ on $\{x < L_\infty\}$. Recalling Corollary 3.14, we already know that L^{-1} is a subordinator killed at rate $\Phi(0)$. Hence, we may easily identify, for $\alpha, \beta \geq 0$, $\kappa(\alpha, \beta) = \Phi(0) + \phi(\alpha, \beta)$, where

$$\phi(\alpha, \beta) = [\Phi(\alpha) - \Phi(0)] + \beta$$

is the Laplace exponent of a bivariate subordinator. Note in particular that $L_\infty < \infty$ if and only if $\Phi(0) > 0$ if and only if $\psi'(0+) = \mathbb{E}(X_1) \in [-\infty, 0)$. Moreover, on account of the fact that L^{-1} is the first-passage process, this occurs if and only if $\mathbb{P}(\limsup_{t\uparrow\infty} X_t < \infty) = 1$.

In the special case that X is a Brownian motion with drift ρ, we know explicitly that $\psi(\theta) = \rho\theta + \frac{1}{2}\theta^2$, $\theta \geq 0$, and hence $\Phi(\alpha) = -\rho + \sqrt{\rho^2 + 2\alpha}$, $\alpha \geq 0$. Inverse local time can then be identified precisely as an inverse Gaussian process (killed at rate $2|\rho|$ if $\rho < 0$).

We close this section by making the important remark that the brief introduction to excursion theory offered here has not paid fair dues to its general setting. Foundational work on excursion theory can be found in Itô (1970) and Maisonneuve (1975). This theory can be applied to a much more general class of Markov processes than just Lévy processes. Recall that $\overline{X} - X$ is a Markov process and hence one may consider L as the local time at 0 of this process. In general, it is possible to identify excursions of well-defined Markov processes from individual points in their state space with the help of local time. The reader interested in a comprehensive account should refer to the detailed but nonetheless approachable account given in Chap. IV of Bertoin (1996a) or Blumenthal (1992).

6.3 Excursions

In Sect. 4.6, we gave an explanation of the Pollaczek–Khintchine formula by decomposing the dual of the path of the Lévy processes considered there in terms of excursions from the maximum. Clearly, this decomposition relied heavily on the fact that the number of new maxima over any finite time horizon is finite. That is to say, 0 is irregular for $[0, \infty)$ and the local time at the maximum is a step function, as in case 1 listed at the end of Sect. 6.1. Now that we have established the concept of local time at the maximum for any Lévy process, we can give the general decomposition of the path of a Lévy process in terms of its excursions from the maximum.

Definition 6.12 For each moment of local time $t > 0$, we define

$$\epsilon_t = \begin{cases} \{X_{L_{t-}^{-1}+s} - X_{L_{t-}^{-1}} : 0 < s \leq L_t^{-1} - L_{t-}^{-1}\} & \text{if } L_{t-}^{-1} < L_t^{-1} \\ \partial & \text{if } L_{t-}^{-1} = L_t^{-1}, \end{cases}$$

where we take $L_{0-}^{-1} = 0$ and ∂ is some "dummy" state. Note that, for each fixed $t > 0$, when $L_{t-}^{-1} < L_t^{-1}$, the object ϵ_t is a stochastic process and hence is double indexed with $\epsilon_t(s) = X_{L_{t-}^{-1}+s} - X_{L_{t-}^{-1}}$ for $0 < s \leq L_t^{-1} - L_{t-}^{-1}$. When $\epsilon_t \neq \partial$, we refer to it as the excursion (from the maximum) associated with local time t.

Note also that, for t such that $\epsilon_t \neq \partial$, ϵ_t has paths that are right-continuous with left limits and, with the exception of its terminal value (in the case that $L_t^{-1} < \infty$), is valued in $(-\infty, 0)$.

Definition 6.13 Let $\mathcal{E}$ be the space of excursions of X from its running supremum, that is, the space of mappings which are right-continuous with left limits satisfying

$$\begin{aligned} &\epsilon : (0, \zeta) \to (-\infty, 0) \quad \text{for some } \zeta \in (0, \infty] \\ &\epsilon : \{\zeta\} \to [0, \infty) \qquad\quad \text{if } \zeta < \infty, \end{aligned}$$

where $\zeta = \zeta(\epsilon)$ is the excursion length. Write $h = h(\epsilon)$ for the terminal value of the excursion, so that $h(\epsilon) = \epsilon(\zeta)$. Finally, let $\overline{\epsilon} = -\inf_{s \in (0,\zeta)} \epsilon(s)$ for the excursion height.

We will shortly state the fundamental result of excursion theory, which relates the process $\{(t, \epsilon_t) : t \leq L_\infty \text{ and } \epsilon_t \neq \partial\}$ to a Poisson point process on $[0, \infty) \times \mathcal{E}$. This process has not yet been discussed in this text and so we devote a little time to its definition first. Recall that, in Chap. 2, the existence of a Poisson random measure on an arbitrary sigma-finite measure space $(S, \mathcal{S}, \eta)$ was proved in Theorem 2.4. If we reconsider the proof of Theorem 2.4, what was in fact shown was the existence in S of a random set of points, each of which is assigned a unit mass, thereby defining the Poisson random measure N. It is the supporting random set of points that we call a *Poisson point process* on $(S, \mathcal{S}, \eta)$ (or sometimes the Poisson point process on S with intensity η). In the case that $S = [0, \infty) \times \mathcal{E}$, we may think of the associated Poisson point process as a process of $\mathcal{E}$-valued points appearing in time.

Theorem 6.14 *There exists a sigma-algebra Σ and σ-finite measure n such that $(\mathcal{E}, \Sigma, n)$ is a measure space and Σ is rich enough to contain sets of the form*

$$\{\epsilon \in \mathcal{E} : \zeta(\epsilon) \in A,\ \overline{\epsilon} \in B,\ h(\epsilon) \in C\},$$

where, for a given $\epsilon \in \mathcal{E}$, $\zeta(\epsilon)$, $\overline{\epsilon}$ and $h(\epsilon)$ were all given in Definition 6.13, *and A, B and C are Borel sets in* $[0, \infty]$.

(i) *If* $\mathbb{P}(\limsup_{t\uparrow\infty} X_t = \infty) = 1$, *then* $\{(t, \epsilon_t) : t \geq 0$ *and* $\epsilon_t \neq \partial\}$ *is a Poisson point process on* $([0, \infty) \times \mathcal{E}, \mathcal{B}[0, \infty) \times \Sigma, \mathrm{d}t \times \mathrm{d}n)$.
(ii) *If* $\mathbb{P}(\limsup_{t\uparrow\infty} X_t < \infty) = 1$, *then* $\{(t, \epsilon_t) : t \leq L_\infty$ *and* $\epsilon_t \neq \partial\}$ *is a Poisson point process on* $([0, \infty) \times \mathcal{E}, \mathcal{B}[0, \infty) \times \Sigma, \mathrm{d}t \times \mathrm{d}n)$ *stopped at the first arrival of an excursion in* $\mathcal{E}_\infty := \{\epsilon \in \mathcal{E} : \zeta(\epsilon) = \infty\}$.

We offer no proof for this result as it goes beyond the scope of this book. We refer instead to Chap. VI in Bertoin (1996a), where a rigorous treatment is given. However, the intuition behind this theorem lies with the observation that, for each $t > 0$, by Lemma 6.8, L_{t-}^{-1} is a stopping time and hence, by Theorem 3.1, the evolution of $X_{L_{t-}^{-1}+s} - X_{L_{t-}^{-1}}$ in the time interval $(L_{t-}^{-1}, L_t^{-1}]$ is independent of $\mathcal{F}_{L_{t-}^{-1}}$. As alluded to earlier, this means that the paths of X may be decomposed into the juxtaposition of independent excursions from the maximum. The case that the drift coefficient, $\mathtt{a}$, of L^{-1} is strictly positive is the case of a bounded variation Lévy process with 0 irregular for $(-\infty, 0)$. Hence L is a local time that is proportional to the Lebesgue measure of the time that $X = \overline{X}$. In this case, excursions from the maximum are interlaced by moments of real time where X can be described as *drifting at its maximum*. If there is a last maximum, then the process of excursions is stopped at the first arrival of an excursion with infinite length; i.e. stopped at the first arrival of an excursion in $\mathcal{E}_\infty$.

Theorem 6.14 generalises the statement of Theorem 6.9. To see why, suppose that we write

$$\Lambda(\mathrm{d}x, \mathrm{d}y) = n\big(\zeta(\epsilon) \in \mathrm{d}x, h(\epsilon) \in \mathrm{d}y\big), \quad x, y > 0. \tag{6.11}$$

On $\{t < L_\infty\}$, the jumps of the ladder process (L^{-1}, H) form a Poisson point process on $[0, \infty) \times (0, \infty)^2$, with intensity measure $\mathrm{d}t \times \Lambda(\mathrm{d}x, \mathrm{d}y)$. We can write L_t^{-1} as the sum of the Lebesgue measure of the time X spends drifting at the maximum (if at all) together with the jumps L^{-1} makes (due to excursions from the maximum). Hence, if N is the counting measure associated with the Poisson point process of excursions, then, on $\{L_\infty > t\}$,

$$\begin{aligned} L_t^{-1} &= \int_0^{L_t^{-1}} \mathbf{1}_{(\epsilon_s = \partial)} \mathrm{d}s + \int_{[0,t]} \int_{\mathcal{E}} \zeta(\epsilon) N(\mathrm{d}s \times \mathrm{d}\epsilon) \\ &= \int_0^{L_t^{-1}} \mathbf{1}_{(\overline{X}_s = X_s)} \mathrm{d}s + \int_{[0,t]} \int_{\mathcal{E}} \zeta(\epsilon) N(\mathrm{d}s \times \mathrm{d}\epsilon) \\ &= \mathtt{a}t + \int_{[0,t]} \int_{(0,\infty)} x N_{L^{-1}}(\mathrm{d}s \times \mathrm{d}x). \end{aligned} \tag{6.12}$$

We can also write the ladder height process, H, in terms of a drift, say $\mathtt{b} \geq 0$, and its jumps, which are given by the terminal values of excursions. Hence, on $\{t < L_\infty\}$,

$$H_t = \mathtt{b}t + \int_{[0,t]} \int_{\mathcal{E}} h(\epsilon) N(\mathrm{d}s \times \mathrm{d}\epsilon). \tag{6.13}$$

Also, we can see that $\mathbb{P}(L_\infty > t)$ is the probability that, in the process of excursions, the first arrival in $\mathcal{E}_\infty$ is after time t. Written in terms of the Poisson point process of excursions, we see that

$$\mathbb{P}(L_\infty > t) = \mathbb{P}\big(N([0, t] \times \mathcal{E}_\infty) = 0\big) = \mathrm{e}^{-n(\mathcal{E}_\infty)t}, \quad t \geq 0.$$

This reinforces the earlier conclusion that L_∞ is exponentially distributed and we equate the parameters

$$\kappa(0,0)=n(\mathcal{E}_\infty). \tag{6.14}$$

6.4 The Wiener–Hopf Factorisation

A fundamental aspect of the theory of Lévy processes is a set of conclusions which, in modern times, are loosely referred to as *the Wiener–Hopf factorisation*. Historically, the identities around which the Wiener–Hopf factorisation is centred are the culmination of a number of works, initiated from within the theory of random walks. These include Baxter (1958), Spitzer (1956, 1957, 1960a, 1960b, 1964), Port (1963), Feller (1971), Borovkov (1976), Percheskii and Rogozin (1969), Gusak and Korolyuk (1969), Greenwood and Pitman (1980b), Fristedt (1974) and many others. The analytical roots of the so-called Wiener–Hopf method go much further back than these probabilistic references (see Sect. 6.7). The importance of the Wiener–Hopf factorisation is that it gives us information concerning the characteristics of the ascending and descending ladder processes. As indicated earlier, we shall use this knowledge in later chapters to consider a number of applications, as well as to extract some generic results concerning coarse and fine path properties of Lévy process.

In this section, we treat the Wiener–Hopf factorisation following closely the presentation of Greenwood and Pitman (1980a, 1980b), which relies heavily on the decomposition of the path of a Lévy process in terms of excursions from the maximum. Examples of the Wiener–Hopf factorisation will be treated in Sect. 6.5.

We begin by recalling that, for $\alpha,\beta\geq 0$, the Laplace exponents $\kappa(\alpha,\beta)$ and $\widehat{\kappa}(\alpha,\beta)$ are defined, respectively, by,

$$\mathbb{E}\big(\mathrm{e}^{-\alpha L_1^{-1}-\beta H_1}\mathbf{1}_{(1<L_\infty)}\big)=\mathrm{e}^{-\kappa(\alpha,\beta)}\quad\text{and}\quad\mathbb{E}\big(\mathrm{e}^{-\alpha\widehat{L}_1^{-1}-\beta\widehat{H}_1}\mathbf{1}_{(1<\widehat{L}_\infty)}\big)=\mathrm{e}^{-\widehat{\kappa}(\alpha,\beta)}.$$

Further, on account of Theorems 6.9 and 6.14,

$$\kappa(\alpha,\beta)=q+\phi(\alpha,\beta),\quad \alpha,\beta\geq 0, \tag{6.15}$$

where ϕ is the Laplace exponent of a bivariate subordinator and $q=n(\mathcal{E}_\infty)\geq 0$. The exponent ϕ can be written in the form

$$\phi(\alpha,\beta)=\alpha\mathtt{a}+\beta\mathtt{b}+\int_{(0,\infty)^2}\big(1-\mathrm{e}^{-\alpha x-\beta y}\big)\Lambda(\mathrm{d}x,\mathrm{d}y), \tag{6.16}$$

where the constant $\mathtt{a}$ was identified in Corollary 6.7, $\mathtt{b}$ is some non-negative constant representing the drift of H and $\Lambda(\mathrm{d}x,\mathrm{d}y)$ is given in terms of the excursion measure n in (6.11). It is also important to remark that both $\kappa(\alpha,\beta)$ and $\widehat{\kappa}(\alpha,\beta)$ can be analytically extended in α and β to $\mathbb{C}^+:=\{z\in\mathbb{C}:\Re z\geq 0\}$.

The next theorem gives the collection of statements which are known as the Wiener–Hopf factorisation. We need to introduce some additional notation first. As in earlier chapters, we shall understand $\mathbf{e}_p$ to be an independent random variable which is exponentially distributed with mean $1/p$. Further, we define

$$\overline{G}_t = \sup\{s < t : \overline{X}_s = X_s\}$$

and

$$\underline{G}_t = \sup\{s < t : \underline{X}_s = X_s\}.$$

An important fact concerning the definition of $\overline{G}_t$, which is responsible for the first sentence in the statement of the Wiener–Hopf factorisation (Theorem 6.15 below), is the following: If X is not a compound Poisson process, then its maxima are obtained at unique times. To see this, first suppose that 0 is regular for $[0, \infty)$. Since we have excluded compound Poisson processes, then this implies that 0 is regular for $(0, \infty)$. In this case, for any stopping time T such that $\overline{X}_T = X_T$, it follows by the strong Markov property and regularity that $\overline{X}_{T+u} > \overline{X}_T$ for all $u > 0$. In particular, we may consider the stopping times L_t^{-1}, for $t \geq 0$, which run through all the times when X visits its maximum. If the aforementioned regularity fails, then since X is assumed not to be a compound Poisson process, 0 must be regular for $(-\infty, 0)$. Hence, the conclusions of the previous case apply to $-X$. However, over finite time horizons $-X$ has the same law as X time reversed. In particular, the path of X over any finite time horizon when time reversed has new maxima which are obtained at unique times. This implies that X itself cannot touch the same maximum at two different times.

As mentioned earlier, if X is a compound Poisson process with an appropriate jump distribution, then it is possible that X visits the same maxima at distinct ladder times.

Theorem 6.15 (The Wiener–Hopf factorisation) *Suppose that X is any Lévy process other than a compound Poisson process. As usual, denote by $\mathbf{e}_p$ an independent and exponentially distributed random variable with parameter $p > 0$.*

(i) *The pairs*

$$(\overline{G}_{\mathbf{e}_p}, \overline{X}_{\mathbf{e}_p}) \quad \textit{and} \quad (\mathbf{e}_p - \overline{G}_{\mathbf{e}_p}, \overline{X}_{\mathbf{e}_p} - X_{\mathbf{e}_p})$$

are independent and infinitely divisible, yielding the factorisation

$$\frac{p}{p - \mathrm{i}\vartheta + \Psi(\theta)} = \Psi_p^+(\vartheta, \theta) \cdot \Psi_p^-(\vartheta, \theta), \tag{6.17}$$

where $\theta, \vartheta \in \mathbb{R}$,

$$\Psi_p^+(\vartheta, \theta) = \mathbb{E}\big(\mathrm{e}^{\mathrm{i}\vartheta \overline{G}_{\mathbf{e}_p} + \mathrm{i}\theta \overline{X}_{\mathbf{e}_p}}\big) \quad \textit{and} \quad \Psi_p^-(\vartheta, \theta) = \mathbb{E}\big(\mathrm{e}^{\mathrm{i}\vartheta \underline{G}_{\mathbf{e}_p} + \mathrm{i}\theta \underline{X}_{\mathbf{e}_p}}\big).$$

Here, the pair $\Psi_p^+(\vartheta, \theta)$ and $\Psi_p^-(\vartheta, \theta)$ are called the Wiener–Hopf factors.

(ii) *Via analytical extension, the Wiener–Hopf factors may be identified from the Laplace transforms*

$$\mathbb{E}\big(e^{-\alpha\overline{G}_{\mathbf{e}_p}-\beta\overline{X}_{\mathbf{e}_p}}\big)=\frac{\kappa(p,0)}{\kappa(p+\alpha,\beta)}\quad\textit{and}\quad\mathbb{E}\big(e^{-\alpha\underline{G}_{\mathbf{e}_p}+\beta\underline{X}_{\mathbf{e}_p}}\big)=\frac{\widehat{\kappa}(p,0)}{\widehat{\kappa}(p+\alpha,\beta)}\tag{6.18}$$

for $\alpha,\beta\in\mathbb{C}^+$.

(iii) *The Laplace exponents* $\kappa(\alpha,\beta)$ *and* $\widehat{\kappa}(\alpha,\beta)$ *may also be identified in terms of the law of X by:*

$$\kappa(\alpha,\beta)=k\exp\bigg(\int_0^\infty\int_{(0,\infty)}\big(e^{-t}-e^{-\alpha t-\beta x}\big)\frac{1}{t}\mathbb{P}(X_t\in dx)dt\bigg)\tag{6.19}$$

and

$$\widehat{\kappa}(\alpha,\beta)=\widehat{k}\exp\bigg(\int_0^\infty\int_{(-\infty,0)}\big(e^{-t}-e^{-\alpha t+\beta x}\big)\frac{1}{t}\mathbb{P}(X_t\in dx)dt\bigg),\tag{6.20}$$

where $\alpha,\beta\geq 0$ *and* k *and* $\widehat{k}$ *are strictly positive constants.*

(iv) *By setting* $\vartheta=0$ *and taking limits as p tends to zero in* (6.17), *we obtain*

$$k\widehat{k}\Psi(\theta)=\kappa(0,-i\theta)\widehat{\kappa}(0,i\theta).\tag{6.21}$$

Let us now make some notes concerning this theorem. Firstly, there are a number of unidentified constants in the given expressions. To some extent, these constants are meaningless since they are dependent on the normalisation chosen in the definition of local time (cf. Definition 6.1). In this context, local time is nothing other than an artificial clock to measure the intrinsic time spent at the maximum. Naturally, a different choice of local time will induce a different inverse local time and, hence, a different ladder height process. Nonetheless the range of the bivariate ladder process will be the same as it will always correspond to the range of the real times and positions of the new maxima of the underlying Lévy process. In this respect, we may always normalise the choice of k and $\widehat{k}$ so that, for example $k\widehat{k}=1$.

Secondly, the exclusion of the compound Poisson processes from the statement of the theorem is not to say that a Wiener–Hopf factorisation for this class of Lévy processes does not exist. The case of the compound Poisson process is essentially the case of the random walk and has some subtle differences which we shall come back to later on.

The proof of Theorem 6.15 we shall give makes use of a simple fact about infinitely divisible distributions, as well as the fundamental properties of the Poisson point processes describing the excursions of X. We give these facts in the following two preparatory lemmas. For the first, it may be useful to recall Exercise 2.10.

Lemma 6.16 *Suppose that* $\mathbf{X}=\{\mathbf{X}_t:t\geq 0\}$ *is any d-dimensional Lévy process with characteristic exponent* $\Psi(\theta)=-\log\mathbb{E}(e^{i\theta\cdot\mathbf{X}_1})$, *for* $\theta\in\mathbb{R}^d$. *Then the pair*

$(\mathbf{e}_p, \mathbf{X}_{\mathbf{e}_p})$ has a $(d+1)$-dimensional infinitely divisible distribution with Lévy–Khintchine exponent given by

$$\mathbb{E}\left(\mathrm{e}^{\mathrm{i}\vartheta \mathbf{e}_p + \mathrm{i}\theta \mathbf{X}_{\mathbf{e}_p}}\right) = \exp\left\{-\int_0^\infty \int_{\mathbb{R}^d} \left(1 - \mathrm{e}^{\mathrm{i}\vartheta t + \mathrm{i}\theta\cdot x}\right)\frac{1}{t}\mathrm{e}^{-pt}\mathbb{P}(\mathbf{X}_t \in \mathrm{d}x)\mathrm{d}t\right\}$$
$$= \frac{p}{p - \mathrm{i}\vartheta + \Psi(\theta)},$$

for $\theta \in \mathbb{R}^d$ and $\vartheta \in \mathbb{R}$.

Proof Recall from the Lévy–Khintchine formula (cf. Exercise 2.10) that $\Re\Psi(\theta) = \theta\cdot\mathbf{A}\theta/2 + \int_{\mathbb{R}^d}(1-\cos\theta\cdot x)\Pi(\mathrm{d}x)$, where $\mathbf{A}$ is a $d\times d$ Gaussian correlation matrix, Π is the Lévy measure on $\mathbb{R}^d$ and $\theta \in \mathbb{R}^d$. Hence $\Re\Psi(\theta) \geq 0$ for all $\theta \in \mathbb{R}^d$ and we have

$$\mathbb{E}\left(\mathrm{e}^{\mathrm{i}\vartheta \mathbf{e}_p + \mathrm{i}\theta \mathbf{X}_{\mathbf{e}_p}}\right) = \int_0^\infty p\mathrm{e}^{-pt + \mathrm{i}\vartheta t - \Psi(\theta)t}\mathrm{d}t = \frac{p}{p - \mathrm{i}\vartheta + \Psi(\theta)},$$

for all $\theta \in \mathbb{R}^d$ and $\vartheta \in \mathbb{R}$. On the other hand, using the Frullani integral in Lemma 1.7, we see that

$$\exp\left\{-\int_0^\infty \int_{\mathbb{R}^d} \left(1 - \mathrm{e}^{\mathrm{i}\vartheta t + \mathrm{i}\theta\cdot x}\right)\frac{1}{t}\mathrm{e}^{-pt}\mathbb{P}(\mathbf{X}_t \in \mathrm{d}x)\mathrm{d}t\right\}$$
$$= \exp\left\{-\int_0^\infty \left(1 - \mathrm{e}^{-(\Psi(\theta) - \mathrm{i}\vartheta)t}\right)\frac{1}{t}\mathrm{e}^{-pt}\mathrm{d}t\right\}$$
$$= \frac{p}{p - \mathrm{i}\vartheta + \Psi(\theta)},$$

for $\theta \in \mathbb{R}^d$ and $\vartheta \in \mathbb{R}$. The result now follows. □

Although the next result is stated for the Poisson point process of excursions, $\{(t, \epsilon_t) : t \geq 0 \text{ and } \epsilon_t \neq \partial\}$, the measure space $(\mathcal{E}, \Sigma, n)$ can be replaced by any σ-finite measure space.

Lemma 6.17 *Suppose that $\{(t, \epsilon_t)\}$ is a Poisson point process on $([0,\infty) \times \mathcal{E}, \mathcal{B}[0,\infty) \times \Sigma, \mathrm{d}t \times \mathrm{d}n)$. Choose $A \in \Sigma$ such that $n(A) < \infty$ and define*

$$\sigma^A = \inf\{t > 0 : \epsilon_t \in A\}.$$

(i) *The random time σ^A is exponentially distributed with parameter $n(A)$.*
(ii) *The process $\{(t, \epsilon_t) : t < \sigma^A\}$ is equal in law to a Poisson point process on $[0,\infty) \times \mathcal{E}\backslash A$ with intensity $\mathrm{d}t \times \mathrm{d}n'$, where $n'(\mathrm{d}\epsilon) = n(\mathrm{d}\epsilon \cap \mathcal{E}\backslash A)$, which is stopped at an independent exponential time with parameter $n(A)$.*
(iii) *The process $\{(t, \epsilon_t) : t < \sigma^A\}$ is independent of ϵ_{σ^A}.*

Proof Let $S_1 = [0,\infty) \times A$ and $S_2 = [0,\infty) \times (\mathcal{E}\backslash A)$. Suppose that N is the Poisson random measure associated with the given Poisson point process. All three conclusions follow from Corollary 2.5 applied to the restriction of N to the disjoint sets S_1 and S_2, say $N^{(1)}$ and $N^{(2)}$, respectively.

Specifically, for (i), note that $\mathbb{P}(\sigma^A > t) = \mathbb{P}(N^{(1)}([0,t] \times A) = 0) = \mathrm{e}^{-n(A)t}$ as $N^{(1)}$ has intensity $\mathrm{d}t \times n(\mathrm{d}\epsilon \cap A)$. For (ii) and (iii), it suffices to note that $N^{(2)}$ has intensity $\mathrm{d}t \times n(\mathrm{d}\epsilon \cap \mathcal{E}\backslash A)$, that

$$\{(t,\epsilon_t) \in [0,\infty) \times \mathcal{E} : t < \sigma^A\} = \{(t,\epsilon_t) \in [0,\infty) \times (\mathcal{E}\backslash A) : t < \sigma^A\}$$

and that the first arrival in A is a point belonging to the process $N^{(1)}$, which is independent of $N^{(2)}$. □

Since $\{\sigma^A \le t\} = \{N([0,t] \times A) \ge 1\}$, it is easily seen that σ^A is a stopping time with respect to the filtration $\mathbb{G} = \{\mathcal{G}_t : t \ge 0\}$, where

$$\mathcal{G}_t = \sigma\big(N(U \times V) : U \in \mathcal{B}[0,t] \text{ and } V \in \Sigma\big).$$

In the case that $\mathcal{E}$ is the space of excursions, one may take $\mathbb{G} = \mathbb{F}$.

Now we are ready to give the proof of the Wiener–Hopf factorisation.

Proof of Theorem 6.15 (i) The crux of the first part of the Wiener–Hopf factorisation lies with the following important observation. Consider the Poisson point process of marked excursions on

$$\big([0,\infty) \times \mathcal{E} \times [0,\infty), \mathcal{B}[0,\infty) \times \Sigma \times \mathcal{B}[0,\infty), \mathrm{d}t \times \mathrm{d}n \times \mathrm{d}\eta\big),$$

where $\eta(\mathrm{d}x) = p\mathrm{e}^{-px}\mathrm{d}x$ for $x \ge 0$. That is to say, consider a Poisson point process whose points are described by $\{(t, \epsilon_t, \mathbf{e}_p^{(t)}) : t \le L_\infty \text{ and } \epsilon_t \ne \partial\}$, where $\mathbf{e}_p^{(t)}$ is an independent copy of an exponentially distributed random variable if t is such that $\epsilon_t \ne \partial$, and otherwise, $\mathbf{e}_p^{(t)} := \partial$. The Poisson point process of unmarked excursions is then obtained as a projection on to $[0,\infty) \times \mathcal{E}$. Sampling the Lévy process X up to an independent exponentially distributed random time $\mathbf{e}_p$ corresponds to sampling the Poisson process of excursions up to time $L_{\mathbf{e}_p}$; that is, $\{(t,\epsilon_t) : t \le L_{\mathbf{e}_p} \text{ and } \epsilon_t \ne \partial\}$. In turn, we claim that this process is equal in law to the projection on to $[0,\infty) \times \mathcal{E}$ of

$$\big\{\big(t, \epsilon_t, \mathbf{e}_p^{(t)}\big) : t \le \sigma_1 \wedge \sigma_2 \text{ and } \epsilon_t \ne \partial\big\}, \tag{6.22}$$

where

$$\sigma_1 := \inf\left\{t > 0 : \int_0^{L_t^{-1}} \mathbf{1}_{(\overline{X}_s = X_s)} \mathrm{d}s > \mathbf{e}_p\right\}$$

(with the usual understanding that $\inf \emptyset = \infty$) and

$$\sigma_2 := \inf\big\{t > 0 : \zeta(\epsilon_t) > \mathbf{e}_p^{(t)}\big\}.$$

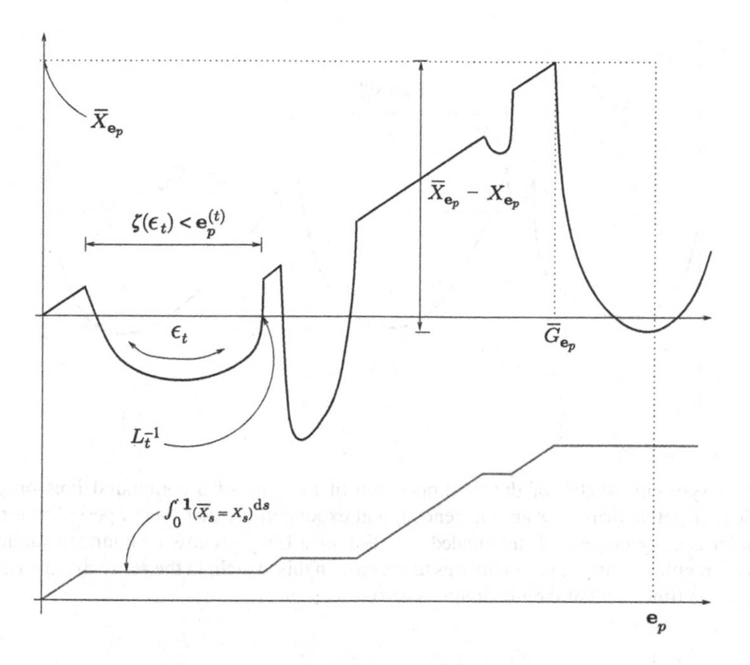

Fig. 6.2 A symbolic sketch of the decomposition of the path of a compound Poisson process with strictly positive drift over an independent and exponentially distributed period of time. The situation for bounded variation Lévy processes for which 0 is irregular for $(-\infty, 0)$ is analogous to the case in this sketch, in the sense that the Lebesgue measure of the time spent at the maximum over any finite time horizon is strictly positive.

(Recall that $\zeta(\epsilon_t)$ is the duration of the excursion indexed by local time t.) A formal proof of this claim would require the use of some additional mathematical tools. However, for the sake of brevity, we shall lean instead on an intuitive explanation.

We recall that the path of the Lévy process up to time $\mathbf{e}_p$ is the independent juxtaposition of excursions interlaced with moments of real time when $X = \overline{X}$ (which accumulate positive Lebesgue measure when $\mathrm{a} > 0$). The event $\{t < L_{\mathbf{e}_p}\}$ corresponds to the event that there are at least t units of local time for $\mathbf{e}_p$ units of real time. By the lack-of-memory property, this is equivalent to the intersection of two events. The first is that the total amount of real time spent at the maximum, when the local time at the maximum is equal to t, has survived independent exponential killing at rate p. The second is that, when the local time at the maximum is equal to t, each of the excursion lengths have survived independent exponential killing at rate p. This idea is easier to visualise when one considers the case that X is a compound Poisson process with strictly positive or strictly negative drift; see Figs. 6.2 and 6.3.

The times σ_1 and σ_2 are independent and, further, σ_2 is of the type of stopping time considered in Lemma 6.17, with $A = \{\zeta(\epsilon) > \mathbf{e}_p\}$, for the Poisson point process (6.22). From each of the three statements given in Lemma 6.17, we respectively deduce three facts concerning the Poisson point process (6.22).

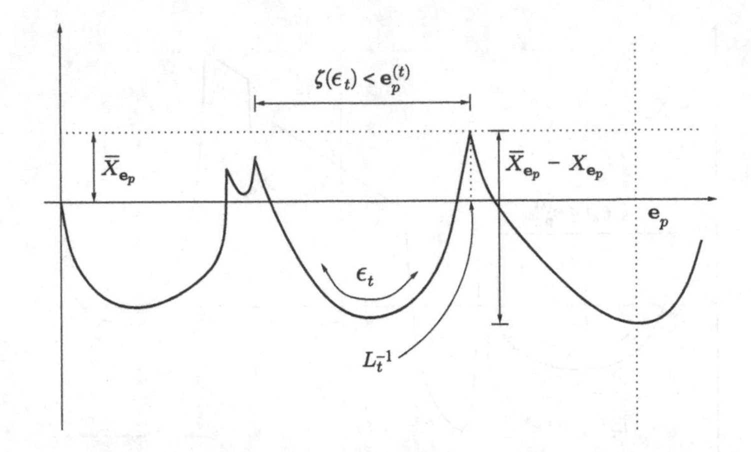

Fig. 6.3 A symbolic sketch of the decomposition of the path of a compound Poisson process with strictly negative drift over an independent and exponentially distributed period of time. The situation for a Lévy process of unbounded variation or a Lévy process of bounded variation for which 0 is irregular for $[0,\infty)$ is analogous to the case in this sketch, in the sense that the Lebesgue measure of the time spent at the maximum is zero.

(1) Since $\int_0^{L_t^{-1}} \mathbf{1}_{(\overline{X}_s = X_s)} \mathrm{d}s = \mathrm{a}t$, we have

$$\mathbb{P}(\sigma_1 > t) = \mathbb{P}\left(\int_0^{L_t^{-1}} \mathbf{1}_{(\overline{X}_s = X_s)} \mathrm{d}s < \mathbf{e}_p\right) = \mathrm{e}^{-\mathrm{a}pt}, \quad t \geq 0.$$

As mentioned earlier, if the constant $\mathrm{a} = 0$, then we have that $\sigma_1 = \infty$. Further, with the help of Lemma 6.17 (i), we also have that

$$\begin{aligned}
&\mathbb{P}(\sigma_2 > t)\\
&\quad = \exp\left\{-t \int_0^\infty p\mathrm{e}^{-px} \mathrm{d}x \cdot n\big(\zeta(\epsilon) > x\big)\right\}\\
&\quad = \exp\left\{-t \int_0^\infty p\mathrm{e}^{-px} \mathrm{d}x \cdot \big[n\big(\infty > \zeta(\epsilon) > x\big) + n\big(\zeta(\epsilon) = \infty\big)\big]\right\}\\
&\quad = \exp\left\{-n(\mathcal{E}_\infty)t - t \int_{(0,\infty)} \big(1 - \mathrm{e}^{-px}\big) n\big(\zeta(\epsilon) \in \mathrm{d}x\big)\right\},
\end{aligned}$$

where we recall that $\mathcal{E}_\infty = \{\epsilon \in \mathcal{E} : \zeta(\epsilon) = \infty\}$. As σ_1 and σ_2 are independent and exponentially distributed, it follows[1] that

$$\mathbb{P}(\sigma_1 \wedge \sigma_2 > t) = \exp\left\{-t\left(n(\mathcal{E}_\infty) + \mathrm{a}p + \int_{(0,\infty)} \big(1 - \mathrm{e}^{-px}\big) n\big(\zeta(\epsilon) \in \mathrm{d}x\big)\right)\right\}.$$

[1] Recall that the minimum of two independent exponential random variables is again exponentially distributed with the sum of their rates.

However, recall from (6.11) and (6.14) that $\kappa(0,0) = n(\mathcal{E}_\infty)$ and $\Lambda(\mathrm{d}x, [0,\infty)) = n(\zeta(\epsilon) \in \mathrm{d}x)$, and hence the exponent above is equal to $\kappa(p,0)$, where κ is given by (6.15) and (6.16).

(2) From Lemma 6.17 (ii) and the observation (1) above, we see that the Poisson point process (6.22) is equal in law to a Poisson point process on $[0,\infty) \times \mathcal{E} \times [0,\infty)$ with intensity

$$\mathrm{d}t \times n\big(\mathrm{d}\epsilon; \zeta(\epsilon) < x\big) \times \eta(\mathrm{d}x), \tag{6.23}$$

which is stopped at an independent time which is exponentially distributed with parameter $\kappa(p,0)$.

(3) Lemma 6.17 (iii) tells us that, on the event $\{\sigma_2 < \sigma_1\}$, the process

$$\big\{\big(t, \epsilon_t, \mathbf{e}_p^{(t)}\big) : t < \sigma_1 \wedge \sigma_2 \text{ and } \epsilon_t \neq \partial\big\} \tag{6.24}$$

is independent of $\epsilon_{\sigma_2} = \epsilon_{\sigma_1\wedge\sigma_2}$. On the other hand, when $\sigma_1 < \sigma_2$, since $\partial = \epsilon_{\sigma_1} = \epsilon_{\sigma_1\wedge\sigma_2}$, we conclude that $\epsilon_{\sigma_1\wedge\sigma_2}$, and indeed $\mathbf{e}_p^{(\sigma_1\wedge\sigma_2)}$, are independent of (6.24).

Now note, with the help of (6.12) and (6.13), that

$$(\overline{G}_{\mathbf{e}_p}, \overline{X}_{\mathbf{e}_p}) \overset{d}{=} \big(L^{-1}_{(\sigma_1\wedge\sigma_2)-}, H_{(\sigma_1\wedge\sigma_2)-}\big), \tag{6.25}$$

where

$$L^{-1}_{(\sigma_1\wedge\sigma_2)-} = \mathrm{a}(\sigma_1\wedge\sigma_2) + \int_{[0,\sigma_1\wedge\sigma_2)}\int_{\mathcal{E}} \zeta(\epsilon_t) N(\mathrm{d}t \times \mathrm{d}\epsilon) \tag{6.26}$$

and

$$H_{(\sigma_1\wedge\sigma_2)-} = \mathrm{b}(\sigma_1\wedge\sigma_2) + \int_{[0,\sigma_1\wedge\sigma_2)}\int_{\mathcal{E}} h(\epsilon_t) N(\mathrm{d}t \times \mathrm{d}\epsilon). \tag{6.27}$$

From point (3) above, the right-hand sides of (6.26) and (6.27) are independent of the excursion $\epsilon_{\sigma_1\wedge\sigma_2}$. Moreover, simultaneously on the same probability spaces referred to in (6.25), the pair $(\mathbf{e}_p - \overline{G}_{\mathbf{e}_p}, X_{\mathbf{e}_p} - \overline{X}_{\mathbf{e}_p})$ is equal in law to

$$\big(\mathbf{e}_p^{(\sigma_1\wedge\sigma_2)}, \epsilon_{\sigma_1\wedge\sigma_2}\big(\mathbf{e}_p^{(\sigma_1\wedge\sigma_2)}\big)\big)\mathbf{1}_{(\sigma_2<\sigma_1)} + (0,0)\mathbf{1}_{(\sigma_1<\sigma_2)}.$$

See Figs. 6.2 and 6.3. In conclusion, $(\overline{G}_{\mathbf{e}_p}, \overline{X}_{\mathbf{e}_p})$ is independent of $(\mathbf{e}_p - \overline{G}_{\mathbf{e}_p}, \overline{X}_{\mathbf{e}_p} - X_{\mathbf{e}_p})$.

From point (2), the process $\{(L_t^{-1}, H_t) : t < \sigma_1 \wedge \sigma_2\}$ behaves like a subordinator with characteristic measure

$$\int_0^\infty p\mathrm{e}^{-pt}\mathrm{d}t \cdot n\big(\zeta(\epsilon) \in \mathrm{d}x, h(\epsilon) \in \mathrm{d}y, x < t\big) = \mathrm{e}^{-px}\Lambda(\mathrm{d}x, \mathrm{d}y)$$

and drift (a, b), which is stopped at an independent exponentially distributed time with parameter $\kappa(p,0)$. Suppose that we denote this subordinator $(\mathbb{L}^{-1}, \mathbb{H}) =$

$\{(\mathbb{L}_t^{-1}, \mathbb{H}_t) : t \geq 0\}$. Then

$$(\mathbb{L}_{\mathbf{e}_\chi}^{-1}, \mathbb{H}_{\mathbf{e}_\chi}) \stackrel{d}{=} (\overline{G}_{\mathbf{e}_p}, \overline{X}_{\mathbf{e}_p}),$$

where $\mathbf{e}_\chi$ is an independent exponential random variable with parameter $\chi = \kappa(p, 0)$. From Lemma 6.16, we also see that $(\overline{G}_{\mathbf{e}_p}, \overline{X}_{\mathbf{e}_p})$ is infinitely divisible. Now note that, by appealing to the Duality Lemma and the fact that maxima are attained at unique times (recall the discussion preceding the statement of Theorem 6.15), one sees that

$$(\mathbf{e}_p - \overline{G}_{\mathbf{e}_p}, \overline{X}_{\mathbf{e}_p} - X_{\mathbf{e}_p}) \stackrel{d}{=} (\underline{G}_{\mathbf{e}_p}, -\underline{X}_{\mathbf{e}_p}). \tag{6.28}$$

(This is also seen, for example, in Figs. 6.2 and 6.3 by rotating them about 180°.) For reasons similar to those given above, the pair $(\underline{G}_{\mathbf{e}_p}, -\underline{X}_{\mathbf{e}_p})$ must also be infinitely divisible. The factorisation (6.17) now follows. □

Proof of Theorem 6.15 (ii) From the proof of part (i), the bivariate subordinator $(\mathbb{L}^{-1}, \mathbb{H})$ has Laplace exponent equal to

$$\mathrm{a}\alpha + \mathrm{b}\beta + \int_{(0,\infty)^2} (1 - \mathrm{e}^{-\alpha x - \beta y})\mathrm{e}^{-px} \Lambda(\mathrm{d}x, \mathrm{d}y) = \kappa(\alpha + p, \beta) - \kappa(p, 0),$$

for $\alpha, \beta \geq 0$, where the equality follows from (6.15) and (6.16). Hence, from the second equality in the statement of Lemma 6.16,

$$\begin{aligned} \mathbb{E}\big(\mathrm{e}^{-\alpha \overline{G}_{\mathbf{e}_p} - \beta \overline{X}_{\mathbf{e}_p}}\big) &= \mathbb{E}\big(\mathrm{e}^{-\alpha \mathbb{L}_{\mathbf{e}_\chi}^{-1} - \beta \mathbb{H}_{\mathbf{e}_\chi}}\big) \\ &= \frac{\chi}{\kappa(\alpha + p, \beta) - \kappa(p, 0) + \chi} \\ &= \frac{\kappa(p, 0)}{\kappa(\alpha + p, \beta)}. \end{aligned} \tag{6.29}$$

Part (ii) follows from (6.29) by analytically extending the identity from $\alpha, \beta \geq 0$ to $\mathbb{C}^+$. □

Proof of Theorem 6.15 (iii) According to Lemma 6.16, the bivariate random variable $(\mathbf{e}_p, X_{\mathbf{e}_p})$ is infinitely divisible and has Lévy measure on $(0, \infty) \times \mathbb{R}$ given by

$$\pi(\mathrm{d}t, \mathrm{d}x) = \frac{1}{t}\mathrm{e}^{-pt}\mathbb{P}(X_t \in \mathrm{d}x)\mathrm{d}t.$$

Since, by part (i), we can write $(\mathbf{e}_p, X_{\mathbf{e}_p})$ as the independent sum

$$(\overline{G}_{\mathbf{e}_p}, \overline{X}_{\mathbf{e}_p}) + (\mathbf{e}_p - \overline{G}_{\mathbf{e}_p}, X_{\mathbf{e}_p} - \overline{X}_{\mathbf{e}_p}),$$

it follows that $\pi = \pi^+ + \pi^-$ where π^+ and π^- are the Lévy measures of $(\overline{G}_{\mathbf{e}_p}, \overline{X}_{\mathbf{e}_p})$, and $(\mathbf{e}_p - \overline{G}_{\mathbf{e}_p}, X_{\mathbf{e}_p} - \overline{X}_{\mathbf{e}_p})$, respectively. Further, the support of π^+

must be contained in $[0,\infty)\times[0,\infty)$ and the support of π^- must be contained in $[0,\infty)\times(-\infty,0]$, since these are the supports of the distributions of $(\overline{G}_{\mathbf{e}_p},\overline{X}_{\mathbf{e}_p})$ and $(\mathbf{e}_p-\overline{G}_{\mathbf{e}_p},X_{\mathbf{e}_p}-\overline{X}_{\mathbf{e}_p})$, respectively.

As X is not a compound Poisson process, we have that $\mathbb{P}(X_t=0)=0$ for Lebesgue almost all $t>0$.[2] We can now identify π^+ as the restriction of π to $[0,\infty)\times(0,\infty)$ and π^- as the restriction of π to $[0,\infty)\times(-\infty,0)$. Using the Lévy–Khintchine formula (2.29) for a bivariate pair of infinitely divisible random variables (cf. Exercise 2.10), we can identify the Wiener–Hopf factors in the form

$$\Psi_p^+(\vartheta,\theta)=\exp\left\{ik\vartheta+\mathrm{i}\mathrm{k}\theta+\int_0^\infty\int_{(0,\infty)}(\mathrm{e}^{\mathrm{i}\vartheta t+\mathrm{i}\theta x}-1)\frac{1}{t}\mathrm{e}^{-pt}\mathbb{P}(X_t\in\mathrm{d}x)\mathrm{d}t\right\}$$

and

$$\Psi_p^-(\vartheta,\theta)=\exp\left\{-ik\vartheta-\mathrm{i}\mathrm{k}\theta+\int_0^\infty\int_{(-\infty,0)}(\mathrm{e}^{\mathrm{i}\vartheta t+\mathrm{i}\theta x}-1)\frac{1}{t}\mathrm{e}^{-pt}\mathbb{P}(X_t\in\mathrm{d}x)\mathrm{d}t\right\},$$

for some constants $k\geq 0$ and $\mathrm{k}\geq 0$, where $\theta,\vartheta\geq 0$. The identification of Ψ^+ and Ψ^- should also take account of the fact that Ψ^+ extends analytically to the upper half of the complex plane in θ and Ψ^- extends to the lower half of the complex plane in θ. Since $\mathbf{e}_p$ can take arbitrarily small values, then so can $\overline{G}_{\mathbf{e}_p}$ and $\overline{X}_{\mathbf{e}_p}$. In that case, the Lévy–Khintchine exponent of $(\overline{G}_{\mathbf{e}_p},\overline{X}_{\mathbf{e}_p})$ should not contain the drift term $ik\vartheta+\mathrm{i}\mathrm{k}\theta$, i.e. $k=\mathrm{k}=0$ (otherwise the distributions of $\overline{G}_{\mathbf{e}_p}$ and $\overline{X}_{\mathbf{e}_p}$ would have supports bounded strictly away from the origin).

From (6.29), we can now identify $\kappa(\alpha,\beta)$ up to a constant and formula (6.19) follows. Similarly, we may identify the formula given for $\widehat{\kappa}(\alpha,\beta)$. □

Proof of Theorem 6.15 (iv) From the expressions established in part (iii) and Lemma 1.7 for the Frullani integral,

$$\begin{aligned}
&\kappa(p,0)\widehat{\kappa}(p,0)\\
&=k'\exp\left\{\int_0^\infty(\mathrm{e}^{-t}-\mathrm{e}^{-pt})\frac{1}{t}\mathrm{d}t\right\}\\
&=k'\exp\left\{\int_0^\infty(1-\mathrm{e}^{-pt})\mathrm{e}^{-t}\frac{1}{t}\mathrm{d}t-\int_0^\infty(1-\mathrm{e}^{-t})\mathrm{e}^{-pt}\frac{1}{t}\mathrm{d}t\right\}\\
&=k'p,
\end{aligned}\tag{6.30}$$

[2] This statement is intuitively appealing; however it requires a rigorous proof. We refrain from giving it here in order to avoid distraction from the proof at hand. The basic idea is to prove, in the spirit of Theorem 5.4, that, for each $q>0$, the potential measure $U^{(q)}(\mathrm{d}x):=\mathbb{E}(\int_0^\infty\mathrm{e}^{-qt}\mathbf{1}_{(X_t\in\mathrm{d}x)}\mathrm{d}t)$ has no atoms. See for example Proposition I.15 of Bertoin (1996a).

where $k' = k\widehat{k}$. Equation (6.17) now reads

$$\frac{1}{p - \mathrm{i}\vartheta + \Psi(\theta)} = \frac{k'}{\kappa(p - \mathrm{i}\vartheta, -\mathrm{i}\theta) \cdot \widehat{\kappa}(p - \mathrm{i}\vartheta, \mathrm{i}\theta)}.$$

Setting $\vartheta = p = 0$ delivers the required result. □

Let us mention the following corollary, merely as a curiosity in light of the discussion on *special subordinators* in Sect. 5.6. Its proof is an immediate consequence of (6.30).

Corollary 6.18 *The ascending inverse local time process L^{-1} and the descending inverse local time process $\widehat{L}^{-1}$ are conjugate special subordinators.*

We conclude this section with some remarks about the case that X is a compound Poisson process. In this case, most of the proof of Theorem 6.15 goes through as stated. However, the following subtleties need to be taken account of.

In the proof of the part (i) of Theorem 6.15, it is no longer true that (6.28) holds. One needs to be more careful concerning the definition of $\overline{G}_t$ and $\underline{G}_t$. For compound Poisson processes, it is necessary to work with the new definitions

$$\overline{G}_t = \sup\{s < t : X_s = \overline{X}_t\} \quad \text{and} \quad \underline{G}_t^* = \inf\{s < t : X_s = \underline{X}_t\}, \tag{6.31}$$

instead. It was shown in the case that X is not a compound Poisson process that maxima are obtained at distinct times. Hence, the above definitions are consistent with the original definitions of $\overline{G}_t$ and $\underline{G}_t$ outside the class of compound Poisson processes.

Appealing to duality, the statement (6.28) should now be replaced by

$$(\mathbf{e}_p - \overline{G}_{\mathbf{e}_p}, \overline{X}_{\mathbf{e}_p} - X_{\mathbf{e}_p}) \stackrel{d}{=} \big(\underline{G}^*_{\mathbf{e}_p}, -\underline{X}_{\mathbf{e}_p}\big), \tag{6.32}$$

and the factorisation (6.17) requires redefining so that

$$\Psi_p^-(\vartheta, \theta) = \mathbb{E}\big(\mathrm{e}^{\mathrm{i}\vartheta \underline{G}^*_{\mathbf{e}_p} + \mathrm{i}\theta \underline{X}_{\mathbf{e}_p}}\big),$$

for $\theta, \vartheta \in \mathbb{R}$. Further, in the proof of parts (ii) and (iii) of Theorem 6.15, an adjustment is required in the definitions of κ and $\widehat{\kappa}$. Recall that in the decomposition $\pi = \pi^+ + \pi^-$, the respective supports of π^+ and π^- are contained in

$[0,\infty)\times(0,\infty)$ and $[0,\infty)\times(-\infty,0)$. Unlike earlier, we are now faced with the difficulty of assigning the mass given by the probabilities $\mathbb{P}(X_t=0)$ for $t\geq 0$ to one or the other of the integrals that, respectively, define κ and $\widehat{\kappa}$. The way to do this is to first consider the process $X^\varepsilon=\{X^\varepsilon_t : t\geq 0\}$, where

$$X^\varepsilon_t := X_t+\varepsilon t,\quad t\geq 0,$$

and $\varepsilon\in\mathbb{R}$. A little thought reveals that, for each fixed $t\geq 0$, $\lim_{\varepsilon\downarrow 0}\overline{G}^\varepsilon_t=\overline{G}_t$, where $\overline{G}^\varepsilon_t$ is given by (6.31) applied to X^ε and $\overline{G}_t$ is also given by (6.31). Similarly, $\lim_{\varepsilon\downarrow 0}\overline{X}^\varepsilon_t=\overline{X}_t$, where $\overline{X}^\varepsilon=\sup_{s\leq t}X^\varepsilon_s$. Next note that, in the sense of weak convergence of measures,

$$\lim_{\varepsilon\downarrow 0}\frac{1}{t}\mathrm{e}^{-pt}\mathbb{P}\big(X^\varepsilon_t\in\mathrm{d}x\big)\mathrm{d}t\mathbf{1}_{(x>0)}=\frac{1}{t}\mathrm{e}^{-pt}\mathbb{P}(X_t\in\mathrm{d}x)\mathrm{d}t\mathbf{1}_{(x\geq 0)}$$

whilst

$$\lim_{\varepsilon\downarrow 0}\frac{1}{t}\mathrm{e}^{-pt}\mathbb{P}\big(X^\varepsilon_t\in\mathrm{d}x\big)\mathrm{d}t\mathbf{1}_{(x<0)}=\frac{1}{t}\mathrm{e}^{-pt}\mathbb{P}(X_t\in\mathrm{d}x)\mathrm{d}t\mathbf{1}_{(x<0)}.$$

Hence, applying Theorem 6.15 to X^ε and taking limits as $\varepsilon\downarrow 0$ in (6.18) and (6.19), one recovers statements (ii), (iii) and (iv) of Theorem 6.15 for compound Poisson processes, but now with

$$\kappa(\alpha,\beta)=k\exp\left(\int_0^\infty\int_{[0,\infty)}\big(\mathrm{e}^{-t}-\mathrm{e}^{-\alpha t-\beta x}\big)\frac{1}{t}\mathbb{P}(X_t\in\mathrm{d}x)\mathrm{d}t\right)$$

(there is now closure of the interval at zero on the delimiter of the inner integral).

The reader may be curious about what would happen if we considered applying the conclusion of Theorem 6.15 to $X^{-\varepsilon}$ as $\varepsilon\downarrow 0$. In this case, using obvious notation for $\overline{G}_t^{-\varepsilon}$, it would follow that $\lim_{\varepsilon\downarrow 0}\overline{G}_t^{-\varepsilon}=\overline{G}^*_t$, where, now,

$$\overline{G}^*_t=\inf\{s<t : X_s=\overline{X}_t\}.$$

This pertains to another version of the Wiener–Hopf factorisation for compound Poisson processes, which states that

$$\big(\mathbf{e}_p-\overline{G}^*_{\mathbf{e}_p},\overline{X}_{\mathbf{e}_p}-X_{\mathbf{e}_p}\big)\stackrel{d}{=}(\underline{G}_{\mathbf{e}_p},-\underline{X}_{\mathbf{e}_p}),$$

with the new definition

$$\underline{G}_t=\sup\{s<t : X_s=\underline{X}_t\}.$$

Further, we would also have that κ satisfies (6.19) but $\widehat{\kappa}$ satisfies (6.20), with the delimiter $(-\infty, 0)$ replaced by $(-\infty, 0]$.

6.5 Examples of the Wiener–Hopf Factorisation

We finish this chapter by describing some examples for which the Wiener–Hopf factorisation is explicit. Before doing so, let us note that, until renewed interest in this topic around the turn of the millennium, there were frustratingly few known examples of the Wiener–Hopf factorisation for which anything concrete could be said.

6.5.1 Brownian Motion

The simplest example of the Wiener–Hopf factorisation is for a standard Brownian motion $B = \{B_t : t \geq 0\}$. In this case, $\Psi(\theta) = \theta^2/2$, for $\theta \in \mathbb{R}$, and

$$\frac{p}{p - \mathrm{i}\vartheta + \theta^2/2} = \frac{\sqrt{2p}}{\sqrt{2p - 2\mathrm{i}\vartheta} - \mathrm{i}\theta} \cdot \frac{\sqrt{2p}}{\sqrt{2p - 2\mathrm{i}\vartheta} + \mathrm{i}\theta}.$$

From the factorisation (6.17) and the transforms given in (6.18), we can identify

$$\kappa(\alpha, \beta) = \widehat{\kappa}(\alpha, \beta) = \sqrt{2\alpha} + \beta, \tag{6.33}$$

for $\alpha, \beta \geq 0$. The fact that both κ and $\widehat{\kappa}$ have the same expression is obvious by symmetry. Further, (6.33) tells us that L^{-1} is a $\frac{1}{2}$-stable subordinator and H is a unit-rate linear drift. This is to be expected when one reconsiders Example 6.11. In particular, it was shown there that L^{-1} has Laplace exponent $\Phi(\alpha) - \Phi(0)$, where Φ is the inverse of the Lévy–Khintchine exponent of B. For Brownian motion

$$\Phi(q) = \sqrt{2q} = \int_0^\infty (1 - \mathrm{e}^{-qx})(2\pi)^{-1/2} x^{-3/2} \mathrm{d}x, \quad q \geq 0,$$

where the second equality uses Exercise 1.4.

6.5.2 Spectrally Negative Lévy Processes

The previous example could also be seen as a consequence of the following more general analysis for spectrally negative Lévy processes. For such processes, recall that we defined $\psi(\theta) = \log \mathbb{E}(\exp\{\theta X_1\})$, $\theta \geq 0$ to be its Laplace exponent, with right inverse Φ; see Sect. 3.3 and Exercise 3.6. Moreover, we know, from Example 6.2, that we may work with the definition $L = \overline{X}$. We also know, from Example 6.11, that

$$L_x^{-1} = \inf\{s > 0 : \overline{X}_s > x\} = \inf\{s > 0 : X_s > x\} = \tau_x^+, \quad x \geq 0,$$

and

$$H_x = X_{L_x^{-1}} = x, \quad x \geq 0,$$

on $\{x < L_\infty\}$. Hence,

$$\mathbb{E}\big(\mathrm{e}^{-\alpha L_1^{-1} - \beta H_1} \mathbf{1}_{(1 < L_\infty)}\big) = \mathrm{e}^{-\Phi(\alpha) - \beta}, \quad \alpha, \beta \geq 0,$$

showing that we may take

$$\kappa(\alpha, \beta) = \Phi(\alpha) + \beta. \tag{6.34}$$

In that case, taking account of (6.18), for $p \geq 0$ and $\theta, \vartheta \in \mathbb{R}$, one of the Wiener–Hopf factors must be

$$\frac{\Phi(p)}{\Phi(p - \mathrm{i}\vartheta) - \mathrm{i}\theta},$$

and hence, by (6.17), the other factor must be

$$\frac{p}{\Phi(p)} \frac{\Phi(p - \mathrm{i}\vartheta) - \mathrm{i}\theta}{p - \mathrm{i}\vartheta + \Psi(\theta)}.$$

By inspection of the second of the two Laplace transforms in (6.18), we see that

$$\widehat{\kappa}(\alpha, \beta) = \frac{\alpha + \Psi(-\mathrm{i}\beta)}{\Phi(\alpha) - \beta} = \frac{\alpha - \psi(\beta)}{\Phi(\alpha) - \beta}, \quad \alpha, \beta \geq 0, \tag{6.35}$$

where in the second equality, we have used the relation $\psi(\theta) = -\Psi(-\mathrm{i}\theta)$, $\theta \geq 0$, between the Laplace exponent and the Lévy–Khintchine exponent. Given this expression for $\widehat{\kappa}$, there is nothing immediate we can say about the descending ladder process $(\widehat{L}^{-1}, \widehat{H})$. Nonetheless, as we shall see in later chapters, the identification of the Wiener–Hopf factors does form the basis of a semi-explicit account of a number of fluctuation identities for spectrally negative processes. Accordingly the case of spectrally negative Lévy processes turns out to be of significant practical value in a variety of applications, some of which we shall pursue in the remainder of this book.

6.5.3 Stable Processes

Suppose that X is an α-stable process, so that, for each $t > 0$, X_t is equal in distribution to $t^{1/\alpha} X_1$. This has the immediate consequence that for all $t > 0$,

$$\mathbb{P}(X_t \geq 0) = \rho,$$

for some $\rho \in [0, 1]$ known as the *positivity parameter*. It is possible to compute ρ in terms of the original parameters; see Zolotarev (1986), who showed that

$$\rho = \frac{1}{2} + \frac{1}{\pi\alpha} \arctan\left(\beta \tan \frac{\pi\alpha}{2}\right),$$

for $\alpha \in (0, 1) \cup (1, 2)$ and $\beta \in [-1, 1]$. For $\alpha = 1$ and $\eta = 0$, we have $\rho = 1/2$. We exclude the cases $\rho = 1$ and $\rho = 0$ in the subsequent discussion as these correspond, respectively, to the cases that X and $-X$ are subordinators with $\alpha \in (0, 1)$.

Note now from (6.19) that, for $\lambda \geq 0$,

$$\begin{aligned}
\kappa(\lambda, 0) &= k \exp\left(\int_0^\infty \int_{[0,\infty)} \left(e^{-t} - e^{-\lambda t}\right) \frac{1}{t} \mathbb{P}(X_t \in dx) dt\right) \\
&= k \exp\left(\int_0^\infty \left(e^{-t} - e^{-\lambda t}\right) \frac{\rho}{t} dt\right) \\
&= k\lambda^\rho, \qquad (6.36)
\end{aligned}$$

where in the final equality, we have used the Frullani integral from Lemma 1.7. This tells us directly that the process L^{-1} is a stable subordinator. We can proceed further and calculate, for $\lambda \geq 0$,

$$\begin{aligned}
\kappa(0, \lambda) &= k \exp\left(\int_0^\infty \int_{[0,\infty)} \left(e^{-t} - e^{-\lambda x}\right) \frac{1}{t} \mathbb{P}(X_t \in dx) dt\right) \\
&= k \exp\left(\int_0^\infty \frac{1}{t} \mathbb{E}\left(\left(e^{-t} - e^{-\lambda X_t}\right) \mathbf{1}_{(X_t \geq 0)}\right) dt\right) \\
&= k \exp\left(\int_0^\infty \frac{1}{t} \mathbb{E}\left(\left(e^{-t} - e^{-\lambda t^{1/\alpha} X_1}\right) \mathbf{1}_{(X_1 \geq 0)}\right) dt\right) \\
&= k \exp\left(\int_0^\infty \frac{1}{s} \mathbb{E}\left(\left(e^{-s\lambda^{-\alpha}} - e^{-s^{1/\alpha} X_1}\right) \mathbf{1}_{(X_1 \geq 0)}\right) ds\right) \\
&= k \exp\left(\int_0^\infty \frac{1}{s} \mathbb{E}\left(\left(e^{-s} - e^{-X_s}\right) \mathbf{1}_{(X_s \geq 0)}\right) ds\right) \\
&\quad \times \exp\left(-\int_0^\infty \frac{\rho}{s} \left(e^{-s} - e^{-s\lambda^{-\alpha}}\right) ds\right) \\
&= \kappa(0, 1) \times \exp\left(-\int_0^\infty \frac{\rho}{s} \left(e^{-s} - e^{-s\lambda^{-\alpha}}\right) ds\right) \\
&= \kappa(0, 1)\lambda^{\alpha\rho}, \qquad (6.37)
\end{aligned}$$

where in the third and fifth equality, we have used the fact that $s^{1/\alpha} X_1$ is equal in distribution to X_s. In the final equality, we have again used the Frullani integral. The term $\kappa(0, 1)$ is just a constant and, hence, we deduce that the ascending ladder height process is also a stable subordinator of index $\alpha\rho$. It is now immediate, from (6.21), that the descending ladder height process is a stable subordinator of index $\alpha(1-\rho)$, which is consistent with the fact that $\mathbb{P}(X_t \le 0) = 1-\rho$. This necessarily implies that $0 < \alpha\rho \le 1$ and $0 < \alpha(1-\rho) \le 1$ when $\rho \in (0,1)$. The extreme cases $\alpha\rho = 1$ and $\alpha(1-\rho) = 1$ correspond to spectrally negative and spectrally positive processes, respectively. For example, when $\beta = -1$ and $\alpha \in (1,2)$, we have a spectrally negative process of unbounded variation. It is easily checked in this situation that $\rho = 1/\alpha$ and hence, from the calculation above, $\kappa(0,\lambda) = \text{const.} \times \lambda$. This is consistent with earlier established facts for spectrally negative Lévy processes. Note that $\kappa(0,0) = \widehat{\kappa}(0,0) = 0$, showing that the killing rates in the ascending and descending ladder height processes are equal to zero. Hence,

$$\limsup_{t\uparrow\infty} X_t = -\liminf_{t\uparrow\infty} X_t = \infty$$

almost surely.

Unfortunately, it is not as easy to establish a convenient closed form expression for the bivariate exponent κ. The only known results in this direction are lead by the work of Doney (1987), who deals with a set of parameter values of α and β which are dense in the full parameter range $(0,2)$ and $[-1,1]$, respectively.[3] More recently Bernyk et al. (2008), Doney and Savov (2010), Kuznetsov (2011) and Graczyk and Jakubowski (2011) have made some significant improvements on Doney's original contribution. The expressions involved are quite complicated and we refrain from including them here.

6.5.4 Meromorphic Lévy Processes

One example where one would expect to be able develop the Wiener–Hopf factors is the case that X is the difference of two independent compound Poisson processes with exponentially distributed jumps. The reason why this example should be, up to a certain point, analytically tractable boils down to the fact that the ladder height processes must be a (possibly-killed) compound Poisson subordinators with exponentially distributed jumps. Moreover, the ladder height processes must be independent of their corresponding ladder time process. This is obvious when one considers that if a new maximum occurs then it is achieved by an exponentially distributed jump and then, by the lack-of-memory property, the overshoot beyond the previous maximum must again be exponentially distributed and independent of when it happens.[4] This heuristic reasoning is still valid even if we add in an independent Gaussian component or a linear drift. However, in that case, there is also the possibility that, for example, a new maximum is achieved continuously. This would result in the ascending ladder height gaining an additional linear drift. This information would be sufficient to develop further the factors $\Psi_p^+(0,\theta)$ and $\Psi_p^-(0,\theta)$.

---

[3]For related results see Bingham (1971, 1972, 1973b).

[4]The details can be found in Example (c), Chap. XVIII.3 of Feller (1971).

From an analytical point of view, the reason why this special class of processes can be handled is because the characteristic exponent necessarily takes a rational form, namely the ratio of two polynomials of finite degree, from which a factorisation can be forced. Starting with the early work of Borovkov (1976) and Feller (1971), various authors have tried to generalise this idea by replacing the exponentially distributed jumps by jumps whose common distribution have a Laplace transform which is rational, or indeed, by jumps whose distribution belongs to the so-called phase-type class; cf. Asmussen et al. (2004), Pistorius (2006) and Mordecki (2008). Contained in both of the aforementioned families of distributions are jump distributions that are finite mixtures of exponential densities.

In this section, we shall present another possible generalisation which has some overlap with all of the aforementioned families of Lévy processes; the so-called *meromorphic Lévy processes*. These processes were introduced by Kuznetsov (2010a, 2010b) and Kuznetsov et al. (2012).

Definition 6.19 A Lévy process is said to belong to the *meromorphic class* if its Lévy measure Π is absolutely continuous, with density given by

$$\pi(x) = \sum_{i=1}^{\infty} a_i \rho_i \mathrm{e}^{-\rho_i x} \mathbf{1}_{\{x>0\}} + \sum_{i=1}^{\infty} \hat{a}_i \hat{\rho}_i \mathrm{e}^{\hat{\rho}_i x} \mathbf{1}_{\{x<0\}}. \tag{6.38}$$

Here, the constants $a_i, \hat{a}_i, \rho_i, \hat{\rho}_i$ are non-negative, ρ_i and $\hat{\rho}_i$ are arranged in increasing order with $\lim_{n\uparrow\infty} \rho_n = \lim_{n\uparrow\infty} \hat{\rho}_n = \infty$ and they satisfy the summability condition

$$\sum_{i=1}^{\infty} a_i \rho_i^{-2} + \sum_{i=1}^{\infty} \hat{a}_i \hat{\rho}_i^{-2} < \infty. \tag{6.39}$$

Let us pursue a number of remarks concerning this definition. First note that the summability condition (6.39) is sufficient to ensure the integrability condition $\int_{\mathbb{R}} (1 \wedge x^2)\pi(x)\mathrm{d}x < \infty$ is satisfied. In fact it is easily confirmed, with the help of Fubini's Theorem, that it guarantees the stronger condition $\int_{\mathbb{R}} x^2 \pi(x)\mathrm{d}x < \infty$. It is not automatic that π is the density of a finite measure and, hence, not all meromorphic Lévy processes have a compound Poisson jump structure. Finite activity does occur, however, if the number of summands in both sums of (6.38) are finite. Next, note that the non-negativity of the constants a_i and $\hat{a}_i$ allows for the possibility that one or both of the sums in the density π have a finite number of summands. The following theorem, lifted from Kuznetsov et al. (2012), shows some convenient properties that follow from this definition.

Theorem 6.20 *Any meromorphic Lévy process, X, has the following properties:*

(i) *The characteristic exponent $\Psi(z)$ is a meromorphic function which has poles at points $\{-\mathrm{i}\rho_n, \mathrm{i}\hat{\rho}_n\}_{n\geq 1}$, where ρ_n and $\hat{\rho}_n$ are positive real numbers.*

(ii) *For* $p \geq 0$, *the function* $p + \Psi(z)$ *has roots at points* $\{-\mathrm{i}\zeta_n(p), \mathrm{i}\hat{\zeta}_n(p)\}_{n\geq 1}$ *where* $\zeta_n(p)$ *and* $\hat{\zeta}_n(p)$ *are non-negative real numbers (strictly positive if* $p > 0$*).*

(iii) *The roots and poles of* $p + \Psi(\mathrm{i}z)$ *satisfy the interlacing condition*

$$\cdots < -\rho_2 < -\zeta_2(p) < -\rho_1 < -\zeta_1(p) < 0 < \hat{\zeta}_1(p) < \hat{\rho}_1 < \hat{\zeta}_2(p) < \hat{\rho}_2 < \cdots$$

(iv) *The spatial Wiener–Hopf factors are expressed as convergent infinite products,*

$$\Psi_p^+(0, \mathrm{i}z) = \mathbb{E}\big[\mathrm{e}^{-z\overline{X}_{\mathbf{e}_p}}\big] = \prod_{n\geq 1} \frac{1 + \frac{z}{\rho_n}}{1 + \frac{z}{\zeta_n(p)}}$$

and

$$\Psi_p^-(0, -\mathrm{i}z) = \mathbb{E}\big[\mathrm{e}^{z\underline{X}_{\mathbf{e}_p}}\big] = \prod_{n\geq 1} \frac{1 + \frac{z}{\hat{\rho}_n}}{1 + \frac{z}{\hat{\zeta}_n(p)}},$$

for $z \geq 0$.

Conversely, any Lévy process with the above properties belongs to the meromorphic class.

As one might expect, the proof is somewhat technical, relying predominantly on complex analytic techniques. We omit it for the sake of brevity. Part (iv) of the above theorem leads quickly to the following corollary with the help of straightforward residue calculus or a partial fraction expansion.

Corollary 6.21 *For all* $x \geq 0$,

$$\mathbb{P}(\overline{X}_{\mathbf{e}_p} \in \mathrm{d}x) = \mathrm{c}_0\delta_0(\mathrm{d}x) + \sum_{n\geq 1} \mathrm{c}_n\zeta_n(p)\mathrm{e}^{-\zeta_n(p)x}, \tag{6.40}$$

where

$$\mathrm{c}_0 := \prod_{n=1}^{\infty} \frac{\zeta_n(p)}{\rho_n} \quad \text{and} \quad \mathrm{c}_n := \left(1 - \frac{\zeta_n(p)}{\rho_n}\right) \prod_{\substack{k\geq 1\\ k\neq n}} \frac{1 - \frac{\zeta_n(p)}{\rho_k}}{1 - \frac{\zeta_n(p)}{\zeta_k(p)}}.$$

From this corollary, it is straightforward to deduce a similar expression for the law of $-\underline{X}_{\mathbf{e}_p}$ using duality.

This corollary also exemplifies a numerical point of interest concerning meromorphic Lévy processes. Knowing the numerical values of even a finite number of the roots and poles would be sufficient to approximate the products and sums in (6.40) by making obvious truncations in the sums and products. The reader is referred to Kuznetsov (2010a) for further details of numerical methods.

There are a number of specific examples of meromorphic Lévy processes (found in the above-mentioned works of Kuznetsov and co-authors) for which the Lévy density and/or the characteristic exponent can be expressed in a more suitable closed form. The most notable of these is the so-called β-class[5] of Lévy processes, found in the landmark paper of Kuznetsov (2010a).

The characteristic exponent is given by

$$\begin{aligned}\Psi(\theta) = \mathrm{i}a\theta + \frac{1}{2}\sigma^2\theta^2 + \frac{c_1}{\beta_1}\left\{\mathrm{B}(\alpha_1, 1-\lambda_1) - \mathrm{B}\left(\alpha_1 - \frac{\mathrm{i}\theta}{\beta_1}, 1-\lambda_1\right)\right\} \\ + \frac{c_2}{\beta_2}\left\{\mathrm{B}(\alpha_2, 1-\lambda_2) - \mathrm{B}\left(\alpha_2 + \frac{\mathrm{i}\theta}{\beta_2}, 1-\lambda_2\right)\right\}, \quad \theta \in \mathbb{R},\end{aligned}$$

where $\mathrm{B}(x, y) = \Gamma(x)\Gamma(y)/\Gamma(x+y)$ is the beta function, with parameter ranges $a \in \mathbb{R}$, $\sigma^2 \geq 0$, $c_i \geq 0$, $\alpha_i > 0$, $\beta_i > 0$ and $\lambda_i \in (0,3)\setminus\{1,2\}$, for $i = 1, 2$. The corresponding Lévy measure, Π, has density

$$\pi(x) = c_1\frac{\mathrm{e}^{-\alpha_1\beta_1 x}}{(1-\mathrm{e}^{-\beta_1 x})^{\lambda_1}}\mathbf{1}_{\{x>0\}} + c_2\frac{\mathrm{e}^{\alpha_2\beta_2 x}}{(1-\mathrm{e}^{\beta_2 x})^{\lambda_2}}\mathbf{1}_{\{x<0\}}.$$

To see why a Lévy process in the β-class is also a meromorphic Lévy process, one may expand the expression for the Lévy density above on the positive and negative half-lines using the generalised binomial formula. The large number of parameters allows one to choose Lévy processes within the β-class that have paths that are both of unbounded variation (when at least one of the conditions $\sigma \neq 0$, $\lambda_1 \in (2,3)$ or $\lambda_2 \in (2,3)$ holds) and bounded variation (when all of the conditions $\sigma = 0$, $\lambda_1 \in (0,2)$ and $\lambda_2 \in (0,2)$ hold) as well as having infinite and finite activity in the jump component (accordingly as both $\lambda_1, \lambda_2 \in (1,2)\cup(2,3)$ or $\lambda_1, \lambda_2 \in (0,1)$).

6.6 Vigon's Theory of Philanthropy and More Examples

At the level of a spatial decomposition, the Wiener–Hopf factorisation expresses a fundamental relationship between the underlying Lévy process, its ascending ladder height processes and its descending ladder height processes. In his seminal Ph.D. thesis, Vigon (2002a) gives a remarkably simple and precise characterisation of what kind of ladder height processes belong together in a Wiener–Hopf factorisation. This is what Vigon colourfully calls *the problem of friends*.[6] In this section, we give a brief outline of his solution to this problem using so-called *philanthropy*. A direct consequence of Vigon's theory of philanthropy is that it gives a simple recipe for generating countless examples of explicit Wiener–Hopf factorisations.

[5] The β-class of Lévy processes is also referred to as the β-family of Lévy processes. Processes in this class are also called β-processes or β-Lévy processes.

[6] Le problème des amis.

Consider any two (killed) subordinators H and $\widehat{H}$. Let us write the Laplace exponent of the (killed) subordinator H in the form

$$\varphi(u) = \eta + \delta u + \int_{(0,\infty)} \left(1 - \mathrm{e}^{-ux}\right)\Upsilon(\mathrm{d}x), \quad u \geq 0,$$

where $\eta \geq 0$, $\delta \geq 0$ and the measure Υ satisfies $\int_{(0,\infty)}(1 \wedge x)\Upsilon(\mathrm{d}x) < \infty$. Recall that H is killed if and only if $\eta > 0$. Moreover, we shall use the symbols $\widehat{\varphi}, \widehat{\eta}, \widehat{\delta}$ and $\widehat{\Upsilon}$ in the obvious way for the process $\widehat{H}$.

Vigon says that H and $\widehat{H}$ are *friends* if there exists a Lévy process with characteristic exponent Ψ and a constant $p \geq 0$ such that, for $\theta \in \mathbb{R}$,

$$p + \Psi(\theta) = \varphi(-\mathrm{i}\theta)\widehat{\varphi}(\mathrm{i}\theta).$$

In particular, if H and $\widehat{H}$ are friends, then necessarily $p = \eta\widehat{\eta}$. Moreover, if at most one of the two friends is killed (i.e. $\eta\widehat{\eta} = 0$), then their friendship constitutes a spatial Wiener–Hopf factorisation (in the sense of part (iv) of Theorem 6.15). The following theorem, given without proof, characterises Π, the Lévy measure associated with Ψ.

Theorem 6.22 (Vigon's theorem of friends) *Suppose that H and $\widehat{H}$ are friends. Then Υ (resp. $\widehat{\Upsilon}$) is absolutely continuous with density υ (resp. $\widehat{\upsilon}$) if $\widehat{\delta} > 0$ (resp. $\delta > 0$), in which case, it has a version that is right-continuous with left limits. Moreover, for all $x > 0$,*

$$\Pi(x,\infty) = \int_{(0,\infty)} \widehat{\Upsilon}(u,\infty)\Upsilon(x + \mathrm{d}u) + \widehat{\delta}\upsilon(x) + \widehat{\eta}\Upsilon(x,\infty) \tag{6.41}$$

and similarly,

$$\Pi(-\infty,-x) = \int_{(0,\infty)} \Upsilon(u,\infty)\widehat{\Upsilon}(x + \mathrm{d}u) + \delta\widehat{\upsilon}(x) + \eta\widehat{\Upsilon}(x,\infty). \tag{6.42}$$

Here, we understand the term $\widehat{\delta}\upsilon(x) = 0$ (resp. $\delta\widehat{\upsilon}(x) = 0$) when $\widehat{\delta} = 0$ (resp. $\delta = 0$).

Conversely, suppose that Υ (resp. $\widehat{\Upsilon}$) is absolutely continuous with density υ (resp. $\widehat{\upsilon}$) whenever $\widehat{\delta} > 0$ (resp. $\delta > 0$). If the expressions given on the right-hand side of (6.41) and (6.42) are both non-increasing, then H and $\widehat{H}$ are friends.

The "converse" part of this theorem is particularly interesting as it gives criteria with which one could potentially *engineer* a spatial Wiener–Hopf factorisation, by first choosing the factors and then characterising the associated Lévy process. The criterion to check for any two factors φ and $\widehat{\varphi}$, namely the non-increasingness of (6.41) and (6.42), is not particularly convenient. With a view to a more convenient criteria, Vigon introduces the concept of philanthropy. A (killed) subordinator is called a *philanthropist* if its Lévy measure is absolutely continuous with non-increasing density. Vigon's strange choice of terminology now becomes clear with the following mathematically and linguistically elegant result.

Theorem 6.23 (Vigon's theorem of philanthropy) *Any two philanthropists are friends.*

On a final note, it is obvious from this last theorem that if at most one of the philanthropists is killed, then their friendship constitutes a spatial Wiener–Hopf factorisation.

We conclude with two examples, the second of which will prove to be of particular pertinence later on when considering the theory of so-called *scale functions* for spectrally negative Lévy processes, in Chap. 9.

6.6.1 Hypergeometric Lévy Processes

Kyprianou et al. (2010a) and Kuznetsov et al. (2011) propose choosing the two philanthropists φ and $\widehat{\varphi}$ from the family of subordinators which belong to the class of β-Lévy processes. The resulting friendship defines a class of processes which are called *hypergeometric Lévy processes*. In particular, we have

$$\varphi(u) = \eta + \delta u + \frac{c}{\beta}\left\{ \mathrm{B}(1-\alpha+\gamma, -\gamma) - \mathrm{B}\left(1-\alpha+\gamma+\frac{u}{\beta}, -\gamma\right)\right\},$$

for $u \geq 0$, where

$$\Upsilon(\mathrm{d}x) = c\frac{\mathrm{e}^{\alpha\beta x}}{(\mathrm{e}^{\beta x}-1)^{1+\gamma}}\mathrm{d}x \quad \text{on } (0,\infty),$$

with $\gamma \in (-\infty, 0) \cup (0, 1)$, $\beta, c > 0$ and $1 - \alpha + \gamma > 0$. A similar expression may be taken for $\widehat{\varphi}$ but with the parameters $(\eta, \delta, \alpha, \beta, \gamma, c)$ replaced by different parameters, $(\widehat{\eta}, \widehat{\delta}, \widehat{\alpha}, \widehat{\beta}, \widehat{\gamma}, \widehat{c})$, satisfying the same constraints. Recall that the case $p = \eta\widehat{\eta} > 0$ corresponds to the case that the Lévy process is killed at an independent and exponentially distributed random time, with rate p.

By appealing to a special set of parameters, Kuznetsov and Pardo (2012) showed that

$$\Psi(\theta) = \frac{\Gamma(1-\mathrm{a}+\gamma-\mathrm{i}\theta)}{\Gamma(1-\mathrm{a}-\mathrm{i}\theta)}\frac{\Gamma(\widehat{\mathrm{a}}+\widehat{\gamma}+\mathrm{i}\theta)}{\Gamma(\widehat{\mathrm{a}}+\mathrm{i}\theta)}, \quad \theta \in \mathbb{R},$$

where

$$\mathrm{a} \leq 1, \quad \gamma \in (0,1), \quad \widehat{\mathrm{a}} \geq 0, \quad \widehat{\gamma} \in (0,1),$$

is a convenient subclass of hypergeometric Lévy processes, for which the associated Lévy measure can be computed precisely. Indeed, setting

$$\mathrm{k} = 1 - \mathrm{a} + \gamma + \widehat{\mathrm{a}} + \widehat{\gamma},$$

they showed that the Lévy measure is absolutely continuous with density

$$\pi(x) = \begin{cases} -\frac{\Gamma(\mathrm{k})}{\Gamma(\mathrm{k}-\widehat{\gamma})\Gamma(-\gamma)} \mathrm{e}^{-(1-\mathrm{a}+\gamma)x} {}_2F_1(1+\gamma, \mathrm{k}; \mathrm{k}-\widehat{\gamma}; \mathrm{e}^{-x}) & \text{if } x > 0 \\ -\frac{\Gamma(\mathrm{k})}{\Gamma(\mathrm{k}-\gamma)\Gamma(-\widehat{\gamma})} \mathrm{e}^{(\widehat{\mathrm{a}}+\widehat{\gamma})x} {}_2F_1(1+\widehat{\gamma}, \mathrm{k}; \mathrm{k}-\gamma; \mathrm{e}^{x}) & \text{if } x < 0. \end{cases} \tag{6.43}$$

Here, ${}_2F_1$ is the Gauss hypergeometric function, satisfying

$${}_2F_1(a, b; c; z) = \sum_{n \geq 0} \frac{(a)_n (b)_n}{(c)_n} \frac{z^n}{n!},$$

where $z \in \mathbb{C}$ such that $|z| < 1$ and $(x)_n = \Gamma(x+n)/\Gamma(x)$.

More can be said about a given Lévy process, X, chosen from this subclass. If $\mathrm{a} < 1$ and $\widehat{\mathrm{a}} > 0$, then X is killed at rate

$$p = \Psi(0) = \frac{\Gamma(1-\mathrm{a}+\gamma)}{\Gamma(1-\mathrm{a})} \frac{\Gamma(\widehat{\mathrm{a}}+\widehat{\gamma})}{\Gamma(\widehat{\mathrm{a}})}.$$

When $\mathrm{a} = 1$ and $\widehat{\mathrm{a}} > 0$ (resp. $\mathrm{a} < 1$ and $\widehat{\mathrm{a}} = 0$), the process X drifts to ∞ (resp. $-\infty$). Moreover, when $\mathrm{a} = 1$ and $\widehat{\mathrm{a}} = 0$, then X is oscillating. Finally, the process X has no Gaussian component and it has paths of bounded variation (resp. unbounded variation) when $\gamma + \widehat{\gamma} < 1$ (resp. $1 \leq \gamma + \widehat{\gamma} \leq 2$).

6.6.2 Spectrally Negative Lévy Processes Revisited

From Sect. 6.5.2, we know that the ascending ladder height process of a spectrally negative Lévy process, X, must be a (possibly-killed) linear drift. In the language of Vigon, this means that one of the two friends involved in the Wiener–Hopf factorisation of a spectrally negative Lévy process necessarily satisfies

$$\varphi(u) = \Phi(0) + u, \quad u \geq 0,$$

where Φ is the right inverse of the Laplace exponent of X. Vigon's theorem of friends thus tells us that the descending ladder height process has a Lévy measure $\widehat{\Upsilon}$ which is absolutely continuous and, together with its density $\widehat{\upsilon}$, satisfies

$$\Pi(-\infty, -x) = \widehat{\upsilon}(x) + \Phi(0)\widehat{\Upsilon}(x, \infty), \tag{6.44}$$

for $x > 0$. Noting that $\widehat{\upsilon}(x) = -\mathrm{d}\widehat{\Upsilon}(x, \infty)/\mathrm{d}x$, we can treat (6.44) as a first order differential equation. Using standard techniques, we can solve this differential equation and obtain

$$\widehat{\Upsilon}(x, \infty) = \mathrm{e}^{\Phi(0)x} \int_x^\infty \mathrm{e}^{-\Phi(0)y} \Pi(-\infty, -y)\mathrm{d}y.$$

Conversely, Vigon's theorem of philanthropy says that we may always choose the descending ladder height process so that $\widehat{\Upsilon}$ is absolutely continuous with non-increasing density, say $\widehat{\upsilon}$. In that case, (6.44) must follow.

6.7 Brief Remarks on the Term "Wiener–Hopf"

Having now completed our exposition of the Wiener–Hopf factorisation, the reader may feel somewhat confused as to the association of the name "Wiener–Hopf" with Theorem 6.15. Indeed, in our presentation, we have made no reference to works of Wiener or Hopf. The connection between Theorem 6.15 and these two mathematicians lies in their analytic study of the solutions to integral equations of the form

$$Q(x) = \int_0^\infty Q(y) f(x-y) \mathrm{d}y, \quad x > 0, \tag{6.45}$$

where $f : \mathbb{R} \to [0, \infty)$ is a pre-specified kernel; see Wiener and Hopf (1931), Payley and Wiener (1934) and Hopf (1934). If one considers a compound Poisson process X which has the property that $\limsup_{t\uparrow\infty} X_t < \infty$, then the strong Markov property implies that $\overline{X}_\infty$ is equal in distribution to $(\xi + \overline{X}_\infty) \vee 0$, where ξ is independent of $\overline{X}_\infty$ and has the same distribution as the jumps of X. If the aforesaid jump distribution has density f, then one shows easily that $H(x) = \mathbb{P}(\overline{X}_\infty \leq x)$ satisfies

$$H(x) = \int_{-\infty}^x H(x-y) f(y) \mathrm{d}y = \int_0^\infty f(x-y) H(y) \mathrm{d}y,$$

and hence one obtains immediately the existence of a solution to (6.45) for the given f. This observation dates back to the work of Spitzer (1957).

Embedded in the complex analytic techniques used to analyse (6.45) and generalisations thereof by Wiener, Hopf and many others that followed are factorisations of operators (which can take the form of Fourier transforms). In the probabilistic setting here, this is manifested in the form of the independence seen in Theorem 6.15 (i) and the way this is used to identify the factors Ψ^+ and Ψ^-, in conjunction with analytic extension, in the proof of part (iii) of the same theorem. The full extent of the analytic Wiener–Hopf factorisation technique goes far beyond the current setting and we make no attempt to expose it here.[7] The name "Wiener–Hopf" factorisation thus honours the somewhat obscure analytical origins of what may, otherwise, be considered as a sophisticated path decomposition of a Lévy process.

Exercises

6.1 Give an example of a Lévy process which has bounded variation with zero drift for which 0 is regular for both $(0, \infty)$ and $(-\infty, 0)$. Give an example of a Lévy process of bounded variation and zero drift for which 0 is only regular for $(0, \infty)$.

[7] The interested reader may consider looking up Noble (1958), Busbridge (1960) and Chandrasekhar (1960).

6.2 Suppose that X is a spectrally negative Lévy process of unbounded variation with Laplace exponent ψ and recall the definition $\tau_x^+ = \inf\{t > 0 : X_t > x\}$. Recall also that the process $\tau^+ := \{\tau_x^+ : x \geq 0\}$ is a (possibly-killed) subordinator (see Corollary 3.14) with Laplace exponent Φ, the right inverse of ψ.

(i) Suppose that δ is the drift of the process τ^+. Show that $\delta = 0$.

(ii) Deduce that

$$\lim_{x \downarrow 0} \frac{\tau_x^+}{x} = 0$$

almost surely, and hence that

$$\limsup_{t \downarrow 0} \frac{X_t}{t} = \infty$$

almost surely. Conclude that 0 is regular for $(0, \infty)$ and hence that the jump measure of τ^+ cannot be finite.

(iii) From the Wiener–Hopf factorisation of X show that

$$\lim_{\theta \uparrow \infty} \mathbb{E}\big(\mathrm{e}^{\theta \underline{X}_{\mathbf{e}_q}}\big) = 0,$$

and hence use this to give an alternative proof that 0 is regular for $(0, \infty)$.

6.3 Fix $\rho \in (0, 1]$. Show that a compound Poisson subordinator with jump rate $\lambda\rho$, killed at an independent and exponentially distributed time with parameter $\lambda(1-\rho)$, is equal in law to a compound Poisson subordinator killed after an independent number of jumps, which is distributed geometrically with parameter $1 - \rho$.

6.4 Show that the only processes for which

$$\int_0^\infty \mathbf{1}_{(\overline{X}_t = X_t)} \mathrm{d}t > 0 \quad \text{and} \quad \int_0^\infty \mathbf{1}_{(X_t = \underline{X}_t)} \mathrm{d}t > 0$$

almost surely are compound Poisson processes.

6.5 Suppose that X is spectrally negative with characteristic triple (a, σ, Π) and that $\mathbb{E}(X_t) > 0$. (Recall that, in general, $\mathbb{E}(X_t) \in [-\infty, \infty)$.)

(i) Show that

$$\int_{-\infty}^{-1} \Pi(-\infty, x)\mathrm{d}x < \infty.$$

(ii) Using Theorem 6.15 (iv), deduce that, up to a constant,

$$\widehat{\kappa}(0, \mathrm{i}\theta) = \left(-a + \int_{(-\infty,-1)} x\Pi(\mathrm{d}x)\right)$$

$$-\frac{1}{2}\mathrm{i}\theta\sigma^2 + \int_{(-\infty,0)} (1 - \mathrm{e}^{\mathrm{i}\theta x})\Pi(-\infty, x)\mathrm{d}x.$$

Hence deduce that there exists a choice of local time at the maximum for which the descending ladder height process has jump measure given by $\Pi(-\infty, -x)\mathrm{d}x$ on $(0, \infty)$, drift $\sigma^2/2$ and is killed at rate $\mathbb{E}(X_1)$.

6.6 Suppose that X is a spectrally negative stable process of index $1 < \alpha < 2$.

(i) Deduce, with the help of Theorem 3.12, that up to a multiplicative constant

$$\kappa(\theta, 0) = \theta^{1/\alpha}, \quad \theta \geq 0,$$

and hence that $\mathbb{P}(X_t \geq 0) = 1/\alpha$ for all $t \geq 0$.

(ii) By reconsidering the Wiener–Hopf factorisation, show that, for each $t \geq 0$ and $\theta \geq 0$,

$$\mathbb{E}\big(\mathrm{e}^{-\theta \overline{X}_t}\big) = \sum_{n=0}^{\infty} \frac{(-\theta t^{1/\alpha})^n}{\Gamma(1 + n/\alpha)}.$$

This identity is taken from Bingham (1971, 1972).

6.7 (The second factorisation identity) In this exercise, we derive what is commonly called the *second factorisation identity*, which is due to Percheskii and Rogozin (1969). It uses the Laplace exponents κ and $\widehat{\kappa}$ to give an identity concerning the problem of first passage above a fixed level $x \in \mathbb{R}$. The derivation we use here makes use of calculations in Darling et al. (1972) and Alili and Kyprianou (2005). We shall use the derivation of this identity later to solve some optimal stopping problems.

Define as usual

$$\tau_x^+ = \inf\{t > 0 : X_t > x\}, \quad x \geq 0,$$

where X is any Lévy process.

(i) Using the same technique as in Exercise 5.7, prove that, for all $\alpha > 0$, $\beta \geq 0$ and $x \in \mathbb{R}$, we have

$$\mathbb{E}\big(\mathrm{e}^{-\alpha\tau_x^+ - \beta X_{\tau_x^+}} \mathbf{1}_{(\tau_x^+ < \infty)}\big) = \frac{\mathbb{E}(\mathrm{e}^{-\beta \overline{X}_{\mathbf{e}_\alpha}} \mathbf{1}_{(\overline{X}_{\mathbf{e}_\alpha} > x)})}{\mathbb{E}(\mathrm{e}^{-\beta \overline{X}_{\mathbf{e}_\alpha}})}. \tag{6.46}$$

Note that the identity is still true when $\alpha = 0$ if $\mathbb{P}(\overline{X}_\infty < \infty) = 1$.

(ii) Establish the second factorisation identity as follows: If X is not a subordinator then, for $\alpha, \beta \geq 0$,

$$\int_0^\infty \mathrm{e}^{-qx}\mathbb{E}\big(\mathrm{e}^{-\alpha\tau_x^+ - \beta(X_{\tau_x^+} - x)} \mathbf{1}_{(\tau_x^+ < \infty)}\big)\mathrm{d}x = \frac{\kappa(\alpha, q) - \kappa(\alpha, \beta)}{(q - \beta)\kappa(\alpha, q)}.$$

6.8 Suppose that X is any Lévy process which is not a subordinator and $\mathbf{e}_p$ is an independent random variable which is exponentially distributed with parameter $p > 0$. Note that 0 is regular for $(0, \infty)$ if and only if $\mathbb{P}(\overline{X}_{\mathbf{e}_p} = 0) = 0$.

(i) Use the Wiener–Hopf factorisation to show that 0 is regular for $(0, \infty)$ if and only if

$$\int_0^1 \frac{1}{t}\mathbb{P}(X_t > 0)\mathrm{d}t = \infty.$$

(ii) Now noting that 0 is irregular for $[0, \infty)$ if and only if $\mathbb{P}(\overline{G}_{\mathbf{e}_p} = 0) > 0$, show that 0 is regular for $[0, \infty)$ if and only if

$$\int_0^1 \frac{1}{t}\mathbb{P}(X_t \geq 0)\mathrm{d}t = \infty.$$

6.9 This exercise gives the random walk analogue of the Wiener–Hopf factorisation. In fact, this is the original setting of the Wiener–Hopf factorisation. We give the formulation in Greenwood and Pitman (1980a). However, one may also consult Feller (1971) and Borovkov (1976) for other accounts.

Suppose that, under P, $S = \{S_n : n \geq 0\}$ is a random walk with $S_0 = 0$ and increment distribution F. We assume that S can jump both upwards and downwards, in other words $\min\{F(-\infty, 0), F(0, \infty)\} > 0$ and that F has no atoms. Denote by $\mathbf{\Gamma}_p$ an independent random variable which has a geometric distribution with parameter $p \in (0, 1)$ and let

$$G = \min\left\{k = 0, 1, \ldots, \mathbf{\Gamma}_p : S_k = \max_{j=1,\ldots,\mathbf{\Gamma}_p} S_j\right\}.$$

Note that S_G is the last maximum over times $\{0, 1, \ldots, \mathbf{\Gamma}_p\}$. Define $N = \inf\{n > 0 : S_n > 0\}$ the first-passage time into $(0, \infty)$, or equivalently the first strict ladder time. Our aim is to characterise the joint laws (G, S_G) and (N, S_N) in terms of F, the basic data of the random walk.

(i) Show that (even without the restriction that $\min\{F(0, \infty), F(-\infty, 0)\} > 0$),

$$E\left(s^{\mathbf{\Gamma}_p}\mathrm{e}^{\mathrm{i}\theta S_{\mathbf{\Gamma}_p}}\right) = \exp\left\{-\int_{\mathbb{R}}\sum_{n=1}^{\infty}(1 - s^n\mathrm{e}^{\mathrm{i}\theta x})q^n\frac{1}{n}F^{*n}(\mathrm{d}x)\right\}$$

where $0 < s \leq 1$, $\theta \in \mathbb{R}$, $q = 1 - p$ and E is expectation under P. Deduce that the pair $(\mathbf{\Gamma}_p, S_{\mathbf{\Gamma}_p})$ is infinitely divisible.

(ii) Let ν be an independent random variable which is geometrically distributed on $\{0, 1, 2, \ldots\}$ with parameter $P(N > \mathbf{\Gamma}_p)$. Using a path decomposition in terms of excursions from the maximum, show that the pair (G, S_G) is equal in distribution to the component-wise sum of ν independent copies of (N, S_N) conditioned on the event $\{N \leq \mathbf{\Gamma}_p\}$, and hence it is an infinitely divisible two-dimensional random variable.

(iii) Show that (G, S_G) and $(\mathbf{\Gamma}_p - G, S_{\mathbf{\Gamma}_p} - S_G)$ are independent. Further, show that the latter pair is equal in distribution to (D, S_D), where

$$D = \max\Big\{k = 0, 1, \ldots, \mathbf{\Gamma}_p : S_k = \min_{j=1,\ldots,\mathbf{\Gamma}_p} S_j\Big\}.$$

(iv) Deduce that

$$E\big(s^G e^{i\theta S_G}\big) = \exp\bigg\{-\int_{(0,\infty)} \sum_{n=1}^{\infty} \big(1 - s^n e^{i\theta x}\big) q^n \frac{1}{n} F^{*n}(dx)\bigg\},$$

for $0 < s \leq 1$ and $\theta \in \mathbb{R}$. Note, when $s = 1$, this identity was established by Spitzer (1956).

(v) Show that

$$E\big(s^G e^{i\theta S_G}\big) = \frac{P(\Gamma_p < N)}{1 - E((qs)^N e^{i\theta S_N})}$$

and hence deduce the Spitzer–Baxter identity

$$\frac{1}{1 - E(s^N e^{i\theta S_N})} = \exp\bigg\{\int_{(0,\infty)} \sum_{n=1}^{\infty} s^n e^{i\theta x} \frac{1}{n} F^{*n}(dx)\bigg\}.$$

See, for example, Bingham (2001).

6.10 Suppose that X is a spectrally negative Lévy process with Laplace exponent ψ whose right inverse is denoted by Φ.

(i) Use the Frullani integral to show that, for $\lambda, q > 0$,

$$\frac{\Phi(q)}{\Phi(q) + \lambda} = \exp\bigg\{\int_0^\infty dx \int_{[0,\infty)} \big(e^{-\lambda x} - 1\big) \frac{e^{-qt}}{x} \mathbb{P}\big(\tau_x^+ \in dt\big)\bigg\},$$

where $\tau_x^+ = \inf\{t > 0 : X_t > x\}$.

(ii) Next use Theorem 6.15 to show that, for $q, \lambda \geq 0$,

$$\frac{\Phi(q)}{\Phi(q) + \lambda} = \exp\bigg\{\int_0^\infty dt \int_{[0,\infty)} \big(e^{-\lambda x} - 1\big) \frac{e^{-qt}}{t} \mathbb{P}(X_t \in dx)\bigg\}.$$

(iii) Hence deduce *Kendall's identity*, that

$$t\mathbb{P}\big(\tau_x^+ \in dt\big)dx = x\mathbb{P}(X_t \in dx)dt$$

on $[0, \infty) \times [0, \infty)$.

Chapter 7
Lévy Processes at First Passage

This chapter is devoted to studying how the Wiener–Hopf factorisation can be used to characterise the behaviour of any Lévy process at first passage over a fixed level. The case of a subordinator will be excluded throughout this chapter, as this has been dealt with in Chap. 5. Nonetheless, the analysis of how subordinators make first passage will play a crucial role in understanding the case of a general Lévy process.

To some extent, the results we present on the first-passage problem suffer from a lack of analytical explicitness. This is due to the same symptoms present in our understanding of the Wiener–Hopf factorisation. Nonetheless there is sufficient mathematical structure to establish qualitative statements concerning the characterisation of the first-passage problem. This becomes more apparent when looking at asymptotic properties of the established characterisations.

7.1 Drifting and Oscillating

For any Lévy process, X, define as usual

$$\tau_x^+ = \inf\{t > 0 : X_t > x\},$$

for $x \in \mathbb{R}$. In this section, we shall establish precisely when $\mathbb{P}(\tau_x^+ < \infty)$ is strictly less than one. Further, we shall give sufficient conditions under which the first-passage probability decays exponentially as $x \uparrow \infty$; that is to say, we handle the case of Cramér's estimate.

Suppose now that $H = \{H_t : t \geq 0\}$ is the ascending ladder height process of X. If

$$T_x^+ = \inf\{t > 0 ; H_t > x\},$$

then quite clearly

$$\mathbb{P}(\tau_x^+ < \infty) = \mathbb{P}(T_x^+ < \infty). \tag{7.1}$$

Recall from Theorem 6.9 that the process H has the law of a subordinator, possibly killed at an independent and exponentially distributed time. The criterion for killing

A.E. Kyprianou, *Fluctuations of Lévy Processes with Applications*, Universitext,
DOI 10.1007/978-3-642-37632-0_7,

is that $\mathbb{P}(\limsup_{t\uparrow\infty} X_t < \infty) = 1$. Suppose this equality fails. Then the probability on the right-hand side of (7.1) is equal to 1. If on the other hand there is killing, then, since killing can occur at arbitrarily small times with positive probability, we have $\mathbb{P}(T_x^+ < \infty) < 1$ for all $x > 0$. In conclusion, we know that

$$\mathbb{P}\big(\tau_x^+ < \infty\big) < 1 \quad \text{for all } x > 0 \quad \Leftrightarrow \quad \mathbb{P}\Big(\limsup_{t\uparrow\infty} X_t < \infty\Big) = 1. \tag{7.2}$$

We devote the remainder of this section to establishing conditions under which $\mathbb{P}(\limsup_{t\uparrow\infty} X_t < \infty) = 1$.

Theorem 7.1 *Suppose that X is a Lévy process.*

(i) *If* $\int_1^\infty t^{-1}\mathbb{P}(X_t \ge 0)\mathrm{d}t < \infty$, *then*

$$\lim_{t\uparrow\infty} X_t = -\infty$$

almost surely and X is said to drift to $-\infty$.

(ii) *If* $\int_1^\infty t^{-1}\mathbb{P}(X_t \le 0)\mathrm{d}t < \infty$, *then*

$$\lim_{t\uparrow\infty} X_t = \infty$$

almost surely and X is said to drift to ∞.

(iii) *If both the integral tests in* (i) *and* (ii) *fail,*[1] *then*

$$\limsup_{t\uparrow\infty} X_t = -\limsup_{t\uparrow\infty} X_t = \infty$$

almost surely and X is said to oscillate.

Proof We follow a similar proof to the one given in Bertoin (1996a).

(i) From Theorem 6.15 (see also the discussion at the end of Sect. 6.4 concerning the adjusted definitions of $\overline{G}_\infty$ and $\underline{G}_\infty$ for the case of compound Poisson processes), we have, for all $\alpha \ge 0$,

$$\mathbb{E}\big(\mathrm{e}^{-\alpha \overline{G}_{\mathbf{e}_p}}\big) = \exp\left\{-\int_0^\infty \big(1-\mathrm{e}^{-\alpha t}\big)\frac{1}{t}\mathrm{e}^{-pt}\mathbb{P}(X_t \ge 0)\mathrm{d}t\right\}. \tag{7.3}$$

Letting p tend to zero in (7.3), and applying the Dominated Convergence Theorem on the left-hand side and the Monotone Convergence Theorem on the right-hand side, we see that

$$\mathbb{E}\big(\mathrm{e}^{-\alpha \overline{G}_\infty}\big) = \exp\left\{-\int_0^\infty \big(1-\mathrm{e}^{-\alpha t}\big)\frac{1}{t}\mathbb{P}(X_t \ge 0)\mathrm{d}t\right\}. \tag{7.4}$$

[1] Note that $\int_1^\infty t^{-1}\mathbb{P}(X_t \ge 0)\mathrm{d}t + \int_1^\infty t^{-1}\mathbb{P}(X_t \le 0)\mathrm{d}t \ge \int_1^\infty t^{-1}\mathrm{d}t = \infty$ and hence at least one of the integral tests in (i) or (ii) fails.

If $\int_1^\infty t^{-1}\mathbb{P}(X_t \geq 0)\mathrm{d}t < \infty$, then since $0 \leq (1 - \mathrm{e}^{-\alpha t}) \leq 1 \wedge t$ for all sufficiently small α, we see that, for the same range of α,

$$\begin{aligned}\int_0^\infty \big(1 - \mathrm{e}^{-\alpha t}\big)\frac{1}{t}\mathbb{P}(X_t \geq 0)\mathrm{d}t &\leq \int_0^\infty (1 \wedge t)\frac{1}{t}\mathbb{P}(X_t \geq 0)\mathrm{d}t \\ &\leq \int_1^\infty \frac{1}{t}\mathbb{P}(X_t \geq 0)\mathrm{d}t + \int_0^1 \mathbb{P}(X_t \geq 0)\mathrm{d}t < \infty.\end{aligned}$$

Hence, once again appealing to the Dominated Convergence Theorem, taking α to zero in (7.4), it follows that

$$\lim_{\alpha \downarrow 0} \int_0^\infty \big(1 - \mathrm{e}^{-\alpha t}\big)\frac{1}{t}\mathbb{P}(X_t \geq 0)\mathrm{d}t = 0,$$

and, therefore, that $\mathbb{P}(\overline{G}_\infty < \infty) = 1$. This implies that $\mathbb{P}(\overline{X}_\infty < \infty) = 1$.

Now noting that $\int_1^\infty t^{-1}\mathbb{P}(X_t \geq 0)\mathrm{d}t = \int_1^\infty t^{-1}(1 - \mathbb{P}(X_t < 0))\mathrm{d}t < \infty$, since $\int_1^\infty t^{-1}\mathrm{d}t = \infty$, we are forced to conclude that

$$\int_1^\infty \frac{1}{t}\mathbb{P}(X_t < 0)\mathrm{d}t = \infty.$$

The Wiener–Hopf factorisation also gives us

$$\mathbb{E}\big(\mathrm{e}^{-\alpha \underline{G}_{\mathbf{e}_p}}\big) = \exp\left\{-\int_0^\infty \big(1 - \mathrm{e}^{-\alpha t}\big)\frac{1}{t}\mathrm{e}^{-pt}\mathbb{P}(X_t \leq 0)\mathrm{d}t\right\}.$$

Taking limits as $p \downarrow 0$ and noting that

$$\int_0^\infty \big(1 - \mathrm{e}^{-\alpha t}\big)\frac{1}{t}\mathbb{P}(X_t \leq 0)\mathrm{d}t \geq k\int_1^\infty \frac{1}{t}\mathbb{P}(X_t \leq 0)\mathrm{d}t = \infty,$$

for some appropriate constant $k > 0$, we get $\mathbb{P}(\underline{G}_\infty = \infty) = 1$. Equivalently, we have $\mathbb{P}(\underline{X}_\infty = -\infty) = 1$.

We have proved that $\limsup_{t\uparrow\infty} X_t < \infty$ and $\liminf_{t\uparrow\infty} X_t = -\infty$ almost surely. This means that

$$\tau_{-x}^- := \inf\{t > 0 : X_t < -x\}$$

is almost surely finite, for each $x > 0$. Note that

$$\{X_t > x/2 \text{ for some } t > 0\} = \{\overline{X}_\infty > x/2\},$$

and hence, since $\mathbb{P}(\overline{X}_\infty < \infty) = 1$, for each $1 > \varepsilon > 0$, there exists an $x_\varepsilon > 0$ such that, for all $x > x_\varepsilon$,

$$\mathbb{P}(X_t > x/2 \text{ for some } t > 0) < \varepsilon.$$

Since τ^-_{-x} is a stopping time which is almost surely finite, we can use the previous estimate together with the strong Markov property and conclude that, for all $x > x_\varepsilon$,

$$\mathbb{P}\big(X_t > -x/2 \text{ for some } t > \tau^-_{-x}\big) \\ \leq \mathbb{P}(X_t > x/2 \text{ for some } t > 0) < \varepsilon.$$

This gives us the uniform estimate for $x > x_\varepsilon$,

$$\mathbb{P}\Big(\limsup_{t\uparrow\infty} X_t \leq -x/2\Big) \geq \mathbb{P}\big(X_t \leq -x/2 \text{ for all } t > \tau^-_{-x}\big) \geq 1-\varepsilon.$$

Since both x may be taken arbitrarily large and ε may be taken arbitrarily close to 0, the proof of part (i) is complete.

(ii) The second part follows from the first part applied to $-X$.

(iii) The argument in (i) shows that if

$$\int_1^\infty t^{-1}\mathbb{P}(X_t \leq 0)\mathrm{d}t = \int_1^\infty t^{-1}\mathbb{P}(X_t \geq 0)\mathrm{d}t = \infty,$$

then $-\underline{X}_\infty = \overline{X}_\infty = \infty$ almost surely and the assertion follows. □

Whilst the last theorem shows that there are only three types of asymptotic behaviour, the integral tests which help to distinguish between the three cases are not particularly user friendly. What would be more appropriate is a criterion in terms of the triple (a, σ, Π). This was provided by Chung and Fuchs (1951) and Erickson (1973) for random walks; see also Bertoin (1997a). To state their criteria, recall with the help of Theorem 3.8 and Exercise 3.3 that the mean of X_1 is well defined if and only if $\mathbb{E}(X_1^+) < \infty$ or $\mathbb{E}(X_1^-) < \infty$ which occurs if and only if

$$\int_{(1,\infty)} x\Pi(\mathrm{d}x) < \infty \quad \text{or} \quad \int_{(-\infty,-1)} |x|\Pi(\mathrm{d}x) < \infty.$$

When both the above integrals are infinite the mean $\mathbb{E}(X_1)$ is undefined.

Theorem 7.2 *Suppose that X is a Lévy process with characteristic measure Π.*

(i) *If $\mathbb{E}(X_1)$ is defined and valued in $[-\infty, 0)$, or if $\mathbb{E}(X_1)$ is undefined and*

$$\int_{(1,\infty)} \frac{x\Pi(\mathrm{d}x)}{\int_0^x \Pi(-\infty,-y)\mathrm{d}y} < \infty,$$

then $\lim_{t\uparrow\infty} X_t/t = c_-$, where $c_- = \mathbb{E}(X_1)$ in the first case and $c_- = -\infty$ in the second case. In particular, in both cases,

$$\lim_{t\uparrow\infty} X_t = -\infty.$$

(ii) *If $\mathbb{E}(X_1)$ is defined and valued in $(0, \infty]$, or if $\mathbb{E}(X_1)$ is undefined and*

$$\int_{(-\infty,-1)} \frac{|x|\Pi(\mathrm{d}x)}{\int_0^{|x|} \Pi(y,\infty)\mathrm{d}y} < \infty,$$

then $\lim_{t\uparrow\infty} X_t/t = c_+$, *where* $c_+ = \mathbb{E}(X_1)$ *in the first case and* $c_+ = \infty$ *in the second case. In particular, in both cases,*

$$\lim_{t\uparrow\infty} X_t = \infty.$$

(iii) *If* $\mathbb{E}(X_1)$ *is defined and equal to zero, or if* $\mathbb{E}(X_1)$ *is undefined and both of the integral tests in part* (i) *and* (ii) *fail, then* $\lim_{t\uparrow\infty} X_t/t = 0$ *in the first case and* $\limsup_{t\uparrow\infty} X_t/t = -\liminf_{t\uparrow\infty} X_t/t = \infty$ *in the second case. Moreover, in both cases,*

$$\limsup_{t\uparrow\infty} X_t = -\liminf_{t\uparrow\infty} X_t = \infty.$$

We give no proof here of this important result, although one may consult Exercise 7.2 for related results, which lean on the classical Strong Law of Large Numbers.

It is interesting to compare the integral tests in the above theorem with those of Theorem 6.5. It would seem that the issue of regularity of 0 for the half-line may be seen as the "small time" analogue of drifting or oscillating. There is no known formal path-wise connection, however.

In the case that X is spectrally negative, thanks to the finiteness and convexity of its Laplace exponent, $\psi(\theta) := \log \mathbb{E}(\mathrm{e}^{\theta X_1})$ on $\theta \geq 0$ (see Exercise 3.5), one always has that $\mathbb{E}(X_1) \in [-\infty, \infty)$. Hence, the asymptotic behaviour of a spectrally negative Lévy process can always be determined from its mean, or equivalently from $\psi'(0+)$. See Exercise 7.3, which shows how to derive this conclusion from Theorem 7.1 and the Wiener–Hopf factorisation.

On account of the dichotomy of drifting to $\pm\infty$ and oscillating, we may now revise the statement (7.2) to

$$\mathbb{P}(\tau_x^+ < \infty) < 1 \quad \text{for all } x > 0 \quad \Leftrightarrow \quad \mathbb{P}\Big(\lim_{t\uparrow\infty} X_t = -\infty\Big) = 1.$$

We close this section by making some brief remarks on the link between drifting and oscillating, and another closely related dichotomy known as *transience* and *recurrence*. The latter dichotomy is often discussed within the more general context of potential theory for Markov processes. See for example Sect. I.4 and Chap. II of Bertoin (1996a).

Definition 7.3 A Lévy process, X, is said to be transient if, for all $a > 0$,

$$\mathbb{P}\left(\int_0^\infty \mathbf{1}_{(|X_t|<a)} \mathrm{d}t < \infty\right) = 1,$$

and recurrent if, for all $a > 0$,

$$\mathbb{P}\left(\int_0^\infty \mathbf{1}_{(|X_t|<a)} \mathrm{d}t = \infty\right) = 1.$$

In the previous definition, the requirements for transience and recurrence may appear quite strong as, in principle, the relevant probabilities could be valued in $(0,1)$. However, the events in the definition belong to the tail sigma-algebra $\bigcap_{t\in\mathbb{Q}\cap[0,\infty)}\sigma(X_s : s \geq t)$. Hence, according to Kolmogorov's zero-one law, they can only have probabilities equal to zero or one. Nonetheless, we could argue that $\mathbb{P}(\int_0^\infty \mathbf{1}_{(|X_t|<a)}\mathrm{d}t = \infty) = 0$ for small a, but $\mathbb{P}(\int_0^\infty \mathbf{1}_{(|X_t|<a)}\mathrm{d}t = \infty) = 1$ for large a. It turns out that Lévy processes always adhere to one of the two cases given in the definition above, as is confirmed by the following classic analytic dichotomy, due to Port and Stone (1971a, 1971b).[2]

Theorem 7.4 *Suppose that X is a Lévy process with characteristic exponent Ψ, then it is transient if and only if, for some sufficiently small $\varepsilon > 0$,*

$$\int_{(-\varepsilon,\varepsilon)} \Re\left(\frac{1}{\Psi(\theta)}\right)\mathrm{d}\theta < \infty,$$

and otherwise it is recurrent.

Probabilistic reasoning also leads to the following interpretation of the dichotomy.

Theorem 7.5 *Let X be any Lévy process.*

(i) *We have transience if and only if*

$$\lim_{t\uparrow\infty} |X_t| = \infty$$

almost surely.

(ii) *If X is not a compound Poisson process, then we have recurrence if and only if, for all $x \in \mathbb{R}$,*

$$\liminf_{t\uparrow\infty} |X_t - x| = 0 \tag{7.5}$$

almost surely.

The reason for the exclusion of compound Poisson processes in part (ii) can be seen when one considers the following example. Take X to be a compound Poisson process, where the jump distribution is supported on a lattice, say $\delta\mathbb{Z}$ for some $\delta > 0$. In that case, it is clear that the set of points visited will be a subset of $\delta\mathbb{Z}$ and (7.5) no longer makes sense.

By definition, a process which is recurrent cannot drift to ∞ or $-\infty$, and therefore must oscillate. Whilst it is clear that a process drifting to ∞ or $-\infty$ is transient, an oscillating process may not necessarily be recurrent. Indeed, it is possible to construct an example of a transient process which oscillates. Inspired by

[2] Theorem 7.4 is built on the foundational, but weaker, result of Chung and Fuchs (1951). See also Kingman (1964).

similar remarks for random walks in Feller (1971), one finds such an example in the form of a symmetric stable process of index $0 < \alpha < 1$. Note that symmetry dictates that the parameter β in the definition of a stable process is necessarily equal to zero. Up to a multiplicative constant, the characteristic exponent for this process is simply $\Psi(\theta) = |\theta|^{\alpha}$. According to the integral test in Theorem 7.4, members of this class of processes are transient. Nonetheless, since by symmetry $\mathbb{P}(X_t \geq 0) = 1/2 = \mathbb{P}(X_t \leq 0)$, it is clear from Theorem 7.1 that X oscillates. In contrast, note that for a one-dimensional linear Brownian motion, the conditions of oscillation and recurrence coincide as do the definitions of transience and drifting to $\pm\infty$. Intuitively speaking, the reason for this difference is because symmetric stable processes with $\alpha \in (0, 1)$ do not have a well-defined mean at each fixed time, whereas Brownian motion has zero mean at each fixed time.

7.2 Cramér's Estimate

In this section, we extend the classical result of Cramér that was presented earlier in Theorem 1.10 for the case of a general Lévy process. The Lévy process we consider may have a relatively general jump structure (in particular, positive jumps are permitted). We follow the treatment of Bertoin and Doney (1994a). Roughly speaking, our aim is to show that, under suitable conditions, there exists a constant $\nu > 0$ so that $\mathrm{e}^{\nu x}\mathbb{P}(\tau_x^+ < \infty)$ has a limit as $x \uparrow \infty$. The result is formulated as follows.

Theorem 7.6 *Suppose that X is a Lévy process which does not have monotone paths. Assume that*

(i) $\lim_{t\uparrow\infty} X_t = -\infty$,
(ii) *there exists a $\nu \in (0, \infty)$ such that $\psi(\nu) = 0$, where $\psi(\theta) = \log \mathbb{E}(\exp\{\theta X_1\})$ is the Laplace exponent of X and*
(iii) *the support of Π is not lattice if $\Pi(\mathbb{R}) < \infty$.*

Then

$$\lim_{x\uparrow\infty} \mathrm{e}^{\nu x}\mathbb{P}\big(\tau_x^+ < \infty\big) = \kappa(0, 0)\left(\nu \frac{\partial \kappa(0, \beta)}{\partial \beta}\bigg|_{\beta=-\nu}\right)^{-1}, \tag{7.6}$$

where the limit is interpreted to be zero if the derivative on the right-hand side is infinite.

Note that condition (ii) implies the existence of $\mathbb{E}(X_1)$ and, on account of the conclusion in Theorem 7.2, condition (i) implies, further, that $\mathbb{E}(X_1) < 0$. We know that if the moment generating function of X_1 exists in the positive half-line, then it must be convex there (this may be shown using arguments similar to those in Exercise 3.5, or alternatively note the remarks in the proof of Theorem 3.9). Conditions (i) and (ii) therefore also imply that the function $\psi(\theta)$ is negative for $\theta \in (0, \nu)$ and equal to zero at the end points of this interval. Condition (ii) is known as *Cramér's condition*. Essentially Theorem 7.6, known as *Cramér's estimate*, says that the existence of exponential moments of a Lévy process which drifts to $-\infty$ implies an exponentially decaying tail of the distribution of its global maximum. Indeed, note that $\mathbb{P}(\tau_x^+ < \infty) = \mathbb{P}(\overline{X}_\infty > x)$. Since renewal theory will play a predominant role in the proof, the third condition of Theorem 7.6 is simply for convenience, allowing the use of the Renewal Theorem without running into the special case of lattice supports. Nonetheless, it is possible to remove condition (iii). See Bertoin and Doney (1994a) for further details.

Proof of Theorem 7.6 The proof is long and we break it into steps.

Step 1. Define the potential measure for the ascending ladder height process on Borel sets $A \in [0,\infty)$ by

$$U(A) = \mathbb{E}\left(\int_0^\infty \mathbf{1}_{(H_t \in A)} \mathrm{d}t\right),$$

where $H = \{H_t : t \geq 0\}$ is the ascending ladder height process. Let $L = \{L_t : t \geq 0\}$ be the local time of X at its running maximum and define $T_x^+ = \inf\{t > 0 : H_t > x\}$. Applying the strong Markov property at this stopping time, we get

$$\begin{aligned} U(x,\infty) &= \mathbb{E}\left(\int_0^\infty \mathbf{1}_{(H_t > x)} \mathrm{d}t;\, T_x^+ < L_\infty\right) \\ &= \mathbb{P}(T_x^+ < L_\infty)\mathbb{E}\left(\int_{T_x^+}^{L_\infty} \mathbf{1}_{(H_t > x)} \mathrm{d}t \,\middle|\, H_s : s \leq T_x^+\right) \\ &= \mathbb{P}(T_x^+ < L_\infty)\mathbb{E}\left(\int_0^{L_\infty} \mathbf{1}_{(H_t \geq 0)} \mathrm{d}t\right) \\ &= \mathbb{P}(T_x^+ < L_\infty)\mathbb{E}(L_\infty). \end{aligned} \tag{7.7}$$

Since $\lim_{t\uparrow\infty} X_t = -\infty$, we know that L_∞ is exponentially distributed with some parameter which is recovered from the joint Laplace exponent $\kappa(\alpha,\beta)$ by setting $\alpha = \beta = 0$. Note also that $\mathbb{P}(T_x^+ < L_\infty) = \mathbb{P}(\overline{X}_\infty > x) = \mathbb{P}(\tau_x^+ < \infty)$. Hence, (7.7) now takes the form

$$\kappa(0,0)U(x,\infty) = \mathbb{P}(\tau_x^+ < \infty). \tag{7.8}$$

Step 2. In order to complete the proof, we need to establish a precise asymptotic for $\mathrm{e}^{\nu x}U(x,\infty)$. To this end, we shall show, via a change of measure, that in fact $U_\nu(\mathrm{d}x) := \mathrm{e}^{\nu x}U(\mathrm{d}x)$ on $(0,\infty)$ is a potential measure.[3] In that case, the Key Renewal Theorem 5.1 (ii) will help us clarify the required asymptotic.

[3] Recall the definition in (5.2).

Since $\psi(\nu)=0$, we know (cf. Chap. 3) that $\{\exp\{\nu X_t\}: t\geq 0\}$ is a martingale with unit mean. Hence, it can be used to define a change of measure via

$$\left.\frac{\mathrm{d}\mathbb{P}^\nu}{\mathrm{d}\mathbb{P}}\right|_{\mathcal{F}_t}=\mathrm{e}^{\nu X_t},\quad t\geq 0,$$

which, by Theorem 3.9, keeps the process X within the class of Lévy processes. From Theorem 6.8, we know that L_t^{-1} is a stopping time and, hence, we have, with the help of Corollary 3.11, that for all $x\geq 0$,

$$\begin{aligned}\mathbb{P}^\nu\big(H_t\in\mathrm{d}x, L_t^{-1}<s\big)&=\mathbb{E}\big(\mathrm{e}^{\nu X_{L_t^{-1}}}; H_t\in\mathrm{d}x, L_t^{-1}<s\big)\\&=\mathrm{e}^{\nu x}\mathbb{P}\big(H_t\in\mathrm{d}x, L_t^{-1}<s\big).\end{aligned}$$

Appealing to monotone convergence and taking $s\uparrow\infty$,

$$\mathbb{P}^\nu(H_t\in\mathrm{d}x)=\mathrm{e}^{\nu x}\mathbb{P}(H_t\in\mathrm{d}x). \tag{7.9}$$

Now note that, on $[0,\infty)$,

$$U_\nu(\mathrm{d}x)=\mathrm{e}^{\nu x}U(\mathrm{d}x)=\int_0^\infty \mathbb{P}^\nu(H_t\in\mathrm{d}x)\mathrm{d}t.$$

The final equality shows that $U_\nu(\mathrm{d}x)$ is equal to the potential measure of the ascending ladder height process H under $\mathbb{P}^\nu$. According to Lemma 5.2, $U_\nu(\mathrm{d}x)$ is equal to a renewal measure providing that H is a subordinator under $\mathbb{P}^\nu$ (as opposed to a killed subordinator). This is proved in the next step.

Step 3. A similar argument to the one above yields

$$\mathbb{P}^\nu(\widehat{H}_t\in\mathrm{d}x)=\mathrm{e}^{-\nu x}\mathbb{P}(\widehat{H}_t\in\mathrm{d}x),$$

where, now, $\widehat{H}$ is the descending ladder height process. From the last two equalities, we can easily deduce that the Laplace exponent, $\widehat{\kappa}^\nu$, of the descending ladder process under the measure $\mathbb{P}^\nu$ satisfies

$$\widehat{\kappa}^\nu(0,\beta)=\widehat{\kappa}(0,\beta+\nu).$$

This shows, in particular, that $\widehat{\kappa}^\nu(0,0)=\widehat{\kappa}(0,\nu)>0$. This is the exponential rate with which the local time $\widehat{L}_\infty$ is distributed under $\mathbb{P}^\nu$. We therefore have $\mathbb{P}^\nu(\liminf_{t\uparrow\infty}X_t>-\infty)=1$. By Theorem 7.1, this is equivalent to $\mathbb{P}^\nu(\lim_{t\uparrow\infty}X_t=\infty)=1$. We now have, as required in the previous step, that H, under $\mathbb{P}^\nu$, is a subordinator without killing.

Step 4. We would like to use the Renewal Theorem in conjunction with $U_\nu(\mathrm{d}x)$. Note from Lemma 5.2 that the underlying distribution of this renewal measure is given by $F(\mathrm{d}x)=U_\nu^{(1)}(\mathrm{d}x)$ on $[0,\infty)$. In order to calculate its mean, we need to reconsider briefly some properties of κ^ν.

From (7.9), we deduce that $\kappa(0,\beta)<\infty$, for $\beta\geq -\nu$. Recall that convexity of ψ on $(0,\infty)$ (see the proof of Theorem 3.9) implies that it is also finite on $[0,\nu]$. We may now appeal to analytic extension and conclude from Theorem 6.15 (iv) that

$$\Psi(\theta-\mathrm{i}\beta)=-\psi(\beta+\mathrm{i}\theta)=k'\kappa(0,-\beta-\mathrm{i}\theta)\widehat{\kappa}(0,\beta+\mathrm{i}\theta),$$

for some $k'>0$, any $\beta\in[0,\nu]$ and $\theta\in\mathbb{R}$. Now setting $\beta=\nu$ and $\theta=0$, we deduce further that

$$-\psi(\nu)=0=k'\kappa(0,-\nu)\widehat{\kappa}(0,\nu).$$

Since $k'\widehat{\kappa}(0,\nu)>0$, we conclude that $\kappa(0,-\nu)=0$.

We may now compute the mean of the distribution F:

$$\begin{aligned}\mu&=\int_{[0,\infty)}xU_\nu^{(1)}(\mathrm{d}x)\\&=\int_0^\infty \mathrm{d}t\cdot \mathrm{e}^{-t}\int_{[0,\infty)}x\mathbb{P}^\nu(H_t\in \mathrm{d}x)\\&=\int_0^\infty \mathrm{d}t\cdot \mathrm{e}^{-t}\mathbb{E}\big(H_t\mathrm{e}^{\nu H_t}\big)\\&=\int_0^\infty \mathrm{d}t\cdot \mathrm{e}^{-t-\kappa(0,-\nu)t}\left.\frac{\partial\kappa(0,\beta)}{\partial\beta}\right|_{\beta=-\nu}\\&=\left.\frac{\partial\kappa(0,\beta)}{\partial\beta}\right|_{\beta=-\nu},\end{aligned}$$

which is possibly infinite in value.

Finally, appealing to the Key Renewal Theorem 5.1 (ii), we have that $U_\nu(\mathrm{d}x)$ converges weakly as a measure to $\mu^{-1}\mathrm{d}x$. Hence, it now follows from (7.8) that

$$\begin{aligned}\lim_{x\uparrow\infty}\mathrm{e}^{\nu x}\mathbb{P}\big(\tau_x^+<\infty\big)&=\kappa(0,0)\lim_{x\uparrow\infty}\int_x^\infty \mathrm{e}^{-\nu(y-x)}U_\nu(\mathrm{d}y)\\&=\kappa(0,0)\lim_{x\uparrow\infty}\int_0^\infty \mathrm{e}^{-\nu z}U_\nu(x+\mathrm{d}z)\\&=\frac{\kappa(0,0)}{\mu}\int_0^\infty \mathrm{e}^{-\nu z}\mathrm{d}z\\&=\frac{\kappa(0,0)}{\nu\mu},\end{aligned}$$

where the limit is understood to be zero if $\mu=\infty$. □

Let us close this section by making a couple of remarks. Firstly, in the case where X is spectrally negative, the Laplace exponent $\psi(\theta)$ is finite on $\theta\geq 0$. When

$\psi'(0+) < 0$, condition (i) of Theorem 7.6 holds. In that case, we know already from Theorem 3.12 that

$$\mathbb{E}\big(\mathrm{e}^{\Phi(q)x - q\tau_x^+}\mathbf{1}_{(\tau_x^+<\infty)}\big) = 1,$$

where Φ is the right inverse of ψ. Taking $q \downarrow 0$, we recover

$$\mathrm{e}^{\Phi(0)x}\mathbb{P}\big(\tau_x^+ < \infty\big) = 1,$$

for all $x \geq 0$, which is a stronger statement than that of Theorem 7.6. Taking account of the fact that the Wiener–Hopf factorisation gives $\kappa(\alpha, \beta) = \beta + \Phi(\alpha)$, for $\alpha, \beta \geq 0$, one may also check for consistency that the constant on the right-hand side of (7.6) is equal to 1.

Secondly, when X is any spectrally positive Lévy process of bounded variation, it is a straightforward exercise to show that formula (7.8) can be rewritten to give the Pollaczek–Khintchine formula, consistently with the one given in (4.20). The point here is that, by irregularity of the upper half-line, the ascending ladder height process H, whose potential measure is U, is equal in law to a killed compound Poisson subordinator whose jumps have the integrated tail distribution given in (4.16).

7.3 A Quintuple Law at First Passage

In this section, we shall give a quantitative account of how a general Lévy process undergoes first passage over a fixed barrier on the event that it jumps clear over it. There will be a number of parallels between the analysis here and the analysis in Chap. 5 concerning first passage of a subordinator. Since subordinators have already been dealt with, they are excluded from the following discussion.

Recall the notation from Chap. 6

$$\overline{G}_t = \sup\{s < t : \overline{X}_s = X_s\}$$

and our standard notation, already used in this chapter,

$$\tau_x^+ = \inf\{t > 0 : X_t > x\}.$$

The centrepiece of this section will concern a quintuple law at first passage, involving

- $\overline{G}_{\tau_x^+-}$: the time of the last maximum prior to first passage,
- $\tau_x^+ - \overline{G}_{\tau_x^+-}$: the length of the excursion making first passage,
- $X_{\tau_x^+} - x$: the overshoot at first passage,
- $x - X_{\tau_x^+-}$: the undershoot at first passage,
- $x - \overline{X}_{\tau_x^+-}$: the undershoot of the last maximum at first passage.

In order to state the main result of this section, let us introduce some more notation. Recall from Chap. 6 that, for $\alpha, \beta \geq 0$, $\kappa(\alpha, \beta)$ is the Laplace exponent of the ascending ladder process (L^{-1}, H); see (6.8). Associated with $\kappa(\alpha,\beta)$ is the bivariate potential measure

$$\mathcal{U}(\mathrm{d}s,\mathrm{d}x)=\int_0^\infty \mathrm{d}t\cdot\mathbb{P}\big(L_t^{-1}\in \mathrm{d}s, H_t\in\mathrm{d}x\big),\quad x,s\geq 0.$$

On taking a bivariate Laplace transform, we find, with the help of Fubini's Theorem, that

$$\int_{[0,\infty)^2}\mathrm{e}^{-\alpha s-\beta x}\mathcal{U}(\mathrm{d}s,\mathrm{d}x)=\int_0^\infty \mathrm{d}t\cdot\mathbb{E}\big(\mathrm{e}^{-\alpha L_t^{-1}-\beta H_t}\big)=\frac{1}{\kappa(\alpha,\beta)},\qquad(7.10)$$

for $\alpha,\beta\geq 0$. Since L can only be defined up to a multiplicative constant, this affects the exponent κ, which in turn affects the measure $\mathcal{U}$. To see precisely how, suppose that $\mathcal{L}=cL$, where L is some choice of local time at the maximum (and hence so is $\mathcal{L}$). It is easily checked that $\mathcal{L}_t^{-1}=L_{t/c}^{-1}$ and if $\mathcal{H}$ is the ladder height process associated with $\mathcal{L}$, then $\mathcal{H}_t=X_{\mathcal{L}_t^{-1}}=X_{L_{t/c}^{-1}}=H_{t/c}$. If $\mathcal{U}^*$ is the measure associated with $\mathcal{L}$ instead of L, then we see that

$$\mathcal{U}^*(\mathrm{d}s,\mathrm{d}x)=\int_0^\infty \mathrm{d}t\cdot\mathbb{P}\big(L_{t/c}^{-1}\in\mathrm{d}s, H_{t/c}\in\mathrm{d}x\big)=c\mathcal{U}(\mathrm{d}s,\mathrm{d}x),\quad s,x\geq 0,$$

where the final equality follows by the substitution $u=t/c$ in the integral.

We shall define the bivariate measure $\widehat{\mathcal{U}}$ on $[0,\infty)^2$ in the obvious way, using the descending ladder process $(\widehat{L}^{-1},\widehat{H})$.

The following main result is due to Doney and Kyprianou (2005), although similar ideas to those used in the proof can be found in Spitzer (1964), Borovkov (1976) and Bertoin (1996a).

Theorem 7.7 *Suppose that X is not a compound Poisson process. Then there exists a normalisation of local time at the maximum such that, for each $x>0$, we have on $u>0$, $v\geq y$, $y\in[0,x]$, $s,t\geq 0$,*

$$\begin{aligned}&\mathbb{P}\big(\tau_x^+-\overline{G}_{\tau_x^+-}\in\mathrm{d}t,\overline{G}_{\tau_x^+-}\in\mathrm{d}s,X_{\tau_x^+}-x\in\mathrm{d}u,x-X_{\tau_x^+-}\in\mathrm{d}v,x-\overline{X}_{\tau_x^+-}\in\mathrm{d}y\big)\\&\quad=\mathcal{U}(\mathrm{d}s,x-\mathrm{d}y)\widehat{\mathcal{U}}(\mathrm{d}t,\mathrm{d}v-y)\Pi(\mathrm{d}u+v),\end{aligned}$$

where Π is the Lévy measure of X.

Before going to the proof, let us give some intuition behind the statement of this result with the help of Fig. 7.1. Roughly speaking the event on the left-hand side of the quintuple law requires that the time-space point $(s,x-y)$ belongs to the range of the ascending ladder process, before going into the final excursion that crosses the level x. Recall that excursions, when indexed by local time at the maximum, form a Poisson point process. This means that the behaviour of the last excursion is independent of the preceding ones, and hence the quintuple law factorises according the laws of the last excursion and the preceding excursions. The first factor,

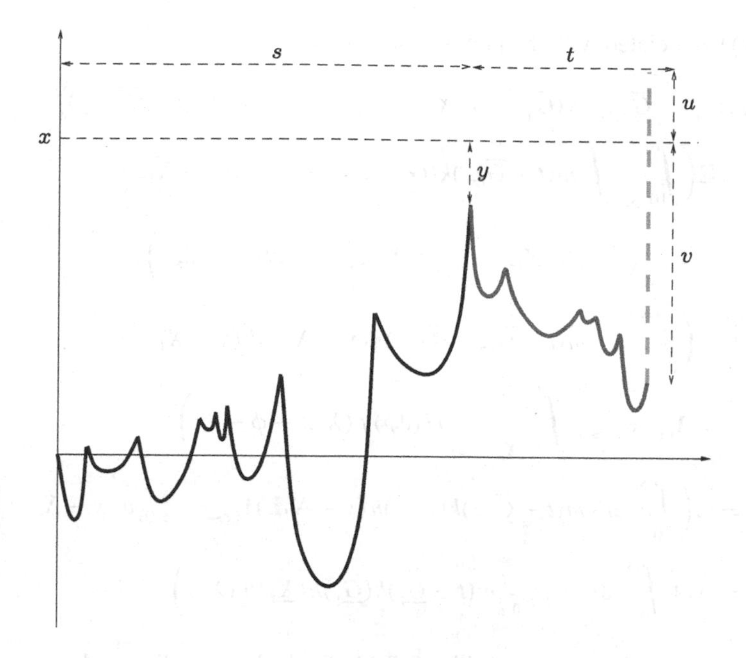

Fig. 7.1 A symbolic description of the quantities involved in the quintuple law.

$\mathcal{U}(\mathrm{d}s, x - \mathrm{d}y)$, thus measures the aforementioned event for the ascending ladder process. To measure the behaviour of the final excursion, one should look at it rotated about 180°. In the rotated excursion, one starts with a jump of size $u + v$, which is measured by $\Pi(\mathrm{d}u + v)$. The remaining path of the rotated excursion must meet the last ascending ladder height with one of its own descending ladder points. By the Duality Lemma 3.4, 180° rotation of a finite segment of path of a Lévy process produces a path with the same law as the original process. Hence in the rotated excursion, independently of the initial jump of size $u + v$, the path descends to time-space ladder point $(t, v - y)$, and this has measure $\widehat{\mathcal{U}}(\mathrm{d}t, \mathrm{d}v - y)$.

Proof of Theorem 7.7 We prove the result in three steps.

Step 1. Let us suppose that m, k, f, g and h are all positive, continuous functions with compact support satisfying $f(0) = g(0) = h(0) = 0$. We prove in this step that

$$\mathbb{E}\big(m(\tau_x^+ - \overline{G}_{\tau_x^+-})k(\overline{G}_{\tau_x^+-})f(X_{\tau_x^+} - x)g(x - X_{\tau_x^+-})h(x - \overline{X}_{\tau_x^+-})\big)$$

$$= \widehat{\mathbb{E}}_x\left(\int_0^{\tau_0^-} m(t - \underline{G}_t)k(\underline{G}_t)h(\underline{X}_t)w(X_t)\mathrm{d}t\right), \tag{7.11}$$

where $w(z) = g(z)\int_{(z,\infty)}\Pi(\mathrm{d}u)f(u - z)$ and $\widehat{\mathbb{E}}_x$ is expectation under the law, $\widehat{\mathbb{P}}_x$, of $-X$ initiated from position $-X_0 = x$.

The proof of (7.11) follows by an application of the compensation formula (cf. Theorem 4.4) applied to the Poisson random measure, N (with intensity measure

$\mathrm{d}t\Pi(\mathrm{d}x))$ associated with the jumps of X. We have

$$\begin{aligned}
&\mathbb{E}\big(m(\tau_x^+ - \overline{G}_{\tau_x^+ -})k(\overline{G}_{\tau_x^+ -})f(X_{\tau_x^+} - x)g(x - X_{\tau_x^+ -})h(x - \overline{X}_{\tau_x^+ -})\big)\\
&\quad = \mathbb{E}\bigg(\int_{[0,\infty)}\int_{\mathbb{R}} m(t - \overline{G}_{t-})k(\overline{G}_{t-})g(x - X_{t-})h(x - \overline{X}_{t-})\\
&\qquad \times \mathbf{1}_{(x-\overline{X}_{t-}>0)} f(X_{t-} + z - x)\mathbf{1}_{(z>x-X_{t-})}N(\mathrm{d}t\times \mathrm{d}z)\bigg)\\
&\quad = \mathbb{E}\bigg(\int_0^\infty \mathrm{d}t\cdot m(t - \overline{G}_{t-})k(\overline{G}_{t-})g(x - X_{t-})h(x - \overline{X}_{t-})\\
&\qquad \times \mathbf{1}_{(x-\overline{X}_{t-}>0)}\int_{(x-X_{t-},\infty)} \Pi(\mathrm{d}\phi)f(X_{t-}+\phi - x)\bigg)\\
&\quad = \mathbb{E}\bigg(\int_0^\infty \mathrm{d}t\cdot m(t - \overline{G}_{t-})k(\overline{G}_{t-})h(x - \overline{X}_{t-})\mathbf{1}_{(x-\overline{X}_{t-}>0)}w(x - X_{t-})\bigg)\\
&\quad = \widehat{\mathbb{E}}_x\bigg(\int_0^\infty \mathrm{d}t\cdot \mathbf{1}_{(t<\tau_0^-)}m(t - \underline{G}_t)k(\underline{G}_t)h(\underline{X}_t)w(X_t)\bigg),
\end{aligned}$$

which is equal to the right-hand side of (7.11). In the last equality, we have rewritten the previous expectation in terms of the path of $-X$. Note that the condition $f(0) = g(0) = h(0) = 0$ has been used implicitly to exclude from the calculation averaging over the event $\{X_{\tau_x^+} = x\}$.

Step 2. Next, we prove that

$$\begin{aligned}
&\mathbb{E}_x\bigg(\int_0^{\tau_0^-} m(t - \underline{G}_t)k(\underline{G}_t)h(\underline{X}_t)w(X_t)\mathrm{d}t\bigg)\\
&\quad = \int_{[0,\infty)}\int_{[0,\infty)} \mathcal{U}(\mathrm{d}t, \mathrm{d}\phi)\\
&\qquad \cdot \int_{[0,\infty)}\int_{[0,x]} \widehat{\mathcal{U}}(\mathrm{d}s, \mathrm{d}\theta)m(t)k(s)h(x-\theta)w(x+\phi-\theta).
\end{aligned} \tag{7.12}$$

(In the next step, we will apply this identity to the process $-X$, which amounts to swapping throughout the roles of $\mathbb{E}_x$ and $\widehat{\mathbb{E}}_x$, and $\mathcal{U}$ and $\widehat{\mathcal{U}}$.)

For $q > 0$,

$$\begin{aligned}
&\mathbb{E}_x\bigg(\int_0^{\tau_0^-} \mathrm{d}t\cdot m(t - \underline{G}_t)k(\underline{G}_t)h(\underline{X}_t)w(X_t)\mathrm{e}^{-qt}\bigg)\\
&\quad = q^{-1}\mathbb{E}_x\big(m(\mathbf{e}_q - \underline{G}_{\mathbf{e}_q})k(\underline{G}_{\mathbf{e}_q})h(\underline{X}_{\mathbf{e}_q})w(X_{\mathbf{e}_q} - \underline{X}_{\mathbf{e}_q} + \underline{X}_{\mathbf{e}_q}); \mathbf{e}_q < \tau_0^-\big)\\
&\quad = q^{-1}\int_{[0,\infty)}\int_{[0,x]} \mathbb{P}(\underline{G}_{\mathbf{e}_q}\in \mathrm{d}s, -\underline{X}_{\mathbf{e}_q}\in \mathrm{d}\theta)k(s)
\end{aligned}$$

$$\cdot \int_{[0,\infty)} \int_{[0,\infty)} \mathbb{P}(\mathbf{e}_q - \underline{G}_{\mathbf{e}_q} \in \mathrm{d}t, X_{\mathbf{e}_q} - \underline{X}_{\mathbf{e}_q} \in \mathrm{d}\phi) m(t) h(x-\theta) w(x+\phi-\theta)$$

$$= q^{-1} \int_{[0,\infty)} \int_{[0,x]} \mathbb{P}(\underline{G}_{\mathbf{e}_q} \in \mathrm{d}s, -\underline{X}_{\mathbf{e}_q} \in \mathrm{d}\theta) k(s)$$

$$\cdot \int_{[0,\infty)} \int_{[0,\infty)} \mathbb{P}(\overline{G}_{\mathbf{e}_q} \in \mathrm{d}t, \overline{X}_{\mathbf{e}_q} \in \mathrm{d}\phi) m(t) h(x-\theta) w(x+\phi-\theta), \tag{7.13}$$

where the Wiener–Hopf factorisation[4] and duality have been used in the second and third equalities, respectively. Further, it is also known from the Wiener–Hopf factorisation, Theorem 6.15, that, for $q > 0$ and $\alpha, \beta \geq 0$,

$$\frac{1}{\kappa(q,0)} \mathbb{E}\left(\mathrm{e}^{-\alpha \overline{G}_{\mathbf{e}_q} - \beta \overline{X}_{\mathbf{e}_q}}\right) = \frac{1}{\kappa(\alpha+q,\beta)},$$

and hence, recalling (7.10), it follows from the Continuity Theorem for Laplace transforms (cf. Theorem 2a in Chap. XIII.1 of Feller (1971)) that, for $t, \phi \geq 0$,

$$\lim_{q \downarrow 0} \frac{1}{\kappa(q,0)} \mathbb{P}(\overline{G}_{\mathbf{e}_q} \in \mathrm{d}t, \overline{X}_{\mathbf{e}_q} \in \mathrm{d}\phi) = \mathcal{U}(\mathrm{d}t, \mathrm{d}\phi), \tag{7.14}$$

in the sense of vague convergence. A similar convergence holds for

$$\mathbb{P}(\underline{G}_{\mathbf{e}_q} \in \mathrm{d}s, -\underline{X}_{\mathbf{e}_q} \in \mathrm{d}\theta) / \widehat{\kappa}(q,0), \quad s, \theta \geq 0.$$

Equality (7.12) thus follows by splitting the divisor q into the product $\kappa(q,0) \times \widehat{\kappa}(q,0)$ (this factorisation was observed in the proof of Theorem 6.15 (iv)) and taking limits in (7.13). In general, $q = k\kappa(q,0)\widehat{\kappa}(q,0)$ for some $k > 0$, which depends on the normalisation of local time (at the maximum). It is thus at this point in the argument that we require a suitable normalisation of local time at the maximum in order to have $k = 1$.

Step 3. We combine the conclusions of steps 1 and 2 (where step 2 is applied to $-X$) to conclude that

$$\mathbb{E}\big(m(\tau_x^+ - \overline{G}_{\tau_x^+-}) k(\overline{G}_{\tau_x^+-}) f(X_{\tau_x^+} - x) g(x - X_{\tau_x^+-}) h(x - \overline{X}_{\tau_x^+-})\big)$$

$$= \int_{u>0, y \in [0,x], 0<y\leq v, s\geq 0, t\geq 0} m(t) k(s) f(u) g(v) h(y)$$

$$\mathbb{P}\big(\tau_x^+ - \overline{G}_{\tau_x^+-} \in \mathrm{d}t, \overline{G}_{\tau_x^+-} \in \mathrm{d}s, X_{\tau_x^+} - x \in \mathrm{d}u, x - X_{\tau_x^+-} \in \mathrm{d}v, x - \overline{X}_{\tau_x^+-} \in \mathrm{d}y\big)$$

$$= \int_{[0,\infty)} \int_{[0,\infty)} \widehat{\mathcal{U}}(\mathrm{d}t, \mathrm{d}\phi) \int_{[0,\infty)} \int_{[0,x]} \mathcal{U}(\mathrm{d}s, \mathrm{d}\theta) m(t) k(s)$$

$$\cdot h(x-\theta) g(x+\phi-\theta) \int_{(x+\phi-\theta,\infty)} \Pi(\mathrm{d}\eta) f\big(\eta - (x+\phi-\theta)\big).$$

---

[4] Specifically we use the independence of the pairs $(\underline{G}_{\mathbf{e}_q}, \underline{X}_{\mathbf{e}_q})$ and $(\mathbf{e}_q - \underline{G}_{\mathbf{e}_q}, X_{\mathbf{e}_q} - \underline{X}_{\mathbf{e}_q})$.

Substituting $y = x - \theta$, then $y + \phi = v$ and finally $\eta = v + u$ in the right-hand side above yields

$$\mathbb{E}\big(m(\tau_x^+ - \overline{G}_{\tau_x^+-})k(\overline{G}_{\tau_x^+-})f(X_{\tau_x^+} - x)g(x - X_{\tau_x^+-})h(x - \overline{X}_{\tau_x^+-})\big)$$
$$= \int_{[0,\infty)}\int_{[0,x]} \mathcal{U}(\mathrm{d}s, x - \mathrm{d}y)\int_{[0,\infty)}\int_{[y,\infty)} \widehat{\mathcal{U}}(\mathrm{d}t, \mathrm{d}v - y)$$
$$\cdot \int_{(0,\infty)} \Pi(\mathrm{d}u + v)m(t)k(s)f(u)g(v)h(y),$$

and the statement of the theorem follows. □

The case of a compound Poisson process has been excluded from the statement of the theorem on account of the additional subtleties that occur in connection with the ascending and descending ladder height processes and their definitions in the weak or strict sense. (Recall the discussion of weak and strict ladder processes in Sect. 6.1.) Nonetheless, the result is still valid, provided one takes the bivariate renewal measure $\mathcal{U}$ as that of the *weak (resp. strict)* ascending ladder process and $\widehat{\mathcal{U}}$ is taken as the bivariate renewal measure of the *strict (resp. weak)* descending ladder process.

To be realistic, the quintuple law in general does not necessarily bring us closer to explicit formulae for special examples of Lévy processes. Indeed, for this to be the case, we would need to know some explicit examples of the pairs $\mathcal{U}$ and $\widehat{\mathcal{U}}$. Ultimately, this boils down to knowing explicit examples of the Wiener–Hopf factorisation. Nonetheless, there are examples where one may make reasonable progress in making these formulae more explicit. We consider here two cases: stable processes, dealt with in Exercise 7.4, and spectrally positive processes.

For any spectrally positive Lévy process X, let $U(\mathrm{d}x) = \int_{[0,\infty)} \mathcal{U}(\mathrm{d}s, \mathrm{d}x)$. Using the Wiener–Hopf factorisation in Sect. 6.5.2, which gives an expression for $\kappa(\alpha, \beta)$, we can deduce from the Laplace transform (7.10) that

$$\int_{[0,\infty)} \mathrm{e}^{-\beta x} U(\mathrm{d}x) = \frac{\beta - \Phi(0)}{\psi(\beta)}, \tag{7.15}$$

where Φ is the right inverse of the Laplace exponent ψ of $-X$. Using obvious notation, it is also clear from (7.10) that since $\widehat{\kappa}(0, \beta) = \Phi(0) + \beta$, $\beta \geq 0$, we may identify $\widehat{U}(\mathrm{d}x) = \mathrm{e}^{-\Phi(0)x}\mathrm{d}x$, $x \geq 0$.

The quintuple law for spectrally positive Lévy processes marginalises to the triple law

$$\mathbb{P}(X_{\tau_x^+} - x \in \mathrm{d}u, x - X_{\tau_x^+-} \in \mathrm{d}v, x - \overline{X}_{\tau_x^+-} \in \mathrm{d}y)$$
$$= \mathrm{e}^{-\Phi(0)(v-y)} U(x - \mathrm{d}y)\Pi(\mathrm{d}u + v)\mathrm{d}v \tag{7.16}$$

for $y \in [0, x]$, $v \geq y$ and $u > 0$. If we assume further that $\liminf_{t\uparrow\infty} X_t = -\infty$, then we know that $\Phi(0) = 0$ and the right-hand side of (7.16) is written in terms of Π and the inverse Laplace transform of $\beta/\psi(\beta)$.

7.4 The Jump Measure of the Ascending Ladder Height Process

Recall that basic information concerning the ladder height process, H, is captured from its Laplace exponent $\kappa(0,\beta)$, which itself is embedded in the Wiener–Hopf factorisation. In this section, we shall show that the quintuple law (which heuristically contains a similar amount of information to the Wiener–Hopf factorisation) allows us to gain some additional insight into the analytical form of the jump measure of the ascending ladder height. The next result is due to Vigon (2002b).

Theorem 7.8 *Suppose that X is a Lévy process which is not a compound Poisson process and whose Lévy measure is denoted by Π. Suppose, further, that Π_H is the jump measure associated with the ascending ladder height process of X. Then, for all $y>0$ and a suitable normalisation of local time at the maximum,*

$$\Pi_H(y,\infty)=\int_{[0,\infty)}\widehat{U}(\mathrm{d}z)\Pi(z+y,\infty),\quad y>0,$$

where $\widehat{U}(\mathrm{d}z)=\int_{[0,\infty)}\widehat{\mathcal{U}}(\mathrm{d}s,\mathrm{d}z)=\mathbb{E}(\int_0^\infty \mathbf{1}_{(\widehat{H}_t\in\mathrm{d}z)}\mathrm{d}t)$, $z\geq 0$.

Proof The result follows from the joint law of the overshoot and undershoot of the maximum of X at first passage over some $x>0$, as given by the quintuple law, by comparing it against the overshoot and undershoot of the process H at the same level.

Recall $T_x^+=\inf\{t>0: H_t>x\}$ and use again the definition $U(\mathrm{d}x)=\int_{[0,\infty)}\mathcal{U}(\mathrm{d}s,\mathrm{d}x)$. Note that since the range of $\overline{X}$ is the same as the range of H, it follows that $H_{T_x^+-}=\overline{X}_{\tau_x^+-}$. Hence, from Theorem 5.6, we have

$$\begin{aligned}\mathbb{P}&(X_{\tau_x^+}-x\in\mathrm{d}u,x-\overline{X}_{\tau_x^+-}\in\mathrm{d}y)\\&=\mathbb{P}(H_{T_x^+}-x\in\mathrm{d}u,x-H_{T_x^+-}\in\mathrm{d}y)\\&=U(x-\mathrm{d}y)\Pi_H(\mathrm{d}u+y),\end{aligned}\tag{7.17}$$

for $u>0$ and $y\in[0,x]$. On the other hand, the quintuple law gives

$$\begin{aligned}\mathbb{P}&(X_{\tau_x^+}-x\in\mathrm{d}u,x-\overline{X}_{\tau_x^+-}\in\mathrm{d}y)\\&=U(x-\mathrm{d}y)\int_{[y,\infty)}\widehat{U}(\mathrm{d}v-y)\Pi(\mathrm{d}u+v),\end{aligned}\tag{7.18}$$

for $u>0$ and $y\in[0,x]$. Equating the right-hand sides of (7.17) and (7.18) implies that

$$\Pi_H(\mathrm{d}u+y)=\int_{[y,\infty)}\widehat{U}(\mathrm{d}v-y)\Pi(\mathrm{d}u+v),\quad u>0.$$

Integrating over $u>0$, the statement of the theorem easily follows. □

Similar techniques allow one to make a more general statement concerning the bivariate jump measure of the ascending ladder process (L^{-1}, H). This is done in Exercise 7.5. As in the previous theorem, the expression for this jump measure still suffers from a lack of explicitness due to the involvement of the quantity $\widehat{\mathcal{U}}$. If one considers the case of a spectrally positive Lévy process then the situation in Theorem 7.8 becomes somewhat more favourable for Π_H.

Corollary 7.9 *Under the conditions of Theorem 7.8, if X is spectrally positive, then*

$$\Pi_H(y,\infty) = \int_0^\infty e^{-\Phi(0)z}\Pi(z+y,\infty)dz, \quad y>0$$

where Φ is the right inverse of the Laplace exponent ψ of $-X$.

Proof Taking into account the remarks in the final paragraph of Sect. 7.3 the result follows easily. □

Note in particular that if the spectrally positive process in the above corollary has the property that $\liminf_{t\uparrow\infty} X_t = -\infty$, then $\Phi(0) = 0$ and hence, for $x > 0$,

$$\Pi_H(dx) = \Pi(x,\infty)dx. \tag{7.19}$$

The same conclusion was drawn in Sect. 6.6.2 and Exercise 6.5, appealing there to the Wiener–Hopf factorisation.

7.5 Creeping

As with the case of a subordinator, one may talk of a Lévy process *creeping* over a fixed level $x > 0$. To be precise, a Lévy process *creeps upwards* over the level x when

$$\mathbb{P}(X_{\tau_x^+} = x) > 0. \tag{7.20}$$

The class of Lévy processes which creep upwards over (at least) one point can easily be seen to be non-empty by simply considering any spectrally negative Lévy process. By definition, any spectrally negative Lévy process has the property that, for all $x \geq 0$,

$$\mathbb{P}\big(X_{\tau_x^+} = x \,|\, \tau_x^+ < \infty\big) = 1.$$

From the above, (7.20) easily follows when we recall from Theorem 3.12 that $\mathbb{P}(\tau_x^+ < \infty) = e^{-\Phi(0)x} > 0$, where Φ is the right inverse of the Laplace exponent of X.

Lemma 7.10 *Suppose that X is a Lévy process but not a compound Poisson process. Then X creeps upwards over some (and then all) $x > 0$ if and only if*

$$\lim_{\beta\uparrow\infty} \frac{\kappa(0,\beta)}{\beta} > 0. \tag{7.21}$$

Proof The key to understanding when an arbitrary Lévy process creeps upwards is embedded within the problem of whether a subordinator creeps upwards. Indeed, since the range of $\{\overline{X}_t : t \geq 0\}$ agrees with the range of the ascending ladder process $H := \{H_t : t \geq 0\}$, it follows that X creeps across $x > 0$ if and only if H does. For this reason, it also follows that if a Lévy process creeps over some $x > 0$, then it will creep over all $x > 0$, provided H has the same behaviour. Let us now split the discussion into two cases, according to the regularity of 0 for $(0, \infty)$.

Suppose that 0 is regular for $(0, \infty)$. The (possibly-killed) subordinator H cannot have a compound Poisson process jump structure by the assumption of regularity. We are then within the scope of Theorem 5.9, which tells us that there is creeping if and only if the underlying subordinator has a strictly positive drift. The presence of a strictly positive drift coefficient is identified from the Laplace exponent of H, $\kappa(0,\beta)$, by taking the limit given in the statement of the lemma (recall Exercise 2.11). In other words, there is creeping if and only if (7.21) holds.

Suppose now that 0 is irregular for $(0, \infty)$. This has the consequence that the ascending ladder height must be a (possibly-killed) compound Poisson process subordinator. Creeping of H (and hence X) is therefore ruled out. □

In the final case of the proof above, it is interesting to ask whether the ascending ladder height process has atoms in its Lévy measure. Indeed, if such an atom were present at, say, $x_0 > 0$, then H (and hence X) would ascend to level x_0 with positive probability.[5] It turns out that no such atoms can exist. To see why, recall that it was assumed that X is not a compound Poisson process. Hence, when 0 is irregular for $(0, \infty)$, we must have that 0 is regular for $(-\infty, 0)$, and the descending ladder height process, $\widehat{H}$, cannot be a compound Poisson subordinator. According to Theorem 7.8, we know that

$$\Pi_H(\mathrm{d}x) = \int_{[0,\infty)} \widehat{U}(\mathrm{d}v)\Pi(\mathrm{d}x + v).$$

Theorem 5.4 (i) shows that $\widehat{U}$ has no atoms on $(0, \infty)$, as $\widehat{H}$ is not a compound Poisson process. It follows that Π_H has no atoms. In conclusion, whilst H is a compound Poisson process, its Lévy measure has no atoms and therefore H cannot hit specified points.

[5] Recall from the discussion at the end of Sect. 5.3 that, formally speaking, a compound Poisson subordinator cannot creep, despite the fact that a given point may lie in its range with positive probability.

The criterion given in Lemma 7.10 is not particularly useful for determining whether a process can creep upwards or not. Ideally, we would like to establish a criterion in terms of the components of the Lévy–Khintchine exponent. The following result does precisely this.

Theorem 7.11 *Suppose that X is a Lévy process which is not a compound Poisson process and not a Lévy process having no positive jumps. Then X creeps upwards if and only if one of the following three situations occurs*:

(i) *X has bounded variation with Lévy–Khintchine exponent*

$$\Psi(\theta) = -\mathrm{i}\theta\delta + \int_{\mathbb{R}\setminus\{0\}} \left(1 - \mathrm{e}^{\mathrm{i}\theta x}\right)\Pi(\mathrm{d}x)$$

and $\delta > 0$,

(ii) *X has a Gaussian component,*

(iii) *X has unbounded variation, has no Gaussian component and its Lévy measure Π satisfies*

$$\int_0^1 \frac{x\Pi(x,\infty)}{\int_0^x \int_y^1 \Pi(-1,-u)\mathrm{d}u\mathrm{d}y}\mathrm{d}x < \infty.$$

Note that spectrally negative Lévy processes are excluded as they obviously creep upwards, as discussed earlier.

Elements of the proof of parts (i) and (ii) appear in Exercise 7.6. The precise formulation and proof of part (iii) remained a challenging open problem until recently, when it was resolved by Vigon (2002b). We do not give details of the proof, which requires a deep analytical understanding of the Wiener–Hopf factorisation and goes far beyond the scope of this text. A recent, more probabilistic proof is given in Chap. 6 of Doney (2007).

We close this section by making some remarks on the difference between a Lévy process X creeping over x and *hitting the point x*. Formally speaking, we say that X hits the point x if $\mathbb{P}(\tau^{\{x\}} < \infty) > 0$, where

$$\tau^{\{x\}} = \inf\{t > 0 : X_t = x\},$$

with the usual convention that $\inf\emptyset = \infty$. Clearly, if X creeps over x (either upwards or downwards), then it must hit x. When X is a subordinator, the converse is also obviously true, providing the Lévy measure has infinite mass. However, if X is not a subordinator, then it can be shown that the converse is not necessarily true. The following result, due to Kesten (1969) and Bretagnolle (1971), gives a complete characterisation of the range of a Lévy process.

Theorem 7.12 *Suppose that X is not a compound Poisson process. Let*

$$C := \left\{x \in \mathbb{R} : \mathbb{P}\left(\tau^{\{x\}} < \infty\right) > 0\right\}$$

be the set of points that a Lévy process can hit. Then $C \neq \emptyset$ if and only if

$$\int_{\mathbb{R}} \Re\left(\frac{1}{1+\Psi(u)}\right) du < \infty. \tag{7.22}$$

Moreover,

(i) *If $\sigma > 0$, then (7.22) is satisfied and $C = \mathbb{R}$.*
(ii) *If $\sigma = 0$, X is of unbounded variation and (7.22) is satisfied, then $C = \mathbb{R}$.*
(iii) *If X is of bounded variation, then (7.22) is satisfied if and only if $\delta \neq 0$, where δ is the drift in the representation (2.22) of its Lévy–Khintchine exponent Ψ. In that case, $C = \mathbb{R}$, unless X or $-X$ is a subordinator, and then $C = (0, \infty)$ or $C = (-\infty, 0)$, respectively.*

From this characterisation, one may deduce that, for example, a symmetric α-stable process where $\alpha \in (1, 2)$ cannot creep and yet $C = \mathbb{R}$. In order to hit a given point, say $x \in \mathbb{R}$, a stable process in this class must approach the point by crossing above and below it infinitely often in such a way that x is an accumulation point in its range. See Exercise 7.6 for details.

7.6 Regular Variation and Infinite Divisibility

It has been pointed out at several points earlier in this chapter that, to some extent, the quintuple law lacks a degree of explicitness which would otherwise give it far greater practical value. In Sect. 7.7, we shall give some indication of how the quintuple law gives some analytical advantage when studying the asymptotic behaviour of the first-passage problem, as the crossing threshold tends to infinity. We need to make a short digression first into the behaviour of infinitely divisible random variables whose Lévy measures have regularly varying tails.

Recall from Definition 5.12 that a measurable function $f : [0, \infty) \to (0, \infty)$ is regularly varying at infinity with index $\rho \in \mathbb{R}$ (written $f \in \mathcal{R}_\infty(\rho)$) if, for all $\lambda > 0$,

$$\lim_{x\uparrow\infty} \frac{f(\lambda x)}{f(x)} = \lambda^\rho.$$

Moreover, when $\rho = 0$, we say that f is slowly varying at infinity (written $f \in \mathcal{R}_\infty$). Let us suppose that H is a random variable valued on $[0, \infty)$ which is infinitely divisible with Lévy measure Π_H.

Throughout this section, we shall suppose that $\Pi_H(\cdot, \infty) \in \mathcal{R}_\infty(-\alpha)$ for some $\alpha > 0$.

Our interest here is to understand how this assumed tail behaviour of Π_H reflects on the tail behaviour of the distribution of the random variable H. We do this with a sequence of lemmas. The reader may skip their proofs at no cost to the understanding of their application in Sect. 7.7. The first of these lemmas is taken from Chap. VIII.8 of Feller (1971).

Lemma 7.13 *Define the probability measure*

$$\nu(\mathrm{d}x) = \frac{\Pi_H(\mathrm{d}x)}{\Pi_H(1,\infty)}\mathbf{1}_{(x>1)}.$$

Then using the usual notation ν^{*n} for the n-fold convolution of ν with itself, we have that

$$\nu^{*n}(x,\infty) \sim n\nu(x,\infty) \tag{7.23}$$

as $x \uparrow \infty$ for each $n = 2, 3, \ldots$.

Proof The result follows by proving a slightly more general result. Suppose that F_1 and F_2 are distribution functions on $[0,\infty)$, such that $F_i(x,\infty) \sim x^{-\alpha}L_i(x)$ for $i = 1, 2$, as $x \uparrow \infty$, where L_1 and L_2 are slowly varying at infinity. Then

$$(F_1 * F_2)(x,\infty) \sim x^{-\alpha}\big(L_1(x) + L_2(x)\big) \tag{7.24}$$

as $x \uparrow \infty$. One may then argue that (7.23) clearly holds for $n = 2$ and hence, by induction, it holds for all integers $n \geq 2$.

To prove (7.24), let Y_1 and Y_2 be independent random variables with distributions F_1 and F_2. Fix $\delta > 0$ and write $x' = x(1+\delta)$. The event $\{Y_1 + Y_2 > x\}$ contains the event $\{Y_1 > x'\} \cup \{Y_2 > x'\}$, and hence

$$F_1 * F_2(x,\infty) \geq F_1\big(x',\infty\big) + F_2\big(x',\infty\big).$$

On the other hand, set $1/2 > \delta > 0$. If $x'' = (1-\delta)x$, then the event $\{Y_1 + Y_2 > x\}$ is a subset of the event $\{Y_1 > x''\} \cup \{Y_2 > x''\} \cup \{\min(Y_1, Y_2) > \delta x\}$. On account of the assumptions made on F_1 and F_2, it is clear that, as $x \uparrow \infty$, $\mathbb{P}(\min(Y_1, Y_2) > \delta x) = \mathbb{P}(Y_1 > \delta x)^2$ and the latter is of considerably smaller order than $\mathbb{P}(Y_i > x'')$, for each $i = 1, 2$. It follows that, as $x \uparrow \infty$,

$$F_1 * F_2(x,\infty) \leq (1+\varepsilon)\big(F_1\big(x'',\infty\big) + F_2\big(x'',\infty\big)\big),$$

for all small $\varepsilon > 0$. The two inequalities for $F_1 * F_2$ together with the assumed regular variation imply that

$$\begin{aligned}(1+\delta)^{-\alpha} &\leq \liminf_{x\uparrow\infty} \frac{F_1 * F_2(x,\infty)}{x^{-\alpha}(L_1(x) + L_2(x))} \\ &\leq \limsup_{x\uparrow\infty} \frac{F_1 * F_2(x,\infty)}{x^{-\alpha}(L_1(x) + L_2(x))} \leq (1+\varepsilon)(1-\delta)^{-\alpha}.\end{aligned}$$

Since δ and ε may be made arbitrarily small, the required result follows. □

Any distribution on $[0,\infty)$ which fulfils the condition (7.23) belongs to a larger class of distributions known as *subexponential*.[6] This class was introduced by Chistyakov (1964) within the context of branching processes. The following lemma, due to Kesten, thus gives a general property of all subexponential distributions.

Lemma 7.14 *Suppose that Y is any random variable whose distribution G, satisfying $G(x) > 0$ for all $x > 0$, has the same asymptotic convolution properties as (7.23). Then, given any $\varepsilon > 0$, there exists a constant $C > 0$ (which depends on ε) such that*

$$\frac{G^{*n}(x,\infty)}{G(x,\infty)} \leq C(1+\varepsilon)^n,$$

for all $n \in \{1,2,\ldots\}$ and $x > 0$.

Proof The proof is by induction. Suppose that for each $n = 1,2,\ldots,$

$$\xi_n := \sup_{x\geq 0} \frac{G^{*n}(x,\infty)}{G(x,\infty)}.$$

It is clear that $\xi_1 \leq 1$. Next note that $1 - G^{*(n+1)} = 1 - G + G*(1-G^{*n})$. Then, for any $0 < T < \infty$ and $x > 0$,

$$\begin{aligned}
\xi_{n+1} &\leq 1 + \sup_{0\leq x\leq T} \int_0^x \frac{1-G^{*n}(x-y)}{1-G(x)} G(dy) \\
&\quad + \sup_{x>T} \int_0^x \frac{1-G^{*n}(x-y)}{1-G(x-y)} \frac{1-G(x-y)}{1-G(x)} G(dy) \\
&\leq 1 + \frac{1}{1-G(T)} + \xi_n \sup_{x>T} \frac{G(x)-G^{*2}(x)}{1-G(x)}.
\end{aligned}$$

Since G satisfies (7.23), given any $\varepsilon > 0$, we can choose $T > 0$ such that

$$\xi_{n+1} \leq 1 + \frac{1}{1-G(T)} + \xi_n(1+\varepsilon).$$

Iterating, we find, after some straightforward algebra, that

$$\xi_{n+1} \leq \left(\frac{2-G(T)}{1-G(T)}\right)\frac{1}{\varepsilon}(1+\varepsilon)^{n+1},$$

which establishes the claim with the obvious choice of C. □

In the next lemma, we use the asymptotic behaviour in Lemma 7.13 and the uniform bounds in Lemma 7.14 to show that the distribution of H must also have

[6] Formally speaking, any distribution F on $[0,\infty)$ is subexponential if, when X_1 and X_2 are independent random variables with distribution F, $\mathbb{P}(X_1 + X_n > x) \sim 2\mathbb{P}(X_1 > x)$, as $x \uparrow \infty$.

regularly varying tails. The result is due to Embrechts et al. (1979). Recall that we are assuming throughout that $\Pi_H(\cdot,\infty) \in \mathcal{R}_\infty(-\alpha)$ for some $\alpha > 0$.

Lemma 7.15 *As $x \uparrow \infty$, we have*

$$\mathbb{P}(H > x) \sim \Pi_H(x,\infty),$$

which implies that the tail of the distribution of H belongs to $\mathcal{R}_\infty(-\alpha)$.

Proof The relationship between the distribution of H and Π_H is expressed via the Lévy–Khintchine formula. In this case, since H is $[0,\infty)$-valued, we may consider instead its Laplace exponent $\Phi(\theta) := -\log \mathbb{E}(\mathrm{e}^{-\theta H})$, which, from the Lévy–Khintchine formula, satisfies

$$\begin{aligned}\Phi(\theta) &= \delta\theta + \int_{(0,\infty)} (1-\mathrm{e}^{-\theta x})\Pi_H(\mathrm{d}x)\\ &= \delta\theta + \int_{(0,1]} (1-\mathrm{e}^{-\theta x})\Pi_H(\mathrm{d}x) \qquad (7.25)\\ &\quad + \int_{(1,\infty)} (1-\mathrm{e}^{-\theta x})\Pi_H(\mathrm{d}x), \quad \theta \geq 0. \qquad (7.26)\end{aligned}$$

The second equality above allows the random variable H to be seen as equal in distribution to the independent sum of two infinitely divisible random variables, say H_1 and H_2, where H_1 has Laplace exponent given by (7.25) and H_2 has Laplace exponent given by (7.26). According to Theorem 3.6, $\mathbb{E}(\mathrm{e}^{\lambda H_1}) < \infty$ for any $\lambda > 0$, because, trivially, $\int_{(1,\infty)} \mathrm{e}^{\lambda x}\Pi_{H_1}(\mathrm{d}x) < \infty$ where $\Pi_{H_1}(\mathrm{d}x) = \Pi_H(\mathrm{d}x)\mathbf{1}_{(x\in(0,1])}$. It follows that one may upper estimate the tail of H_1 by any exponentially decaying function. Specifically, with the help of the Markov inequality, $\mathbb{P}(H_1 > x) \leq \mathbb{E}(\mathrm{e}^{\lambda H_1})\mathrm{e}^{-\lambda x}$, for any $\lambda > 0$.

On the other hand, by assumption, the tail of the measure $\Pi_{H_2}(\mathrm{d}x) = \Pi_H(\mathrm{d}x)\mathbf{1}_{(x>1)}$ belongs to $\mathcal{R}_\infty(-\alpha)$. Since Π_{H_2} necessarily has finite total mass, we may consider H_2 as the distribution at time 1 of a compound Poisson process with rate $\eta := \Pi_H(1,\infty)$ and jump distribution ν (defined in Lemma 7.13). We know that

$$\mathbb{P}(H_2 > x) = \mathrm{e}^{-\eta}\sum_{k\geq 0}\frac{\eta^k}{k!}\nu^{*k}(x,\infty), \quad x > 0,$$

where, as usual, we interpret $\nu^{*0}(\mathrm{d}x) = \delta_0(\mathrm{d}x)$ (so in fact the first term of the above sum is equal to zero). Next, use the conclusion of Lemma 7.14 with dominated convergence to establish that the limits

$$\lim_{x\uparrow\infty}\frac{\mathbb{P}(H_2 > x)}{\Pi_H(x,\infty)/\eta} = \lim_{x\uparrow\infty}\mathrm{e}^{-\eta}\sum_{k\geq 1}\frac{\eta^k}{k!}\frac{\nu^{*k}(x,\infty)}{\nu(x,\infty)}$$

exist. The conclusion of Lemma 7.13 allows the computation of the limiting sum explicitly. That is to say $\sum_{k\geq 1}\eta^k/(k-1)! = \eta e^\eta$. In conclusion, we have

$$\lim_{x\uparrow\infty}\frac{\mathbb{P}(H_2 > x)}{\Pi_H(x,\infty)} = 1.$$

The proof of this lemma is thus completed once we show that

$$\lim_{x\uparrow\infty}\frac{\mathbb{P}(H_1 + H_2 > x)}{\mathbb{P}(H_2 > x)} = 1. \tag{7.27}$$

However, this fact follows by reconsidering the proof of Lemma 7.13. If in this proof one takes F_i as the distribution of H_i for $i = 1, 2$, then with the slight difference that F_1 has exponentially decaying tails, one may follow the proof step by step to deduce that the above limit holds. Intuitively speaking, the tails of H_1 are considerably lighter than those of H_2 and hence, for large x, the event whose probability is in the numerator of (7.27) occurs due to a large observation of H_2. The details are left as an exercise to the reader. □

7.7 Asymptotic Behaviour at First Passage

In this section, we give the promised example of how to use the quintuple law to obtain precise analytic statements concerning the asymptotic behaviour of the first-passage problem, under assumptions of regular variation. The following theorem, due to Asmussen and Klüppelberg (1996) and Klüppelberg and Kyprianou (2006), is our main objective.

Theorem 7.16 *If X is any spectrally positive Lévy process with mean $\mathbb{E}(X_1) < 0$ and $\Pi(\cdot,\infty) \in \mathcal{R}_\infty(-(\alpha+1))$, for some $\alpha \in (0,\infty)$, then we have the following asymptotic behaviour:*

(i) *As $x \uparrow \infty$, we have*

$$\mathbb{P}\big(\tau_x^+ < \infty\big) \sim \frac{1}{|\mathbb{E}(X_1)|}\int_x^\infty \Pi(y,\infty)\mathrm{d}y,$$

and consequently, the first-passage probability belongs to $\mathcal{R}_\infty(-\alpha)$. (Note that convexity of the Laplace exponent of $-X$ dictates that $|\mathbb{E}(X_1)| < \infty$ when $\mathbb{E}(X_1) < 0$.)

(ii) *For all $u, v > 0$,*

$$\lim_{x\uparrow\infty}\mathbb{P}\left(\frac{X_{\tau_x^+} - x}{x/\alpha} > u, \frac{-X_{\tau_x^+ -}}{x/\alpha} > v \,\middle|\, \tau_x^+ < \infty\right) = \left(1 + \frac{v+u}{\alpha}\right)^{-\alpha}. \tag{7.28}$$

Part (i) of the above theorem shows that when the so-called *Cramér condition* appearing in Theorem 7.6 fails, conditions may exist where one may still gain information about the asymptotic behaviour of the first-passage probability. Part (ii) shows that, with rescaling, the joint law of the overshoot and undershoot converges to a non-trivial distribution. In fact, the limiting distribution takes the form of a bivariate generalised Pareto distribution (cf. Definition 3.4.9 in Embrechts et al. (1997)). The result in part (ii) is also reminiscent of the following extraction from extreme value theory. It is known that a distribution, F, is in the domain of attraction of a generalised Pareto distribution if $F(\cdot,\infty)$ is regularly varying at infinity with index $-\alpha$, for some $\alpha>0$. In that case, we have

$$\lim_{x\uparrow\infty}\frac{F(x+xu/\alpha,\infty)}{F(x,\infty)}=\left(1+\frac{u}{\alpha}\right)^{-\alpha},$$

for $\alpha>0$ and $u>0$.

Generalised Pareto distributions have heavy tails in the sense that their moment generating functions do not exist on the positive half of the real axis. Roughly speaking, this means that there is a good chance to observe relatively large values when sampling from this distribution.

Proof of Theorem 7.16 (i) Following the logic that leads to (7.8), we have that

$$\mathbb{P}(\tau_x^+<\infty)=qU(x,\infty)=q\int_0^\infty \mathbb{P}(H_t>x)\mathrm{d}t,$$

where $q=\kappa(0,0)>0$ is the killing rate of the ascending ladder process. Write $[t]$ for the integer part of t and note, with the help of Lemma 7.14, that, for $x>0$,

$$\frac{\mathbb{P}(H_t>x)}{\mathbb{P}(H_1>x)}\le\frac{\mathbb{P}(H_{[t]+1}>x)}{\mathbb{P}(H_1>x)}\le C(1+\varepsilon)^{[t]+1}\mathrm{e}^{-q[t]}.$$

(To see where the exponential term on the right-hand side comes from, recall that H is equal in law to a subordinator killed independently at rate q.) Now appealing to the Dominated Convergence Theorem, we have

$$\lim_{x\uparrow\infty}\frac{\mathbb{P}(\tau_x^+<\infty)}{\mathbb{P}(H_1>x)}=q\int_0^\infty \mathrm{d}t\cdot\lim_{x\uparrow\infty}\frac{\mathbb{P}(H_t>x)}{\mathbb{P}(H_1>x)}. \tag{7.29}$$

In order to deal with the limit on the right-hand side above, we shall use the fact that $\mathbb{P}(H_t>x)=\mathrm{e}^{-qt}\mathbb{P}(\mathcal{H}_t>x)$, where $\mathcal{H}_t$ is an infinitely divisible random variable. To be more specific, one may think of $\{\mathcal{H}_t:t\ge0\}$ as a subordinator which, when killed at an independent and exponentially distributed time with parameter q, has the same law as $\{H_t:t\ge0\}$. Associated to the random variable $\mathcal{H}_t$ is its Lévy measure, which necessarily takes the form $t\Pi_H$. By Lemma 7.15, it follows that

$$\lim_{x\uparrow\infty}\frac{\mathbb{P}(H_t>x)}{\mathbb{P}(H_1>x)}=t\mathrm{e}^{-q(t-1)}.$$

Hence, referring back to (7.29) and Lemma 7.15, we have that

$$\lim_{x\uparrow\infty}\frac{\mathbb{P}(\tau_x^+<\infty)}{\Pi_H(x,\infty)}=q\int_0^\infty t\mathrm{e}^{-qt}\mathrm{d}t=\frac{1}{q}. \tag{7.30}$$

On the other hand, taking account of exponential killing, one easily computes

$$U(\infty) = \int_0^\infty \mathbb{P}(H_t < \infty)\mathrm{d}t = \int_0^\infty \mathrm{e}^{-qt}\mathrm{d}t = \frac{1}{q}.$$

Since $\psi'(0+) > 0$, we have $\Phi(0) = 0$, where Φ is the right inverse of ψ. From (7.15), we may thus identify $U(\infty) = \lim_{\beta\downarrow 0} \beta/\psi(\beta) = 1/\psi'(0+)$, where $\psi(\beta) = \log\mathbb{E}(\mathrm{e}^{-\beta X_1})$. In particular, this implies $q = |\mathbb{E}(X_1)|$. Moreover, from Corollary 7.9, we have that $\Pi_H(\mathrm{d}x) = \Pi(x,\infty)\mathrm{d}x$, $x > 0$. Putting the pieces together in (7.30) completes the proof of part (i).

(ii) Applying the quintuple law in marginalised form, we have

$$\mathbb{P}\big(X_{\tau_x^+} - x > u^*, x - X_{\tau_x^+ -} > v^*\big) = \int_0^x U(x - \mathrm{d}y) \int_{[v^*\vee y,\infty)} \mathrm{d}z \Pi\big(u^* + z, \infty\big) \tag{7.31}$$

for $u^*, v^* > 0$. As noted in the proof of part (i), we also have

$$\Pi_H(u,\infty) = \int_u^\infty \Pi(z,\infty)\mathrm{d}z.$$

Choosing $u^* = ux/\alpha$ and $v^* = x + vx/\alpha$, we find that

$$\mathbb{P}\left(\frac{X_{\tau_x^+} - x}{x/\alpha} > u, \frac{-X_{\tau_x^+ -}}{x/\alpha} > v\right) = U(x)\Pi_H\big(x + x(v+u)/\alpha, \infty\big). \tag{7.32}$$

From part (i), if the limit exists then it holds that

$$\lim_{x\uparrow\infty} \mathbb{P}\left(\frac{X_{\tau_x^+} - x}{x/\alpha} > u, \frac{-X_{\tau_x^+ -}}{x/\alpha} > v \,\middle|\, \tau_x < \infty\right) = \lim_{x\uparrow\infty} \frac{U(x)}{U(\infty)} \frac{\Pi_H(x + x(v+u)/\alpha, \infty)}{\Pi_H(x,\infty)}. \tag{7.33}$$

Since, by assumption, $\Pi(\cdot,\infty) \in \mathcal{R}_\infty(-(\alpha+1))$, the Monotone Density Theorem 5.14 implies that $\Pi_H(\cdot,\infty)$ is regularly varying with index $-\alpha$. Hence, the limit in (7.33) exists and, in particular, (7.28) holds, thus concluding the proof. □

Exercises

7.1 (Moments of the supremum) Fix $n = 1, 2, \ldots$ and suppose that

$$\int_{(1,\infty)} x^n \Pi(\mathrm{d}x) < \infty \tag{7.34}$$

(or equivalently $\mathbb{E}((\max\{X_1, 0\})^n) < \infty$ by Exercise 3.3).

(i) Suppose that X^K is the Lévy process with the same characteristics as X except that the measure Π is replaced by Π^K, where, for some $K > 0$,

$$\Pi^K(\mathrm{d}x) = \Pi(\mathrm{d}x)\mathbf{1}_{(x>-K)} + \delta_{-K}(\mathrm{d}x)\Pi(-\infty, -K).$$

In other words, the paths of X^K are an adjustment of the paths of X in that all negative jumps of magnitude K or greater are replaced by a negative jump of magnitude precisely K.

Deduce that $\mathbb{E}(|X_t^K|^n) < \infty$ for all $t \geq 0$ and that the descending ladder height process of X^K has moments of all orders.

(ii) Use the Wiener–Hopf factorisation, together with a Maclaurin expansion up to order n, to deduce that

$$\mathbb{E}\big(\overline{X}_{\mathbf{e}_q}^n\big) < \infty$$

holds for any $q > 0$.

(iii) Now suppose that $q = 0$, $\limsup_{t\uparrow\infty} X_t < \infty$ and that, for $n = 2, 3, \ldots$, (7.34) holds. By adapting the arguments above, show that

$$\mathbb{E}\big(\overline{X}_\infty^{n-1}\big) < \infty.$$

(iv) Suppose now that X is a spectrally positive Lévy process which has paths of bounded variation and which drifts to $-\infty$. Use the Pollaczek-Khintchine formula discussed in Sect. 4.6 to deduce that, even if

$$\int_{(1,\infty)} x^2\Pi(\mathrm{d}x) < \infty,$$

it is not necessarily the case that $\mathbb{E}(\overline{X}_\infty^2) < \infty$.

7.2 (The Strong Law of Large Numbers for Lévy processes) Suppose that X is a Lévy process such that $\mathbb{E}|X_1| < \infty$. For $n \geq 0$, let $Y_n = \sup_{t\in[n,n+1]} |X_t - X_n|$. Clearly, this is a sequence of independent and identically distributed random variables.

(i) Use the previous exercise to show that $\mathbb{E}(Y_n) < \infty$.

(ii) Use the classical Strong Law of Large Numbers to deduce that $\lim_{n\uparrow\infty} n^{-1} Y_n = 0$ almost surely.

(iii) Prove that

$$\lim_{t\uparrow\infty} \frac{X_t}{t} = \mathbb{E}(X_1)$$

almost surely.

(iv) Now suppose that $\mathbb{E}(X_1) = \infty$. Show that

$$\lim_{t\uparrow\infty} \frac{X_t}{t} = \infty.$$

(v) Finally, suppose that $\mathbb{E}(X_1)$ is undefined but $\lim_{t\uparrow\infty} X_t = \infty$. Show that the same conclusion as in part (iv) holds.

Hint: in the last two parts, consider truncating the Lévy measure on $(0,\infty)$.

7.3 The idea of this exercise is to recover the conclusion of Theorem 7.2 for a spectrally negative Lévy process, X, using Theorem 7.1 and the Wiener–Hopf factorisation. As usual, the Laplace exponent of X is denoted ψ and its right inverse is Φ.

(i) Using one of the Wiener–Hopf factors, show that

$$\mathbb{E}\big(e^{\beta \underline{X}_\infty}\mathbf{1}_{(-\underline{X}_\infty<\infty)}\big) = \begin{cases} 0 & \text{if } \psi'(0) < 0 \\ \psi'(0+)\beta/\psi(\beta) & \text{if } \psi'(0) \geq 0. \end{cases}$$

(ii) Using the other Wiener–Hopf factor, show that

$$\mathbb{E}\big(e^{-\beta \overline{X}_\infty}\mathbf{1}_{(\overline{X}_\infty<\infty)}\big) = \begin{cases} \Phi(0)/(\beta+\Phi(0)) & \text{if } \psi'(0) < 0 \\ 0 & \text{if } \psi'(0) \geq 0. \end{cases}$$

(iii) Deduce from Theorem 7.1 that $\lim_{t\uparrow\infty} X_t = \infty$ when $\mathbb{E}(X_1) > 0$, $\lim_{t\uparrow\infty} X_t = -\infty$ when $\mathbb{E}(X_1) < 0$, and $\limsup_{t\uparrow\infty} X_t = -\liminf_{t\uparrow\infty} X_t = \infty$ when $\mathbb{E}(X_1) = 0$.

(iv) Show that a spectrally negative stable process of index $\alpha \in (1,2)$ necessarily oscillates.

7.4 Let X be a stable process with index $\alpha \in (0,2)$ which has both positive and negative jumps. Let $\rho = \mathbb{P}(X_t \geq 0)$.

(i) Explain why such processes cannot creep upwards. If, further, it experiences negative jumps, explain why it cannot creep downwards either.

(ii) Suppose that $U(dx) = \int_{[0,\infty)} \mathcal{U}(dx, ds)$ for $x \geq 0$. Show that (up to a multiplicative constant)

$$U(dx) = \frac{x^{\alpha\rho-1}}{\Gamma(\alpha\rho)}dx,$$

for $x \geq 0$.

Hint: reconsider Exercise 5.8.

(iii) Show that, for $y \in [0,x]$, $v \geq y$ and $u > 0$,

$$\begin{aligned}&\mathbb{P}(X_{\tau_x^+} - x \in du, x - X_{\tau_x^+-} \in dv, x - \overline{X}_{\tau_x^+-} \in dy) \\ &= c \cdot \frac{(x-y)^{\alpha\rho-1}(v-y)^{\alpha(1-\rho)-1}}{(v+u)^{1+\alpha}} dy dv du,\end{aligned}$$

where c is a strictly positive constant.

(iv) Explain why the constant c must normalise the above triple law to a probability distribution. Show that

$$c = \frac{\sin \alpha\rho\pi}{\pi} \frac{\Gamma(\alpha+1)}{\Gamma(\alpha\rho)\Gamma(\alpha(1-\rho))}.$$

7.5 Suppose that X is a Lévy process (but not a compound Poisson process) with jump measure Π and ascending ladder process (L^{-1}, H), whose jump measure is denoted by $\mathbf{\Pi}(dt, dh)$.

(i) Using the conclusion of Exercise 5.6, show that, up to a multiplicative constant,

$$\mathbf{\Pi}(\mathrm{d}t, \mathrm{d}h) = \int_{[0,\infty)} \widehat{\mathcal{U}}(\mathrm{d}t, \mathrm{d}\theta)\Pi(\mathrm{d}h+\theta), \quad t, h > 0.$$

(ii) Show, further, that if X is spectrally positive, then

$$\mathbf{\Pi}(\mathrm{d}t, \mathrm{d}h) = \int_0^\infty \mathrm{d}\theta \cdot \mathbb{P}\big(\widehat{L}_\theta^{-1} \in \mathrm{d}t\big)\Pi(\mathrm{d}h+\theta), \quad t, h > 0,$$

where $\widehat{L}$ is the local time of X at its minimum.

7.6 Here, we deduce some statements about creeping and hitting points.

(i) Show that

$$\lim_{|\theta|\uparrow\infty} \frac{\Psi(\theta)}{\theta^2} = \frac{\sigma^2}{2},$$

where σ is the Gaussian coefficient of Ψ. With the help of the Wiener–Hopf factorisation, prove that a Lévy process creeps both upwards and downwards if and only if it has a Gaussian component.

(ii) Show that a Lévy process of bounded variation with Lévy–Khintchine representation

$$\Psi(\theta) = -i\theta\delta + \int_{\mathbb{R}\setminus\{0\}} (1 - \mathrm{e}^{i\theta x})\Pi(\mathrm{d}x), \quad \theta \in \mathbb{R},$$

creeps upwards if $\delta > 0$.

(iii) Show that any Lévy process for which 0 is irregular for $(0, \infty)$ cannot creep upwards.

(iv) Show that a spectrally negative Lévy process with no Gaussian component cannot creep downwards.

(v) Use part (i) to show that a symmetric α-stable process with $\alpha \in (1, 2)$ cannot creep. Use the integral test (7.22) to deduce that this Lévy process can hit points.

7.7 This exercise concerns an example where an explicit characterisation of the two-sided exit problem can be obtained. The result is due to Rogozin (1972).

Suppose that X is an α-stable process with both positive and negative jumps[7] and index $\alpha \in (0,2)$. From the discussion in Sect. 6.5.3, we know that the positivity parameter satisfies $\rho \in (0,1)$ and that both $\alpha\rho$ and $\alpha(1-\rho)$ are valued in $(0,1)$.

(i) With the help of the conclusion of Exercise 5.8 (ii), show that

$$\mathbb{P}_x(X_{\tau_1^+} \le 1+y) = \Phi_{\alpha\rho}\left(\frac{y}{1-x}\right),$$

for $x \le 1$, and

$$\mathbb{P}_x(-X_{\tau_0^-} \le y) = \Phi_{\alpha(1-\rho)}\left(\frac{y}{x}\right),$$

for $x \ge 0$, where

$$\Phi_q(u) = \begin{cases} \frac{\sin \pi q}{\pi} \int_0^u t^{-q}(1+t)^{-1}\mathrm{d}t & \text{for } u \ge 0 \\ 0 & \text{for } u < 0. \end{cases}$$

Hint: it will be helpful to prove that

$$\int_0^{1/(1+\theta)} u^{\alpha-1}(1-u)^{-(\alpha+1)}\mathrm{d}v = \frac{\theta^{-\alpha}}{\alpha}$$

for any $\theta > 0$.

(ii) Let

$$r(x,y) = \mathbb{P}_x\left(X_{\tau_1^+} \le 1+y;\, \tau_1^+ < \tau_0^-\right)$$

and

$$l(x,y) = \mathbb{P}_x\left(X_{\tau_0^-} \ge -y;\, \tau_1^+ > \tau_0^-\right),$$

where $x \in (0,1)$ and $y \ge 0$. Show that the following system of equations hold:

$$r(x,y) = \Phi_{\alpha\rho}\left(\frac{y}{1-x}\right) - \int_{(0,\infty)} \Phi_{\alpha\rho}\left(\frac{y}{1+z}\right) l(x,\mathrm{d}z)$$

and

$$l(x,y) = \Phi_{\alpha(1-\rho)}\left(\frac{y}{x}\right) - \int_{(0,\infty)} \Phi_{\alpha(1-\rho)}\left(\frac{y}{1+z}\right) r(x,\mathrm{d}z),$$

for $x \in (0,1)$ and $y \ge 0$.

(iii) Assuming the above system of equations has a unique solution, show that

$$r(x,y) = \frac{\sin \pi\alpha\rho}{\pi}(1-x)^{\alpha\rho} x^{\alpha(1-\rho)} \int_0^y t^{-\alpha\rho}(t+1)^{-\alpha(1-\rho)}(t+1-x)^{-1}\mathrm{d}t,$$

for $x \in (0,1)$ and $y \ge 1$. Write down a similar expression for $l(x,y)$.

---

[7] The case that X or $-X$ is spectrally negative is dealt with later in Exercise 8.11.

(iv) Now consider the integral

$$\int_0^\infty t^{-\alpha\rho}(t+1)^{-\alpha(1-\rho)}(t+1-x)^{-1}\mathrm{d}t,$$

where $0 < x < 1$. By performing the change of variable $(t+1-x)^{-1} = (1-x)^{-1}u$, differentiating the resulting integral in the variable x and then applying a further change of variable $(1/u - 1) = (1-x)^{-1}z$, show that[8]

$$\mathbb{P}_x(\tau_1^+ < \tau_0^-) = \frac{\Gamma(\alpha)}{\Gamma(\alpha\rho)\Gamma(\alpha(1-\rho))}\int_0^x z^{\alpha(1-\rho)-1}(1-z)^{\alpha\rho-1}\mathrm{d}z,$$

for $x \in (0,1)$. Write down a similar expression for $\mathbb{P}_z(\tau_0^- < \tau_1^+)$.

7.8 Suppose that X is any Lévy process. The following problem is taken from Kyprianou et al. (2010a) and gives an identity which allows one to convert distributional statements about overshoot and undershoot at first passage into distributional statements about overshoot, undershoot and undershoot of the last maximum at first passage. Define the following quantities:

$$\mathcal{U}_x = x - \overline{X}_{\tau_x^+-}, \qquad \mathcal{V}_x = x - X_{\tau_x^+-}, \qquad \mathcal{O}_x = X_{\tau_x^+} - x, \quad x > 0.$$

Prove that

$$\mathbb{P}(\mathcal{U}_x > u, \mathcal{O}_x > w, \mathcal{V}_x > v) = \mathbb{P}(\mathcal{O}_{x-u} > w+u, \mathcal{V}_{x-u} > v-u).$$

Hint: A simple sketch will prove to be very useful.

7.9 Suppose that X is a Lévy process with jump measure Π and ascending ladder height H, satisfying $\mathbb{E}(H_1) < \infty$. Suppose, moreover, that the drift coefficient of H is written $\gamma \geq 0$ and the descending ladder height process has potential function denoted by $\widehat{U}$. Show that, for $y, z \geq 0$,

$$\lim_{x\uparrow\infty}\mathbb{P}(X_{\tau_x^+} - x \in \mathrm{d}y,\, x - X_{\tau_x^+-} - x \in \mathrm{d}z)$$
$$= \frac{1}{\mathbb{E}(H_1)}\big(\widehat{U}(z)\Pi(z+\mathrm{d}y)\mathrm{d}z + \gamma\delta_0(\mathrm{d}y)\delta_0(\mathrm{d}z)\big),$$

in the sense of vague convergence.

7.10 In this exercise, we shall consider the expected occupation measure of a Lévy process before first entry into $(-\infty, 0)$. Related computations can be found in the proof of Theorem 7.7. The original result is due to Silverstein (1980).

---

[8] There is a typographic error in Lemma 3 of Rogozin (1972) for the two-sided exit formula. In the notation of that paper, the roles of q and $(1-q)$ should be exchanged and the upper delimiter in the integral should be x and not ∞.

Suppose that X is any Lévy process (other than a compound Poisson process) and recall that $H = \{H_t : t \geq 0\}$ denotes the ascending ladder height process. Let

$$U(x) = \mathbb{E}\left(\int_0^\infty \mathbf{1}_{(H_t \leq x)} \mathrm{d}t\right), \quad x \geq 0,$$

and define $\widehat{U}$ in the obvious way, with the help of the descending ladder height process. Appealing to the techniques used in Step 2 of the proof Theorem 7.7, show that, up to a multiplicative constant, for positive, bounded and measurable f,

$$\mathbb{E}_x\left(\int_0^{\tau_0^-} f(X_t)\mathrm{d}t\right) = \int_{[0,\infty)} U(\mathrm{d}y) \int_{[0,x]} \widehat{U}(\mathrm{d}z) f(x+y-z), \quad x \geq 0.$$

Chapter 8
Exit Problems for Spectrally Negative Processes

In this chapter, we consider in more detail the special case of spectrally negative Lévy processes. As we have already seen in a number of examples from previous chapters, Lévy processes which have jumps in only one direction turn out to offer a significant advantage for many calculations. We devote our time in this chapter, initially, to gathering facts about spectrally negative processes from earlier chapters, and then to an ensemble of fluctuation identities which are semi-explicit in terms of a class of functions known as scale functions, whose properties we shall also explore.

8.1 Basic Properties Reviewed

Let us gather what we have already established in previous chapters together with other easily derived facts.

The Laplace exponent. Rather than working with the Lévy–Khintchine characteristic exponent, it is preferable to work with the Laplace exponent,

$$\psi(\lambda) := \frac{1}{t} \log \mathbb{E}\big(\mathrm{e}^{\lambda X_t}\big) = -\Psi(-\mathrm{i}\lambda), \tag{8.1}$$

which is finite at least for all $\lambda \geq 0$. The function $\psi : [0, \infty) \to \mathbb{R}$ is zero at zero and tends to infinity at infinity. Further, it is infinitely differentiable and strictly convex on $(0, \infty)$. In particular, $\psi'(0+) = \mathbb{E}(X_1) \in [-\infty, \infty)$. Define the right inverse

$$\Phi(q) = \sup\big\{\lambda \geq 0 : \psi(\lambda) = q\big\},$$

for each $q \geq 0$. If $\psi'(0+) \geq 0$, then $\lambda = 0$ is the unique solution to $\psi(\lambda) = 0$ on $[0, \infty)$ and otherwise there are two solutions, with $\lambda = \Phi(0) > 0$ the larger of the two. The other is $\lambda = 0$ (see Fig. 3.3).

A.E. Kyprianou, *Fluctuations of Lévy Processes with Applications*, Universitext,
DOI 10.1007/978-3-642-37632-0_8, © Springer-Verlag Berlin Heidelberg 2014

First passage upwards. The first-passage time above a level $x > 0$ has been defined by $\tau_x^+ = \inf\{t > 0 : X_t > x\}$. From Theorem 3.12, we know that, for each $q \geq 0$,

$$\mathbb{E}\big(\mathrm{e}^{-q\tau_x^+}\mathbf{1}_{(\tau_x^+<\infty)}\big) = \mathrm{e}^{-\Phi(q)x}.$$

Further, the process $\{\tau_x^+ : x \geq 0\}$ is a subordinator with Laplace exponent $\Phi(q) - \Phi(0)$, killed at rate $\Phi(0)$.

Path variation. Given the triple (a, σ, Π) as in Theorem 1.6, where, now, the measure Π is necessarily concentrated on $(-\infty, 0)$, we may always write

$$\psi(\lambda) = -a\lambda + \frac{1}{2}\sigma^2\lambda^2 + \int_{(-\infty,0)} \big(\mathrm{e}^{\lambda x} - 1 - \lambda x\mathbf{1}_{(x>-1)}\big)\Pi(\mathrm{d}x), \qquad (8.2)$$

for $\lambda \geq 0$. When X has bounded variation we may always write

$$\psi(\lambda) = \delta\lambda - \int_{(-\infty,0)} \big(1 - \mathrm{e}^{\lambda x}\big)\Pi(\mathrm{d}x), \qquad (8.3)$$

where necessarily

$$\delta = -a - \int_{(-1,0)} x\Pi(\mathrm{d}x)$$

is strictly positive. Hence, a spectrally negative Lévy process of bounded variation must always take the form of a strictly positive drift minus a pure jump subordinator. Note that, if $\delta \leq 0$, then we would see the Laplace exponent of a decreasing subordinator, which is excluded from the definition of a spectrally negative Lévy process.

Regularity. From Theorem 6.5 (i) and (ii) one sees immediately that 0 is regular for $(0, \infty)$ for X, irrespective of path variation. Further, by considering the process $-X$, we can see from the same theorem that 0 is regular for $(-\infty, 0)$ for X if and only if X has unbounded variation. Said another way, 0 is regular for both $(0, \infty)$ and $(-\infty, 0)$ if and only if it has unbounded variation.

Creeping. We know from Corollary 3.13 and the fact that there are no positive jumps that

$$\mathbb{P}\big(X_{\tau_x^+} = x|\tau_x^+ < \infty\big) = 1.$$

Hence spectrally negative Lévy processes necessarily creep upwards. It was shown, however, in Exercise 7.6 that they creep downwards if and only if $\sigma > 0$.

Wiener–Hopf factorisation. In Chap. 6, we identified, up to a multiplicative constant,

$$\kappa(\alpha, \beta) = \Phi(\alpha) + \beta \quad \text{and} \quad \widehat{\kappa}(\alpha, \beta) = \frac{\alpha - \psi(\beta)}{\Phi(\alpha) - \beta},$$

for $\alpha, \beta \geq 0$. Appropriate choices of local time at the maximum and minimum allow the multiplicative constants to be taken as equal to unity. From Theorem 6.15 (ii) this leads to

$$\mathbb{E}\big(e^{-\beta \overline{X}_{\mathbf{e}_p}}\big) = \frac{\Phi(p)}{\Phi(p)+\beta} \quad \text{and} \quad \mathbb{E}\big(e^{\beta \underline{X}_{\mathbf{e}_p}}\big) = \frac{p}{\Phi(p)} \frac{\Phi(p)-\beta}{p-\psi(\beta)}, \tag{8.4}$$

where $\mathbf{e}_p$ is an independent and exponentially distributed random variable with parameter $p \geq 0$. The first of these two expressions shows that $\overline{X}_{\mathbf{e}_p}$ is exponentially distributed with parameter $\Phi(p)$. Note that, when $p = 0$ in the last statement, we employ our usual convention that an exponential variable with parameter zero is infinite with probability one.

Drifting and oscillating. From Theorem 7.2 or Exercise 7.3, we have the following asymptotic behaviour for X. The process drifts to infinity if and only if $\psi'(0+) > 0$, oscillates if and only if $\psi'(0+) = 0$ and drifts to minus infinity if and only if $\psi'(0+) < 0$.

Exponential change of measure. From Exercise 1.5, we know that, for each $c \geq 0$,

$$\{e^{cX_t - \psi(c)t} : t \geq 0\}$$

is a martingale. For each $c \geq 0$, define the change of measure

$$\left.\frac{d\mathbb{P}^c}{d\mathbb{P}}\right|_{\mathcal{F}_t} = e^{cX_t - \psi(c)t}. \tag{8.5}$$

When X is a Brownian motion this is the same change of measure that appears in the most elementary form of the Cameron–Martin–Girsanov Theorem. In that case, we know that the effect of the change of measure makes $(X, \mathbb{P}^c)$ equal in law to a Brownian motion with drift c. In Sect. 3.3, we showed that, if $(X, \mathbb{P})$ is a spectrally negative Lévy process, then $(X, \mathbb{P}^c)$ is also a spectrally negative Lévy process. Moreover, we showed that its Laplace exponent, $\psi_c(\lambda)$, is given by

$$\begin{aligned}\psi_c(\lambda) &= \psi(\lambda + c) - \psi(c) \\ &= \left(\sigma^2 c - a + \int_{(-\infty,0)} x\big(e^{cx} - 1\big)\mathbf{1}_{(x>-1)}\Pi(dx)\right)\lambda \\ &\quad + \frac{1}{2}\sigma^2\lambda^2 + \int_{(-\infty,0)} \big(e^{\lambda x} - 1 - \lambda x \mathbf{1}_{(x>-1)}\big)e^{cx}\Pi(dx),\end{aligned} \tag{8.6}$$

for $\lambda \geq -c$.

When we set $c = \Phi(p)$ for $p \geq 0$ we discover that $\psi_{\Phi(p)}(\lambda) = \psi(\lambda + \Phi(p)) - p$, and hence $\psi'_{\Phi(p)}(0) = \psi'(\Phi(p)) \geq 0$ on account of the strict convexity of ψ. In particular, $(X, \mathbb{P}^{\Phi(p)})$ always drifts to infinity for $p > 0$. Roughly speaking, the effect of the change of measure has been to change the characteristics of X to those of a spectrally negative Lévy process with the same Gaussian coefficient, an exponentially tilted Lévy measure and an adjusted linear drift. Note also that $(X,\mathbb{P})$ is of bounded variation if and only if $(X,\mathbb{P}^c)$ is of bounded variation. This statement is clear when $\sigma>0$. When $\sigma=0$ it is justified by noting that $\int_{(-1,0)}|x|\Pi(\mathrm{d}x)<\infty$ if and only if $\int_{(-1,0)}|x|\mathrm{e}^{cx}\Pi(\mathrm{d}x)<\infty$. In the case that X is of bounded variation and we write the Laplace exponent in the form (8.3), we also see from the second equality of (8.6) that

$$\psi_c(\lambda)=\delta\lambda-\int_{(-\infty,0)}(1-\mathrm{e}^{\lambda x})\mathrm{e}^{cx}\Pi(\mathrm{d}x),\quad \lambda\geq -c.$$

Hence, under $\mathbb{P}^c$, the process retains the same drift and only the Lévy measure is exponentially tilted.

8.2 The One-Sided and Two-Sided Exit Problems

In this section, we shall develop semi-explicit identities concerning exiting from a half-line and a strip. Recall that $\mathbb{P}_x$ and $\mathbb{E}_x$ are shorthand for $\mathbb{P}(\cdot|X_0=x)$ and $\mathbb{E}(\cdot|X_0=x)$, respectively, and for the special case that $x=0$, we keep with our old notation, so that $\mathbb{P}_0=\mathbb{P}$ and $\mathbb{E}_0=\mathbb{E}$, unless we wish to emphasise the fact that $X_0=0$. Recall also that

$$\tau_x^+=\inf\{t>0:X_t>x\}\quad\text{and}\quad \tau_x^-=\inf\{t>0:X_t<x\},\tag{8.7}$$

for all $x\in\mathbb{R}$. The main results of this section are the following.

Theorem 8.1 (One- and two-sided exit formulae) *There exist a family of functions* $W^{(q)}:\mathbb{R}\to[0,\infty)$ *and*

$$Z^{(q)}(x):=1+q\int_0^x W^{(q)}(y)\mathrm{d}y,\quad \textit{for } x\in\mathbb{R},$$

defined for each $q\geq 0$, *such that the following hold (for short we shall write* $W^{(0)}=W$).

(i) *For any* $q\geq 0$, *we have* $W^{(q)}(x)=0$ *for* $x<0$ *and* $W^{(q)}$ *is characterised on* $[0,\infty)$ *as a strictly increasing and continuous function whose Laplace transform satisfies*

$$\int_0^\infty \mathrm{e}^{-\beta x}W^{(q)}(x)\mathrm{d}x=\frac{1}{\psi(\beta)-q}\quad \textit{for } \beta>\Phi(q).\tag{8.8}$$

(ii) *For any* $x\in\mathbb{R}$ *and* $q\geq 0$,

$$\mathbb{E}_x\big(\mathrm{e}^{-q\tau_0^-}\mathbf{1}_{(\tau_0^-<\infty)}\big)=Z^{(q)}(x)-\frac{q}{\Phi(q)}W^{(q)}(x),\tag{8.9}$$

where we understand $q/\Phi(q)$ in the limiting sense for $q=0$, so that

$$\mathbb{P}_x(\tau_0^- < \infty) = \begin{cases} 1-\psi'(0+)W(x) & \text{if } \psi'(0+) \geq 0 \\ 1 & \text{if } \psi'(0+) < 0 \end{cases}. \tag{8.10}$$

(iii) *For any $x \leq a$ and $q \geq 0$,*

$$\mathbb{E}_x\left(\mathrm{e}^{-q\tau_a^+}\mathbf{1}_{(\tau_0^- > \tau_a^+)}\right) = \frac{W^{(q)}(x)}{W^{(q)}(a)}, \tag{8.11}$$

and

$$\mathbb{E}_x\left(\mathrm{e}^{-q\tau_0^-}\mathbf{1}_{(\tau_0^- < \tau_a^+)}\right) = Z^{(q)}(x) - Z^{(q)}(a)\frac{W^{(q)}(x)}{W^{(q)}(a)}. \tag{8.12}$$

Note that (8.10) should agree with the Pollaczek–Khintchine formula (1.15) when X is taken as the Cramér–Lundberg risk process discussed in Chap. 1. Exercise 8.3 handles the details.

The name "scale function" for W was first used by Bertoin (1992) to reflect the analogous role it plays in (8.11) to scale functions for diffusions. In keeping with existing literature, we will refer to the functions $W^{(q)}$ and $Z^{(q)}$ as the q-scale functions.[1]

Identity (8.9) appears in the form of its Fourier transform in Emery (1973) and, for the case that Π is finite and $\sigma=0$, in Korolyuk (1975a). Identity (8.11) first appeared for the case $q=0$ in Zolotarev (1964), followed by Takács (1966) and then, with a short proof, in Rogers (1990). The case $q>0$ is found in Korolyuk (1975a) for the case that Π is finite and $\sigma=0$, in Bertoin (1996b) for the case of a purely asymmetric stable process and again for a general spectrally negative Lévy process in Bertoin (1997b) (who referred to a method used for the case $q=0$ in Bertoin (1996a)). See also Doney (2007) for further remarks on this identity. Finally (8.12) was proved for the case that Π is finite and $\sigma=0$ by Korolyuk (1974, 1975a); see Bertoin (1997b) for the general case.

Proof of Theorem 8.1 (8.11) We prove (8.11) for the case that $\psi'(0+)>0$ and $q=0$, then for the case that $q>0$ (with no restriction on $\psi'(0+)$). Finally the case that $\psi'(0+)\leq 0$ and $q=0$ is handled by passing to the limit as q tends to zero.

Assume that $\psi'(0+)>0$ so that $-\underline{X}_\infty$ is $\mathbb{P}$-almost surely finite. Now define the non-decreasing function

$$W(x) = \mathbb{P}_x(\underline{X}_\infty \geq 0).$$

A simple argument using the law of total probability and the strong Markov property now yields for $x \in [0,a)$

[1] One may also argue that the terminology "scale function" is inappropriate as the mentioned analogy breaks down in a number of other respects.

$$
\begin{aligned}
\mathbb{P}_x(\underline{X}_\infty \geq 0) &= \mathbb{E}_x\big(\mathbb{P}_x(\underline{X}_\infty \geq 0|\mathcal{F}_{\tau_a^+})\big) \\
&= \mathbb{E}_x\big(\mathbf{1}_{(\tau_a^+<\tau_0^-)}\mathbb{P}_a(\underline{X}_\infty \geq 0)\big) + \mathbb{E}_x\big(\mathbf{1}_{(\tau_a^+>\tau_0^-)}\mathbb{P}_{X_{\tau_0^-}}(\underline{X}_\infty \geq 0)\big) \\
&= \mathbb{P}_a(\underline{X}_\infty \geq 0)\mathbb{P}_x\big(\tau_a^+ < \tau_0^-\big).
\end{aligned}
$$

To justify that the second term in the second equality disappears, note the following. If X has no Gaussian component, then it cannot creep downwards, implying that $X_{\tau_0^-} < 0$, and then we use that $\mathbb{P}_x(\underline{X}_\infty \geq 0) = 0$ for $x < 0$. If X has a Gaussian component, then $X_{\tau_0^-} \leq 0$ and we need to additionally know that $\mathbb{P}(\underline{X}_\infty \geq 0) = 0$. However, since 0 is regular for $(-\infty, 0)$ and $(0, \infty)$, it follows that $\underline{X}_\infty < 0$ $\mathbb{P}$-almost surely, which is the same as $\mathbb{P}(\underline{X}_\infty \geq 0) = 0$.

We now have

$$
\mathbb{P}_x\big(\tau_a^+ < \tau_0^-\big) = \frac{W(x)}{W(a)}, \quad x \geq 0, \tag{8.13}
$$

which proves (8.11) for the case $\psi'(0+) > 0$ and $q = 0$. It is trivial, but nonetheless useful for later, to note that the same equality holds even when $x < 0$ since both sides are equal to zero there.

Now assume that $q > 0$ or that $q = 0$ and $\psi'(0+) < 0$. In these cases, by the convexity of ψ, we know that $\Phi(q) > 0$ and hence $\psi'_{\Phi(q)}(0) = \psi'(\Phi(q)) > 0$ (again by convexity), which implies that under $\mathbb{P}^{\Phi(q)}$, the process X drifts to infinity. For $(X, \mathbb{P}^{\Phi(q)})$, we have already established the existence of a 0-scale function $W_{\Phi(q)}(x) = \mathbb{P}_x^{\Phi(q)}(\underline{X}_\infty \geq 0)$, which fulfils the relation

$$
\mathbb{P}_x^{\Phi(q)}\big(\tau_a^+ < \tau_0^-\big) = \frac{W_{\Phi(q)}(x)}{W_{\Phi(q)}(a)}. \tag{8.14}
$$

However, by definition of $\mathbb{P}^{\Phi(q)}$, we also have that

$$
\begin{aligned}
\mathbb{P}_x^{\Phi(q)}\big(\tau_a^+ < \tau_0^-\big) &= \mathbb{E}_x\big(\mathrm{e}^{\Phi(q)(X_{\tau_a^+}-x)-q\tau_a^+}\mathbf{1}_{(\tau_a^+<\tau_0^-)}\big) \\
&= \mathrm{e}^{\Phi(q)(a-x)}\mathbb{E}_x\big(\mathrm{e}^{-q\tau_a^+}\mathbf{1}_{(\tau_a^+<\tau_0^-)}\big).
\end{aligned} \tag{8.15}
$$

Combining (8.14) and (8.15) gives

$$
\mathbb{E}_x\big(\mathrm{e}^{-q\tau_a^+}\mathbf{1}_{(\tau_a^+<\tau_0^-)}\big) = \mathrm{e}^{-\Phi(q)(a-x)}\frac{W_{\Phi(q)}(x)}{W_{\Phi(q)}(a)} = \frac{W^{(q)}(x)}{W^{(q)}(a)}, \tag{8.16}
$$

where $W^{(q)}(x) = \mathrm{e}^{\Phi(q)x}W_{\Phi(q)}(x)$. Clearly $W^{(q)}$ is identically zero on $(-\infty, 0)$ and non-decreasing.

Consider now the final case that $\psi'(0+) = 0$ and $q = 0$. Since the limit as $q \downarrow 0$ on the left-hand side of (8.16) exists, the same is true of the right-hand side. By choosing an arbitrary $b > a$, we can thus define, $W(x) = \lim_{q\downarrow 0} W^{(q)}(x)/W^{(q)}(b)$ for each $x \leq a$. Consequently,

$$
\begin{aligned}
W(x) &= \lim_{q\downarrow 0} \frac{W^{(q)}(x)}{W^{(q)}(b)} \\
&= \lim_{q\downarrow 0} \mathbb{E}_x\left(\mathrm{e}^{-q\tau_a^+}\mathbf{1}_{(\tau_a^+<\tau_0^-)}\right)\frac{W^{(q)}(a)}{W^{(q)}(b)} \\
&= \mathbb{P}_x\left(\tau_a^+ < \tau_0^-\right)W(a). \qquad (8.17)
\end{aligned}
$$

Again it is clear that W is identically zero on $(-\infty, 0)$ and non-decreasing.

It is important to note for the remaining parts of the proof that the definition of $W^{(q)}$ we have given above may be taken up to any multiplicative constant without affecting the validity of the arguments.[2] □

Proof of Theorem 8.1 (i) Suppose again that X is assumed to drift to infinity so that $\psi'(0+) > 0$. First consider the case that $q = 0$. Recalling that the definition of W in (8.11) may be taken up to a multiplicative constant, let us work with

$$
W(x) = \frac{1}{\psi'(0+)}\mathbb{P}_x(\underline{X}_\infty \geq 0). \qquad (8.18)
$$

We may take limits in the second Wiener–Hopf factor given in (8.4) to deduce that

$$
\mathbb{E}(\mathrm{e}^{\beta \underline{X}_\infty}) = \psi'(0+)\frac{\beta}{\psi(\beta)}
$$

for $\beta > 0$. Integrating by parts, we also see that

$$
\begin{aligned}
\mathbb{E}(\mathrm{e}^{\beta \underline{X}_\infty}) &= \int_{[0,\infty)} \mathrm{e}^{-\beta x}\mathbb{P}(-\underline{X}_\infty \in \mathrm{d}x) \\
&= \mathbb{P}(-\underline{X}_\infty = 0) + \int_{(0,\infty)} \mathrm{e}^{-\beta x}\,\mathrm{d}\mathbb{P}\big(-\underline{X}_\infty \in (0,x]\big) \\
&= \int_0^\infty \mathbb{P}(-\underline{X}_\infty = 0)\beta\,\mathrm{e}^{-\beta x}\,\mathrm{d}x + \beta\int_0^\infty \mathrm{e}^{-\beta x}\mathbb{P}\big(-\underline{X}_\infty \in (0,x]\big)\mathrm{d}x \\
&= \beta\int_0^\infty \mathrm{e}^{-\beta x}\mathbb{P}(-\underline{X}_\infty \leq x)\mathrm{d}x \\
&= \beta\int_0^\infty \mathrm{e}^{-\beta x}\mathbb{P}_x(\underline{X}_\infty \geq 0)\mathrm{d}x,
\end{aligned}
$$

and hence

$$
\int_0^\infty \mathrm{e}^{-\beta x}W(x)\,\mathrm{d}x = \frac{1}{\psi(\beta)} \qquad (8.19)
$$

for all $\beta > 0 = \Phi(0)$.

Now for the case that $q > 0$ or that $q = 0$ and $\psi'(0+) < 0$. Take, as before, $W^{(q)}(x) = \mathrm{e}^{\Phi(q)x}W_{\Phi(q)}(x)$. As remarked earlier, X under $\mathbb{P}^{\Phi(q)}$ drifts to infinity,

---

[2] This also justifies the terminology "scale function".

and hence, using the conclusion from the previous paragraph together with (8.6), we have

$$\begin{aligned}\int_0^\infty \mathrm{e}^{-\beta x} W^{(q)}(x)\mathrm{d}x &= \int_0^\infty \mathrm{e}^{-(\beta-\Phi(q))x} W_{\Phi(q)}(x)\mathrm{d}x \\ &= \frac{1}{\psi_{\Phi(q)}(\beta-\Phi(q))} \\ &= \frac{1}{\psi(\beta)-q},\end{aligned}$$

provided $\beta - \Phi(q) > 0$. Since $W^{(q)}$ is an increasing function, we work with the measure $W^{(q)}(\mathrm{d}x)$ on $[0,\infty)$, associated with the distribution $W^{(q)}(a,b] := W^{(q)}(b) - W^{(q)}(a)$ for $-\infty < a \le b < \infty$. Integration by parts gives a characterisation of the measure $W^{(q)}$,

$$\begin{aligned}\int_{[0,\infty)} \mathrm{e}^{-\beta x} W^{(q)}(\mathrm{d}x) &= W^{(q)}(0) + \int_{(0,\infty)} \mathrm{e}^{-\beta x}\, \mathrm{d}W^{(q)}(0,x] \\ &= \int_0^\infty \beta\, \mathrm{e}^{-\beta x} W^{(q)}(0)\mathrm{d}x + \int_0^\infty \beta\, \mathrm{e}^{-\beta x} W^{(q)}(0,x]\mathrm{d}x \\ &= \frac{\beta}{\psi(\beta)-q}\end{aligned} \tag{8.20}$$

for $\beta > \Phi(q)$.

For the case that $q = 0$ and $\psi'(0+) = 0$ one may appeal to the Extended Continuity Theorem for Laplace Transforms (see Feller (1971), Theorem XIII.1.2a) to deduce that, since

$$\lim_{q\downarrow 0} \int_{[0,\infty)} \mathrm{e}^{-\beta x} W^{(q)}(\mathrm{d}x) = \lim_{q\downarrow 0} \frac{\beta}{\psi(\beta)-q} = \frac{\beta}{\psi(\beta)},$$

there exists a measure W^* such that, in the sense of vague convergence, $W^*(\mathrm{d}x) = \lim_{q\downarrow 0} W^{(q)}(\mathrm{d}x)$ and

$$\int_{[0,\infty)} \mathrm{e}^{-\beta x} W^*(\mathrm{d}x) = \frac{\beta}{\psi(\beta)}.$$

Clearly $W^*(x) := W^*[0,x]$ is a multiple of W given in (8.17), so we may define $W = W^*$. Integration by parts now shows that (8.19) holds again.

Next, we turn to continuity and strict monotonicity of $W^{(q)}$. The argument is taken from Bertoin (1996a). Recall that $\{(t,\epsilon_t) : t \ge 0 \text{ and } \epsilon_t \neq \partial\}$ is the Poisson point process of excursions on $[0,\infty) \times \mathcal{E}$, with intensity $\mathrm{d}t \times \mathrm{d}n$, decomposing the path of X. Write $\overline{\epsilon}$ for the height of each excursion $\epsilon \in \mathcal{E}$; see Definition 6.13. For spectrally negative Lévy processes, we work with the definition of local time

$L=\overline{X}$. Hence, for $0\le x<a$, $L_{\tau^+_{a-x}}=\overline{X}_{\tau^+_{a-x}}=a-x$. Therefore it holds that

$$\{\underline{X}_{\tau^+_{a-x}}\ge -x\}=\{\forall t\le a-x \text{ and } \epsilon_t\ne\partial, \overline{\epsilon}_t\le t+x\}.$$

It follows with the help of (8.13) that

$$\begin{aligned}\frac{W(x)}{W(a)}&=\mathbb{P}_x(\underline{X}_{\tau^+_a}\ge 0)\\&=\mathbb{P}(\underline{X}_{\tau^+_{a-x}}\ge -x)\\&=\mathbb{P}(\forall t\le a-x \text{ and } \epsilon_t\ne\partial, \overline{\epsilon}_t\le t+x)\\&=\mathbb{P}\big(N(A)=0\big),\end{aligned}\tag{8.21}$$

where N is the Poisson random measure associated with the process of excursions and $A=\{(t,\epsilon_t): t\le a-x \text{ and } \overline{\epsilon}_t>t+x\}$. Since $N(A)$ is Poisson distributed with parameter $\int \mathbf{1}_A\, n(\mathrm{d}\epsilon)\mathrm{d}t=\int_0^{a-x} n(\overline{\epsilon}>t+x)\mathrm{d}t=\int_x^a n(\overline{\epsilon}>t)\mathrm{d}t$, we have that

$$\frac{W(x)}{W(a)}=\exp\bigg\{-\int_x^a n(\overline{\epsilon}>t)\mathrm{d}t\bigg\}.\tag{8.22}$$

Since a may be chosen arbitrarily large, continuity and strict monotonicity follow from (8.22). Continuity of W also guarantees that it is uniquely defined via its Laplace transform on $[0,\infty)$. From the definition

$$W^{(q)}(x)=\mathrm{e}^{\Phi(q)x}W_{\Phi(q)}(x),\tag{8.23}$$

the properties of continuity, uniqueness and strict monotonicity carry over to the case $q>0$. □

Proof of Theorem 8.1 (ii) Using the Laplace transform of $W^{(q)}(x)$ (given in (8.8)), as well as the Laplace–Stieltjes transform (8.20), we can interpret the second Wiener–Hopf factor in (8.4) as saying that, for $x\ge 0$,

$$\mathbb{P}(-\underline{X}_{\mathbf{e}_q}\in\mathrm{d}x)=\frac{q}{\Phi(q)}W^{(q)}(\mathrm{d}x)-qW^{(q)}(x)\mathrm{d}x,\tag{8.24}$$

and hence, for $x\ge 0$,

$$\begin{aligned}\mathbb{E}_x\big(\mathrm{e}^{-q\tau_0^-}\mathbf{1}_{(\tau_0^-<\infty)}\big)&=\mathbb{P}_x\big(\mathbf{e}_q>\tau_0^-\big)\\&=\mathbb{P}_x(\underline{X}_{\mathbf{e}_q}<0)\\&=\mathbb{P}(-\underline{X}_{\mathbf{e}_q}>x)\\&=1-\mathbb{P}(-\underline{X}_{\mathbf{e}_q}\le x)\end{aligned}$$

$$= 1 + q\int_0^x W^{(q)}(y)\mathrm{d}y - \frac{q}{\Phi(q)}W^{(q)}(x)$$

$$= Z^{(q)}(x) - \frac{q}{\Phi(q)}W^{(q)}(x). \tag{8.25}$$

Note that since $Z^{(q)}(x) = 1$ and $W^{(q)}(x) = 0$ for all $x \in (-\infty, 0)$, the statement is valid for all $x \in \mathbb{R}$. The proof is now complete for the case that $q > 0$.

Finally, we have that $\lim_{q\downarrow 0} q/\Phi(q) = \lim_{q\downarrow 0} \psi(\Phi(q))/\Phi(q)$. If $\psi'(0+) \geq 0$. i.e. the process drifts to infinity or oscillates, then $\Phi(0) = 0$ and the limit is equal to $\psi'(0+)$. Otherwise, when $\Phi(0) > 0$, the aforementioned limit is zero. The proof is thus completed by taking the limit in q in (8.9). □

Proof of Theorem 8.1 (8.12) Fix $q > 0$. We have for $x \geq 0$,

$$\mathbb{E}_x\big(\mathrm{e}^{-q\tau_0^-}\mathbf{1}_{(\tau_0^- < \tau_a^+)}\big) = \mathbb{E}_x\big(\mathrm{e}^{-q\tau_0^-}\mathbf{1}_{(\tau_0^- < \infty)}\big) - \mathbb{E}_x\big(\mathrm{e}^{-q\tau_0^-}\mathbf{1}_{(\tau_a^+ < \tau_0^-)}\big).$$

Applying the strong Markov property at τ_a^+ and using the fact that X creeps upwards, we also have that

$$\mathbb{E}_x\big(\mathrm{e}^{-q\tau_0^-}\mathbf{1}_{(\tau_a^+ < \tau_0^-)}\big) = \mathbb{E}_x\big(\mathrm{e}^{-q\tau_a^+}\mathbf{1}_{(\tau_a^+ < \tau_0^-)}\big)\mathbb{E}_a\big(\mathrm{e}^{-q\tau_0^-}\mathbf{1}_{(\tau_0^- < \infty)}\big).$$

Appealing to (8.9) and (8.11) we now have that

$$\mathbb{E}_x\big(\mathrm{e}^{-q\tau_0^-}\mathbf{1}_{(\tau_0^- < \tau_a^+)}\big) = Z^{(q)}(x) - \frac{q}{\Phi(q)}W^{(q)}(x) - \frac{W^{(q)}(x)}{W^{(q)}(a)}\left(Z^{(q)}(a) - \frac{q}{\Phi(q)}W^{(q)}(a)\right),$$

and the required result follows in the case that $q > 0$. The case that $q = 0$ is again dealt with by taking limits as $q \downarrow 0$. □

8.3 The Scale Functions $W^{(q)}$ and $Z^{(q)}$

Let us explore a little further the analytical properties of the functions $W^{(q)}$ and $Z^{(q)}$. As an abuse of notation, let us write $W^{(q)} \in C^1(0, \infty)$ to mean the restriction of $W^{(q)}$ to $(0, \infty)$ belongs to $C^1(0, \infty)$.

Lemma 8.2 *For all $q \geq 0$, the function $W^{(q)}$ has left and right derivatives on $(0, \infty)$, which agree if and only if the measure $n(\overline{\epsilon} \in \mathrm{d}x)$ has no atoms. In that case, $W^{(q)} \in C^1(0, \infty)$.*

Proof Since $W^{(q)}(x) := \mathrm{e}^{\Phi(q)x}W_{\Phi(q)}(x)$, it suffices to prove the result for $q = 0$. However, in this case, we identified in Eq. (8.22),

$$W(x) = W(a)\exp\left\{-\int_x^a n(\overline{\epsilon} > t)\mathrm{d}t\right\},$$

for any arbitrary $a > x$. It follows then that the left and right first derivatives exist and are given by

$$W'_-(x) = n(\overline{\epsilon} \geq x)W(x) \quad \text{and} \quad W'_+(x) = n(\overline{\epsilon} > x)W(x). \tag{8.26}$$

Since W is continuous, W' exists if and only if $n(\overline{\epsilon} \in \mathrm{d}x)$ has no atoms as claimed. In that case it is clear that it also belongs to the class $C^1(0,\infty)$. □

Although the proof is a little technical, it can be shown that $n(\overline{\epsilon} \in \mathrm{d}x)$ has no atoms if X is a process of unbounded variation. If X has bounded variation then it is very easy to construct an example where $n(\overline{\epsilon} \in \mathrm{d}x)$ has at least one atom. Consider for example the case of a compound Poisson process with positive drift and negative jumps whose distribution has an atom at unity. An excursion may therefore begin with a jump of size one. Since thereafter the process may fail to jump again before reaching its previous maximum, we see the excursion measure of heights must have at least an atom at 1, i.e. $n(\overline{\epsilon} = 1) > 0$. In fact, it can be shown in the case of bounded variation paths that $n(\overline{\epsilon} \in \mathrm{d}x)$ has no atoms if and only if the Lévy measure Π is atomless. See Exercise 8.4.

Next, we look at how $W^{(q)}$ and $Z^{(q)}$ extend analytically in the parameter q. This will turn out to be important in some of the exercises at the end of this chapter. The following result is found in Bertoin (1997b).

Lemma 8.3 *For each $x \geq 0$, the function $q \mapsto W^{(q)}(x)$ may be analytically extended in q to $\mathbb{C}$.*

Proof For a fixed choice of $q > 0$ and $\beta > \Phi(q)$ (so that $0 < q/\psi(\beta) < 1$),

$$\begin{aligned}\int_0^\infty \mathrm{e}^{-\beta x} W^{(q)}(x)\mathrm{d}x &= \frac{1}{\psi(\beta) - q} \\ &= \frac{1}{\psi(\beta)} \frac{1}{1 - q/\psi(\beta)} \\ &= \frac{1}{\psi(\beta)} \sum_{k\geq 0} q^k \frac{1}{\psi(\beta)^k}.\end{aligned} \tag{8.27}$$

Next, we claim that

$$\sum_{k\geq 0} q^k W^{*(k+1)}(x)$$

converges for each $x \geq 0$ where $W^{*(k+1)}$ is the $(k+1)$-th convolution of W with itself. This is easily deduced once one has the estimate

$$W^{*(k+1)}(x) \leq \frac{x^k}{k!} W(x)^{k+1}, \tag{8.28}$$

for $k \geq 0$ and $x \geq 0$, which itself can easily be proved by induction. Indeed, (8.28) holds trivially for $k = 0$ and if (8.28) holds for $k \geq 0$, then by monotonicity of W,

$$\begin{aligned} W^{*(k+1)}(x) &\leq \int_0^x \frac{y^{k-1}}{(k-1)!} W(y)^k W(x-y) \mathrm{d}y \\ &\leq \frac{1}{(k-1)!} W(x)^{k+1} \int_0^x y^{k-1} \mathrm{d}y \\ &= \frac{x^k}{k!} W(x)^{k+1}. \end{aligned}$$

Returning to (8.27), we may now apply Fubini's Theorem (justified by the assumption that $\beta > \Phi(q)$) and deduce that

$$\begin{aligned} \int_0^\infty \mathrm{e}^{-\beta x} W^{(q)}(x) \mathrm{d}x &= \sum_{k \geq 0} q^k \frac{1}{\psi(\beta)^{k+1}} \\ &= \sum_{k \geq 0} q^k \int_0^\infty \mathrm{e}^{-\beta x} W^{*(k+1)}(x) \mathrm{d}x \\ &= \int_0^\infty \mathrm{e}^{-\beta x} \sum_{k \geq 0} q^k W^{*(k+1)}(x) \mathrm{d}x. \end{aligned}$$

Thanks to continuity of W and $W^{(q)}$, we have that

$$W^{(q)}(x) = \sum_{k \geq 0} q^k W^{*(k+1)}(x). \tag{8.29}$$

Now noting that $\sum_{k \geq 0} q^k W^{*(k+1)}(x)$ converges for all $q \in \mathbb{C}$, we may extend the definition of $W^{(q)}$ for each fixed $x \geq 0$ by the equality given in (8.29). □

Suppose that, for each $c \geq 0$, we call $W_c^{(q)}$ the function fulfilling the definitions given in Theorem 8.1 but with respect to the measure $\mathbb{P}^c$. The previous lemma allows us to establish the following relationship for $W_c^{(q)}$ with different values of q and c.

Lemma 8.4 *For any $q \in \mathbb{C}$ and $c \in \mathbb{R}$ such that $\psi(c) < \infty$, we have*

$$W^{(q)}(x) = \mathrm{e}^{cx} W_c^{(q-\psi(c))}(x) \tag{8.30}$$

for all $x \geq 0$.

Proof For a given $c \in \mathbb{R}$ such that $\psi(c) < \infty$, the identity (8.30) holds for $q \geq 0$ and $q - \psi(c) \geq 0$ on account of both left- and right-hand sides being continuous functions with the same Laplace transform. By Lemma 8.3, both left- and right-hand sides of (8.30) are analytic in q for each fixed $x \geq 0$. The Identity Theorem for analytic functions thus implies that they are equal for all $q \in \mathbb{C}$. □

Unfortunately, a convenient relation such as (8.30) cannot be given for $Z^{(q)}$. Nonetheless, we do have the following obvious corollary.

Corollary 8.5 *For each $x > 0$ the function $q \mapsto Z^{(q)}(x)$ may be analytically extended to $q \in \mathbb{C}$.*

The final lemma of this section shows that a discontinuity of $W^{(q)}$ at zero may occur even when $W^{(q)}$ belongs to $C^1(0, \infty)$.

Lemma 8.6 *For all $q \geq 0$, $W^{(q)}(0) = 0$ if and only if X has unbounded variation. Otherwise, when X has bounded variation, it is equal to $1/\delta$, where $\delta > 0$ is the drift.*

Proof Recall the second identity in (8.4). Note that for all $q > 0$,

$$\begin{aligned} W^{(q)}(0) &= \lim_{\beta \uparrow \infty} \int_0^\infty \beta \, \mathrm{e}^{-\beta x} W^{(q)}(x) \mathrm{d}x \\ &= \lim_{\beta \uparrow \infty} \frac{\beta}{\psi(\beta) - q} \\ &= \lim_{\beta \uparrow \infty} \frac{\beta - \Phi(q)}{\psi(\beta) - q} \\ &= \frac{\Phi(q)}{q} \lim_{\beta \uparrow \infty} \mathbb{E}\big(\mathrm{e}^{\beta \underline{X}_{\mathbf{e}_q}}\big) \\ &= \frac{\Phi(q)}{q} \mathbb{P}(\underline{X}_{\mathbf{e}_q} = 0). \end{aligned}$$

Now recall that $\mathbb{P}(\underline{X}_{\mathbf{e}_q} = 0) > 0$ if and only if 0 is irregular for $(-\infty, 0)$, which was shown earlier to be equivalent to the case that X has paths of bounded variation. The above calculation also shows that

$$W^{(q)}(0) = \lim_{\beta \uparrow \infty} \frac{\beta}{\psi(\beta) - q} = \lim_{\beta \uparrow \infty} \frac{\beta}{\psi(\beta)},$$

which in turn is equal to $1/\delta$ by Exercise 2.11.

To deal with the case that $q = 0$, note from (8.29) that for any $p > 0$, $W^{(p)}(0) = W(0)$. □

Returning to (8.11), we see that the conclusion of the previous lemma indicates that, precisely when X has bounded variation,

$$\mathbb{P}_0\big(\tau_a^+ < \tau_0^-\big) = \frac{W(0)}{W(a)} > 0. \tag{8.31}$$

Note that the stopping time τ_0^- is defined by strict first passage. Hence when X has the property that 0 is irregular for $(-\infty, 0)$, it takes an almost surely positive amount of time to exit the half-line $[0, \infty)$. Since the aforementioned irregularity is equivalent to bounded variation for this class of Lévy processes, we see that (8.31) intuitively makes sense.

8.4 Potential Measures

In this section, we give an example of how scale functions may be used to describe potential measures associated with the one- and two-sided exit problems. This gives the opportunity to study the overshoot distributions at first passage below a level. Many of the calculations in this section concerning potential measures are reproduced from Bertoin (1997a).

To introduce the idea of potential measures and their relevance in this context, fix $a > 0$ and suppose that

$$\tau = \tau_a^+ \wedge \tau_0^-.$$

A computation in the spirit of Theorem 5.6 and Lemma 5.8, with the help of the Compensation Formula (Theorem 4.4), gives, for $x \in [0, a]$, A any Borel set in $[0, a)$ and B any Borel set in $(-\infty, 0)$,

$$\begin{aligned}
&\mathbb{P}_x(X_\tau \in B, X_{\tau-} \in A) \\
&= \mathbb{E}_x\left(\int_{[0,\infty)}\int_{(-\infty,0)} \mathbf{1}_{(\overline{X}_{t-}\le a, \underline{X}_{t-}\ge 0, X_{t-}\in A)}\mathbf{1}_{(y\in B-X_{t-})}N(\mathrm{d}t\times\mathrm{d}y)\right) \\
&= \mathbb{E}_x\left(\int_0^\infty \mathbf{1}_{(t<\tau)}\Pi(B-X_t)\mathbf{1}_{(X_t\in A)}\mathrm{d}t\right) \\
&= \int_A \Pi(B-y)U(a,x,\mathrm{d}y),
\end{aligned} \tag{8.32}$$

where N is the Poisson random measure associated with the jumps of X and

$$U(a, x, \mathrm{d}y) := \int_0^\infty \mathbb{P}_x(X_t \in \mathrm{d}y, \tau > t)\mathrm{d}t.$$

The above is called the *potential measure of X killed on exiting* $[0, a]$ when issued from x. It is also known as the *resolvent* measure. More generally, we can work with the q-potential measure, where

$$U^{(q)}(a, x, \mathrm{d}y) := \int_0^\infty \mathrm{e}^{-qt}\,\mathbb{P}_x(X_t \in \mathrm{d}y, \tau > t)\mathrm{d}t,$$

for $q \ge 0$, with the agreement that $U^{(0)} = U$. If, for each $x \in [0, a]$, a density of $U^{(q)}(a, x, \mathrm{d}y)$ exists with respect to Lebesgue measure, then we call it the *potential density* and denote it by $u^{(q)}(a,x,y)$ (with $u^{(0)}=u$). It turns out that, for a spectrally negative process, not only does a potential density exist, but also we can write it in semi-explicit terms. This is the subject of the next theorem, which is due to Suprun (1976) and later Bertoin (1997a). Note, in the statement of the result, it is implicitly understood that $W^{(q)}(z)$ is identically zero for $z<0$.

Theorem 8.7 *Suppose, for $q\geq 0$, that $U^{(q)}(a,x,\mathrm{d}y)$ is the q-potential measure of a spectrally negative Lévy process killed on exiting $[0,a]$ where $x,y\in[0,a]$. Then it has a density $u^{(q)}(a,x,y)$ given by*

$$u^{(q)}(a,x,y)=\frac{W^{(q)}(x)W^{(q)}(a-y)}{W^{(q)}(a)}-W^{(q)}(x-y). \tag{8.33}$$

Proof We start by noting that for all $x,y\geq 0$ and $q>0$,

$$R^{(q)}(x,\mathrm{d}y):=\int_0^\infty \mathrm{e}^{-qt}\,\mathbb{P}_x\big(X_t\in\mathrm{d}y,\tau_0^->t\big)\mathrm{d}t=\frac{1}{q}\mathbb{P}_x(X_{\mathbf{e}_q}\in\mathrm{d}y,\underline{X}_{\mathbf{e}_q}\geq 0),$$

where $\mathbf{e}_q$ is an independent, exponentially distributed random variable with parameter $q>0$. Recall, one may think of $R^{(q)}$ as the q-potential measure of the process X when killed on exiting $[0,\infty)$.

Appealing to the Wiener–Hopf factorisation, specifically that $X_{\mathbf{e}_q}-\underline{X}_{\mathbf{e}_q}$ is independent of $\underline{X}_{\mathbf{e}_q}$, we have that

$$\begin{aligned}R^{(q)}(x,\mathrm{d}y)&=\frac{1}{q}\mathbb{P}\big((X_{\mathbf{e}_q}-\underline{X}_{\mathbf{e}_q})+\underline{X}_{\mathbf{e}_q}\in\mathrm{d}y-x,-\underline{X}_{\mathbf{e}_q}\leq x\big)\\&=\frac{1}{q}\int_{[x-y,x]}\mathbb{P}(-\underline{X}_{\mathbf{e}_q}\in\mathrm{d}z)\mathbb{P}(X_{\mathbf{e}_q}-\underline{X}_{\mathbf{e}_q}\in\mathrm{d}y-x+z).\end{aligned}$$

By duality, $X_{\mathbf{e}_q}-\underline{X}_{\mathbf{e}_q}$ is equal in distribution to $\overline{X}_{\mathbf{e}_q}$, which itself is exponentially distributed with parameter $\Phi(q)$. In addition, the law of $-\underline{X}_{\mathbf{e}_q}$ has been identified in (8.24). We may therefore develop the expression for $R^{(q)}(x,\mathrm{d}y)$ as follows:

$$R^{(q)}(x,\mathrm{d}y)=\left\{\int_{[x-y,x]}\left(\frac{1}{\Phi(q)}W^{(q)}(\mathrm{d}z)-W^{(q)}(z)\mathrm{d}z\right)\Phi(q)\mathrm{e}^{-\Phi(q)(y-x+z)}\right\}\mathrm{d}y.$$

This shows that there exists a density, $r^{(q)}(x,y)$, for the measure $R^{(q)}(x,\mathrm{d}y)$. Now applying integration by parts to the integral in the last equality, we have that

$$r^{(q)}(x,y)=\mathrm{e}^{-\Phi(q)y}W^{(q)}(x)-W^{(q)}(x-y).$$

Finally, we may use the above established facts to compute the potential density $u^{(q)}$ as follows. First note that, with the help of the strong Markov property,

$$\begin{aligned}
qU^{(q)}(a, x, \mathrm{d}y) &= \mathbb{P}_x(X_{\mathbf{e}_q} \in \mathrm{d}y, \underline{X}_{\mathbf{e}_q} \geq 0, \overline{X}_{\mathbf{e}_q} \leq a) \\
&= \mathbb{P}_x(X_{\mathbf{e}_q} \in \mathrm{d}y, \underline{X}_{\mathbf{e}_q} \geq 0) \\
&\quad - \mathbb{P}_x(X_{\mathbf{e}_q} \in \mathrm{d}y, \underline{X}_{\mathbf{e}_q} \geq 0, \overline{X}_{\mathbf{e}_q} > a) \\
&= \mathbb{P}_x(X_{\mathbf{e}_q} \in \mathrm{d}y, \underline{X}_{\mathbf{e}_q} \geq 0) \\
&\quad - \mathbb{P}_x(X_\tau = a, \tau < \mathbf{e}_q)\mathbb{P}_a(X_{\mathbf{e}_q} \in \mathrm{d}y, \underline{X}_{\mathbf{e}_q} \geq 0).
\end{aligned}$$

The first and third of the three probabilities on the right-hand side above have been computed in the previous paragraph, the second probability is equal to

$$\mathbb{E}_x\big(\mathrm{e}^{-q\tau_a^+}\mathbf{1}_{(\tau_a^+ < \tau_0^-)}\big) = \frac{W^{(q)}(x)}{W^{(q)}(a)}.$$

In conclusion, we have that $U^{(q)}(a, x, \mathrm{d}y)$ has a density

$$r^{(q)}(x, y) - \frac{W^{(q)}(x)}{W^{(q)}(a)} r^{(q)}(a, y),$$

which, after a short amount of algebra, can be shown to be equal to the right-hand side of (8.33).

To complete the proof when $q = 0$, one may take limits in (8.33), noting that the right-hand side is analytic and hence continuous in q for fixed values x, a, y. The right-hand side of (8.33) tends to $u(a, x, y)$ by monotone convergence of $U^{(q)}$ as $q \downarrow 0$. □

The above proof contains the following corollary.

Corollary 8.8 *For $q \geq 0$, the q-potential measure of a spectrally negative Lévy process killed on exiting $[0, \infty)$ has density given by*

$$r^{(q)}(x, y) = \mathrm{e}^{-\Phi(q)y} W^{(q)}(x) - W^{(q)}(x - y),$$

for $x, y \geq 0$.

Define further the q-potential measure of X without killing by

$$\Theta^{(q)}(x, \mathrm{d}y) = \int_0^\infty \mathrm{e}^{-qt}\mathbb{P}_x(X_t \in \mathrm{d}y)\mathrm{d}t,$$

for $x, y \in \mathbb{R}$. Note, by spatial homogeneity, we have that $\Theta^{(q)}(x, \mathrm{d}y) = \Theta^{(q)}(0, \mathrm{d}y - x)$. If $\Theta^{(q)}(x, \mathrm{d}y)$ has a density, then we may always write it in the form $\theta^{(q)}(y - x)$ for some function $\theta^{(q)}$. The following corollary was established in Bingham (1975).

Corollary 8.9 *For $q > 0$, the q-potential density of a spectrally negative Lévy process is given by*

$$\theta^{(q)}(z) = \Phi'(q)\mathrm{e}^{-\Phi(q)z} - W^{(q)}(-z),$$

for all $z \in \mathbb{R}$.

Proof The result is obtained from Corollary 8.8 by considering the effect of moving the killing barrier to an arbitrary large distance from the initial point. Formally, with the help of spatial homogeneity,

$$\theta^{(q)}(z) = \lim_{x\uparrow\infty} r^{(q)}(x, x+z) = \lim_{x\uparrow\infty} \mathrm{e}^{-\Phi(q)(x+z)} W^{(q)}(x) - W^{(q)}(-z).$$

Note, however, that, from the proof of Theorem 8.1 (iii), we identified $W^{(q)}(x) = \mathrm{e}^{\Phi(q)x} W_{\Phi(q)}(x)$ where

$$\int_0^\infty \mathrm{e}^{-\theta x} W_{\Phi(q)}(x)\mathrm{d}x = \frac{1}{\psi_{\Phi(q)}(\theta)}.$$

It follows that

$$\theta^{(q)}(z) = \mathrm{e}^{-\Phi(q)z} W_{\Phi(q)}(\infty) - W^{(q)}(-z).$$

Note that $(X, \mathbb{P}^{\Phi(q)})$ drifts to infinity and hence $W_{\Phi(q)}(\infty) < \infty$. Since $W_{\Phi(q)}$ is a continuous function on $(0,\infty)$, we have that

$$W_{\Phi(q)}(\infty) = \lim_{\theta\downarrow 0} \int_0^\infty \theta\, \mathrm{e}^{-\theta x} W_{\Phi(q)}(x)\mathrm{d}x = \lim_{\theta\downarrow 0} \frac{\theta}{\psi_{\Phi(q)}(\theta)} = \frac{1}{\psi'_{\Phi(q)}(0+)}.$$

As $\psi(\Phi(q)) = q$, differentiation of this equality implies that the right-hand side above is equal to $\Phi'(q)$ and the proof is complete. □

To conclude this section, let us now return to (8.32). Recall that $\tau = \tau_a^+ \wedge \tau_0^-$. The above results now show that for $z \in (-\infty, 0)$ and $y \in (0, a]$,

$$\begin{aligned}&\mathbb{P}_x(X_\tau \in \mathrm{d}z, X_{\tau-} \in \mathrm{d}y)\\ &\quad = \Pi(\mathrm{d}z - y)\left\{\frac{W(x)W(a-y) - W(a)W(x-y)}{W(a)}\right\}\mathrm{d}y. \qquad (8.34)\end{aligned}$$

Similarly, in the limiting case when a tends to infinity,

$$\begin{aligned}&\mathbb{P}_x(X_{\tau_0^-} \in \mathrm{d}z, X_{\tau_0^- -} \in \mathrm{d}y)\\ &\quad = \Pi(\mathrm{d}z - y)\{\mathrm{e}^{-\Phi(0)y} W(x) - W(x-y)\}\mathrm{d}y. \qquad (8.35)\end{aligned}$$

8.5 Identities for Reflected Processes

In this final section, we give further support to the idea that the functions $W^{(q)}$ and $Z^{(q)}$ play a central role in many fluctuation identities concerning spectrally negative Lévy processes. We give a brief account of their appearance in a number of identities for spectrally negative Lévy processes reflected at either their supremum or their infimum.

We begin by reiterating what we mean by a Lévy process reflected at its supremum or reflected at its infimum. Fix $x \geq 0$. Then the process

$$\overline{Y}_t^x := (x \vee \overline{X}_t) - X_t, \quad t \geq 0$$

is called the process reflected at its supremum (with initial value x) and the process

$$\underline{Y}_t^x := X_t - \big(\underline{X}_t \wedge (-x)\big), \quad t \geq 0$$

is called the process reflected at its infimum (with initial value x).

For such processes, we may consider the exit times

$$\overline{\sigma}_a^x = \inf\{t > 0 : \overline{Y}_t^x > a\} \quad \text{and} \quad \underline{\sigma}_a^x = \inf\{t > 0 : \underline{Y}_t^x > a\}$$

for levels $a > 0$. In the spirit of Theorem 8.1, we have the following result.

Theorem 8.10 *Let X be a spectrally negative Lévy process with Lévy measure Π. Fix $a > 0$. We have,*

(i) *for $x \in [0, a]$ and $\theta \in \mathbb{R}$ such that $\psi(\theta) < \infty$,*

$$\mathbb{E}\big(\mathrm{e}^{-q\overline{\sigma}_a^x - \theta \overline{Y}_{\overline{\sigma}_a^x}^x}\big) = \mathrm{e}^{-\theta x}\left(Z_\theta^{(p)}(a - x) - W_\theta^{(p)}(a - x)\frac{pW_\theta^{(p)}(a) + \theta Z_\theta^{(p)}(a)}{W_\theta^{(p)\prime}(a) + \theta W_\theta^{(p)}(a)}\right),$$

where $p = q - \psi(\theta)$ and $W_\theta^{(q)\prime}(a)$ is understood to be the right derivative of $W_\theta^{(q)}$ at a. Further,

(ii) *for $x \in [0, a]$,*

$$\mathbb{E}\big(\mathrm{e}^{-q\underline{\sigma}_a^x}\big) = \frac{Z^{(q)}(x)}{Z^{(q)}(a)}.$$

Part (i) was proved[3] in Avram et al. (2004) and part (ii) in Pistorius (2004). Their proofs turn out to be quite complicated, requiring the need for a theory which is slightly beyond the scope of this text, namely, Itô's excursion theory. Doney (2005, 2007) gives another proof of the above theorem, again based on excursion theory. Part (ii) for processes of bounded variation is proved in Exercise 8.10.

It turns out that it is also possible to say something about the q-potential measures of $\overline{Y}^x$ and $\underline{Y}^x$ with killing at first passage over a specified level $a > 0$. These potentials are defined, respectively, by

---

[3] See also the note at the end of this chapter.

$$\overline{U}^{(q)}(a,x,\mathrm{d}y)=\int_0^\infty \mathrm{e}^{-qt}\mathbb{P}\big(\overline{Y}_t^x\in \mathrm{d}y, \overline{\sigma}_a^x>t\big)\mathrm{d}t,$$

for $x, y \in [0,a]$, and

$$\underline{U}^{(q)}(a,x,\mathrm{d}y)=\int_0^\infty \mathrm{e}^{-qt}\mathbb{P}\big(\underline{Y}_t^x\in \mathrm{d}y, \underline{\sigma}_a^x>t\big)\mathrm{d}t,$$

for $x, y \in [0,a]$. The following results are due to Pistorius (2004). Alternative proofs are also given in Doney (2005, 2007). Once again, we offer no proofs here on account of their difficulty.

Theorem 8.11 *Fix $a>0$ and $q\geq 0$.*

(i) *For $x, y\in[0,a]$,*

$$\begin{aligned}\overline{U}^{(q)}(a,x,\mathrm{d}y) &= \left(W^{(q)}(a-x)\frac{W^{(q)}(0)}{W^{(q)\prime}(a)}\right)\delta_0(\mathrm{d}y)\\ &\quad+\left(W^{(q)}(a-x)\frac{W^{(q)\prime}(y)}{W^{(q)\prime}(a)}-W^{(q)}(y-x)\right)\mathrm{d}y.\end{aligned}$$

(ii) *For $x, y\in[0,a]$, the measure $\underline{U}^{(q)}(a,x,\mathrm{d}y)$ has a density given by*

$$\underline{u}^{(q)}(a,x,y)=W^{(q)}(a-y)\frac{Z^{(q)}(x)}{Z^{(q)}(a)}-W^{(q)}(x-y).$$

As in Theorem 8.10, we take $W^{(q)\prime}$ to mean the right derivative. Note in particular that when the underlying Lévy process is of unbounded variation, the q-potential for $\overline{Y}^x$ killed on first passage above a is absolutely continuous with respect to Lebesgue measure and otherwise it has an atom at zero. A little thought reveals that the atom in the bounded variation case appears as a consequence of the accumulation of Lebesgue measure at the maximum of X; see Theorem 6.7.

On a final note, we emphasise that there exists an additional body of literature, written in Russian and Ukrainian by members of the Kiev school of probability, which considers the type of boundary problems described above for spectrally one-sided Lévy processes using a so-called "potential method", developed in Korolyuk (1974). For example, Theorem 8.10 (i) can be found for the case that Π has finite total mass and $\sigma=0$ in Korolyuk (1975a, 1975b) and Bratiychuk and Gusak (1991). The reader is also referred to Korolyuk et al. (1976) and Korolyuk and Borovskich (1981) and references therein.[4]

[4] I am grateful to Professors V.S. Korolyuk and M.S. Bratiychuk for bringing this literature to my attention.

Exercises

8.1 Suppose that X is a spectrally negative Lévy process with Laplace exponent ψ such that $\psi'(0+) < 0$. Show that, for $t \geq 0$ and any A in $\mathcal{F}_t$,

$$\lim_{x\uparrow\infty} \mathbb{P}\big(A|\tau_x^+ < \infty\big) = \mathbb{P}^{\Phi(0)}(A),$$

where, as usual, Φ is the right inverse of ψ.

8.2 Suppose that X is a spectrally negative stable process with index $\alpha \in (1,2)$ and assume, without loss of generality, that its Laplace exponent is given by $\psi(\theta) = \theta^\alpha$, for $\theta \geq 0$ (cf. Exercise 3.7).

(i) Show that, for $q > 0$ and $\beta > q^{1/\alpha}$,

$$\int_0^\infty \mathrm{e}^{-\beta x}\overline{W}^{(q)}(x)\mathrm{d}x = \frac{1}{\beta(\beta^\alpha - q)} = \sum_{n\geq 1} q^{n-1}\beta^{-\alpha n-1},$$

where $\overline{W}^{(q)}(x) = \int_0^x W^{(q)}(y)\mathrm{d}y$.

(ii) Conclude that, for $x \geq 0$

$$Z^{(q)}(x) = \sum_{n\geq 0} q^n \frac{x^{\alpha n}}{\Gamma(\alpha n + 1)}.$$

Note that the right-hand side above is also equal to $\mathrm{E}_{\alpha,1}(qx^\alpha)$ where $\mathrm{E}_{\alpha,1}(\cdot)$ is the Mittag–Leffler function defined in (5.30).

(iii) Deduce that, for $q \geq 0$,

$$W^{(q)}(x) = \alpha x^{\alpha-1}\mathrm{E}'_{\alpha,1}\big(qx^\alpha\big),$$

for $x \geq 0$.

(iv) Show that, for standard Brownian motion,

$$W^{(q)}(x) = \sqrt{\frac{2}{q}}\sinh\big(\sqrt{2q}x\big) \quad \text{and} \quad Z^{(q)}(x) = \cosh\big(\sqrt{2q}x\big),$$

for $x \geq 0$ and $q \geq 0$.

(v) Suppose now that X is a tempered stable spectrally negative Lévy process, with Laplace exponent given by $\psi(\theta) = (\theta + c)^\alpha - c^\alpha$, where $c \geq 0$ and $\alpha \in (1,2)$. Show that, for $q \geq 0$,

$$W^{(q)}(x) = \mathrm{e}^{-cx}\alpha x^{\alpha-1}\mathrm{E}'_{\alpha,1}\big((q + c^\alpha)x^\alpha\big).$$

8.3 Suppose that X is a spectrally negative Lévy process of bounded variation such that $\lim_{t\uparrow\infty} X_t = \infty$. For convenience, write $X_t = \delta t - S_t$ where $S = \{S_t : t \geq 0\}$ is a subordinator with jump measure Υ and no drift.

(i) Show that, necessarily, $\delta^{-1}\int_0^\infty \Upsilon(y,\infty)\mathrm{d}y < 1$.
(ii) Show that the scale function, W, satisfies

$$\int_{[0,\infty)} \mathrm{e}^{-\beta x} W(\mathrm{d}x) = \frac{1}{\delta - \int_0^\infty \mathrm{e}^{-\beta y}\Upsilon(y,\infty)\mathrm{d}y}$$

and deduce that

$$W(\mathrm{d}x) = \frac{1}{\delta}\sum_{n\geq 0} \nu^{*n}(\mathrm{d}x),$$

where $\nu(\mathrm{d}x) = \delta^{-1}\Upsilon(x,\infty)\mathrm{d}x$ and, as usual, we understand $\nu^{*0}(\mathrm{d}x) = \delta_0(\mathrm{d}x)$.
(iii) Suppose that S is a compound Poisson process with rate $\lambda > 0$ and jump distribution which is exponential, with parameter $\mu > 0$. Show that

$$W(x) = \frac{1}{\delta}\left(1 + \frac{\lambda}{\delta\mu - \lambda}\left(1 - \mathrm{e}^{-(\mu - \delta^{-1}\lambda)x}\right)\right),$$

for $x \geq 0$.

8.4 It is known that, when X has paths of bounded variation, and accordingly its Laplace exponent is written in the form (8.3), the excursion measure, n, satisfies

$$n(\overline{\epsilon} > a) = \frac{1}{\delta}\int_{(-\infty,0)} \Pi(\mathrm{d}x)\mathbb{P}_{-x}\left(\tau_{-a}^- < \tau_0^+\right), \tag{8.36}$$

for $a > 0$. See for example formula (20) of Pistorius (2004).

(i) Use (8.36) to show that

$$n(\overline{\epsilon} > a) = \frac{1}{\delta}\Pi(-\infty,-a) + \frac{1}{\delta}\int_{[-a,0)} \Pi(\mathrm{d}x)\left(1 - \frac{W(x+a)}{W(a)}\right).$$

(ii) Deduce that

$$n(\overline{\epsilon} = a) = \frac{1}{\delta}\frac{W(0)}{W(a)}\Pi(\{-a\})$$

and hence conclude that $W \in C^1(0,\infty)$ if and only if Π has no atoms.
(iii) Use part (ii), together with (8.23), to show further that $W^{(q)} \in C^1(0,\infty)$, for all $q \geq 0$, if and only if Π has no atoms.

8.5 Let X be any spectrally negative Lévy process with Laplace exponent ψ.

(i) Use (8.12) and (8.9) to establish that, for each $q \geq 0$,

$$\lim_{x\uparrow\infty} \frac{Z^{(q)}(x)}{W^{(q)}(x)} = \frac{q}{\Phi(q)},$$

where the right-hand side is understood in the limiting sense when $q=0$. In addition, show that

$$\lim_{a\uparrow\infty}\frac{W^{(q)}(a-x)}{W^{(q)}(a)}=\mathrm{e}^{-\Phi(q)x}.$$

(ii) Taking account of a possible atom at the origin, write down the Laplace transform of $W^{(q)}(\mathrm{d}x)$ on $[0,\infty)$ and show that, if X has unbounded variation, then $W^{(q)\prime}(0+)=2/\sigma^2$, where σ is the Gaussian coefficient in the Lévy–Itô decomposition and it is understood that $1/0=\infty$. If, however, X has bounded variation, then the right derivative of $W^{(q)}$ at zero (with an abuse of notation, also written here as $W^{(q)\prime}(0+)$), satisfies

$$W^{(q)\prime}(0+)=\frac{\Pi(-\infty,0)+q}{\delta^2},$$

where δ is the drift coefficient and it is understood that the right-hand side is infinite if $\Pi(-\infty,0)=\infty$.

8.6 Suppose that X is a spectrally negative Lévy process. Using the results of Chap. 5, show, with the help of the Wiener–Hopf factorisation and scale functions, that

$$\mathbb{P}\big(X_{\tau_x^-}=x,\ \tau_x^-<\infty\big)=\frac{\sigma^2}{2}\big[W'(-x)-\Phi(0)W(-x)\big],$$

for all $x\le 0$. As usual, W is the scale function, Φ is the inverse of the Laplace exponent, ψ, of X and σ is the Gaussian coefficient.

8.7 This exercise deals with first hitting of points below zero of spectrally negative Lévy processes, following the work of Doney (1991). For each $x>0$, define

$$T(-x)=\inf\{t>0:X_t=-x\},$$

where X is a spectrally negative Lévy process with Laplace exponent ψ and right inverse Φ.

(i) Show that, for all $c\ge 0$ and $q\ge 0$,

$$\Phi_c(q)=\Phi\big(q+\psi(c)\big)-c.$$

(ii) Show, for $x>0$, $c\ge 0$ and $p\ge\psi(c)\vee 0$, that

$$\mathbb{E}\big(\mathrm{e}^{-p\tau_{-x}^-+c(X_{\tau_{-x}^-}+x)}\mathbf{1}_{(\tau_{-x}^-<\infty)}\big)=\mathrm{e}^{cx}\left(Z_c^{(q)}(x)-\frac{q}{\Phi_c(q)}W_c^{(q)}(x)\right),$$

where $q=p-\psi(c)$. Use analytic extension to justify that the above identity is in fact valid for all $x>0$, $c\ge 0$ and $p\ge 0$.

(iii) By noting that $T(-x) \geq \tau_{-x}^{-}$, condition on $\mathcal{F}_{\tau_{-x}^{-}}$ to deduce that, for $p, u \geq 0$,

$$\begin{aligned}&\mathbb{E}\big(\mathrm{e}^{-pT(-x)-u(T(-x)-\tau_{-x}^{-})}\mathbf{1}_{(T(-x)<\infty)}\big)\\&\quad=\mathbb{E}\big(\mathrm{e}^{-p\tau_{-x}^{-}+\Phi(p+u)(X_{\tau_{-x}^{-}}+x)}\mathbf{1}_{(\tau_{-x}^{-}<\infty)}\big).\end{aligned}$$

(iv) By taking a limit as $u \downarrow 0$ in part (iii) and making use of the identity in part (ii), deduce that

$$\mathbb{E}\big(\mathrm{e}^{-pT(-x)}\mathbf{1}_{(T(-x)<\infty)}\big)=\mathrm{e}^{\Phi(p)x}-\psi'\big(\Phi(p)\big)W^{(p)}(x)$$

and hence by taking limits again as $x \downarrow 0$,

$$\mathbb{E}\big(\mathrm{e}^{-pT(0)}\mathbf{1}_{(T(0)<\infty)}\big)=\begin{cases}1-\psi'(\Phi(p))\frac{1}{\delta} & \text{if } X \text{ has bounded variation}\\ 1 & \text{if } X \text{ has unbounded variation,}\end{cases}$$

where δ is the drift term in the Laplace exponent in the case that X has bounded variation paths.

8.8 Again relying on Doney (1991), we shall make the following application of part (iii) of the previous exercise. Suppose that $B = \{B_t : t \geq 0\}$ is a Brownian motion. Denote

$$\sigma = \inf\{t > 0 : B_t = \overline{B}_t = t\}.$$

(i) Suppose that X is a descending stable-$\frac{1}{2}$ subordinator with upward unit drift. Show that

$$\mathbb{P}(\sigma < \infty) = \mathbb{P}\big(T(0) < \infty\big),$$

where $T(0)$ is defined in Exercise 8.7.

(ii) Deduce from part (i) that $\mathbb{P}(\sigma < \infty) = \frac{1}{2}$.

8.9 This exercise is based on the results of Chiu and Yin (2005) and Baurdoux (2009). Suppose that X is any spectrally negative Lévy process with Laplace exponent ψ, satisfying $\lim_{t\uparrow\infty} X_t = \infty$. Recall that this necessarily implies that $\psi'(0+) > 0$. Define for each $x \in \mathbb{R}$,

$$\Lambda_0 = \sup\{t > 0 : X_t < 0\}.$$

Here, we work with the definition $\sup\emptyset = 0$ so that the event $\{\Lambda_0 = 0\}$ corresponds to the event that X never enters $(-\infty, 0)$.

(i) Using the almost surely equivalent events $\{\Lambda_0 < t\} = \{X_t \geq 0, \inf_{s\geq t} X_s \geq 0\}$ and the Markov property, show that for each $q > 0$ and $y \in \mathbb{R}$

$$\mathbb{E}_y\big(\mathrm{e}^{-q\Lambda_0}\big)=q\int_0^\infty \theta^{(q)}(x-y)\mathbb{P}_x(\underline{X}_\infty \geq 0)\mathrm{d}x,$$

where $\theta^{(q)}$ is the q-potential density of X.

(ii) Hence show that for $y \le 0$,

$$\mathbb{E}_y(\mathrm{e}^{-q\Lambda_0}) = \psi'(0+)\Phi'(q)\mathrm{e}^{\Phi(q)y},$$

where Φ is the right inverse of ψ and, in particular,

$$\mathbb{P}(\Lambda_0 = 0) = \begin{cases} \psi'(0+)/\delta & \text{if } X \text{ has bounded variation with drift } \delta \\ 0 & \text{if } X \text{ has unbounded variation.} \end{cases}$$

(iii) Suppose now that $y > 0$. Use again the strong Markov property to deduce that, for $q > 0$,

$$\mathbb{E}_y(\mathrm{e}^{-q\Lambda_0}\mathbf{1}_{(\Lambda_0>0)}) = \psi'(0+)\Phi'(q)\mathbb{E}_y(\mathrm{e}^{-q\tau_0^- + \Phi(q)X_{\tau_0^-}}\mathbf{1}_{(\tau_0^-<\infty)}).$$

(iv) Deduce that, for $y > 0$ and $q > 0$,

$$\mathbb{E}_y(\mathrm{e}^{-q\Lambda_0}\mathbf{1}_{(\Lambda_0>0)}) = \psi'(0+)\Phi'(q)\mathrm{e}^{\Phi(q)y} - \psi'(0+)W^{(q)}(y).$$

8.10 (Proof of Theorem 8.10 (ii) with Bounded Variation) Adopt the setting of Theorem 8.10 (ii). It may be assumed that $\underline{\sigma}_a^x$ is a stopping time with respect to the filtration $\mathbb{F}$ (recall that in our standard notation, this is the filtration generated by the underlying Lévy process X, which satisfies the usual conditions of completion and right continuity).

(i) Show that for any $x \in (0, a]$,

$$\mathbb{E}(\mathrm{e}^{-q\underline{\sigma}_a^x}) = \mathbb{E}_x(\mathrm{e}^{-q\tau_0^-}\mathbf{1}_{(\tau_0^-<\tau_a^+)})\mathbb{E}(\mathrm{e}^{-q\underline{\sigma}_a^0}) + \mathbb{E}_x(\mathrm{e}^{-q\tau_a^+}\mathbf{1}_{(\tau_a^+<\tau_0^-)}).$$

(ii) By taking limits as x tends to zero in part (i), deduce that

$$\mathbb{E}(\mathrm{e}^{-q\underline{\sigma}_a^x}) = \frac{Z^{(q)}(x)}{Z^{(q)}(a)},$$

for all $x \in [0, a]$.
Hint: recall that $W^{(q)}(0) > 0$ if X has paths of bounded variation.

(iii) The following application comes from Dube et al. (2004). Let W be a general storage process, as described at the beginning of Chap. 4. Now suppose that this storage process has a limited capacity, say $c > 0$. This means that, when the workload exceeds c units, the excess of work is removed and dumped. Prove that the Laplace transform (with parameter $q > 0$) of the first time for the workload of this storage process to become zero, when started from $0 < x < c$, is given by $Z^{(q)}(c - x)/Z^{(q)}(c)$, where $Z^{(q)}$ is the scale function associated with the underlying Lévy process driving W.

8.11 Suppose that X is a spectrally negative α-stable process for $\alpha \in (1, 2)$. We are interested in establishing the distribution of the overshoot below the origin when the process, starting from $x \in (0,1)$, first exits this interval from below. In principle one could attempt to invert the formula given in Exercise 8.7 (ii). However, the following technique, taken from Rogozin (1972), offers a more straightforward method. It will be helpful to first review Exercise 7.7.

(i) Show that

$$\mathbb{P}_x\big(-X_{\tau_0^-} \le y;\, \tau_0^- < \tau_1^+\big) = \Phi_{\alpha-1}\left(\frac{y}{x}\right) - \mathbb{P}_x\big(\tau_1^+ < \tau_0^-\big)\Phi_{\alpha-1}(y),$$

where $\Phi_{\alpha-1}$ was defined in Exercise 7.7.

(ii) Hence, deduce that

$$\mathbb{P}_x\big(-X_{\tau_0^-} \le y;\, \tau_0^- < \tau_1^+\big) = \frac{\sin\pi(\alpha-1)}{\pi} x^{\alpha-1}(1-x)\int_0^y t^{-(\alpha-1)}(t+1)^{-1}(t+x)^{-1}\mathrm{d}t.$$

(iii) Finally let us consider the problem of first entry *into* the strip $(-1,1)$; cf. Port (1967). Let

$$\tau_{(-1,1)} = \inf\{t > 0 : X_t \in (-1,1)\}.$$

Show that the hitting distribution of $(-1,1)$ is given by

$$\mathbb{P}_x(X_{\tau_{(-1,1)}} \in \mathrm{d}y) = \frac{\sin\pi(\alpha-1)}{\pi}(x-1)^{\alpha-1}(1-y)^{1-\alpha}(x-y)^{-1}\mathrm{d}y + \delta_{-1}(\mathrm{d}y)\frac{\sin\pi(\alpha-1)}{\pi}\int_0^{\frac{x-1}{x+1}} t^{\alpha-2}(1-t)^{1-\alpha}\,\mathrm{d}t,$$

for $x \ge 1$ and $y \in (-1,1)$, where $\delta_{-1}(\mathrm{d}y)$ is the Dirac unit point mass at -1. What is the corresponding formula when $x \le -1$?

8.12 Fix $a \in (0,\infty]$ and $q \ge 0$. Show that

$$\mathrm{e}^{-q(t\wedge\tau_a^+\wedge\tau_0^-)}W^{(q)}(X_{t\wedge\tau_a^+\wedge\tau_0^-}) \quad \text{and} \quad \mathrm{e}^{-q(t\wedge\tau_a^+\wedge\tau_0^-)}Z^{(q)}(X_{t\wedge\tau_a^+\wedge\tau_0^-}), \quad t \ge 0,$$

are martingales.

Chapter 9
More on Scale Functions

In the previous chapter, we saw that it is possible to develop many fluctuation identities for spectrally negative Lévy processes in terms of scale functions. In this chapter, we continue in this vein and look in greater detail at the relationship between scale functions and potential measures of subordinators through the Wiener–Hopf factorisation. This will allow us to extract a number of additional analytical properties for scale functions as well as to offer a method for generating many examples of spectrally negative Lévy processes whose scale functions can be computed explicitly. A large part of this chapter is based on Hubalek and Kyprianou (2010) and Kyprianou and Rivero (2008).

9.1 The Wiener–Hopf Factorisation Revisited

Henceforth, we shall assume, as in the previous chapter, that X is a spectrally negative Lévy process with characteristic triple (a, σ, Π) and Laplace exponent ψ, whose right inverse function is denoted by Φ. Suppose temporarily that we denote its characteristic exponent by Ψ. According to (8.1),

$$\psi(\lambda) = -\Psi(-\mathrm{i}\lambda),$$

for all $\lambda \geq 0$. Taking account of Theorem 6.15 and Sect. 6.5.2, it is not difficult to see that, up to a multiplicative constant, for all $\theta \in \mathbb{R}$,

$$\Psi(\theta) = \big(\Phi(0) - \mathrm{i}\theta\big)\phi(\mathrm{i}\theta),$$

where ϕ is the Laplace exponent of the descending ladder height subordinator. This leads to the factorisation identity

$$\psi(\lambda) = \big(\lambda - \Phi(0)\big)\phi(\lambda), \tag{9.1}$$

for all $\lambda \geq 0$. Note that, in a similar manner to the computations in Exercise 6.5, formula (9.1) can also be proved by a direct manipulation of the expression for the

A.E. Kyprianou, *Fluctuations of Lévy Processes with Applications*, Universitext,
DOI 10.1007/978-3-642-37632-0_9, © Springer-Verlag Berlin Heidelberg 2014

Laplace exponent given in (8.2). However, in that case, one may only identify ϕ as the Laplace exponent of a (possibly-killed) subordinator, rather than, specifically, as the Laplace exponent of the descending ladder height process. Either way, the exponent $\phi(\lambda)$ must take the form

$$\phi(\lambda) = \kappa + \delta\lambda + \int_{(0,\infty)} (1 - e^{-\lambda x})\Upsilon(dx), \tag{9.2}$$

where $\kappa, \delta \geq 0$ and Υ is a measure concentrated on $(0, \infty)$, which satisfies $\int_{(0,\infty)}(1 \wedge x)\Upsilon(dx) < \infty$.

Lemma 9.1 *We have that*

$$\Upsilon(x, \infty) = e^{\Phi(0)x} \int_x^\infty e^{-\Phi(0)u} \Pi(-\infty, -u)du, \quad \textit{for } x > 0, \tag{9.3}$$

$\delta = \sigma^2/2$ *and* $\kappa = \psi'(0+) \vee 0$.

Proof In the case that $\psi'(0+) \geq 0$, equivalently $\Phi(0) = 0$, the result may be easily recovered from Exercise 6.5. To deal with the case that $\psi'(0+) < 0$, equivalently $\Phi(0) > 0$, recall from (8.6) that we may write

$$\psi(\lambda) = \psi_{\Phi(0)}(\lambda - \Phi(0)),$$

where $\lambda \geq -\Phi(0)$. Reviewing (9.1) in light of the above equality, it follows that, for $\lambda \geq 0$,

$$\phi(\lambda) = \phi_{\Phi(0)}(\lambda - \Phi(0)), \tag{9.4}$$

where $\phi_{\Phi(0)}$ plays the role of ϕ in the Wiener–Hopf factorisation of $\psi_{\Phi(0)}$. Using obvious notation, we have that

$$\begin{aligned}
\phi(\lambda) &= \kappa_{\Phi(0)} + \delta_{\Phi(0)}(\lambda - \Phi(0)) + \int_{(0,\infty)} (1 - e^{-(\lambda-\Phi(0))x})\Upsilon_{\Phi(0)}(dx) \\
&= \phi_{\Phi(0)}(-\Phi(0)) + \delta_{\Phi(0)}\lambda + \int_{(0,\infty)} (1 - e^{-\lambda x})e^{\Phi(0)x}\Upsilon_{\Phi(0)}(dx) \\
&= \phi(0) + \delta_{\Phi(0)}\lambda + \int_{(0,\infty)} (1 - e^{-\lambda x})e^{\Phi(0)x}\Upsilon_{\Phi(0)}(dx),
\end{aligned}$$

for $\lambda \geq 0$. This shows, in particular, that $\delta = \delta_{\Phi(0)} = \sigma^2/2$ and, by (9.1), $\kappa = \phi(0) = 0$. Next, note that $\psi'_{\Phi(0)}(0+) = \psi'(\Phi(0)) > 0$ and that, from Theorem 3.9, $\Pi_{\Phi(0)}(dx) = e^{\Phi(0)x}\Pi(dx)$ for $x < 0$. From the first sentence of this proof, we know that (9.3) holds for the spectrally negative Lévy process with Laplace exponent $\psi_{\Phi(0)}$. In other words, $\Upsilon_{\Phi(0)}(x, \infty) = \int_x^\infty \Pi_{\Phi(0)}(-\infty, -u)du$, for $x > 0$. Hence, combining these facts,

$$\begin{aligned}\Upsilon(\mathrm{d}x) &= \mathrm{e}^{\Phi(0)x}\Upsilon_{\Phi(0)}(\mathrm{d}x)\\ &= \mathrm{e}^{\Phi(0)x}\Pi_{\Phi(0)}(-\infty,-x)\mathrm{d}x\\ &= \mathrm{e}^{\Phi(0)x}\Pi(-\infty,-x)+\Phi(0)\mathrm{e}^{\Phi(0)x}\int_x^\infty \mathrm{e}^{-\Phi(0)u}\Pi(-\infty,-u)\mathrm{d}u,\end{aligned}$$

for $x \geq 0$, where the final equality follows from an integration by parts. This agrees with the identity given in the statement of the lemma. □

9.2 Scale Functions and Philanthropy

Suppose we now denote the descending ladder height process associated with X by $\widehat{H} = \{\widehat{H}_t : t \geq 0\}$. In the special case that $\Phi(0) = 0$, that is to say, the process X does not drift to $-\infty$ and its Wiener–Hopf factorisation takes the form $\psi(\lambda) = \lambda\phi(\lambda)$, it can be shown that the scale function, W, describes the potential measure of $\widehat{H}$. Indeed, recall that the potential measure of $\widehat{H}$ is defined by

$$\int_0^\infty \mathbb{P}(\widehat{H}_t \in \mathrm{d}x)\mathrm{d}t, \quad \text{for } x \geq 0. \tag{9.5}$$

Calculating its Laplace transform, we get the identity

$$\int_0^\infty\int_0^\infty \mathrm{e}^{-\lambda x}\mathbb{P}(\widehat{H}_t \in \mathrm{d}x)\mathrm{d}t = \int_0^\infty \mathrm{e}^{-\phi(\lambda)t}\mathrm{d}t = \frac{1}{\phi(\lambda)} = \frac{\lambda}{\psi(\lambda)}, \tag{9.6}$$

where $\lambda \geq 0$. Inverting the Laplace transform on the left-hand side with the help of (8.20), we get the identity

$$W(x) = \int_0^\infty \mathbb{P}(\widehat{H}_t \leq x)\mathrm{d}t, \quad x \geq 0. \tag{9.7}$$

It can be easily shown in a similar fashion that, when $\Phi(0) > 0$, the scale function is related to the potential measure of $\widehat{H}$ by the formula

$$W(x) = \mathrm{e}^{\Phi(0)x}\int_0^x \mathrm{e}^{-\Phi(0)y}\int_0^\infty \mathbb{P}(\widehat{H}_t \in \mathrm{d}y)\mathrm{d}t, \quad x \geq 0. \tag{9.8}$$

This relationship between scale functions and potential measures of subordinators lies at the heart of the approach we shall describe in this section. Key to the method is the fact that one can find in the literature several subordinators for which the potential measure is known explicitly.[1] Should these subordinators turn out to be the descending ladder height process of a spectrally negative Lévy process which

[1] We remind the reader that many examples can be found directly in Schilling et al. (2010) and, as inverse local times, in Borodin and Salminen (2002).

does not drift to $-\infty$, i.e. $\Phi(0)=0$, then this would give an exact expression for its scale function. Said another way, we can build scale functions using the following approach.

Step 1. Choose a subordinator with Laplace exponent ϕ, for which one knows its potential measure, or equivalently, in light of (9.6), for which one can explicitly invert the Laplace transform $1/\phi(\lambda)$.

Step 2. Verify whether the relation

$$\psi(\lambda) := \lambda\phi(\lambda), \quad \lambda \geq 0,$$

defines the Laplace exponent of a spectrally negative Lévy process.

Of course, for this method to be useful, we should first provide necessary and sufficient conditions for a subordinator to be the descending ladder height process of some spectrally negative Lévy process, or equivalently, a verification method for Step 2. Precisely this point is addressed by Vigon's Theorem of Philanthropy 6.23. Indeed, noting that the ascending ladder height of a spectrally negative Lévy process necessarily takes the form of a (possibly-killed) linear drift, the aforesaid theorem tells us that one may take any subordinator with Laplace exponent ϕ, so long as the associated Lévy measure is absolutely continuous with non-increasing density. Moreover, the inclusion of a killing term in ϕ can only occur when there is no killing for the ascending ladder height process. More formally, we have the following theorem, taken from Hubalek and Kyprianou (2010), which can also be easily proved directly from Lemma 6.23.

Theorem 9.2 *Consider a given (killed) subordinator with Lévy triple $(\kappa, \delta, \Upsilon)$ and Laplace exponent given by* (9.2). *Then, for all $\varphi \geq 0$, there exists a spectrally negative Lévy process, X, henceforth referred to as the parent process, with Laplace exponent given by*

$$\psi(\lambda) = (\lambda - \varphi)\phi(\lambda), \tag{9.9}$$

for $\lambda \geq 0$, such that $\varphi\kappa = 0$.

The Lévy triple (a, σ, Π) of the parent process is uniquely identified as follows. The Gaussian coefficient is given by $\sigma = \sqrt{2\delta}$. The Lévy measure is given by

$$\Pi(-\infty, -x) = \varphi\Upsilon(x, \infty) + \frac{\mathrm{d}\Upsilon}{\mathrm{d}x}(x), \quad x > 0. \tag{9.10}$$

Finally

$$a = \int_{(-\infty,-1)} x\Pi(\mathrm{d}x) - \kappa \tag{9.11}$$

if $\varphi = 0$ *and otherwise, when* $\varphi > 0$,

$$a = \frac{1}{2}\sigma^2\varphi + \frac{1}{\varphi}\int_{(-\infty,0)} \left(\mathrm{e}^{\varphi x} - 1 - x\varphi\mathbf{1}_{\{x>-1\}}\right)\Pi(\mathrm{d}x). \tag{9.12}$$

Note that when describing parent processes later on in this text, for practical reasons, we shall prefer to specify the triple (σ, Π, ψ) instead of (a, σ, Π). However both triples provide an equivalent amount of information. It is also worth making an observation for later reference concerning the path variation of the process X for a given descending ladder height process.

Corollary 9.3 *Given a (killed) subordinator satisfying the conditions of the previous theorem,*

(i) *the parent process has paths of unbounded variation if and only if* $\Upsilon(0, \infty) = \infty$ *or* $\delta > 0$, *and*
(ii) *if* $\Upsilon(0, \infty) = \mathrm{c} < \infty$, *then the parent process necessarily decomposes in the form*

$$X_t = (\kappa + \mathrm{c} - \delta\varphi)t + \sqrt{2\delta}B_t - S_t, \tag{9.13}$$

where $B = \{B_t : t \geq 0\}$ *is a Brownian motion,* $S = \{S_t : t \geq 0\}$ *is an independent, driftless subordinator with Lévy measure* ν, *satisfying*

$$\nu(x, \infty) = \varphi\Upsilon(x, \infty) + \frac{\mathrm{d}\Upsilon}{\mathrm{d}x}(x).$$

Proof (i) Recalling the discussion at the beginning of Chap. 8, we know that a spectrally negative Lévy process has paths of bounded variation if and only if 0 is irregular for $(-\infty, 0)$. This is equivalent to the descending ladder height process being a driftless compound Poisson subordinator, which is, in turn, equivalent to either $\Upsilon(0, \infty) = \infty$ or $\delta > 0$. See, for example, the discussion preceding Corollary 4.12.

(ii) Using (9.10), the Laplace exponent of the decomposition (9.13) can be computed as follows, with the help of an integration by parts:

$$\begin{aligned}
&(\kappa + \mathrm{c} - \delta\varphi)\lambda + \delta\lambda^2 - \varphi\lambda\int_0^\infty \mathrm{e}^{-\lambda x}\Upsilon(x, \infty)\mathrm{d}x - \lambda\int_0^\infty \mathrm{e}^{-\lambda x}\frac{\mathrm{d}\Upsilon}{\mathrm{d}x}(x)\mathrm{d}x \\
&= \left(\kappa + \Upsilon(0, \infty) - \delta\varphi\right)\lambda + \delta\lambda^2 - \varphi\int_0^\infty \left(1 - \mathrm{e}^{-\lambda x}\right)\frac{\mathrm{d}\Upsilon}{\mathrm{d}x}(x)\mathrm{d}x \\
&\quad - \lambda\int_0^\infty \mathrm{e}^{-\lambda x}\frac{\mathrm{d}\Upsilon}{\mathrm{d}x}(x)\mathrm{d}x \\
&= (\lambda - \varphi)\left(\kappa + \delta\lambda + \int_0^\infty \left(1 - \mathrm{e}^{-\lambda x}\right)\frac{\mathrm{d}\Upsilon}{\mathrm{d}x}(x)\mathrm{d}x\right).
\end{aligned}$$

This agrees with the Laplace exponent $\psi(\lambda) = (\lambda - \varphi)\phi(\lambda)$ of the parent process constructed in Theorem 9.2. □

Let us illustrate the functionality of the previous two results with some examples.

Example 9.4 Consider a spectrally negative Lévy process which is the parent process of a (killed) tempered stable process, that is to say, a subordinator with Laplace exponent given by

$$\phi(\lambda) = \kappa + c\Gamma(-\alpha)\big((\gamma+\lambda)^\alpha - \gamma^\alpha\big), \quad \lambda \ge 0,$$

where $\alpha \in (-1, 1) \setminus \{0\}$, $\gamma \ge 0$ and $c > 0$. The associated Lévy measure is given by

$$\Upsilon(dx) = cx^{-\alpha-1}\mathrm{e}^{-\gamma x}\mathrm{d}x, \quad x > 0.$$

Recall that, for $\alpha, \beta > 0$ and $x \in \mathbb{R}$,

$$\mathrm{E}_{\alpha,\beta}(x) = \sum_{n\ge 0} \frac{x^n}{\Gamma(n\alpha+\beta)} \tag{9.14}$$

denotes the two-parameter Mittag–Leffler function. The following is a well-known transform for the Mittag–Leffler function:

$$\int_0^\infty \mathrm{e}^{-\theta x} x^{\beta-1} \mathrm{E}_{\alpha,\beta}(\lambda x^\alpha)\mathrm{d}x = \frac{\theta^{\alpha-\beta}}{\theta^\alpha - \lambda}, \tag{9.15}$$

where $\lambda \in \mathbb{R}$ and $|\theta^\alpha/\lambda| > 1$. Together with the well-known rules for Laplace transforms concerning primitives and exponential tilting, it is straightforward to deduce the following expressions for the scale functions associated with the parent process with Laplace exponent given by (9.9) such that $\kappa\varphi = 0$.

If $0 < \alpha < 1$, then

$$W(x) = -\frac{\mathrm{e}^{\varphi x}}{c\Gamma(-\alpha)} \int_0^x \mathrm{e}^{-(\gamma+\varphi)y} y^{\alpha-1} \mathrm{E}_{\alpha,\alpha}\left(\frac{\kappa + c\Gamma(-\alpha)\gamma^\alpha}{c\Gamma(-\alpha)} y^\alpha\right)\mathrm{d}y.$$

If $-1 < \alpha < 0$, then

$$\begin{aligned} W(x) = {} & \frac{\mathrm{e}^{\varphi x}}{\kappa + c\Gamma(-\alpha)\gamma^\alpha} \\ & + \frac{c\Gamma(-\alpha)\mathrm{e}^{\varphi x}}{(\kappa + c\Gamma(-\alpha)\gamma^\alpha)^2} \int_0^x \mathrm{e}^{-(\gamma+\varphi)y} y^{-\alpha-1} \\ & \times \mathrm{E}_{-\alpha,-\alpha}\left(\frac{c\Gamma(-\alpha)}{\kappa + c\Gamma(-\alpha)\gamma^\alpha} y^{-\alpha}\right)\mathrm{d}y. \end{aligned}$$

Example 9.5 Let $c > 0$, $\nu \ge 0$ and $\theta \in (0, 1)$ and ϕ be defined by

$$\phi(\lambda) = \frac{c\lambda\Gamma(\nu+\lambda)}{\Gamma(\nu+\lambda+\theta)}, \quad \lambda \ge 0.$$

In Example 5.26, it was shown that ϕ is the Laplace exponent of some subordinator. Its characteristics are $\kappa = 0, \delta = 0$,

$$\Upsilon(x,\infty) = \frac{c}{\Gamma(\theta)} \mathrm{e}^{-x(\nu+\theta-1)}\left(\mathrm{e}^x - 1\right)^{\theta-1}, \quad x > 0.$$

It is not difficult to show that Υ has a non-increasing density. It follows from Theorem 9.2 that there exists an oscillating spectrally negative Lévy process, say X, whose Laplace exponent is $\psi(\lambda) = \lambda\phi(\lambda)$, $\lambda \geq 0$, with $\sigma = 0$, and Lévy density given by $-\mathrm{d}^2\Upsilon(x,\infty)/\mathrm{d}x^2$. Again, referring back to Example 5.26, and taking account of (9.7), we may identify its associated scale function as

$$W(x) = \frac{\Gamma(\nu+\theta)}{c\Gamma(\nu)} + \frac{\theta}{c\Gamma(1-\theta)} \int_0^x \left\{ \int_y^\infty \frac{\mathrm{e}^{z(1-\nu)}}{(\mathrm{e}^z - 1)^{1+\theta}} \mathrm{d}z \right\} \mathrm{d}y, \quad x \geq 0.$$

An interesting feature of this example is that one may use the fact that ϕ is a special subordinator to develop a second example. Indeed the computation in (5.35) shows

$$\phi^*(\lambda) := \frac{\lambda}{\phi(\lambda)} = \frac{\Gamma(\nu+\theta)}{c\Gamma(\nu)} + \frac{\theta}{c\Gamma(1-\theta)} \int_0^\infty \left(1 - \mathrm{e}^{-\lambda x}\right) \frac{\mathrm{e}^{x(1-\nu)}}{(\mathrm{e}^x - 1)^{1+\theta}} \mathrm{d}x, \quad \lambda \geq 0.$$

On inspection, we immediately see that the Lévy density of ϕ^* is non-increasing and hence

$$\psi^*(\lambda) = \lambda\phi^*(\lambda) = \frac{\lambda^2}{\phi(\lambda)}, \quad \lambda \geq 0,$$

defines the Laplace exponent of a spectrally negative Lévy process. Taking account of the fact that $\phi^*(0) > 0$, that is to say, the subordinator corresponding to ϕ^* is killed, it follows that the parent process, corresponding to ψ^*, drifts to $+\infty$.

Looking again back into Example 5.26, we can quickly deduce from (5.36) that

$$W^*(x) = \frac{c}{\Gamma(\theta)} \int_0^x \mathrm{e}^{-z(\nu+\theta-1)}\left(\mathrm{e}^z - 1\right)^{\theta-1} \mathrm{d}z.$$

The method described in this example can be formalised into a general theory that applies to a large family of subordinators, namely that of special subordinators.

9.3 Special and Conjugate Scale Functions

Recall from Sect. 5.6 that the class of Bernstein functions coincides precisely with the class of Laplace exponents of (possibly-killed) subordinators. That is to say, a general Bernstein function takes the form (9.2). Recall, moreover, that a given

Bernstein function, ϕ, is further called a *special Bernstein function* if

$$\phi(\lambda) = \frac{\lambda}{\phi^*(\lambda)}, \quad \lambda \geq 0, \tag{9.16}$$

where $\phi^*(\lambda)$ is another Bernstein function. In that case, ϕ^* is referred to as *conjugate* to ϕ. Accordingly, a (possibly-killed) subordinator is called a special subordinator if its Laplace exponent is a special Bernstein function.

Suppose that we use obvious notation and write $(\kappa^*, \delta^*, \Upsilon^*)$ for the Lévy triple associated with ϕ^*. Then Theorem 5.19 offers a very concise relationship between the potential measure associate to ϕ and the triple $(\kappa^*, \delta^*, \Upsilon^*)$. Let us denote by $W(\mathrm{d}x)$ the potential measure of ϕ. (It will of course prove to be no coincidence that we have chosen this notation to coincide with the notation for a scale function.) Then we have that W necessarily satisfies

$$W(\mathrm{d}x) = \delta^*\delta_0(\mathrm{d}x) + \left\{\kappa^* + \Upsilon^*(x, \infty)\right\}\mathrm{d}x, \quad \text{for } x \geq 0, \tag{9.17}$$

where $\delta_0(\mathrm{d}x)$ is the Dirac measure at zero. Naturally, if W^* is the potential measure of ϕ^* then we may describe it the same way as on the right-hand side of (9.17), using instead the triple $(\kappa, \delta, \Upsilon)$.

We are interested in constructing a parent process whose descending ladder height process is a special subordinator. The following theorem and corollary are now evident given the above discussion when taken in the light of Theorem 9.2.

Theorem 9.6 *Suppose that ϕ and ϕ^* are a conjugate pair of special Bernstein functions such that Υ is absolutely continuous with non-increasing density. Then there exists a spectrally negative Lévy process that does not drift to $-\infty$, whose Laplace exponent is described by*

$$\psi(\lambda) = \frac{\lambda^2}{\phi^*(\lambda)} = \lambda\phi(\lambda), \quad \textit{for } \lambda \geq 0, \tag{9.18}$$

and whose scale function is a concave function, given by

$$W^*(x) = 1 - \mathrm{e}^{-x} + x\int_x^\infty \frac{\mathrm{e}^{-z}}{z}\mathrm{d}z, \quad x \geq 0. \tag{9.19}$$

The assumptions of the previous theorem only require that both the Lévy and potential measures associated with ϕ have a non-increasing density in $(0, \infty)$. Note, from Theorem 5.19, that the aforementioned condition on the potential measure of ϕ is equivalent to insisting that ϕ is a special subordinator. If, in addition, it is assumed that the potential density is a convex function, that is to say, Υ^* has a non-increasing density, then, in light of the representation (9.19), we can interchange the roles of ϕ and ϕ^*, respectively, in the previous theorem. We thus have the following corollary.

Corollary 9.7 *If ϕ and ϕ^* are a conjugate pair of special Bernstein functions such that both Υ and Υ^* are absolutely continuous with non-increasing densities, then* *there exists a pair of scale functions W and W^*, such that W is concave, its first derivative is a convex function,* (9.19) *is satisfied, and*

$$W^*(x) = \delta + \kappa x + \int_0^x \Upsilon(y, \infty)\mathrm{d}y. \tag{9.20}$$

Moreover, the Laplace exponents of the respective parent processes are given by (9.18) *and*

$$\psi^*(\lambda) = \frac{\lambda^2}{\phi(\lambda)} = \lambda\phi^*(\lambda), \quad \lambda \geq 0. \tag{9.21}$$

For obvious reasons, we shall henceforth refer to the scale functions W and W^* as *conjugate special scale functions*. Similarly, we call their respective parent processes *conjugate parent processes*. The conjugation of W and W^* through the relation (9.16) can also be seen via the convolution relation

$$W * W^*(\mathrm{d}x) = \mathrm{d}x,$$

for $x \geq 0$.

9.4 Tilting and Parent Processes Drifting to $-\infty$

In this section, we present two methods for which, given a scale function and its associated parent process, it is possible to construct further examples of scale functions. We use the same notation as in the previous section.

For the first, let ϕ be a special Bernstein function with representation given by (9.2). Then it is a straightforward computation, in the spirit of Theorem 3.9, to show that, for any $\beta \geq 0$, the function $\phi_\beta(\lambda) = \phi(\lambda + \beta)$, $\lambda \geq 0$, is again a Bernstein function with killing rate $\kappa_\beta = \phi(\beta)$, drift coefficient $\delta_\beta = \delta$ and Lévy measure $\Upsilon_\beta(\mathrm{d}x) = \mathrm{e}^{-\beta x}\Upsilon(\mathrm{d}x)$, $x > 0$. By taking Laplace transforms, it is also straightforward to verify that the potential measure associated with ϕ_β, say W_β, has the same-sized atom at zero as W and a decreasing density in $(0, \infty)$ such that $W_\beta(\mathrm{d}x) = \mathrm{e}^{-\beta x}W'(x)\mathrm{d}x$, for $x > 0$, where W' denotes the density of the potential measure associated with ϕ. This immediately qualifies ϕ_β as a special subordinator thanks to Theorem 5.19. Exercise 9.6 gives an expression for its conjugate, ϕ^*_β.

Note that, if Υ has a non-increasing density, then so does Υ_β. Moreover, if W' is convex (equivalently Υ^* has a non-increasing density) then W'_β is convex (equivalently Υ^*_β has a non-increasing density). We have the following lemma.

Lemma 9.8 *Fix $\beta \geq 0$. If conjugate special Bernstein functions ϕ and ϕ^* exist such that both Υ and Υ^* are absolutely continuous with non-increasing densities, then there exist conjugate parent processes with Laplace exponents*

$$\psi_\beta(\lambda) = \lambda\phi_\beta(\lambda) \text{ and } \psi^*_\beta(\lambda) = \lambda\phi^*_\beta(\lambda), \quad \lambda \geq 0,$$

whose respective scale functions are given by

$$\begin{aligned} W_\beta(x) &= \delta^* + \int_0^x \mathrm{e}^{-\beta y}\big(\Upsilon^*(y,\infty)+\kappa^*\big)\mathrm{d}y \\ &= \mathrm{e}^{-\beta x}W(x) + \beta\int_0^x \mathrm{e}^{-\beta z}W(z)\mathrm{d}z, \quad x \geq 0, \end{aligned} \tag{9.22}$$

and

$$W_\beta^*(x) = \delta + \phi(\beta)x + \int_0^x \left(\int_y^\infty \mathrm{e}^{-\beta z}\Upsilon(\mathrm{d}z)\right)\mathrm{d}y, \quad x \geq 0. \tag{9.23}$$

where we have used obvious notation.

Proof Taking account of (9.19) and (9.20), all of the statements in this lemma follow in a straightforward way from the discussion preceding it. We shall, however, only elaborate on the first equality in (9.22). By invoking the formula in (9.19) for W_β, it suffices to identify the triple $(\kappa_\beta^*, \delta_\beta^*, \Upsilon_\beta^*)$ belonging to ϕ_β^*. Note that, on account of (5.26) and the fact that $\kappa_\beta = \phi_\beta(0) = \phi(\beta) > 0$, it follows that the killing rate κ_β^* must be identically zero. Since $W_\beta(\mathrm{d}x) = \mathrm{e}^{-\beta x}W'(x)\mathrm{d}x$, we also have from Theorem 5.19 that $\Upsilon_\beta^*(x,\infty) = \mathrm{e}^{-\beta x}W'(x) = \mathrm{e}^{-\beta x}(\Upsilon^*(x,\infty)+\kappa^*)$. Finally, to obtain the value of δ_β^*, note from Exercise 2.11 that it suffices to consider the limit of $\phi_\beta(\lambda)/\lambda$ as $\lambda \uparrow \infty$. One readily deduces that $\delta_\beta^* = \delta^*$. □

The second procedure builds on the first to construct examples of scale functions whose parent process may be seen as an auxiliary parent process conditioned to drift to $-\infty$.

Suppose that ϕ is a Bernstein function such that $\kappa = 0$ and its associated Lévy measure, Υ, has a non-increasing density. Fix $\beta > 0$. Theorem 9.2 says that there exists a parent process, say X, that drifts to $-\infty$, such that its Laplace exponent, ψ, can be factorised as

$$\psi(\lambda) = (\lambda - \beta)\phi(\lambda), \quad \lambda \geq 0.$$

Necessarily ψ is a convex function with $\psi(0) = 0 = \psi(\beta)$, so that β is the largest positive solution to the equation $\psi(\lambda) = 0$. From previous discussion, we know that $\phi_\beta(\lambda) := \phi(\lambda+\beta)$ is a Bernstein function with a non-zero killing component and Lévy measure with non-increasing density. Hence, Theorem 9.2 permits us to conclude that $\psi_\beta(\lambda) := \psi(\lambda+\beta) = \lambda\phi_\beta(\lambda)$, $\lambda \geq 0$, is also the Laplace exponent of a parent process. Note in particular that $\psi_\beta'(0+) = \psi'(\beta) > 0$ and hence the aforesaid parent process drifts to $+\infty$.

Now, let W_β be the scale function of the spectrally negative Lévy process with Laplace exponent $\psi_\beta(\lambda)$. It follows from formula (8.23), with $q = 0$, that the 0-scale function of the process with Laplace exponent ψ is related to W_β by

$$W(x) = \mathrm{e}^{\beta x}W_\beta(x), \quad x \geq 0.$$

The above considerations thus lead to the following result, which allows for the construction of a scale function of a parent process which drifts to $-\infty$.

Lemma 9.9 *Suppose that ϕ is a special Bernstein function such that Υ is absolutely continuous with non-increasing density and $\kappa = 0$. Fix $\beta > 0$. Then there exists a parent process with Laplace exponent*

$$\psi(\lambda) = (\lambda - \beta)\phi(\lambda), \quad \lambda \geq 0,$$

whose associated scale function is given by

$$W(x) = \delta^* \mathrm{e}^{\beta x} + \mathrm{e}^{\beta x} \int_0^x \mathrm{e}^{-\beta y} \big(\Upsilon^*(y, \infty) + \kappa^*\big) \mathrm{d}y, \quad x \geq 0,$$

where we have used our usual notation.

9.5 Complete Scale Functions

All the results in the previous two sections require that the conjugate pairs of special Bernstein functions have Lévy measures, Υ and Υ^*, which have non-increasing densities. We have seen earlier in Sect. 5.6 that a natural subclass of Bernstein function, which respects this requirement, is that of the complete Bernstein functions. Indeed, all Bernstein functions in the aforementioned class have the defining property that their Lévy densities, and consequently the Lévy densities of their conjugates, are completely monotone and hence, in particular, non-increasing. We have the following obvious corollary to Theorem 9.6.

Corollary 9.10 *For any conjugate pair of complete Bernstein functions, ϕ and ϕ^*, the pair*

$$\psi(\lambda) = \frac{\lambda^2}{\phi^*(\lambda)} = \lambda\phi(\lambda) \quad \textit{and} \quad \psi^*(\lambda) = \frac{\lambda^2}{\phi(\lambda)} = \lambda\phi^*(\lambda), \quad \lambda \geq 0.$$

defines the Laplace exponents of parent processes with respective scale functions given by (9.19) *and* (9.20).

Scale functions which belong to the parent processes of complete Bernstein functions are, naturally, referred to as *complete scale functions*. Note that we can also easily deduce from Corollary 5.24 that complete scale functions have completely monotone densities. Note also that the scale functions discussed in Examples 9.4 and 9.5 are all complete. Let us conclude this chapter and section with another example of a family of complete scale functions.

Example 9.11 Let $0 < \alpha < \beta \leq 1$, $a, b > 0$ and ϕ be the Bernstein function defined by

$$\phi(\lambda) = a\lambda^{\beta-\alpha} + b\lambda^{\beta}, \quad \lambda \geq 0.$$

When $\alpha < \beta < 1$, ϕ is the Laplace exponent of a subordinator which is obtained as the sum of two independent stable subordinators. One has parameter $\beta - \alpha$ and the other has parameter β, so that the killing and drift term of ϕ are both equal to 0, and its Lévy measure is given by

$$\Upsilon(\mathrm{d}x) = \left(\frac{a(\beta-\alpha)}{\Gamma(1-\beta+\alpha)} x^{-(1+\beta-\alpha)} + \frac{b\beta}{\Gamma(1-\beta)} x^{-(1+\beta)}\right)\mathrm{d}x, \quad x > 0.$$

In the case that $\beta = 1$, ϕ is the Laplace exponent of a stable subordinator with parameter $1 - \alpha$ and a linear drift. In all cases, the Lévy measure Υ has a density which is completely monotone, and thus its potential density, or equivalently the density of the associated scale function W, is completely monotone.

Recall the definition (9.14) of the Mittag–Leffler function $\mathrm{E}_{\alpha,\beta}(x)$ and its associated transformation (9.15). With the help of the latter, the scale function associated with the parent process of ϕ can now be identified via its density on $(0, \infty)$,

$$W'(x) = \frac{1}{b} x^{\beta-1} \mathrm{E}_{\alpha,\beta}\big(-ax^{\alpha}/b\big), \quad x > 0, \tag{9.24}$$

which, by Theorem 5.24, is necessarily a completely monotone function. The parent process has Laplace exponent

$$\psi(\lambda) = \lambda\phi(\lambda) = a\lambda^{\beta-\alpha+1} + b\lambda^{\beta+1}, \quad \lambda \geq 0,$$

and hence is the independent sum of two spectrally negative stable processes with stability indices $\beta + 1$ and $1 + \beta - \alpha$, respectively. It therefore has paths of unbounded variation, which implies that $W(0) = 0$. Integrating (9.24), we thus conclude that

$$W(x) = \frac{1}{b} \int_0^x t^{\beta-1} \mathrm{E}_{\alpha,\beta}\big(-at^{\alpha}/b\big)\mathrm{d}t, \quad x \geq 0.$$

The respective conjugates to ϕ, ψ and W are given by

$$\phi^*(\lambda) = \frac{\lambda}{a\lambda^{\beta-\alpha} + b\lambda^{\beta}}, \qquad \psi^*(\lambda) = \frac{\lambda^2}{a\lambda^{\beta-\alpha} + b\lambda^{\beta}}, \quad \lambda \geq 0,$$

and

$$W^*(x) = \frac{a}{\Gamma(2-\beta+\alpha)} x^{1-\beta+\alpha} + \frac{b}{\Gamma(2-\beta)} x^{1-\beta}, \quad x \geq 0. \tag{9.25}$$

The subordinator with Laplace exponent ϕ^* has zero killing and drift terms and its Lévy measure is obtained by taking the derivative of the expression in (9.24). By Theorem 9.2, the spectrally negative Lévy process with Laplace exponent ψ^*

oscillates, has unbounded variation, has zero Gaussian term and the density of its Lévy measure is obtained from expression in (9.24) together with (9.10).

One may mention here that, by letting $a \downarrow 0$, the Continuity Theorem for Laplace transforms tells us that, for the case $\phi(\lambda) = b\lambda^{\beta}$, the associated parent process has the Laplace exponent of a spectrally negative stable process with stability parameter $1+\beta$, and its scale function is given by

$$W(x) = \frac{1}{b\Gamma(1+\beta)} x^{\beta}, \quad x \geq 0.$$

The associated conjugates are given by

$$\phi^*(\lambda) = b^{-1}\lambda^{1-\beta}, \qquad \psi^*(\lambda) = b^{-1}\lambda^{2-\beta}, \quad \lambda \geq 0,$$

and

$$W^*(x) = \frac{b}{\Gamma(2-\beta)} x^{1-\beta}, \quad x \geq 0.$$

The Lévy measure of the conjugate parent process is given by

$$\Pi^*(-\infty, -x) = \frac{\beta(1-\beta)}{b\Gamma(1+\beta)} x^{\beta-2}, \quad x \geq 0.$$

To complete this example, note that we can also consider the construction in Sect. 9.4. For, $m, a, b > 0$, $0 < \alpha < \beta < 1$, there exists a parent process drifting to $-\infty$, with Laplace exponent

$$\psi(\lambda) = (\lambda - m)\big(a\lambda^{\beta-\alpha} + b\lambda^{\beta}\big), \quad \lambda \geq 0.$$

It follows, from the previous calculations, that the scale function associated with the parent process with this Laplace exponent is given by

$$W(x) = \frac{\mathrm{e}^{mx}}{b} \int_0^x \mathrm{e}^{-mt} t^{\beta-1} \mathrm{E}_{\alpha,\beta}\big(-at^{\alpha}/b\big)\mathrm{d}t, \quad x \geq 0.$$

Exercises

9.1 Suppose we are in the setting of Example 9.4. That is to say, we consider the case of a spectrally negative Lévy process which is the parent process of a subordinator with Laplace exponent

$$\phi(\lambda) = \kappa + c\Gamma(-\alpha)\big((\gamma+\lambda)^{\alpha} - \gamma^{\alpha}\big), \quad \lambda \geq 0,$$

where $\kappa \geq 0$, $c, \gamma > 0$ and $\alpha \in (-1, 1)\backslash\{0\}$.

(i) Suppose that $0<\alpha<1$. Show that for all $q\geq 0$, as $x\downarrow 0$,

$$W^{(q)}(x)\sim -\frac{x^{\alpha}}{c\Gamma(-\alpha)\Gamma(1+\alpha)} \quad\text{and}\quad W^{(q)\prime}(x)\sim -\frac{x^{\alpha-1}}{c\Gamma(-\alpha)\Gamma(\alpha)}.$$

(ii) Suppose that $-1<\alpha<0$. Show that for all $q\geq 0$, as $x\downarrow 0$,

$$W^{(q)}(x)\sim \frac{1}{\kappa+c\gamma^{\alpha}\Gamma(-\alpha)} \quad\text{and}\quad W^{(q)\prime}(x)\sim \frac{cx^{-\alpha-1}}{(\kappa+c\gamma^{\alpha}\Gamma(-\alpha))^2}.$$

(iii) Now suppose that $\kappa=0$ and $\alpha=1/2$. Show that, if the parent process is oscillating, then

$$W(x)=\frac{1}{4c\sqrt{\gamma\pi}}\left[(1+2\gamma x)\mathrm{erfc}(-\sqrt{\gamma x})+2\sqrt{\frac{x\gamma}{\pi}}\mathrm{e}^{-\gamma x}-1\right],$$

where

$$\mathrm{erfc}(x)=\frac{2}{\sqrt{\pi}}\int_x^{\infty}\mathrm{e}^{-t^2}\mathrm{d}t$$

is the complementary error function.

9.2 This exercise is based on computations found in Konstantopoulos et al. (2011). Consider the spectrally negative Lévy process with Laplace exponent

$$\psi(\lambda):=\lambda-\sqrt{2\lambda+c^2}+c, \quad \lambda\geq 0,$$

where $c>0$. Show that, for all $q\geq 0$,

$$W^{(q)}(x)=\frac{\mathrm{e}^{-c^2x/2}}{\eta_1-\eta_2}\left(\eta_1\mathrm{e}^{\eta_1^2x/2}\mathrm{erfc}\left(-\eta_1\sqrt{x/2}\right)-\eta_2\mathrm{e}^{\eta_2^2x/2}\mathrm{erfc}\left(-\eta_2\sqrt{x/2}\right)\right),$$

where

$$\eta_1:=1+\sqrt{(1-c)^2+2q}, \qquad \eta_2:=1-\sqrt{(1-c)^2+2q}. \tag{9.26}$$

9.3 Show that $\psi(\lambda)=\lambda\log(1+\lambda)$, $\lambda\geq 0$, is the Laplace exponent of a spectrally negative Lévy process.

(i) Deduce that its scale function satisfies

$$W(x)=\int_0^x \mathrm{e}^{-y}\left\{\int_0^{\infty}\frac{y^{t-1}}{\Gamma(t)}\mathrm{d}t\right\}\mathrm{d}y, \quad x\geq 0.$$

(ii) Show that W given above is a complete scale function and its conjugate scale function is given by

$$W^*(x) = x\int_0^x \frac{\mathrm{e}^{-z}}{z}\mathrm{d}z + \mathrm{e}^{-x}, \quad x \geq 0.$$

9.4 Suppose that X is a Brownian motion with drift and compound Poisson jumps which are exponentially distributed. That is to say

$$X_t = \sigma B_t + \mu t - \sum_{i=1}^{N_t} \xi_i, \quad t \geq 0,$$

where ξ_i are i.i.d. random variables which are exponentially distributed with parameter $\rho > 0$ and $N = \{N_t : t \geq 0\}$ is an independent Poisson process with intensity $a > 0$.

(i) Show that the Laplace exponent, ψ, of X satisfies

$$\psi(\lambda) = \frac{\sigma^2}{2}\lambda^2 + \mu\lambda - \frac{a\lambda}{\rho+\lambda}, \quad \lambda \in \mathbb{R}\backslash\{-\rho\}.$$

(ii) By considering the behaviour of $\psi(\lambda)$ as $\lambda \to \pm\infty$ and $\lambda \to \rho^{\pm}$, verify that, for every $q > 0$, the equation $\psi(\lambda) = q$ has exactly three real solutions $-\zeta_2, -\zeta_1$ and $\Phi(q)$, which satisfy

$$-\zeta_2 < -\rho < -\zeta_1 < 0 < \Phi(q).$$

(iii) Deduce that, for all $q > 0$,

$$W^{(q)}(x) = \frac{\mathrm{e}^{\Phi(q)x}}{\psi'(\Phi(q))} + \frac{\mathrm{e}^{-\zeta_1 x}}{\psi'(-\zeta_1)} + \frac{\mathrm{e}^{-\zeta_2 x}}{\psi'(-\zeta_2)}, \quad x \geq 0.$$

(iv) More generally, suppose that X is a spectrally negative meromorphic Lévy process. In particular, suppose its Lévy density satisfies

$$\pi(x) = \mathbf{1}_{\{x<0\}} \sum_{j=1}^{\infty} a_j \rho_j \mathrm{e}^{\rho_j x},$$

where the coefficients a_j and ρ_j are positive, ρ_j increase to $+\infty$ as $j \to +\infty$ and

$$\sum_{j\geq 1} \frac{a_j}{\rho_j^2} < \infty.$$

Use Corollary 6.21 to show that, for $q > 0$,

$$W^{(q)}(x) = \frac{\mathrm{e}^{\Phi(q)x}}{\psi'(\Phi(q))} + \sum_{j=1}^{\infty} \frac{\mathrm{e}^{-\zeta_j x}}{\psi'(-\zeta_j)}, \quad x \geq 0,$$

where

$$\cdots < -\rho_2 < -\zeta_2 < -\rho_1 < -\zeta_1 < 0 < \Phi(q)$$

solve $\psi(\lambda) = q$ on $\mathbb{R}$. What happens as $q \downarrow 0$?

(v) Suppose that X does not drift to $-\infty$. Explain why W is a complete scale function and write down an expression for its conjugate.

9.5 Suppose that Π is the Lévy measure of a spectrally negative Lévy process, X, and define, for all $x > 0$, $\overline{\Pi}(x) = \Pi(-\infty, -x)$. Suppose now that $-\overline{\Pi}$ has a completely monotone density on $(0, \infty)$.

(i) Suppose that $\psi'(0+) \geq 0$. Show that the scale function, W, of X has a completely monotone density.

(ii) Now remove the assumption on $\psi'(0+)$. For each $q \geq 0$, use the previous part of the question to show that $W_{\Phi(q)}$, the scale function associated with $(X, \mathbb{P}^{\Phi(q)})$, has a completely monotone density.

(iii) Use part (ii) of the question to deduce that, for any given spectrally negative Lévy process whose Lévy measure has the property that $-\overline{\Pi}$ has a completely monotone density on $(0, \infty)$, for each $q \geq 0$, $W^{(q)\prime}$ is a strictly convex function.

9.6 Suppose that $\phi(\lambda)$ is a special Bernstein function. It was shown in Sect. 9.4 that, for fixed $\beta \geq 0$, $\phi_\beta(\lambda) := \phi(\lambda + \beta)$, $\lambda \geq 0$ is a special Bernstein function. Show that its conjugate, ϕ_β^*, satisfies

$$\phi_\beta^*(\lambda) = \phi^*(\lambda + \beta) - \phi^*(\beta) + \beta \int_0^\infty (1 - \mathrm{e}^{-\lambda x})\mathrm{e}^{-\beta x} W'(x)\mathrm{d}x, \quad \lambda \geq 0.$$

9.7 Use (9.8) and (9.19) to give an alternative proof to Lemma 9.9.

9.8 This example may be considered as an extension of Exercise 5.13, see also Chazal et al. (2012). Fix $\beta > 0$ and suppose that $\psi(\lambda)$, $\lambda \geq 0$, is the Laplace exponent of a spectrally negative Lévy process. Consider the following transformation:

$$\mathcal{T}_\beta \psi(\lambda) = \frac{\lambda}{\lambda + \beta} \psi(\lambda + \beta), \quad \lambda \geq -\beta.$$

(i) Suppose that the Lévy process associated with ψ has Gaussian coefficient σ and Lévy measure Π, concentrated on $(-\infty, 0)$. Show that $\mathcal{T}_\beta \psi$ is also the Laplace exponent of a spectrally negative Lévy process with Gaussian coefficient σ. Moreover, its Lévy measure is given by

$$\mathrm{e}^{\beta x} \Pi(\mathrm{d}x) + \beta \mathrm{e}^{\beta x} \overline{\Pi}(x)\mathrm{d}x \quad \text{on } (-\infty, 0),$$

where $\overline{\Pi}(x) = \Pi(-\infty, -x)$.

(ii) Suppose that W_ψ is the scale function associated with the Laplace exponent ψ. Show that, for $x \geq 0$ and $\beta > 0$,

$$W_{T_\beta \psi}(x) = \mathrm{e}^{-\beta x} W_\psi(x) + \beta \int_0^x \mathrm{e}^{-\beta y} W_\psi(y) \mathrm{d}y.$$

Chapter 10
Ruin Problems and Gerber–Shiu Theory

Recall from Sects. 1.3.1 and 2.7.1 that a natural generalisation of the classical Cramér–Lundberg insurance risk model is a spectrally negative Lévy process; also called a *Lévy insurance risk process*. In this chapter, we shall return to the first-passage problem for Lévy processes, which has already been studied in Chap. 7 and look at the role it plays in a family of problems which have proved to be an extensive topic of research in the actuarial literature. Many of the problems we shall consider are inspired by the longstanding collaborative contributions of Hans Gerber and Elias Shiu, thereby motivating the title of this chapter.

We shall start by reviewing classical results that have already been treated implicitly, if not explicitly, earlier in this book. Largely, this concerns the exact and asymptotic distributions of overshoots and undershoots of the Lévy insurance risk process at ruin. Thereafter, we shall turn our attention to more complex models of insurance risk in which dividends or tax are paid out of the insurance risk process, thereby adjusting its trajectory. In this setting, a number of identities concerning ruin of the resulting adjusted process, as well as the dividends or tax paid out until ruin, are investigated.

Throughout this chapter, X will denote an insurance risk process which will always be assumed to belong to the class of spectrally negative Lévy processes. Unless otherwise stated, we shall also assume throughout this chapter the *security loading condition*

$$\lim_{t\uparrow\infty} X_t = \infty, \tag{10.1}$$

which is equivalent to the assumption $\psi'(0+) > 0$ where, as usual, ψ is the Laplace exponent of X; see (8.1). Many of the technical features of the theory of spectrally negative Lévy processes, for example excursion theory and the theory of scale functions, will inevitably play a central role in our analysis. Accordingly, we shall adopt the same notation as in Chap. 8.

A.E. Kyprianou, *Fluctuations of Lévy Processes with Applications*, Universitext,
DOI 10.1007/978-3-642-37632-0_10, © Springer-Verlag Berlin Heidelberg 2014

10.1 Review of Distributional Properties at Ruin

As alluded to earlier, the ruin problem for the process X is one and the same as the first-passage problem, which has already been studied extensively in Chap. 7 (for the case of $-X$). Let us therefore spend some time in this section gathering together some of the facts that we have already established in previous chapters, for the special setting of a Lévy insurance risk process. To this end, we start by recalling that the ruin time is written as

$$\tau_0^- = \inf\{t > 0 : X_t < 0\},$$

in which case the so-called *deficit at ruin* may be identified as $-X_{\tau_0^-}$ and the *wealth prior to ruin* is identified as $X_{\tau_0^- -}$.

The probability of ruin. Recall that, for each spectrally negative Lévy process, we can define its scale function, W, through the Laplace transform in (8.8). The probability of ruin, when the initial surplus is valued at $x \geq 0$, is given in Theorem 8.1 (ii) by

$$\mathbb{P}_x(\tau_0^- < \infty) = 1 - \psi'(0+)W(x).$$

When X has paths of bounded variation, recall that we may write

$$X_t = \delta t - S_t, \quad t \geq 0, \tag{10.2}$$

where $\delta > 0$, $\{S_t : t \geq 0\}$ is a driftless subordinator and the Lévy measure of which we shall denote by Υ. Note that this class includes the Cramér–Lundberg model by taking $\Upsilon(\cdot) = \lambda F(\cdot)$, where λ is the rate of arrival of claims and $F(\cdot)$ is the claim distribution on $(0, \infty)$. Using Exercise 8.3, one can recover easily the Pollaczek–Khintchine formula

$$\mathbb{P}_x(\tau_0^- = \infty) = \frac{\psi'(0+)}{\delta} \sum_{n \geq 0} \nu^{*k}(x), \quad x \geq 0,$$

where $\nu(\mathrm{d}x) = \delta^{-1}\Upsilon(x, \infty)\mathrm{d}x$ on $(0, \infty)$ and we understand $\nu^{*0}(\mathrm{d}x) = \delta_0(\mathrm{d}x)$.

Cramér's estimate of ruin. Recall that the Laplace exponent, ψ, of X is a convex function on $(0, \infty)$. Theorem 7.6 tells us that, if there exists an $\alpha > 0$ such that $\psi(-\alpha) = 0$ (the so-called *Cramér condition*) then, under mild additional conditions, the ruin probability $\mathbb{P}_x(\tau_0^- < \infty)$ should decay exponentially as a function of x with rate α. To be precise, if we assume that the Lévy measure Π of X does not have lattice support when $\Pi(-\infty, 0) < \infty$, then one easily checks from the statement of Theorem 7.6, with the help of (6.35), that

$$\lim_{x \uparrow \infty} \mathrm{e}^{\alpha x}\mathbb{P}_x(\tau_0^- < \infty) = \frac{\psi'(0)}{|\psi'(-\alpha)|}.$$

Note in particular that, thanks to the convexity of the Laplace exponent ψ and the fact that $\psi(-\alpha) = \psi(0) = 0$, it follows that $\psi'(-\alpha) < 0$.

Heavy-tailed estimates of ruin. Cramér's estimate of ruin requires the existence of an $\alpha > 0$ such that $\mathbb{E}(\mathrm{e}^{-\alpha X_t}) = 1$ for all $t \geq 0$. Hence, it follows from Theorem 3.6 that

$$\int_{(-\infty,-1)} \mathrm{e}^{-\alpha x} \Pi(\mathrm{d}x) < \infty,$$

showing the existence of exponential moments in the Lévy measure. Theorem 7.16 (i) shows us that when the Cramér condition fails, then radically different asymptotics of the ruin probability can occur. Indeed, whenever $\Pi(-\infty, -x)$, $x \geq 0$, is regularly varying at infinity with index $-(\alpha+1)$, for $\alpha > 0$, then

$$\mathbb{P}_x\left(\tau_0^- < \infty\right) \sim \frac{1}{\psi'(0+)} \int_x^\infty \Pi(-\infty, -y)\mathrm{d}y,$$

as $x \uparrow \infty$.

Deficit at ruin. As noted above, the deficit at ruin is nothing other than $-X_{\tau_0^-}$. However, in the proof of Theorem 7.8, we saw that this quantity, under $\mathbb{P}_x$, $x > 0$, can also be identified as $\widehat{H}_{\widehat{T}_x}$, where $\widehat{H} := \{\widehat{H}_t : t \geq 0\}$ is the descending ladder height process of X and $\widehat{T}_x = \inf\{t > 0 : \widehat{H}_t > x\}$. Recall that $\widehat{H}$ is a killed subordinator, where the killing is a consequence of the security loading condition (10.1). If we denote its potential measure by $\widehat{U}$, then, recalling that the security loading condition (10.1) is in force, we have, from the discussion in Sect. 9.2, that $\widehat{U}(\mathrm{d}z) = W(\mathrm{d}x)$. Here W is the scale function associated with X. Moreover, from Corollary 7.9, the Lévy measure of $\widehat{H}$, say $\Pi_{\widehat{H}}(\mathrm{d}x)$ on $x > 0$, satisfies

$$\Pi_{\widehat{H}}(x, \infty) = \int_x^\infty \Pi(z, \infty)\mathrm{d}z, \quad x > 0.$$

In that case, we may appeal to Theorem 5.6 to deduce that, for $u > 0$ and $x \geq 0$,

$$\mathbb{P}_x\left(-X_{\tau_0^-} \in \mathrm{d}u,\ \tau_0^- < \infty\right) = \int_{[0,x]} W(\mathrm{d}z)\Pi(x+u-z, \infty)\mathrm{d}u.$$

Ruin by creeping. We know that any spectrally negative Lévy process creeps downwards if and only if it has a Gaussian component; see the discussion in Sect. 8.1 as well as Exercise 7.6. In that case, if σ is the Gaussian coefficient such that $\sigma^2 > 0$, then in light of the comments concerning deficit at ruin given earlier, we may also write the probability of ruin by creeping in terms of the probability that the descending ladder height process creeps over a level. Indeed, recalling from Exercise 6.5 that the descending ladder height process has a drift if and only if $\sigma^2 > 0$, in which case it is equal to $\sigma^2/2$, we may infer from Theorem 5.9 that, for all $x > 0$,

$$\mathbb{P}_x\left(-X_{\tau_0^-} = 0,\ \tau_0^- < \infty\right) = \frac{\sigma^2}{2} W'(x). \tag{10.3}$$

See also Exercise 8.6. Note that the existence of a first derivative of the scale function W is guaranteed by the fact that the underlying Lévy process has paths of unbounded variation; cf. Lemma 8.2 and the comments thereafter.

With these observations in mind, we may dig deeper into some of the other results in earlier parts of this book and extract similarly relevant statements. One may consider for example the relevance of Theorem 5.7, Exercise 5.10 and Theorem 7.16 (ii) to the asymptotic deficit at ruin as $x \uparrow \infty$. In the next section, we address a topic which is well represented in the actuarial literature. This is the time-penalised joint law of the deficit at ruin and wealth immediately prior to ruin.

10.2 The Gerber–Shiu Measure

Within the setting of the classical Cramér–Lundberg model, Gerber and Shiu (1997, 1998) introduced the *expected discounted penalty function* as follows. If we imagine that $f : [0,\infty)^2 \to [0,\infty)$ is a measurable function such that $f(-X_{\tau_0^-}, X_{\tau_0^- -})$ reflects the economic cost to the insurer at the moment of ruin, then taking account of a discounting force of interest, say $q \geq 0$, the penalty function is given by

$$\mathbb{E}_x\big(\mathrm{e}^{-q\tau_0^-} f(-X_{\tau_0^-}, X_{\tau_0^- -}); \tau_0^- < \infty\big), \tag{10.4}$$

where the initial surplus of the insurance company is $x \geq 0$. More commonly, (10.4) is referred to as the *Gerber–Shiu penalty function*.

Since its introduction into the actuarial literature, there has been an arms race of publications studying the penalty function in settings of ever-increasing generality. Although far from exhaustive, on account of the extent of the relevant literature, a list of key papers includes Dickson (1992, 1993), Gerber and Landry (1998), Lin and Willmot (1999), Cai and Dickson (2002), Cai (2004), Garrido and Morales (2006), Morales (2007) and Yin and Wang (2009). For an encyclopaedic overview, see Asmussen and Albrecher (2010). To some extent, until recently, this literature has evolved disjointly from parallel developments in the theory of Lévy processes. However, Zhou (2005) makes the important observation, from the point of view of Lévy insurance risk processes, that the penalty function can be expressed in a straightforward way in terms of scale functions. Moreover, within the same setting, Biffis and Morales (2010) make the observation that a more general version of the Gerber–Shiu penalty function, which allows the cost function f to take account of the last minimum before ruin, that is $\underline{X}_{\tau_0^- -} = \inf_{s<\tau_0^-} X_s$, can be derived from the quintuple law given in Theorem 7.7, again in terms of scale functions.

Let $f : \mathbb{R}^3 \to [0,\infty)$ be a bounded measurable function such that $f(0,\cdot,\cdot) = 0$ and $x, q \geq 0$. The generalised discounted penalty function associated with f and $q \geq 0$ is given by

$$\phi_f(x,q) = \mathbb{E}_x\big(\mathrm{e}^{-q\tau_0^-} f(-X_{\tau_0^-}, X_{\tau_0^- -}, \underline{X}_{\tau_0^- -}); \tau_0^- < \infty\big). \tag{10.5}$$

Note that the requirement $f(0,\cdot,\cdot) = 0$ simply ensures that the penalty function has no contribution from the event of creeping when downward creeping is possible for

X (that is, the case that there is a Gaussian component). The case of creeping at ruin will shortly be developed separately.

It is more convenient to write the penalty function (10.5) in the form

$$\phi_f(x,q) = \int_{(0,\infty)^3} \mathbf{1}_{(v\geq y)} f(u,v,y)\, K_x^{(q)}(\mathrm{d}u,\mathrm{d}v,\mathrm{d}y),$$

where, for $q, x \geq 0$, $u < 0$, $v > 0$ and $0 < y \leq v \wedge x$, we define

$$K_x^{(q)}(\mathrm{d}u,\mathrm{d}v,\mathrm{d}y) \\ = \mathbb{E}_x\big(\mathrm{e}^{-q\tau_0^-}; X_{\tau_0^-} \in \mathrm{d}u,\ X_{\tau_0^- -} \in \mathrm{d}v,\ \underline{X}_{\tau_0^- -} \in \mathrm{d}y,\ \tau_0^- < \infty\big)$$

to be the *Gerber–Shiu measure.*

Theorem 10.1 *The Gerber–Shiu measure for a Lévy insurance risk process satisfies*

$$K_x^{(q)}(\mathrm{d}u,\mathrm{d}v,\mathrm{d}y) \\ = \mathrm{e}^{-\Phi(q)(v-y)}\big\{W^{(q)\prime}(x-y) - \Phi(q)W^{(q)}(x-y)\big\}\Pi(\mathrm{d}u-v)\mathrm{d}y\mathrm{d}v, \tag{10.6}$$

for $q \geq 0, x > 0, u < 0, v > 0$ *and* $0 < y \leq v \wedge x$.

It is worth mentioning that although the first derivative of $W^{(q)}$ is only defined almost everywhere in general, we use $W^{(q)\prime}$ in (10.6) as the density with respect to Lebesgue measure. Unless otherwise stated, this convention will be applied throughout the remainder of this chapter. Note also that this result does not cover the case that $x = 0$. This is of no consequence when X has paths of unbounded variation as ruin is instantaneous. However, when X has paths of bounded variation, we recall that 0 is irregular for $(-\infty, 0)$, in which case one should expect a non-trivial expression for the Gerber–Shiu measure. This is left to the reader in Exercise 10.2.

Proof of Theorem 10.1 According to the quintuple law in Theorem 7.7, we have, for $u < 0$, $v > 0$ and $0 < y \leq v \wedge x$,

$$\mathbb{P}_x\big(X_{\tau_0^-} \in \mathrm{d}u,\ X_{\tau_0^- -} \in \mathrm{d}v,\ \underline{X}_{\tau_0^- -} \in \mathrm{d}y,\ \tau_0^- < \infty\big) \\ = kW'(x-y)\Pi(\mathrm{d}u-v)\mathrm{d}y\mathrm{d}v, \tag{10.7}$$

where W is the scale function associated with X and k is a strictly positive constant, which depends on the normalisation of the local time of X at its supremum.

We claim that the constant k is unity. Indeed, on the one hand, we have from (8.35) that, for $u < 0$, $v > 0$,

$$
\begin{aligned}
&\mathbb{P}_x\big(X_{\tau_0^-}\in du, X_{\tau_0^- -}\in dv, \tau_0^-<\infty\big)\\
&\quad=\big\{W(x)-W(x-v)\big\}\Pi(du-v)dv.
\end{aligned}
\tag{10.8}
$$

On the other hand, integrating out y in (10.7), we get the same expression as on the right-hand side of (10.8), albeit for the factor k. We are thus forced to conclude that $k=1$.

To complete the proof, we need to develop the expression (10.7) so that it incorporates exponential discounting at rate $q>0$. However, this can be done by considering (10.8) under the measure $\mathbb{P}^{\Phi(q)}$, where we recall that $\Phi(q)=\sup\{\theta\geq 0:\psi(\theta)=q\}$ and $\mathbb{P}^{\Phi(q)}$ is defined through the exponential change of measure described in (8.5). Note in particular that, under $\mathbb{P}^{\Phi(q)}$, the process X is still a Lévy insurance risk process, but now with Laplace exponent $\psi_{\Phi(q)}(\theta)=\psi(\theta+\Phi(q))-q$, $\theta\geq 0$, which still respects the security loading condition, $\psi'_{\Phi(q)}(0+)=\psi'(\Phi(q))>0$. Moreover, the scale function of X under $\mathbb{P}^{\Phi(q)}$, written $W_{\Phi(q)}(x)$, is related to $W^{(q)}$, the q-scale function of X under $\mathbb{P}$, via the relation

$$
W^{(q)}(x)=e^{\Phi(q)x}W_{\Phi(q)}(x),
\tag{10.9}
$$

for $x\in\mathbb{R}$; see Lemma 8.4.

Revisiting the identity (10.7) with $k=1$ but under the law $\mathbb{P}^{\Phi(q)}$ instead, we now have

$$
\begin{aligned}
&\mathbb{E}_x\big(e^{-q\tau_0^-}; X_{\tau_0^-}\in du, X_{\tau_0^- -}\in dv, \underline{X}_{\tau_0^- -}\in dy, \tau_0^-<\infty\big)\\
&\quad=e^{\Phi(q)(x-u)}\mathbb{P}_x^{\Phi(q)}\big(X_{\tau_0^-}\in du, X_{\tau_0^- -}\in dv, \underline{X}_{\tau_0^- -}\in dy, \tau_0^-<\infty\big)\\
&\quad=e^{\Phi(q)(x-u)}W'_{\Phi(q)}(x-y)\Pi_{\Phi(q)}(du-v)dydv,
\end{aligned}
\tag{10.10}
$$

where $\Pi_{\Phi(q)}$ is the Lévy measure associated with $(X,\mathbb{P}^{\Phi(q)})$. From Theorem 3.9 we know that $\Pi_{\Phi(q)}(dx)=e^{\Phi(q)x}\Pi(dx)$ on $(-\infty,0)$. Moreover, the almost-everywhere derivative of (10.9) gives us, for $x\in\mathbb{R}$,

$$
W^{(q)\prime}(x)-\Phi(q)W^{(q)}(x)=e^{\Phi(q)x}W'_{\Phi(q)}(x).
\tag{10.11}
$$

Plugging this back into (10.10), we get

$$
\begin{aligned}
&\mathbb{E}_x\big(e^{-q\tau_0^-}; X_{\tau_0^-}\in du, X_{\tau_0^- -}\in dv, \underline{X}_{\tau_0^- -}\in dy, \tau_0^-<\infty\big)\\
&\quad=e^{-\Phi(q)(v-y)}\big\{W^{(q)\prime}(x-y)-\Phi(q)W^{(q)}(x-y)\big\}\Pi(du-v)dydv,
\end{aligned}
$$

where $u<0$, $v>0$ and $0<y\leq v\wedge x$. Note in particular when $q=0$, recalling that $\Phi(0)=0$ as the process X drifts to ∞, we see agreement with the formula (10.7) and the proof is complete. □

It is a straightforward computation to marginalise the kernel $K_x^{(q)}(du,dv,dy)$ in y to give the bivariate Gerber–Shiu measure specifying the joint distribution of the deficit at ruin and wealth prior to ruin, the classical quantities of interest. With a slight abuse of notation, let us refer to this measure as $K_x^{(q)}(\mathrm{d}u, \mathrm{d}v)$.

Corollary 10.2 *Within the setting of Theorem* 10.1, *we have, for* $q \geq 0$, $x > 0$, $v > 0$ *and* $u < 0$,

$$K_x^{(q)}(\mathrm{d}u, \mathrm{d}v) = \{\mathrm{e}^{-\Phi(q)v} W^{(q)}(x) - W^{(q)}(x - v)\}\Pi(\mathrm{d}u - v)\mathrm{d}v.$$

Note that, when $q = 0$, the measure $K^{(q)}(\mathrm{d}u, \mathrm{d}v, \mathrm{d}y)$, and hence the measure $K^{(q)}(\mathrm{d}u, \mathrm{d}v)$, is not necessarily a probability measure on account of the fact that we have excluded consideration of ruin by creeping in its definition. It was remarked earlier that ruin by creeping occurs if and only if $\sigma^2 > 0$, in which case the probability of this event is given by (10.3). Exercise 10.1 gives an identity for the penalised probability of ruin by creeping $\mathbb{E}_x(\mathrm{e}^{-q\tau_0^-}; X_{\tau_0^-} = X_{\tau_0^- -} = \underline{X}_{\tau_0^-} = 0)$.

10.3 Reflection Strategies

An adaptation of the classical ruin problem was introduced by de Finetti (1957) in which dividends are paid out to shareholders up to the moment of ruin. De Finetti was interested in finding a way of paying out dividends such as to optimise the expected present value of the total income of the shareholders from time zero until ruin. *De Finetti's dividend problem* amounts to solving a control problem which we reproduce here, albeit in the framework of a general Lévy insurance risk process.

Let $\xi = \{\xi_t : t \geq 0\}$ be a dividend strategy consisting of a process with initial value zero, which has paths that are left-continuous, non-negative, non-decreasing and adapted to the filtration of X. The quantity ξ_t thus represents the cumulative dividends paid out up to time $t \geq 0$ by the insurance company, whose risk-process is modelled by X. The aggregate, or *controlled*, value of the risk process, when taking account of dividend strategy ξ, is thus $U^\xi = \{U_t^\xi : t \geq 0\}$, where $U_t^\xi = X_t - \xi_t$, $t \geq 0$. An additional constraint on ξ is that $\xi_{t+} - \xi_t \leq \max\{U_t^\xi, 0\}$ for $t \geq 0$ (i.e. lump sum dividend payments are always smaller than the available reserves).

Let Ξ be the family of dividend strategies, as outlined in the previous paragraph, and, for each $\xi \in \Xi$, write $\sigma^\xi = \inf\{t > 0 : U_t^\xi < 0\}$ for the time at which ruin occurs for the controlled risk process. The expected present value, with discounting at rate $q \geq 0$, associated with the dividend policy ξ is given by

$$v_\xi(x) = \mathbb{E}_x\left(\int_0^{\sigma^\xi} \mathrm{e}^{-qt}\mathrm{d}\xi_t\right),$$

where the risk process has initial capital $x \geq 0$. De Finetti's dividend problem consists of solving the stochastic control problem

$$v^*(x) := \sup_{\xi \in \Xi} v_\xi(x), \quad x \geq 0. \tag{10.12}$$

That is to say, if it exists, to establish a strategy, $\xi^* \in \Xi$, such that $v^* = v_{\xi^*}$.

This problem was considered by Gerber (1969), Gerber (1972) and by Azcue and Muler (2005) for the Cramér–Lundberg model. Thereafter, a string of articles, each one successively improving on the previous, treated the case of a general Lévy insurance risk process; see Avram et al. (2007), Loeffen (2008), Kyprianou et al. (2010b) and Loeffen and Renaud (2010). We shall refrain from giving a complete account of their findings other than to say that under appropriate conditions on the underlying Lévy measure of X, the optimal strategy consists of a so-called *reflection strategy*. Specifically, there exists an $a^* \in [0, \infty)$ such that the optimal strategy, $\xi^* = \{\xi_t^* : t \geq 0\}$, satisfies $\xi_0^* = 0$ and

$$\xi_t^* = (a^* \vee \overline{X}_t) - a^*, \quad t \geq 0.$$

In that case, the ξ^*-controlled risk process, say $U^* = \{U_t^* : t \geq 0\}$, is identical to the process $\{a^* - Y_t : t \geq 0\}$ under $\mathbb{P}_x$, where

$$Y_t = (a^* \vee \overline{X}_t) - X_t, \quad t \geq 0,$$

and $\overline{X}_t = \sup_{s \leq t} X_s$ is the running supremum of the Lévy insurance risk process. Note that Y has earlier been identified as the process X reflected in its supremum, cf. Sect. 8.5, which motivates the name of the strategy ξ^*. For $x \in (0, a^*)$, we may now write

$$v^*(x) = \mathbb{E}_x\left(\int_0^{\sigma^*} \mathrm{e}^{-qt} \mathrm{d}\xi_t^*\right),$$

where $\sigma^* = \inf\{t \geq 0 : U_t^* < 0\} = \inf\{t > 0 : Y_t > a^*\}$. From Loeffen (2008), we know that sufficient (but not necessary) conditions that ensure the reflection strategy is optimal are that the q-scale function, $W^{(q)}$, associated with X is sufficiently smooth and has a convex first derivative. Here, sufficiently smooth means that it is continuously differentiable[1] on $(0, \infty)$ when X has bounded variation paths and, otherwise, it is twice continuously differentiable in $(0, \infty)$. In that case, the optimal threshold is given by

$$a^* = \sup\{c \geq 0 : W^{(q)\prime}(c) \leq W^{(q)\prime}(x) \text{ for all } x \geq 0\}.$$

Exercise 9.5 shows that the above sufficient conditions are met when the function $x \mapsto \Pi(-\infty, -x)$, $x \geq 0$, is completely monotone.

Understanding distributional properties of the random variable $\int_0^{\sigma^*} \mathrm{e}^{-qt} \mathrm{d}\xi_t^*$, that is, the present value of the optimal dividend strategy paid until ruin, is our main focus of interest in this section. Theorem 10.3 below pertains to the work of Gerber (1972), Dickson and Waters (2004), Kyprianou and Palmowski (2007) and Renaud

[1] Recall from the discussion following Lemma 8.2 that $W^{(q)}$ is continuously differentiable when X has paths of unbounded variation and otherwise it is continuously differentiable if and only if the Lévy measure of X has no atoms.

and Zhou (2007). See also Albrecher and Gerber (2011). Specifically, it gives a closed-form expression for the optimal value function below the optimal threshold,

$$v^*(x) = \frac{W^{(q)}(x)}{W_+^{(q)\prime}(a^*)}, \quad x \le a^*, \tag{10.13}$$

where, for all $q \ge 0$, $W_+^{(q)\prime}$ is the right derivative of the scale function.

Theorem 10.3 *Let $a > 0$ and define the process $\xi^a = \{\xi_t^a : t \ge 0\}$ by*

$$\xi_t^a = (a \vee \overline{X}_t) - a, \quad t \ge 0.$$

For $n = 1, 2, \ldots$ and $0 \le x \le a$, we have

$$\mathbb{E}_x\left[\left(\int_0^{\sigma^a} \mathrm{e}^{-qt}\mathrm{d}\xi_t^a\right)^n\right] = n!\frac{W^{(qn)}(x)}{W^{(qn)}(a)}\prod_{k=1}^n \frac{W^{(qk)}(a)}{W_+^{(qk)\prime}(a)},$$

where

$$\sigma^a = \inf\{t > 0 : (a \vee \overline{X}_t) - X_t > a\}.$$

Proof We begin by noting that it suffices to prove the result when $x = a$. Indeed, since ξ^a increases on the set of times that the reflected process Y is equal to zero, equivalently, the set of times at which U^{ξ^a} is equal to a, the strong Markov property for $a - Y$ implies that

$$\begin{aligned}
\mathbb{E}_x\left[\left(\int_0^{\sigma^a} \mathrm{e}^{-qt}\mathrm{d}\xi_t^a\right)^n\right] &= \mathbb{E}_x\left[\left(\int_{\tau_a^+}^{\sigma^a} \mathrm{e}^{-qt}\mathrm{d}\xi_t^a\right)^n \mathbf{1}_{(\tau_a^+ < \tau_0^-)}\right] \\
&= \mathbb{E}_x\left(\mathrm{e}^{-qn\tau_a^+}\mathbf{1}_{(\tau_a^+ < \tau_0^-)}\right)\mathbb{E}_a\left[\left(\int_0^{\sigma^a} \mathrm{e}^{-qt}\mathrm{d}\xi_t^a\right)^n\right] \\
&= \frac{W^{(qn)}(x)}{W^{(qn)}(a)}\mathbb{E}_a\left[\left(\int_0^{\sigma^a} \mathrm{e}^{-qt}\mathrm{d}\xi_t^a\right)^n\right],
\end{aligned} \tag{10.14}$$

where we recall that the stopping times τ_a^+ and τ_0^- for X are given by (8.7) and where the final equality is a consequence of Theorem 8.1 (iii).

To deal with the expectation on the right-hand side of (10.14), let us start by identifying the integral $\int_0^{\sigma^a} \mathrm{e}^{-qt}\mathrm{d}\xi_t^a$ in terms of the process of excursions of X from its supremum, introduced in Sect. 6.3. To start with, note that, under $\mathbb{P}_a$, the random variable $\int_0^{\sigma^a} \mathrm{e}^{-qt}\mathrm{d}\xi_t^a$ is equal in law to $\int_0^{\overline{\sigma}_a} \mathrm{e}^{-qt}\mathrm{d}\overline{X}_t$ under $\mathbb{P}$, where $\overline{\sigma}_a = \inf\{t > 0 : \overline{X}_t - X_t > a\}$. In the notation of Chap. 6, recall that, under $\mathbb{P}$, the local time of X at its maximum, L, may be taken as equal to $\overline{X}$. Hence, after the change of variable $t \mapsto L_t^{-1}$, we are interested in the distribution of the random variable

$$\int_0^\infty \mathbf{1}_{(\sup_{s\le t}\overline{\epsilon}_s \le a)}\mathrm{e}^{-qL_t^{-1}}\mathrm{d}t$$

under $\mathbb{P}$, where $\{(t, \epsilon_t) : t \geq 0 \text{ and } t \neq \partial\}$ is the process of excursions of X from its maximum, indexed by local time, as described in Sect. 6.3. Recall, moreover, that $\overline{\epsilon}_s$ is the supremum of the excursion indexed by local time s. Next, define

$$J_t = \int_t^\infty \mathrm{e}^{-qL_u^{-1}} \mathbf{1}_{(\sup_{s\leq u} \overline{\epsilon}_s \leq a)} \mathrm{d}u.$$

Since,

$$\frac{\mathrm{d}}{\mathrm{d}t} J_t^n = -n J_t^{n-1} \mathrm{e}^{-qL_t^{-1}} \mathbf{1}_{(\sup_{s\leq t} \overline{\epsilon}_s \leq a)},$$

we obtain

$$J_0^n - J_t^n = n \int_0^t \mathrm{e}^{-qL_u^{-1}} \mathbf{1}_{(\sup_{s\leq u} \overline{\epsilon}_s \leq a)} J_u^{n-1} \mathrm{d}u. \tag{10.15}$$

Recall from Lemma 6.8 that L_t^{-1} is a stopping time. Hence, by the strong Markov property (cf. Theorem 3.1) and the fact that the process of excursions of X from the maximum, indexed by local time, forms a Poisson point process (cf. Theorem 6.14), we can write for each $t \geq 0$,

$$J_t = \mathrm{e}^{-qL_t^{-1}} \mathbf{1}_{(\sup_{s\leq t} \overline{\epsilon}_s \leq a)} J_0^*,$$

where J_0^* is independent of $\mathcal{F}_{L_t^{-1}}$ and has the same distribution as J_0. In conclusion, if we let

$$\Psi_n = \mathbb{E}\left(J_0^n\right),$$

then

$$\Psi_n\left(1 - \mathbb{E}\left(\mathrm{e}^{-nqL_t^{-1}} \mathbf{1}_{(\sup_{s\leq t} \overline{\epsilon}_s \leq a)}\right)\right) = n\Psi_{n-1} \int_0^t \mathbb{E}\left(\mathrm{e}^{-nqL_u^{-1}} \mathbf{1}_{(\sup_{s\leq u} \overline{\epsilon}_s \leq a)}\right) \mathrm{d}u. \tag{10.16}$$

Recalling again that L_t^{-1} is a stopping time and appealing to the exponential change of measure in (8.5), we have that

$$\mathbb{E}\left(\mathrm{e}^{-qnL_t^{-1}} \mathbf{1}_{(\sup_{s\leq t} \overline{\epsilon}_s \leq a)}\right) = \mathrm{e}^{-\Phi(qn)t} \mathbb{P}^{\Phi(qn)}\left(\sup_{s\leq t} \overline{\epsilon}_s \leq a\right). \tag{10.17}$$

The process $(X, \mathbb{P}^{\Phi(qn)})$ is still a spectrally negative Lévy process which drifts to $+\infty$. (See for example the discussion at the beginning of Sect. 8.1.) Appealing again to Theorem 6.14, we have, for $t \geq 0$, that

$$\mathbb{P}^{\Phi(qn)}\left(\sup_{s\leq t} \overline{\epsilon}_s \leq a\right) = \mathrm{e}^{-n_{\Phi(nq)}(\overline{\epsilon} > a)t}, \tag{10.18}$$

where $n_{\Phi(nq)}$ is the excursion measure of the Poisson point process of excursions; cf. Sect. 6.3.

Considering (8.26) and (10.11) it is straightforward to show that

$$n_{\Phi(nq)}(\overline{\epsilon} > a) = \frac{W_+^{(qn)\prime}(a)}{W^{(qn)}(a)} - \Phi(qn), \tag{10.19}$$

see (8.26).

Plugging (10.17), (10.18) and (10.19) back into (10.16), we now see the iteration

$$\Psi_n = n\Psi_{n-1}\frac{W^{(qn)}(a)}{W_+^{(qn)\prime}(a)},$$

which yields the desired result. □

10.4 Refraction Strategies

An adaptation of the optimal control problem (10.12) studied by Jeanblanc and Shiryaev (1995), Asmussen and Taksar (1997), Gerber and Shiu (2006b) and Kyprianou et al. (2012b), in the setting of (Lévy) insurance risk processes deals with the case that optimality is sought in a subclass, say Ξ_α, of the admissible strategies Ξ, where $\alpha > 0$ is a fixed parameter. Specifically, Ξ_α denotes the set of dividend strategies $\xi \in \Xi$ such that

$$\xi_t = \int_0^t \ell_s \mathrm{d}s, \quad t \geq 0,$$

where $\ell = \{\ell_t : t \geq 0\}$ is uniformly bounded by α. That is to say, Ξ_α consists of dividend strategies which are absolutely continuous with uniformly bounded density.

Again, we refrain from going into the details of their findings, other than to say that, under appropriate conditions, the optimal strategy, $\xi^\alpha = \{\xi_t^\alpha : t \geq 0\}$, in Ξ_α turns out to satisfy

$$\xi_t^\alpha = \alpha \int_0^t \mathbf{1}_{(U_s > b)} \mathrm{d}s, \quad t \geq 0,$$

for some $b \geq 0$, where $U = \{U_t : t \geq 0\}$ is the controlled Lévy risk process $X - \xi^\alpha$. Each element of the pair (U, ξ^α) cannot be expressed autonomously and we are forced to work within the confines of the stochastic differential equation (SDE)

$$U_t = X_t - \alpha \int_0^t \mathbf{1}_{(U_s > b)} \mathrm{d}s, \quad t \geq 0, \tag{10.20}$$

also written as

$$\mathrm{d}U_t = \mathrm{d}X_t - \alpha \mathbf{1}_{(U_t > b)} \mathrm{d}t, \quad t \geq 0.$$

For reasons that we shall elaborate on later, the process in (10.20) is called a *refracted Lévy process*. It will be the main focus of our attention for the remainder of this section. We are guided largely by Kyprianou and Loeffen (2010) in our presentation.

The very first issue we are confronted with when studying (10.20) is whether a solution to this SDE exists. In order to keep our exposition as mathematically convenient as possible, we shall henceforth make the following assumption.

For the remainder of this section, we restrict ourselves to the case that X is a bounded variation spectrally negative Lévy process and $0 < \alpha < \delta$, where δ is the drift appearing in the decomposition (10.2).

Theorem 10.4 *The SDE* (10.20) *has a unique pathwise solution.*

Proof Start by recalling that all spectrally negative Lévy processes of bounded variation have the property that 0 is irregular for $(-\infty, 0)$ and, moreover, that they do not creep downwards. This means that, as $0 < \alpha < \delta$, the process U, when issued from a point in $[b, \infty)$, behaves as the spectrally negative Lévy process $\{X_t - \alpha t : t \geq 0\}$ until the first moment that it passes below b, which it does by a jump. On the other hand, in $(-\infty, b)$, U behaves like the process X until it first passes above b, which it does continuously.

Define the times T_n and S_n recursively as follows. We set $S_0 = 0$ and, for $n = 1, 2, \ldots$, on the events that $\{S_{n-1} < \infty\}$ and $\{T_n < \infty\}$ respectively, put

$$T_n = \inf\left\{t > S_{n-1} : X_t - \alpha \sum_{i=1}^{n-1} (S_i - T_i) > b\right\},$$

$$S_n = \inf\left\{t > T_n : X_t - \alpha \sum_{i=1}^{n-1} (S_i - T_i) - \alpha(t - T_n) < b\right\}.$$

As usual we use the convention that $\inf \emptyset = \infty$. Since 0 is irregular for $(-\infty, 0)$, the difference between the two consecutive times T_n and S_n is strictly positive. Moreover, both sequences T_n and S_n increase to infinity almost surely.

Now we construct a solution to (10.20), $U = \{U_t : t \geq 0\}$, as follows. The process is issued from $X_0 = x$ and

$$U_t = \begin{cases} X_t - \alpha \sum_{i=1}^{n} (S_i - T_i), & \text{for } t \in [S_n, T_{n+1}) \text{ and } n \geq 0, \\ X_t - \alpha \sum_{i=1}^{n-1} (S_i - T_i) - \alpha(t - T_n), & \text{for } t \in [T_n, S_n) \text{ and } n \geq 1. \end{cases}$$

Note that, for $n = 1, 2, \ldots$, on the events $\{S_{n-1} < \infty\}$ and $\{T_n < \infty\}$, the times T_n and S_n can then be identified as

$$T_n = \inf\{t > S_{n-1} : U_t > b\}, \quad S_n = \inf\{t > T_n : U_t < b\}.$$

Hence

$$U_t = X_t - \alpha \int_0^t \mathbf{1}_{\{U_s > b\}} \mathrm{d}s, \quad t \geq 0.$$

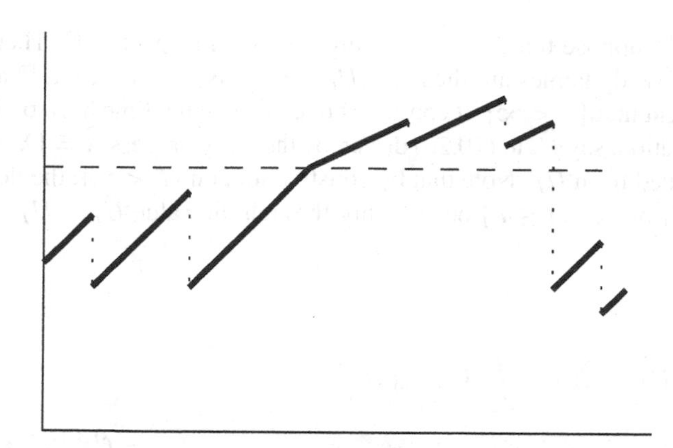

Fig. 10.1 A sample path of U when the driving Lévy process is a Cramér–Lundberg process. Its trajectory "refracts" as it passes continuously above the *horizontal dashed line* at level b.

For uniqueness of this solution, suppose that $\{U_t^{(1)} : t \geq 0\}$ and $\{U_t^{(2)} : t \geq 0\}$ are two pathwise solutions to (10.20). Then, writing

$$\Delta_t = U_t^{(1)} - U_t^{(2)} = -\alpha \int_0^t (\mathbf{1}_{\{U_s^{(1)} > b\}} - \mathbf{1}_{\{U_s^{(2)} > b\}}) \mathrm{d}s,$$

it follows from integration by parts that

$$\Delta_t^2 = -2\alpha \int_0^t \Delta_s (\mathbf{1}_{\{U_s^{(1)} > b\}} - \mathbf{1}_{\{U_s^{(2)} > b\}}) \mathrm{d}s.$$

Thanks to the fact that $\mathbf{1}_{\{x>b\}}$ is an increasing function, it follows from the above representation, that, for all $t \geq 0$, $\Delta_t^2 \leq 0$ and hence $\Delta_t = 0$ almost surely. This concludes the proof of existence and uniqueness amongst the class of pathwise solutions. □

Let us momentarily return to the reason why U is referred to as a refracted Lévy processes. A simple sketch of a realisation of the path of U in the case that X is a Cramér–Lundberg process (see for example Fig. 10.1) gives the impression that the trajectory of U "refracts" each time it passes continuously from $(-\infty, b]$ into (b, ∞), much as a beam of light does when passing from one medium to another.[2]

The construction of the unique pathwise solution described above clearly shows that U is adapted to the natural filtration $\mathbb{F} = \{\mathcal{F}_t : t \geq 0\}$ of X. Conversely, since, for all $t \geq 0$, $X_t = U_t + \alpha \int_0^t \mathbf{1}_{\{U_s > b\}} \mathrm{d}s$, it is also clear that X is adapted to the natural filtration of U. We can use this observation to reason that U is a strong Markov process.

[2]See for example the discussion on p. 80 of Gerber and Shiu (2006b) which also makes reference to "refraction" in the case of compound Poisson jumps. Gerber and Shiu (2006a) also use the terminology "refraction" for the case that X is a linear Brownian motion.

To this end, suppose that T is a stopping time with respect to $\mathbb{F}$. Then define a process $\widehat{U}$ whose dynamics are those of $\{U_t : t \leq T\}$ issued from $x \in \mathbb{R}$ and, given $\mathcal{F}_T$, on the event that $\{T < \infty\}$, it continues to evolve on the time horizon $[T, \infty)$ as the unique solution, say $\widetilde{U}$, to (10.20) driven by the Lévy process $\widetilde{X} = \{X_{T+s} - X_T : s \geq 0\}$ and issued from U_T. Note that by construction, on $\{T < \infty\}$, the dependence of $\{\widehat{U}_t : t \geq T\}$ on $\{\widehat{U}_t : t \leq T\}$ occurs only through the value $\widehat{U}_T = U_T$. Note also that for $t > 0$,

$$
\begin{aligned}
\widehat{U}_{T+t} &= \widetilde{U}_t \\
&= \widehat{U}_T + \widetilde{X}_t - \alpha \int_0^t \mathbf{1}_{\{\widetilde{U}_s > b\}} \mathrm{d}s \\
&= x + X_T - \alpha \int_0^T \mathbf{1}_{\{U_s > b\}} \mathrm{d}s + (X_{T+t} - X_T) - \alpha \int_0^t \mathbf{1}_{\{\widehat{U}_{T+s} > b\}} \mathrm{d}s \\
&= x + X_{T+t} - \alpha \int_0^{T+t} \mathbf{1}_{\{\widehat{U}_s > b\}} \mathrm{d}s,
\end{aligned}
$$

thereby showing that $\widehat{U}$ solves (10.20) issued from x. Since (10.20) has a unique pathwise solution, this solution must be $\widehat{U}$ and therefore possesses the strong Markov property.

Let us now introduce the stopping times for U,

$$
\kappa_a^+ := \inf\{t > 0 : U_t > a\} \quad \text{and} \quad \kappa_0^- := \inf\{t > 0 : U_t < 0\},
$$

where $a > 0$. We are interested in studying the ruin probability

$$
\mathbb{P}_x\big(\kappa_0^- < \infty\big), \tag{10.21}
$$

as well as the expected present value of dividends paid until ruin,

$$
\alpha \mathbb{E}_x \left(\int_0^{\kappa_0^-} \mathrm{e}^{-qt} \mathbf{1}_{\{U_t > b\}} \mathrm{d}t \right). \tag{10.22}
$$

Not unlike our treatment of the analogous objects for X in Theorem 8.1, it turns out to be more convenient to first study the seemingly more complex two-sided exit problem. To this end, let $Y = \{Y_t : t \geq 0\}$, where $Y_t = X_t - \alpha t$ and denote by P_x the law of the process Y when issued from x (with E_x as the associated expectation operator). For each $q \geq 0$, $W^{(q)}$ and $Z^{(q)}$ denote, as usual, the q-scale functions associated with X. We shall write $\mathbb{W}^{(q)}$ for the q-scale function associated with Y. For convenience, we will write

$$
w^{(q)}(x; y) = W^{(q)}(x - y) + \alpha \mathbf{1}_{(x \geq b)} \int_b^x \mathbb{W}^{(q)}(x - z) W^{(q)\prime}(z - y) \mathrm{d}z,
$$

for $x, y \in \mathbb{R}$ and $q \geq 0$. We have two main results concerning the two-sided exit problem, from which more can be said about the quantities (10.21) and (10.22).

Theorem 10.5 *For $q \geq 0$ and $0 \leq x, b \leq a$, we have*

$$\mathbb{E}_x\big(\mathrm{e}^{-q\kappa_a^+}\mathbf{1}_{\{\kappa_a^+<\kappa_0^-\}}\big) = \frac{w^{(q)}(x;0)}{w^{(q)}(a;0)}. \tag{10.23}$$

Theorem 10.6 *For $q \geq 0$ and $0 \leq x, y, b \leq a$,*

$$\begin{aligned}&\int_0^\infty \mathrm{e}^{-qt}\mathbb{P}_x\big(U_t \in \mathrm{d}y,\, t < \kappa_0^- \wedge \kappa_a^+\big)\mathrm{d}t \\ &\quad = \mathbf{1}_{\{y\in[b,a]\}}\left\{\frac{w^{(q)}(x;0)}{w^{(q)}(a;0)}\mathbb{W}^{(q)}(a-y) - \mathbb{W}^{(q)}(x-y)\right\}\mathrm{d}y \\ &\qquad + \mathbf{1}_{\{y\in[0,b)\}}\left\{\frac{w^{(q)}(x;0)}{w^{(q)}(a;0)}w^{(q)}(a;y) - w^{(q)}(x;y)\right\}\mathrm{d}y.\end{aligned} \tag{10.24}$$

Although appealing to relatively straightforward methods, the proofs are quite long, requiring a little patience.

Proof of Theorem 10.5 Write $p(x,\alpha) = \mathbb{E}_x(\mathrm{e}^{-q\kappa_a^+}\mathbf{1}_{\{\kappa_a^+<\kappa_0^-\}})$. Suppose that $x \leq b$. Then, by conditioning on $\mathcal{F}_{\tau_b^+}$, we have

$$p(x,\alpha) = \mathbb{E}_x\big(\mathrm{e}^{-q\tau_b^+}\mathbf{1}_{\{\tau_0^->\tau_b^+\}}\big)p(b,\alpha) = \frac{W^{(q)}(x)}{W^{(q)}(b)}p(b,\alpha), \tag{10.25}$$

where in the last equality, we have used Theorem 8.1 (iii). Suppose now that $b \leq x \leq a$. Using, respectively, that 0 is irregular for $(-\infty, 0)$ for Y, Theorem 8.1 (iii), the strong Markov property (10.25) and the identity in Exercise 10.6, we have

$$\begin{aligned}&p(x,\alpha) \\ &\quad = \mathrm{E}_x\big(\mathrm{e}^{-q\tau_a^+}\mathbf{1}_{\{\tau_b^->\tau_a^+\}}\big) + \mathrm{E}_x\big(\mathrm{e}^{-q\tau_b^-}\mathbf{1}_{\{\tau_b^-<\tau_a^+\}}p(U_{\tau_b^-},\alpha)\big) \\ &\quad = \frac{\mathbb{W}^{(q)}(x-b)}{\mathbb{W}^{(q)}(a-b)} + \frac{p(b,\alpha)}{W^{(q)}(b)}\mathrm{E}_x\big(\mathrm{e}^{-q\tau_b^-}\mathbf{1}_{\{\tau_b^-<\tau_a^+\}}W^{(q)}(Y_{\tau_b^-})\big) \\ &\quad = \frac{\mathbb{W}^{(q)}(x-b)}{\mathbb{W}^{(q)}(a-b)} + \frac{p(b,\alpha)}{W^{(q)}(b)}h(a,b,x),\end{aligned} \tag{10.26}$$

where

$$\begin{aligned}&h(a,b,x) \\ &\quad = \int_0^{a-b}\int_{(-\infty,-y)} W^{(q)}(b+y+\theta) \\ &\qquad \times\left[\frac{\mathbb{W}^{(q)}(x-b)\mathbb{W}^{(q)}(a-b-y)}{\mathbb{W}^{(q)}(a-b)} - \mathbb{W}^{(q)}(x-b-y)\right]\Pi(\mathrm{d}\theta)\mathrm{d}y.\end{aligned}$$

By setting $x=b$ in (10.26) and recalling that $\mathbb{W}^{(q)}(0)=1/(\delta-\alpha)$, we can now solve for $p(b,\alpha)$. Indeed, we have

$$p(b,\alpha)=W^{(q)}(b)\bigg\{(\delta-\alpha)\mathbb{W}^{(q)}(a-b)W^{(q)}(b) - \int_0^{a-b}\int_{(-\infty,-y)} W^{(q)}(b+y+\theta)\mathbb{W}^{(q)}(a-b-y)\Pi(\mathrm{d}\theta)\mathrm{d}y\bigg\}^{-1}. \tag{10.27}$$

Next, we want to simplify the term involving the double integral in the above expression.

To this end, noting that for $\alpha=0$ (the case that there is no refraction), we have, by Theorem 8.1 (iii), that, for all $x\geq 0$,

$$p(b,0)=\mathbb{E}_b\big(\mathrm{e}^{-q\tau_a^+}\mathbf{1}_{\{\tau_0^->\tau_a^+\}}\big)=\frac{W^{(q)}(b)}{W^{(q)}(a)}. \tag{10.28}$$

It follows, by comparing (10.27) (for $\alpha=0$) with (10.28), that

$$\int_0^{a-b}\int_{(-\infty,-y)} W^{(q)}(b+y+\theta)W^{(q)}(a-b-y)\Pi(\mathrm{d}\theta)\mathrm{d}y = \delta W^{(q)}(b)W^{(q)}(a-b)-W^{(q)}(a). \tag{10.29}$$

As $a\geq b$ is taken arbitrarily, we may take Laplace transforms in a on the interval (b,∞) of both sides of the above expression. Denote by $\mathcal{L}_b$ the operator which satisfies $\mathcal{L}_b f[\lambda]:=\int_b^\infty \mathrm{e}^{-\lambda x}f(x)\mathrm{d}x$ and let $\lambda>\Phi(q)$. For the left-hand side of (10.29), we get with the help of Fubini's Theorem

$$\int_b^\infty \mathrm{e}^{-\lambda x}\int_0^\infty\int_{(-\infty,-y)} W^{(q)}(b+y+\theta)W^{(q)}(x-b-y)\mathrm{d}y\Pi(\mathrm{d}\theta)\mathrm{d}x = \frac{\mathrm{e}^{-\lambda b}}{\psi(\lambda)-q}\int_0^\infty\int_{(-\infty,-y)} \mathrm{e}^{-\lambda y}W^{(q)}(b+y+\theta)\Pi(\mathrm{d}\theta)\mathrm{d}y.$$

For the right-hand side of (10.29), we get

$$\int_b^\infty \mathrm{e}^{-\lambda x}\big(W^{(q)}(x-b)\delta W^{(q)}(b)-W^{(q)}(x)\big)\mathrm{d}x = \frac{\mathrm{e}^{-\lambda b}}{\psi(\lambda)-q}\delta W^{(q)}(b)-\int_b^\infty \mathrm{e}^{-\lambda x}W^{(q)}(x)\mathrm{d}x,$$

and so

$$\int_0^\infty\int_{(-\infty,-y)} \mathrm{e}^{-\lambda y}W^{(q)}(b+y+\theta)\Pi(\mathrm{d}\theta)\mathrm{d}y$$

$$= \delta W^{(q)}(b) - \big(\psi(\lambda) - q\big)\mathrm{e}^{\lambda b}\mathcal{L}_b W^{(q)}[\lambda], \qquad (10.30)$$

for $\lambda > \Phi(q)$. Our objective is now to use (10.30) to show that for $q \geq 0$ and $x \geq b$, we have

$$\begin{aligned}\int_0^\infty \int_{(-\infty,-y)} & W^{(q)}(b+y+\theta)\mathbb{W}^{(q)}(x-b-y)\Pi(\mathrm{d}\theta)\mathrm{d}y \\ &= -W^{(q)}(x) + (\delta - \alpha)W^{(q)}(b)\mathbb{W}^{(q)}(x-b) \\ &\quad - \alpha\int_b^x \mathbb{W}^{(q)}(x-y)W^{(q)\prime}(y)\mathrm{d}y. \qquad (10.31)\end{aligned}$$

We will do this by taking Laplace transforms of (10.31) on both sides in x on (b, ∞). To this end note that, by (10.30), it follows, with the help of Fubini's Theorem, that the Laplace transform of the left-hand side of (10.31) equals

$$\begin{aligned}\int_b^\infty \mathrm{e}^{-\lambda x}\int_0^\infty \int_{(-\infty,-y)} & W^{(q)}(b+y+\theta)\mathbb{W}^{(q)}(x-b-y)\Pi(\mathrm{d}\theta)\mathrm{d}y\mathrm{d}x \\ &= \frac{\mathrm{e}^{-\lambda b}}{\psi(\lambda) - \alpha\lambda - q}\Big(\delta W^{(q)}(b) - \big(\psi(\lambda) - q\big)\mathrm{e}^{\lambda b}\mathcal{L}_b W^{(q)}[\lambda]\Big), \qquad (10.32)\end{aligned}$$

where $\lambda > \varphi(q)$ and, for $q \geq 0$, $\varphi(q) = \sup\{\theta \geq 0 : \psi(\theta) - \delta\theta = q\}$. (Note that φ is the right inverse of the Laplace exponent of Y.) Since

$$\mathcal{L}_b\left(\int_b^x f(x-y)g(y)\mathrm{d}y\right)[\lambda] = (\mathcal{L}_0 f)[\lambda](\mathcal{L}_b g)[\lambda]$$

and, for $\lambda > \Phi(q)$,

$$\mathcal{L}_b W^{(q)\prime}[\lambda] = \lambda\mathcal{L}_b W^{(q)}[\lambda] - \mathrm{e}^{-\lambda b}W^{(q)}(b)$$

(which follows from integration by parts), we have that the Laplace transform of the right-hand side of (10.31) is equal to the right-hand side of (10.32), for all sufficiently large λ. Hence (10.31) holds for almost every $x \geq b$. Because both sides of (10.31) are continuous in x, we finally conclude that (10.31) holds for all $x \geq b$.

To complete the proof, it suffices to plug (10.31) and the expression for $h(a, b, x)$ into (10.26) and the desired identity follows after straightforward algebra. □

In anticipation of the proof of Theorem 10.6, we shall note here a particular identity which follows easily from (10.31). That is, for $v \geq u \geq m \geq 0$,

$$\begin{aligned}\int_0^\infty \int_{(-\infty,-z)} & W^{(q)}(z+\theta+m) \\ &\times\left[\frac{\mathbb{W}^{(q)}(v-m-z)}{\mathbb{W}^{(q)}(v-m)}\mathbb{W}^{(q)}(u-m) - \mathbb{W}^{(q)}(u-m-z)\right]\Pi(\mathrm{d}\theta)\mathrm{d}z\end{aligned}$$

$$= -\frac{\mathbb{W}^{(q)}(u-m)}{\mathbb{W}^{(q)}(v-m)}\left(W^{(q)}(v) + \alpha\int_m^v \mathbb{W}^{(q)}(v-z)W^{(q)\prime}(z)\mathrm{d}z\right)$$

$$+ W^{(q)}(u) + \alpha\int_m^u \mathbb{W}^{(q)}(u-z)W^{(q)\prime}(z)\mathrm{d}z. \tag{10.33}$$

Proof of Theorem 10.6 Define for Borel $B \subseteq [0, a]$ and $x, q \geq 0$,

$$V^{(q)}(x, a, B) = \int_0^\infty \mathrm{e}^{-qt}\mathbb{P}_x\big(U_t \in B,\, t < \kappa_0^- \wedge \kappa_a^+\big)\mathrm{d}t.$$

For $x \leq b$, by the strong Markov property, Theorem 8.1 (iii) and Theorem 8.7, we have

$$V^{(q)}(x, a, B) = \mathbb{E}_x\left(\int_0^{\tau_b^+} \mathrm{e}^{-qt}\mathbf{1}_{\{U_t \in B, t<\kappa_a^+\wedge\kappa_0^-\}}\mathrm{d}t\right)$$

$$+ \mathbb{E}_x\left(\int_{\tau_b^+}^\infty \mathrm{e}^{-qt}\mathbf{1}_{\{U_t\in B, t<\kappa_a^+\wedge\kappa_0^-, \tau_b^+<\tau_0^-\}}\mathrm{d}t\right)$$

$$= \mathbb{E}_x\left(\int_0^{\tau_b^+\wedge\tau_0^-} \mathrm{e}^{-qt}\mathbf{1}_{\{X_t\in B\}}\mathrm{d}t\right)$$

$$+ \mathbb{E}_x\left(\mathrm{e}^{-q\tau_b^+}\mathbf{1}_{\{\tau_b^+<\tau_0^-\}}\right)V^{(q)}(b, a, B)$$

$$= \int_B \left(\frac{W^{(q)}(b-y)}{W^{(q)}(b)}W^{(q)}(x) - W^{(q)}(x-y)\right)\mathrm{d}y$$

$$+ \frac{W^{(q)}(x)}{W^{(q)}(b)}V^{(q)}(b, a, B). \tag{10.34}$$

Moreover, for $b \leq x \leq a$, we have, using similar arguments,

$$V^{(q)}(x, a, B)$$

$$= \int_0^\infty \mathrm{e}^{-qt}\mathrm{P}_x\big(Y_t \in B\cap[b, a], t < \tau_b^- \wedge \tau_a^+\big)\mathrm{d}t$$

$$+ \mathrm{E}_x\big(\mathbf{1}_{\{\tau_b^- < \tau_a^+\}}\mathrm{e}^{-q\tau_b^-}V^{(q)}(Y_{\tau_b^-}, a, B)\big)$$

$$= \int_{B\cap[b,a]}\left(\frac{\mathbb{W}^{(q)}(a-z)}{\mathbb{W}^{(q)}(a-b)}\mathbb{W}^{(q)}(x-b) - \mathbb{W}^{(q)}(x-z)\right)\mathrm{d}z$$

$$+ \int_0^\infty \int_{(-\infty,-z)}\left\{\int_B\left[\frac{W^{(q)}(b-y)}{W^{(q)}(b)}W^{(q)}(z+\theta+b)\right.\right.$$

$$\left. - W^{(q)}(z+\theta+b-y)\right]\mathrm{d}y$$

$$+\frac{V^{(q)}(b,a,B)}{W^{(q)}(b)}W^{(q)}(z+\theta+b)\bigg\}$$

$$\times\left[\frac{\mathbb{W}^{(q)}(a-b-z)}{\mathbb{W}^{(q)}(a-b)}\mathbb{W}^{(q)}(x-b)-\mathbb{W}^{(q)}(x-b-z)\right]\Pi(\mathrm{d}\theta)\mathrm{d}z,$$

where in the first equality, we have used the strong Markov property and in the second equality, we have used the identity in Exercise 10.6. Next, we shall apply the identity (10.33) twice in order to simplify the expression for $V^{(q)}(x,a,B)$, $a \geq x \geq b$. We use it once by setting $m=b$, $u=x$, $v=a$ and once by setting $m=b-y$ and $u=x-y$, $v=a-y$ for $y \in [0,b]$. We obtain

$$\begin{aligned}
&V^{(q)}(x,a,B)\\
&=\int_{B\cap[b,a]}\left(\frac{\mathbb{W}^{(q)}(a-z)}{\mathbb{W}^{(q)}(a-b)}\mathbb{W}^{(q)}(x-b)-\mathbb{W}^{(q)}(x-z)\right)\mathrm{d}z\\
&\quad+\int_{B\cap[0,b)}\left\{\frac{W^{(q)}(b-y)}{W^{(q)}(b)}\left(-\frac{\mathbb{W}^{(q)}(x-b)}{\mathbb{W}^{(q)}(a-b)}w^{(q)}(a;0)+w^{(q)}(x;0)\right)\right.\\
&\quad\left.-\left(-\frac{\mathbb{W}^{(q)}(x-b)}{\mathbb{W}^{(q)}(a-b)}w^{(q)}(a;y)+w^{(q)}(x;y)\right)\right\}\mathrm{d}y\\
&\quad+\frac{V^{(q)}(b,a,B)}{W^{(q)}(b)}\left(-\frac{\mathbb{W}^{(q)}(x-b)}{\mathbb{W}^{(q)}(a-b)}w^{(q)}(a;0)+w^{(q)}(x;0)\right).
\end{aligned}\tag{10.35}$$

Setting $x=b$ in (10.35), we get an expression for $V^{(q)}(b,a,B)$ in terms of itself. Solving this and then putting the resulting expression for $V^{(q)}(b,a,B)$ back in (10.34) and (10.35) leads to (10.24) which completes the proof. □

The two expressions we are interested in, namely the ruin probability and the expected present value of dividends paid until ruin, can both be extracted from the identity for the potential measure of U on $[0,\infty)$,

$$\int_0^\infty \mathbb{P}_x\big(U_t \in B, t<\kappa_0^-\big)\mathrm{d}t=\lim_{a\uparrow\infty}V(x,a,B),$$

where B is any Borel set in $[0,\infty)$. Note that the limit is justified by monotone convergence. In order to describe this potential measure, let us introduce some more notation. Recall that φ was defined as the right inverse of the Laplace exponent of Y, so that

$$\varphi(q)=\sup\{\theta\geq 0:\psi(\theta)-\alpha\theta=q\}.$$

Corollary 10.7 *For $x,y,b\geq 0$ and $q\geq 0$*

$$\int_0^\infty \mathrm{e}^{-qt}\mathbb{P}_x\big(U_t\in\mathrm{d}y,\ t<\kappa_0^-\big)\mathrm{d}t$$

$$= \mathbf{1}_{\{y\in[b,\infty)\}}\left\{\frac{w^{(q)}(x;0)}{\alpha\int_b^\infty \mathrm{e}^{-\varphi(q)z}W^{(q)\prime}(z)\mathrm{d}z}\mathrm{e}^{-\varphi(q)y} - \mathbb{W}^{(q)}(x-y)\right\}\mathrm{d}y$$

$$+ \mathbf{1}_{\{y\in[0,b)\}}\left\{\frac{\int_b^\infty \mathrm{e}^{-\varphi(q)z}W^{(q)\prime}(z-y)\mathrm{d}z}{\int_b^\infty \mathrm{e}^{-\varphi(q)z}W^{(q)\prime}(z)\mathrm{d}z}w^{(q)}(x;0) - w^{(q)}(x;y)\right\}\mathrm{d}y. \tag{10.36}$$

Proof Assume that $q > 0$. We begin by recalling from Exercise 8.5 that for all $x, q > 0$,

$$\lim_{a\uparrow\infty}\frac{\mathbb{W}^{(q)}(a-x)}{\mathbb{W}^{(q)}(a)} = \mathrm{e}^{-\varphi(q)x}.$$

Note that, for each $q \geq 0$, $\varphi(q) \geq \Phi(q)$ and hence, appealing to (10.9), it also follows that, for all $q, x > 0$,

$$\lim_{a\uparrow\infty}\frac{W^{(q)}(a-x)}{\mathbb{W}^{(q)}(a)} = 0.$$

For $q > 0$, the result we are after is obtained by dividing the numerator and denominator of each of the first terms in the curly brackets of (10.24) by $\mathbb{W}^{(q)}(a)$ and taking limits as $a \uparrow \infty$, making use of the above two observations. The case that $q = 0$ is handled by taking limits as $q \downarrow 0$ in (10.36). □

Now we are in a position to derive expressions for (10.21) and (10.22).

Corollary 10.8 *For $x \geq 0$, if $\mathbb{E}(X_1) \leq \alpha$ then $\mathbb{P}_x(\kappa_0^- < \infty) = 1$. Otherwise, when $\mathbb{E}(X_1) > \alpha$, we have*

$$\mathbb{P}_x\left(\kappa_0^- < \infty\right) = 1 - \frac{\mathbb{E}(X_1)-\alpha}{1-\alpha W(b)}\left(W(x) + \alpha\mathbf{1}_{(x\geq b)}\int_b^x \mathbb{W}(x-y)W'(y)\mathrm{d}y\right). \tag{10.37}$$

Proof Let $\underline{U}_t = \inf_{s\leq t} U_s$ and, as usual, $\mathbf{e}_q$ denotes an independent and exponentially distributed random variable with mean $1/q$. Note that for $q > 0$,

$$\mathbb{E}_x\left(\mathrm{e}^{-q\kappa_0^-}\mathbf{1}_{\{\kappa_0^-<\infty\}}\right) = 1 - \mathbb{P}_x(\underline{U}_{\mathbf{e}_q} \geq 0) = 1 - q\int_0^\infty \mathrm{e}^{-qt}\mathbb{P}_x\left(U_t \in [0,\infty), t < \kappa_0^-\right)\mathrm{d}t.$$

Computing the integral above from (10.36) is relatively straightforward and gives us, for $x, b \geq 0$ and $q > 0$,

$$\begin{aligned}
&\mathbb{E}_x\left(\mathrm{e}^{-q\kappa_0^-}\mathbf{1}_{\{\kappa_0^-<\infty\}}\right)\\
&\quad = z^{(q)}(x) - \frac{q\int_b^\infty \mathrm{e}^{-\varphi(q)y}W^{(q)}(y)\mathrm{d}y}{\int_b^\infty \mathrm{e}^{-\varphi(q)y}W^{(q)\prime}(y)\mathrm{d}y}w^{(q)}(x;0)\\
&\qquad +q\int_b^x \mathbb{W}^{(q)}(x-z)\mathrm{d}z + q\int_0^b W^{(q)}(x-z)\mathrm{d}z\\
&\qquad -q\int_0^x W^{(q)}(z)\mathrm{d}z - q\alpha\int_b^x \mathbb{W}^{(q)}(x-z)W^{(q)}(z-b)\mathrm{d}z, \qquad (10.38)
\end{aligned}$$

where

$$z^{(q)}(x) = Z^{(q)}(x) + \alpha q\int_b^x \mathbb{W}^{(q)}(x-z)W^{(q)}(z)\mathrm{d}z, \quad x\in\mathbb{R}, q\ge 0.$$

The details of the computation are left to the reader.

Although it is not immediately obvious, it turns out that the last four terms in (10.38) sum to zero. Indeed, in the case that $x\le b$, this observation is straightforward, noting that the two integrals from b to x are identically zero and the second integral may be replaced by $\int_0^x W^{(q)}(x-z)\mathrm{d}z = \int_0^x W^{(q)}(z)\mathrm{d}z$ on account of the fact that $W^{(q)}$ is identically zero on $(-\infty,0)$. In the case that $x>b$, the last four terms of (10.38) can be easily rearranged to be equal to $m(x-b)$, where $m:[0,\infty)\to[0,\infty)$ is the continuous function

$$m(u) = q\int_0^u \mathbb{W}^{(q)}(z)\mathrm{d}z - q\int_0^u W^{(q)}(z)\mathrm{d}z - q\alpha\int_0^u \mathbb{W}^{(q)}(z)W^{(q)}(u-z)\mathrm{d}z.$$

Taking Laplace transforms of m and using (8.8), we easily verify that m is identically zero.

In conclusion, we have that, for $x,b\ge 0$ and $q>0$,

$$\mathbb{E}_x\left(\mathrm{e}^{-q\kappa_0^-}\mathbf{1}_{\{\kappa_0^-<\infty\}}\right) = z^{(q)}(x) - \frac{q\int_b^\infty \mathrm{e}^{-\varphi(q)y}W^{(q)}(y)\mathrm{d}y}{\int_b^\infty \mathrm{e}^{-\varphi(q)y}W^{(q)\prime}(y)\mathrm{d}y}w^{(q)}(x;0).$$

The expression for the ruin probability in (10.37) is obtained by taking limits on the left- and right-hand side above as $q\downarrow 0$. On the left-hand side, thanks to monotone convergence, the limit is equal to $\mathbb{P}_x(\kappa_0^-<\infty)$. Computing the limits on the right-hand side is relatively straightforward, taking account of the fact that

$$\int_0^\infty \mathrm{e}^{-\varphi(q)z}W^{(q)}(z)\mathrm{d}z = \frac{1}{\varphi(q)\alpha} \qquad (10.39)$$

and the fact that

$$\lim_{q\downarrow 0}\frac{q}{\varphi(q)} = \lim_{q\downarrow 0}\frac{\psi(\varphi(q))-\alpha\varphi(q)}{\varphi(q)} = 0\vee\left(\mathbb{E}(X_1)-\alpha\right).$$

The details are again left to the reader. □

Corollary 10.9 *For $x \geq 0$,*

$$\mathbb{E}_x\left(\int_0^{\kappa_0^-} \mathrm{e}^{-qt}\alpha\mathbf{1}_{\{U_t>b\}}\mathrm{d}t\right)$$

$$= -\alpha\int_b^x \mathbb{W}^{(q)}(z-b)\mathrm{d}z + \frac{W^{(q)}(x)+\alpha\mathbf{1}_{(x\geq b)}\int_b^x \mathbb{W}^{(q)}(x-y)W^{(q)\prime}(y)\mathrm{d}y}{\varphi(q)\int_0^\infty \mathrm{e}^{-\varphi(q)y}W^{(q)\prime}(y+b)\mathrm{d}y}.$$

Proof The proof is a simple exercise in integrating the potential measure (10.36) over (b,∞). □

For the sake of completeness, let us finish this section by returning to the discussion at the beginning of the section, concerning the optimal control problem (10.12), and by describing the optimal strategy in a little more detail. We have already indicated that the optimal strategy is one that makes the controlled process a refracted Lévy process, where refraction occurs at some threshold $b \geq 0$. The value function of this "refraction strategy", henceforth denoted by v_b, is given in the previous corollary. If we now let

$$\Lambda(b) = \varphi(q)\int_0^\infty \mathrm{e}^{-\varphi(q)u}W^{(q)\prime}(u+b)\mathrm{d}u,$$

then we have

$$v_b(x) = \frac{W^{(q)}(x)}{\Lambda(b)}, \quad \text{for } x \leq b.$$

The familiarity of the above identity when compared to (10.13) is also mirrored by the description of the optimal value of b, denoted by b^*. Kyprianou et al. (2012b) show that, when $-\Pi(-\infty,-x)$, $x>0$ has a completely monotone density, b^* is the largest argument at which Λ attains its minimum. That is to say,

$$b^* = \sup\{b \geq 0 : \Lambda(b) \leq \Lambda(x) \text{ for all } x \geq 0\}.$$

It is also shown in Kyprianou et al. (2012b) that $b^* \leq a^*$, where we recall that a^* is the optimal threshold for the reflection strategy discussed in Sect. 10.3.

10.5 Perturbed Processes and Tax

In the setting of the classical Cramér–Lundberg risk insurance model, Albrecher and Hipp (2007) introduced the idea of tax payments. More precisely, if $X = \{X_t : t \geq 0\}$ represents the Cramér–Lundberg process and, for all $t \geq 0$, $\overline{X}_t = \sup_{s\leq t} X_s$, then the aforementioned authors study the process

$$X_t - \gamma\overline{X}_t, \quad t \geq 0,$$

where $\gamma \in (0,1)$ is the rate at which tax is paid. Intuitively speaking, since the process X increases whenever $\overline{X}$ increases, it follows that the Cramér–Lundberg process is taxed only when it generates new maxima. This is similar to the case of paying dividends according to a reflection strategy, but the requirement that $\gamma \in (0,1)$ ensures that, in principle, the tax paid does not stop the aggregate process from exploring arbitrarily large values with positive probability.

The above tax model was quickly generalised to the setting that X is a general spectrally negative Lévy process by Albrecher et al. (2008). Finally, Kyprianou and Zhou (2009) and Kyprianou and Ott (2012) extended this model further by allowing the rate at which tax is paid with respect to the process $\overline{X}$ to vary as a function of the current value of $\overline{X}$. Specifically, they consider the so-called *perturbed* spectrally negative Lévy process,

$$U_t = X_t - \int_{(0,t]} \gamma(\overline{X}_u)\,\mathrm{d}\overline{X}_u, \quad t \geq 0, \tag{10.40}$$

where $\gamma : [0,\infty) \to [0,\infty)$ satisfies appropriate conditions. The presentation we shall give here follows the last two references.

We distinguish two regimes, *light-* and *heavy-perturbation* regimes. The first corresponds to the case that $\gamma : [0,\infty) \to [0,1)$ and the second to the case that $\gamma : [0,\infty) \to (1,\infty)$. As alluded to previously, the light-perturbation regime has a similar flavour to paying dividends at a weaker rate than a reflection strategy. In contrast, the heavy-perturbation regime is equivalent to paying dividends at a much stronger rate than a reflection strategy. (The connection with the original motivation to model tax payments is arguably lost.) A little thought reveals that the dividing case $\gamma = 1$ corresponds precisely to a reflection strategy. In principle, it is also possible to consider the more general case that $\gamma : [0,\infty) \to [0,\infty)$ without the aforementioned restrictions, but this is mathematically less convenient than the two main regimes we have already identified.

The key observation, which, with the help of excursion theory, leads to all of the forthcoming results, is that we may write U in the form

$$U_t = A_t - (\overline{X}_t - X_t), \quad t \geq 0, \tag{10.41}$$

where the process $A = \{A_t : t \geq 0\}$ is given by

$$A_t := \overline{X}_t - \int_{(0,t]} \gamma(\overline{X}_u)\,\mathrm{d}\overline{X}_u, \quad t \geq 0. \tag{10.42}$$

Assuming that $X_0 = x$, we may write $A_t = \bar{\gamma}_x(\overline{X}_t)$, where

$$\bar{\gamma}_x(s) := s - \int_x^s \gamma(y)\,\mathrm{d}y = x + \int_x^s \big(1 - \gamma(y)\big)\mathrm{d}y, \quad s \geq x.$$

Noting that $A_t = \int_{(0,t]}(1 - \gamma(\overline{X}_s))\mathrm{d}\overline{X}_s$, $t \geq 0$, we see that, in the light-perturbation (resp. heavy-perturbation) regime, the process A has monotone increasing (resp. decreasing) paths. Let the set $\mathcal{A}$ consist of the points of increase (resp.

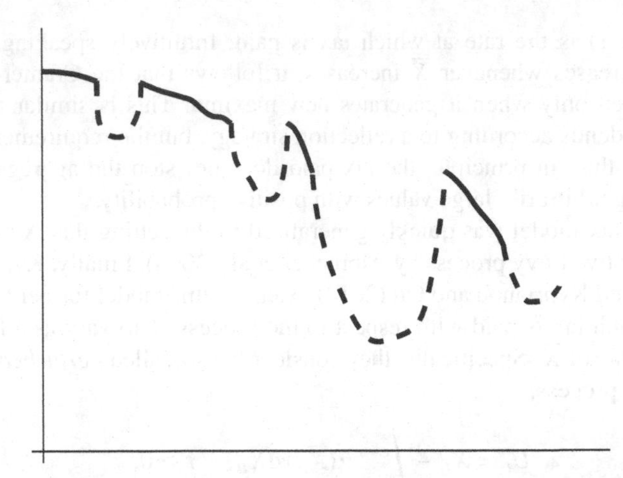

Fig. 10.2 A symbolic representation of the path of U in the case of heavy-perturbation. At times when $\overline{X}$ is increasing, the process U follows the path of $\bar{\gamma}_x$. During the open intervals of time that X executes an excursion away from its previous maximum, the process U undertakes the same excursion, but away from the current value of $A = \bar{\gamma}_x(\overline{X})$.

decrease) times of A. We have that $\mathcal{A}$ is contained in the support of the measure $\mathrm{d}\overline{X}$. If we write $\mathcal{B}$ for the countable union of open intervals of time which correspond to the epochs that the process $\overline{X} - X$ spends away from zero, then $\mathcal{A} \cap \mathcal{B} = \emptyset$. As a consequence, we may interpret (10.41) as a path decomposition in which excursions of X from its maximum (equivalently excursions of $\overline{X} - X$ away from zero) are "hung" off the trajectory of A between its increment (resp. decrement) times. See Fig. 10.2 for a symbolic representation when X is a Cramér–Lundberg process and there is heavy-perturbation.

It is also worth commenting that, in the light-perturbation regime, the process A coincides with $\{\sup_{s\le t} U_s : t \ge 0\}$; cf. Exercise 10.9. Hence, unless it is assumed that

$$\int_x^\infty \big(1 - \gamma(s)\big)\mathrm{d}s = \infty, \tag{10.43}$$

in the light-perturbation regime, the perturbed process U will have an almost surely finite global maximum. In contrast, in the heavy-perturbation regime, the process A coincides with $\{\sup_{s\ge t} U_s : t \ge 0\}$ (see again Exercise 10.9) and hence the process U is always bounded by its initial value x.

Let

$$T_0^- := \inf\{t > 0 : U_t < 0\},$$

where we understand, as usual, $\inf\emptyset := \infty$. We shall also use the stopping time

$$\tau_a^+ = \inf\{t > 0 : X_t > a\} = \inf\{t > 0 : \overline{X}_t > a\}.$$

Note that in the light-perturbation case, the function $\bar{\gamma}_x$ is increasing and hence it has a well-defined inverse, say $\bar{\gamma}_x^{-1}$. In that case, we may write for all values b in the range of $\bar{\gamma}_x$,

$$\tau^+_{\bar{\gamma}_x^{-1}(b)} = T_b^+, \tag{10.44}$$

where $T_b^+ = \inf\{t > 0 : U_t > b\}$.

Theorem 10.10 *Fix $x > 0$ and assume* (10.43) *in the case of the light-perturbation regime. In the case of the heavy-perturbation regime, noting that, $\bar{\gamma}_x$ is monotone decreasing, define*

$$s^*(x) = \inf\{s \geq x : \bar{\gamma}_x(s) < 0\}.$$

Then, for any $q \geq 0$, and $0 \leq x \leq a$ in the case of light-perturbation, resp. $0 \leq x \leq a < s^*(x)$ in the case of heavy-perturbation, we have

$$\mathbb{E}_x\big[\mathrm{e}^{-q\tau_a^+}\mathbf{1}_{\{\tau_a^+ < T_0^-\}}\big] = \exp\bigg(-\int_x^a \frac{W^{(q)\prime}(\bar{\gamma}_x(s))}{W^{(q)}(\bar{\gamma}_x(s))}\,\mathrm{d}s\bigg). \tag{10.45}$$

Taking account of the equivalence (10.44) in the light-perturbation regime, (10.45) can be more conveniently written as

$$\mathbb{E}_x\big[\mathrm{e}^{-qT_a^+}\mathbf{1}_{\{T_a^+ < T_0^-\}}\big] = \exp\bigg(-\int_x^{\bar{\gamma}_x^{-1}(a)} \frac{W^{(q)\prime}(\bar{\gamma}_x(s))}{W^{(q)}(\bar{\gamma}_x(s))}\,\mathrm{d}s\bigg).$$

Proof of Theorem 10.10 The proof does not distinguish between the two different regimes of light- and heavy-perturbation. All that is required in what follows is that $\bar{\gamma}_x^{-1}(a) < \infty$.

Recall from Chap. 6 that $\{(t, \epsilon_t) : t \geq 0 \text{ and } \epsilon_t \neq \partial\}$ is the Poisson point process of excursions on $[0, \infty) \times \mathcal{E}$ with intensity $\mathrm{d}t \times \mathrm{d}n$, indexed by local time. For $x \geq 0$, the connection between local time at zero of $\{\overline{X}_t - X_t : t \geq 0\}$, denoted by $\{L_t : t \geq 0\}$, and real time under $\mathbb{P}_x$ is given by $L_t = \overline{X}_t - x$, $t \geq 0$. Note in particular that, again under $\mathbb{P}_x$, $\tau_a^+ = L_{a-x}^{-1}$. Write $\overline{\epsilon}$ for the height of the canonical excursion $\epsilon \in \mathcal{E}$; see Definition 6.13. Note that, in terms of excursions, the event $\{\tau_a^+ < T_0^-\}$ corresponds precisely to the event

$$\{\overline{\epsilon}_s \leq \bar{\gamma}_x(x+s) \text{ for all } 0 \leq s \leq a-x\}.$$

Using similar reasoning to that found in the computation (8.21), it follows, with the help of the exponential change of measure (8.5) applied at the stopping time L_{a-x}^{-1} and the identity (10.19), that for $x \geq 0$,

$$\begin{aligned}\mathbb{E}_x\big[\mathrm{e}^{-q\tau_a^+}\mathbf{1}_{\{\tau_a^+ < T_0^-\}}\big] &= \mathbb{E}_x\big[\mathrm{e}^{-qL_{a-x}^{-1}}\mathbf{1}_{\{\overline{\epsilon}_s \leq \bar{\gamma}_x(x+s) \text{ for all } 0\leq s\leq a-x\}}\big]\\ &= \mathrm{e}^{-(a-x)\Phi(q)}\mathbb{P}_x^{\Phi(q)}\big(\overline{\epsilon}_s \leq \bar{\gamma}_x(x+s) \text{ for all } 0 \leq s \leq a-x\big)\end{aligned}$$

$$
\begin{aligned}
&= \mathrm{e}^{-(a-x)\Phi(q)} \exp\left(-\int_0^{a-x} n_{\Phi(q)}(\overline{\epsilon} > \bar{\gamma}_x(x+s))\mathrm{d}s\right) \\
&= \exp\left(-\int_0^{a-x} \frac{W^{(q)\prime}(\bar{\gamma}_x(x+s))}{W^{(q)}(\bar{\gamma}_x(x+s))}\mathrm{d}s\right). \qquad (10.46)
\end{aligned}
$$

The required identity follows after a straightforward change of variables in the final integral. □

Theorem 10.10 motivates some interesting observations concerning the event of ruin, $\{T_0^- < \infty\}$. First, suppose that we are in the heavy-perturbation regime and $s^*(x) < \infty$. In that case

$$
\mathbb{P}_x(T_0^- < \infty) \geq \mathbb{P}_x(\tau^+_{s^*(x)} < \infty) \vee \mathbb{P}_x(\tau_0^- < \infty).
$$

Indeed, on the event $\{\tau^+_{s^*(x)} < \infty\}$, we have $\overline{X}_{\tau^+_{s^*(x)}} - X_{\tau^+_{s^*(x)}} = 0$ and hence $U_{\tau^+_{s^*(x)}} = A_{\tau^+_{s^*(x)}} = \bar{\gamma}_x(s^*(x)) = 0$. Moreover, since $U_t \leq X_t$ for all $t \geq 0$, it follows that $\{\tau_0^- < \infty\} \subseteq \{T_0^- < \infty\}$. In the event that $\limsup_{t\uparrow\infty} X_t = \infty$ almost surely, we have $\mathbb{P}_x(\tau^+_{s^*(x)} < \infty) = 1$. Otherwise, it follows that $\mathbb{P}_x(\tau_0^- < \infty) = 1$. Either way, $\mathbb{P}_x(T_0^- < \infty) = 1$.

Remaining in the heavy-perturbation regime, suppose that $s^*(x) = \infty$. Then from (10.45), by taking limits as $a \uparrow \infty$, we get an expression for the ruin probability,

$$
\mathbb{P}_x(T_0^- < \infty) = 1 - \exp\left(-\int_x^\infty \frac{W'(\bar{\gamma}_x(s))}{W(\bar{\gamma}_x(s))}\mathrm{d}s\right). \qquad (10.47)
$$

However, the right-hand side above turns out to be equal to 1. Recalling that $W'(x)/W(x) = n(\overline{\epsilon} > x)$ for almost every $x > 0$, since $n(\overline{\epsilon} > x)$ is non-increasing on $(0, \infty)$ and $\bar{\gamma}_x(s) \leq x$ for all $s \geq 0$, the claim follows.

Finally, in the light-perturbation regime, where necessarily $s^*(x) = \infty$, the reasoning that leads to (10.47) still applies. Exercise 10.10 shows that this probability need not be unity.

Although the perturbed process is almost surely ruined in the heavy-perturbation regime, it is interesting to note that, unlike regular spectrally negative Lévy processes, there are three different ways to become ruined. The first two, i.e. by a jump or creeping downwards (in the presence of a Gaussian component), are properties inherited from the underlying Lévy process. The third way of becoming ruined, which we refer to as *type II* creeping, is the result of continuously passing the origin at the moment in time that an increment in $\overline{X}$ brings U along the curve $\bar{\gamma}_x$ just as it intersects the origin. Said another way, type II creeping corresponds to the event that $\{\tau^+_{s^*(x)} = T_0^-\}$, in which case, as remarked upon above, $U_{\tau^+_{s^*(x)}} = 0$. This can only happen with positive probability if $s^*(x) < \infty$.

The following result is a corollary to Theorem 10.10 on account of the fact that its proof is identical, albeit that one replaces a by $s^*(x)$.

Corollary 10.11 *Fix $x > 0$, and suppose that $\gamma : [0, \infty) \to (1, \infty)$ such that $s^*(x) < \infty$. Then, for all $q \geq 0$,*

$$\mathbb{E}_x\big[\mathrm{e}^{-qT_0^-}1_{\{T_0^-=\tau^+_{s^*(x)}\}}\big] = \exp\bigg(-\int_x^{s^*(x)} \frac{W^{(q)\prime}(\bar{\gamma}_x(s))}{W^{(q)}(\bar{\gamma}_x(s))}\,\mathrm{d}s\bigg).$$

The above corollary tells us that, under mild conditions, the probability of type II creeping is strictly positive if and only if

$$\int_x^{s^*(x)} \frac{W'(\bar{\gamma}_x(s))}{W(\bar{\gamma}_x(s))}\,\mathrm{d}s < \infty.$$

One easily sees that type II creeping may occur in the case that X is a Cramér–Lundberg process. Indeed, if the first jump of X occurs after the time it takes X to climb to a height $s^*(x)$ from the initial position $x > 0$, then type II creeping will trivially occur. One may easily elaborate on this reasoning to deduce that, for all $n \in \mathbb{N}$, type II creeping may occur with positive probability between the n-th and $(n+1)$-th jumps.

Exercise 10.11 deals with a number of scenarios where type II creeping can happen. In particular, under relatively mild assumptions, there will be type II creeping if and only if X has paths of bounded variation.

In the spirit of the Gerber–Shiu-type results presented in the previous sections, our final theorem for perturbed processes (with either light- or heavy-perturbation) considers the present value of dividends (or tax, as appropriate with the interpretation of the perturbation) paid until ruin.

Theorem 10.12 *Fix $x > 0$ and assume* (10.43) *in the case of the light-perturbation regime. Then, for $q \geq 0$,*

$$\mathbb{E}_x\bigg[\int_0^{T_0^-} \mathrm{e}^{-qu}\gamma(\overline{X}_u)\,\mathrm{d}\overline{X}_u\bigg] = \int_x^{s^*(x)} \exp\bigg(-\int_x^t \frac{W^{(q)\prime}(\bar{\gamma}_x(s))}{W^{(q)}(\bar{\gamma}_x(s))}\,\mathrm{d}s\bigg)\gamma(t)\,\mathrm{d}t.$$

Proof Appealing to a straightforward change of variables and Fubini's Theorem, we have

$$\begin{aligned}
\mathbb{E}_x\bigg[\int_0^{T_0^-} \mathrm{e}^{-qu}\gamma(\overline{X}_u)\mathrm{d}\overline{X}_u\bigg] &= \mathbb{E}_x\bigg[\int_0^{s^*(x)} \mathbf{1}_{(u<T_0^-)}\mathrm{e}^{-qu}\gamma(\overline{X}_u)\mathrm{d}\overline{X}_u\bigg] \\
&= \mathbb{E}_x\bigg[\int_x^{s^*(x)} \mathbf{1}_{(\tau_t^+<T_0^-)}\mathrm{e}^{-q\tau_t^+}\gamma(t)\mathrm{d}t\bigg] \\
&= \int_x^{s^*(x)} \mathbb{E}_x\big[\mathrm{e}^{-q\tau_t^+}\mathbf{1}_{(\tau_t^+<T_0^-)}\big]\gamma(t)\mathrm{d}t.
\end{aligned}$$

The proof is completed by taking advantage of the identity in (10.45). □

Exercises

10.1 Using an exponential change of measure together with (10.3), show that, for $x, q \geq 0$,

$$\mathbb{E}_x(\mathrm{e}^{-q\tau_0^-}; X_{\tau_0^-} = 0) = \mathbb{E}_x(\mathrm{e}^{-q\tau_0^-}; X_{\tau_0^-} = X_{\tau_0^- -} = \underline{X}_{\tau_0^-} = 0)$$
$$= \frac{\sigma^2}{2}\{W^{(q)\prime}(x) - \Phi(q)W^{(q)}(x)\},$$

where the right-hand side is understood to be zero when $\sigma = 0$.

10.2 Find an expression for the Gerber–Shiu measure in Theorem 10.1 for the case that X has paths of bounded variation and $x = 0$.

10.3 The following exercise is based on results found in Huzak et al. (2004b). Suppose that X is a Lévy insurance risk process. In particular, we will assume that $X = \sum_{i=1}^n X^{(i)}$, where each of the $X^{(i)}$ are independent spectrally negative Lévy processes with respective Lévy measures, $\Pi^{(i)}$, concentrated on $(-\infty, 0)$. One may think of them as competing risk processes.

(i) With the help of the compensation formula, show that, for $x \geq 0$, $y > 0$, $u < 0$ and $i = 1, \ldots, n$,

$$\mathbb{P}_x(X_{\tau_0^-} \in \mathrm{d}u,\ X_{\tau_0^- -} \in \mathrm{d}y,\ \Delta X_{\tau_0^-} = \Delta X^{(i)}_{\tau_0^-})$$
$$= r(x, y)\Pi^{(i)}(-y + \mathrm{d}u)\mathrm{d}y,$$

where $r(x, y)$ is the potential density of the process killed on first passage into $(-\infty, 0)$ given in Corollary 8.8.

(ii) Suppose now that $x = 0$ and each of the processes $X^{(i)}$ is of bounded variation. Recall that any such spectrally negative Lévy process is the difference of a linear drift and a driftless subordinator. Let δ be the drift of X. Show that for $y > 0$, $u < 0$,

$$\mathbb{P}(X_{\tau_0^-} \in \mathrm{d}u,\ X_{\tau_0^- -} \in \mathrm{d}y,\ \Delta X_{\tau_0^-} = \Delta X^{(i)}_{\tau_0^-})$$
$$= \frac{1}{\delta}\Pi^{(i)}(-y + \mathrm{d}u)\mathrm{d}y.$$

(iii) For each $i = 1, \ldots, n$, let δ_i be the drift of $X^{(i)}$. Note that, necessarily, $\delta = \sum_{i=1}^n \delta_i$. Suppose further that for each $i = 1, \ldots, n$, $\mu_i := \delta_i - \mathbb{E}(X_1^{(i)}) < \infty$. Show that the probability that ruin occurs as a result of a claim from the i-th process when $x = 0$ is equal to μ_i/δ.

10.4 Suppose that $S = \{S_t : t \geq 0\}$ is a subordinator, with Laplace exponent $\Phi(q) = t^{-1}\log \mathbb{E}(\exp\{-qS_t\})$, $t \geq 0$, and $\mathbf{e}_\kappa$ is an independent exponentially distributed random variable with rate $\kappa > 0$.

(i) Use the ideas in the proof of Theorem 10.3 to deduce that

$$\mathbb{E}\left[\left(\int_0^{\mathbf{e}_\kappa} \mathrm{e}^{-qS_t}\mathrm{d}t\right)^n\right] = n!\prod_{k=1}^n \frac{1}{\kappa + \Phi(qk)}.$$

(ii) Explain how part (i) above can be used to rephrase the proof of Theorem 10.3.

10.5 Suppose that X is a Cramér–Lundberg process with premium rate $c > 0$, compound Poisson arrival rate $\lambda > 0$ and claim distribution F with mean value μ. In the notation of Theorem 10.3, define

$$V_a = \mathbb{E}_a\left[\int_0^{\sigma^a} \mathrm{e}^{-qt}\mathrm{d}\xi_t^a\right].$$

(i) By conditioning on the first jump of X, show that

$$V_a = \frac{c}{\lambda + q} + V_a \frac{\lambda}{\lambda + q}\int_{(0,a]} \frac{W^{(q)}(a-y)}{W^{(q)}(a)} F(\mathrm{d}y).$$

(ii) Show by means of taking Laplace transforms that, for all $q \geq 0$ and $a > 0$,

$$cW_+^{(q)\prime}(a) = (\lambda + q)W^{(q)}(a) - \lambda\int_{(0,a]} W^{(q)}(a-y)F(\mathrm{d}y).$$

(iii) Use parts (i) and (ii) to prove Theorem 10.3 in the case that $n = 1$.

10.6 Use reasoning similar to that of the proof of (8.34) to deduce the following result. Let $a > 0$, $x \in [0,a]$, $q \geq 0$ and f, g be positive, bounded measurable functions. Further suppose that either X has no Gaussian component or it has a Gaussian component and $f(0)g(0) = 0$. Then

$$\begin{aligned}&\mathbb{E}_x(\mathrm{e}^{-q\tau_0^-} f(X_{\tau_0^-})g(X_{\tau_0^- -})\mathbf{1}_{\{\tau_0^- < \tau_a^+\}})\\ &= \int_0^a \int_{(-\infty,-y)} f(y+\theta)g(y)\left\{\frac{W^{(q)}(x)W^{(q)}(a-y)}{W^{(q)}(a)} - W^{(q)}(x-y)\right\}\Pi(\mathrm{d}\theta)\mathrm{d}y.\end{aligned}$$

10.7 Show that, for $q \geq 0$ and $0 \leq x, b \leq a$, we have for the refracted process (10.20),

$$\mathbb{E}_x\left(\mathrm{e}^{-q\kappa_0^-}\mathbf{1}_{\{\kappa_0^- < \kappa_a^+\}}\right) = z^{(q)}(x) - z^{(q)}(a)\frac{w^{(q)}(x;0)}{w^{(q)}(a;0)}.$$

10.8 Suppose that X is a spectrally negative Lévy process with bounded variation paths satisfying (10.1). Write, as usual, ψ for its Laplace exponent and Φ for the right inverse of ψ. Thinking of X as a Lévy insurance risk process, we may have the following adjusted definition of ruin. Every time the process X becomes negative, an independent and exponentially distributed clock is started with parameter $q \geq 0$. If the process X recovers and enters $(0, \infty)$ before this clock rings, then the insurance company may continue without becoming ruined. If, however, the process X spends longer below zero than it takes the associated exponential clock to ring, then the process is declared ruined.

(i) Explain why the probability of ruin (according to the new definition) may now be written $1 - V$ where

$$V := \mathbb{E}\left(\mathrm{e}^{-q \int_0^\infty \mathbf{1}_{\{X_s < 0\}} \mathrm{d}s}\right), \quad x \geq 0.$$

(ii) Show that

$$V = \mathbb{E}\left(\mathbf{1}_{\{\tau_0^- < \infty\}} g(X_{\tau_0^-})\right) V + \mathbb{P}(\tau_0^- = \infty),$$

where, for $x \leq 0$, $g(x) = \mathbb{E}_x(\mathrm{e}^{-q\tau_0^+})$. Hence deduce that

$$V = \psi'(0+) \frac{\Phi(q)}{q}.$$

(iii) Now suppose that the drift term of X is denoted δ as in (10.2) and let U be the associated refracted process as in Sect. 10.4, where the threshold for refraction is b and α is the rate of refraction. Using ideas similar to those found in the previous parts of this question, show that, when $\psi'(0+) > \alpha$ and $q \geq 0$,

$$\mathbb{E}_b\left(\mathrm{e}^{-q \int_0^\infty \mathbf{1}_{\{U_s < b\}} \mathrm{d}s}\right) = \frac{(\psi'(0+) - \alpha)\Phi(q)}{q - \alpha\Phi(q)}.$$

10.9 Consider the perturbed spectrally negative Lévy process (10.40). Suppose that we write $\overleftarrow{U}_t = \sup_{s \leq t} U_s$, $t \geq 0$, in the light-perturbation regime and $\overrightarrow{U}_t = \sup_{s \geq t} U_s$, $t \geq 0$, in the heavy-perturbation regime. Show that both of these processes agree with the definition of the process A in (10.42).

10.10 This exercise reproduces the results of Albrecher and Hipp (2007) and Albrecher et al. (2008) for the case of constant light-perturbation. Suppose that U is a perturbed spectrally negative Lévy process with *constant* tax rate $\gamma \in (0, 1)$. Show that, for all $a \geq x$,

$$\mathbb{E}_x\left[\mathrm{e}^{-qT_a^+} \mathbf{1}_{\{T_a^+ < T_0^-\}}\right] = \left(\frac{W^{(q)}(x)}{W^{(q)}(a)}\right)^{1/(1-\gamma)}$$

and give an expression for the probability of ruin. Show also that

$$\mathbb{E}_x\left[\int_0^{T_0^-} \mathrm{e}^{-qu} \gamma(\overline{X}_u)\, \mathrm{d}\overline{X}_u\right] = \frac{\gamma}{1-\gamma} \int_x^\infty \left(\frac{W^{(q)}(x)}{W^{(q)}(u)}\right)^{1/(1-\gamma)} \mathrm{d}u$$

and derive an expression for the last expectation in the case that γ is a constant in $(1, \infty)$.

10.11 This exercise is based on computations found in Kyprianou and Ott (2012). Consider the perturbed process U for the heavy-perturbation regime with $U_0 = x > 0$. Assume that $s^*(x) < \infty$.

(i) Suppose that γ is a continuous function. Show that U exhibits type II creeping under $\mathbb{P}_x$ if and only if X has paths of bounded variation.
(ii) Fix $c > 0$. Define for $s \in [0, c]$,

$$\gamma(s) = 1 + \frac{1}{2}(c - s)^{-\frac{1}{2}}.$$

Show that there exists an $x > 0$ such that $\bar{\gamma}_x(s) = x - \int_x^s \gamma(y)\mathrm{d}y = (c - s)^{1/2}$ with $s^*(x) = c$. Let U be the associated perturbed process such that the underlying Lévy process has a non-zero Gaussian component and $U_0 = x$. Show that type II creeping can occur.

Chapter 11
Applications to Optimal Stopping Problems

The aim of this chapter is to show how some of the established fluctuation identities for (reflected) Lévy processes can be used to solve quite specific, but nonetheless exemplary, optimal stopping problems. To some extent, this will be done in an unsatisfactory way, *without* first giving a thorough account of the general theory of optimal stopping. However, we shall give rigorous proofs relying on the method of "guess and verify". That is to say, our proofs will start with a candidate solution, the choice of which is inspired by intuition, and then we shall prove that this candidate verifies sufficient conditions in order to confirm its status as the actual solution. For a more complete overview of the theory of optimal stopping the reader is referred to the main three texts, Chow et al. (1971), Shiryaev (1978) and Peskir and Shiryaev (2006); see also Chap. 10 of Øksendal (2003) and Chap. 2 of Øksendal and Sulem (2004), as well as the foundational work of Snell (1952) and Dynkin (1963).

The optimal stopping problems we consider in this chapter will be of the form

$$v(x) = \sup_{\tau\in\mathcal{T}} \mathbb{E}_x\big(\mathrm{e}^{-q\tau}G(X_\tau)\big), \quad x\in\mathbb{R}, \tag{11.1}$$

or variants thereof, where $X=\{X_t : t\geq 0\}$ is a Lévy process. Further, G is a non-negative measurable function, $q\geq 0$ and $\mathcal{T}$ is a family of stopping times with respect to the filtration $\mathbb{F}$. Note that, when talking of a solution to (11.1), it is understood that we want to characterise the function v as well as finding a stopping time τ^* such that $v(x)=\mathbb{E}_x(\mathrm{e}^{-q\tau^*}G(X_{\tau^*}))$, for all $x\in\mathbb{R}$.

11.1 Sufficient Conditions for Optimality

Here, we give sufficient conditions under which one may verify that a candidate solution, i.e. a pair (v^*,τ^*), solves the optimal stopping problem (11.1).

A.E. Kyprianou, *Fluctuations of Lévy Processes with Applications*, Universitext,
DOI 10.1007/978-3-642-37632-0_11, © Springer-Verlag Berlin Heidelberg 2014

Lemma 11.1 *Consider the optimal stopping problem* (11.1) *for $q \geq 0$ under the assumption that, for all $x \in \mathbb{R}$,*

$$\mathbb{P}_x\Big(\text{there exists } \lim_{t\uparrow\infty} \mathrm{e}^{-qt} G(X_t) < \infty\Big) = 1. \tag{11.2}$$

Suppose that $\tau^* \in \mathcal{T}$ is a candidate optimal strategy for the optimal stopping problem (11.1) *and let $v^*(x) = \mathbb{E}_x(\mathrm{e}^{-q\tau^*} G(X_{\tau^*}))$, $x \in \mathbb{R}$. Then the pair (v^*, τ^*) is a solution if*

(i) $v^*(x) \geq G(x)$ *for all $x \in \mathbb{R}$,*
(ii) *the process $\{\mathrm{e}^{-qt} v^*(X_t) : t \geq 0\}$ is a right-continuous supermartingale.*

Proof The definition of v^* implies that

$$\sup_{\tau\in\mathcal{T}} \mathbb{E}_x\big(\mathrm{e}^{-q\tau} G(X_\tau)\big) \geq v^*(x),$$

for all $x \in \mathbb{R}$. On the other hand, property (ii) together with Doob's Optional Sampling Theorem[1] imply that, for all $t \geq 0$, $x \in \mathbb{R}$ and $\sigma \in \mathcal{T}$,

$$v^*(x) \geq \mathbb{E}_x\big(\mathrm{e}^{-q(t\wedge\sigma)} v^*(X_{t\wedge\sigma})\big),$$

and hence, by property (i), Fatou's Lemma, the non-negativity of G and assumption (11.2), we have

$$\begin{aligned} v^*(x) &\geq \liminf_{t\uparrow\infty} \mathbb{E}_x\big(\mathrm{e}^{-q(t\wedge\sigma)} G(X_{t\wedge\sigma})\big) \\ &\geq \mathbb{E}_x\big(\liminf_{t\uparrow\infty} \mathrm{e}^{-q(t\wedge\sigma)} G(X_{t\wedge\sigma})\big) \\ &= \mathbb{E}_x\big(\mathrm{e}^{-q\sigma} G(X_\sigma)\big). \end{aligned}$$

As $\sigma \in \mathcal{T}$ is arbitrary, it follows that, for all $x \in \mathbb{R}$,

$$v^*(x) \geq \sup_{\tau\in\mathcal{T}} \mathbb{E}_x\big(\mathrm{e}^{-q\tau} G(X_\tau)\big).$$

In conclusion, it must hold that

$$v^*(x) = \sup_{\tau\in\mathcal{T}} \mathbb{E}_x\big(\mathrm{e}^{-q\tau} G(X_\tau)\big)$$

for all $x \in \mathbb{R}$. □

Note that, when $\mathcal{T}$ contains only almost surely finite stopping times, a brief review of the above proof shows that the condition (11.2) is unnecessary.

When G is a monotone function and $q > 0$, a reasonable class of candidate solutions that one may consider in conjunction with the previous lemma is those based

[1] Right-continuity of paths is implicitly used here.

on first-passage times over a specified threshold. That is, either first passage above a given constant in the case that G is monotone increasing or first passage below a given constant in the case that G is monotone decreasing. A heuristic justification may be given as follows.

Suppose that G is monotone increasing. In order to optimise the value $G(X_\tau)$, one should stop at some time τ for which X_τ is large. On the other hand, this should not happen after too much time on account of the exponential discounting. This suggests that there is a threshold, which may depend on time, over which one should stop X in order to maximise the expected discounted gain. Suppose that by time $t > 0$ one has not reached this threshold. Then, by the Markov property, given $X_t = x$, any stopping time τ which depends only on the continuation of the path of X from the space-time point (x, t) would yield an expected gain $\mathrm{e}^{-qt}\mathbb{E}_x(\mathrm{e}^{-q\tau}G(X_\tau))$. The optimisation of this expression over the aforementioned class of stopping times is essentially the same procedure as in the original problem (11.1). Note that, since X is a Markov process, there is nothing to be gained by considering stopping times which take account of the history of the process $\{X_s : s < t\}$. These arguments suggest that the threshold should not vary with time, and hence a candidate for the optimal strategy takes the form

$$\tau^* = \inf\{t > 0 : X_t \in A\},$$

where $A = [y, \infty)$ or (y, ∞) for some $y \in \mathbb{R}$. Similar heuristic reasoning applies when G is monotone decreasing. The reader should be warned, however, that if one were to try and make these arguments rigorous, one would need to impose more conditions on G than just monotonicity.

When $q = 0$ and G is monotone increasing, it may be optimal to never stop. To avoid this case, we impose the added assumption that $\limsup_{t\uparrow\infty} X_t < \infty$ almost surely. In that case, we may again expect to describe the optimal stopping strategy as first passage above a threshold. The reason for this is that we cannot use a stopping time to stop when the Lévy process is at its all-time maximum.[2] Again, the threshold should be time-invariant due to the Markov property. If $q = 0$ and G is monotone decreasing, then, in light of the aforementioned, we may impose the condition that $\liminf_{t\uparrow\infty} X_t > -\infty$ almost surely and expect to see an optimal strategy consisting of first passage below a time-invariant threshold.

11.2 The McKean Optimal Stopping Problem

This optimal stopping problem is given by

$$v(x) = \sup_{\tau\in\mathcal{T}} \mathbb{E}_x\big(\mathrm{e}^{-q\tau}\big(K - \mathrm{e}^{X_\tau}\big)^+\big), \quad x \in \mathbb{R}, \tag{11.3}$$

[2] See, however, Baurdoux and van Schaik (2012) who investigate the problem of stopping as "close" the maximum as possible in an appropriate sense.

where $\mathcal{T}$ is the set of all $\mathbb{F}$-stopping times, $K > 0$ and, in the current context, we consider the two cases:

$$q > 0 \quad \text{or} \quad q = 0 \quad \text{and} \quad \lim_{t\uparrow\infty} X_t = \infty \quad \text{a.s.} \tag{11.4}$$

The solution to this optimal stopping problem was considered first by McKean (1965) for the case that X is linear Brownian motion. The original motivation for this problem comes from the valuation of the so-called *perpetual American put option.* This is a financial derivative which gives the holder the right, but not the obligation, to sell a risky asset (here modelled by an exponential Lévy process) for a fixed price K, at any time in the future. It turns out that the valuation of this contract boils down to solving (11.3).

In Darling et al. (1972), a solution to a discrete-time analogue of (11.3) was obtained. In that case, the process X is replaced by a random walk. Some years later, and again within the context of the optimal time to sell a risky asset (the pricing of an American put), a number of authors dealt with the solution to (11.3) for a variety of special classes of Lévy processes.[3] Below, we give the solution to (11.3) as presented in Mordecki (2002). The proof we shall give here comes from Alili and Kyprianou (2005) and remains close in nature to the random walk proofs of Darling et al. (1972). More recently, Baurdoux (2013) offers an interesting alternative perspective on our presentation.

Theorem 11.2 *The solution to* (11.3) *under the assumption* (11.4) *is given by*

$$v(x) = \frac{\mathbb{E}((K\mathbb{E}(e^{\underline{X}_{\mathbf{e}_q}}) - e^{x+\underline{X}_{\mathbf{e}_q}})^+)}{\mathbb{E}(e^{\underline{X}_{\mathbf{e}_q}})}$$

and an optimal stopping time is given by

$$\tau^* = \inf\{t > 0 : X_t < x^*\}$$

where

$$x^* = \log K\mathbb{E}\big(e^{\underline{X}_{\mathbf{e}_q}}\big).$$

Here, as usual, $\mathbf{e}_q$ denotes a $1/q$-mean, independent and exponentially distributed random variable, with the understanding that, when $q = 0$, this variable takes the value infinity with probability one. Further, $\underline{X}_t = \inf_{s\le t} X_s$. Note that in the case

[3]Gerber and Shiu (1994) dealt with the case of bounded variation spectrally positive Lévy processes; Boyarchenko and Levendorskii (2002a) handled a class of tempered stable processes; Chan (2004) covers the case of spectrally negative processes; Avram et al. (2002, 2004) deal with spectrally negative Lévy processes again; Asmussen et al. (2004) look at Lévy processes which have phase-type jumps and Chesney and Jeanblanc (2004) again for the spectrally negative case.

that $q = 0$, as we have assumed that $\lim_{t\uparrow\infty} X_t = \infty$, by Theorem 7.1, we know that $|\underline{X}_\infty| < \infty$ almost surely.

Proof of Theorem 11.2 We begin by noting that the assumption (11.2) is trivially satisfied. In view of the remarks following Lemma 11.1, let us define the bounded functions

$$v_y(x) = \mathbb{E}_x\big(\mathrm{e}^{-q\tau_y^-}\big(K - \mathrm{e}^{X_{\tau_y^-}}\big)^+\big), \quad x, y \in \mathbb{R}. \tag{11.5}$$

We shall show that the solution to (11.3) is of the form (11.5), for a suitable choice of $y \le \log K$, by using Lemma 11.1.

According to the conclusion of Exercise 6.7 (i), we have that

$$\mathbb{E}_x\big(\mathrm{e}^{-\alpha\tau_y^- + \beta X_{\tau_y^-}}\mathbf{1}_{(\tau_y^- < \infty)}\big) = \mathrm{e}^{\beta x}\frac{\mathbb{E}(\mathrm{e}^{\beta \underline{X}_{\mathbf{e}_\alpha}}\mathbf{1}_{(-\underline{X}_{\mathbf{e}_\alpha} > x-y)})}{\mathbb{E}(\mathrm{e}^{\beta\underline{X}_{\mathbf{e}_\alpha}})}, \tag{11.6}$$

for $\alpha, \beta \ge 0$ and $x - y \ge 0$, and hence it follows that, for all $x, y \in \mathbb{R}$,

$$v_y(x) = \frac{\mathbb{E}((K\mathbb{E}(\mathrm{e}^{\underline{X}_{\mathbf{e}_q}}) - \mathrm{e}^{x+\underline{X}_{\mathbf{e}_q}})\mathbf{1}_{(-\underline{X}_{\mathbf{e}_q} > x-y)})}{\mathbb{E}(\mathrm{e}^{\underline{X}_{\mathbf{e}_q}})}. \tag{11.7}$$

Lower bound (i). The lower bound $v_y(x) \ge (K - \mathrm{e}^x)^+$, $x \in \mathbb{R}$, is respected if and only if $v_y(x) \ge 0$ and $v_y(x) \ge (K - \mathrm{e}^x)$, for all $x \in \mathbb{R}$. From (11.5), we see that $v_y(x) \ge 0$ always holds, for all $x, y \in \mathbb{R}$. On the other hand, a straightforward manipulation shows that

$$v_y(x) = (K - \mathrm{e}^x) + \frac{\mathbb{E}((\mathrm{e}^{x+\underline{X}_{\mathbf{e}_q}} - K\mathbb{E}(\mathrm{e}^{\underline{X}_{\mathbf{e}_q}}))\mathbf{1}_{(-\underline{X}_{\mathbf{e}_q} \le x-y)})}{\mathbb{E}(\mathrm{e}^{\underline{X}_{\mathbf{e}_q}})}. \tag{11.8}$$

From (11.8), we see that a sufficient condition on y which ensures that $v_y(x) \ge (K - \mathrm{e}^x)$ is

$$\mathrm{e}^y \ge K\mathbb{E}(\mathrm{e}^{\underline{X}_{\mathbf{e}_q}}). \tag{11.9}$$

Supermartingale property (ii). On the event $\{t < \mathbf{e}_q\}$ the identity $\underline{X}_{\mathbf{e}_q} = \underline{X}_t \wedge (X_t + I)$ holds, where conditionally on $\mathcal{F}_t$, I has the same distribution as $\underline{X}_{\mathbf{e}_q}$. In particular, it follows that, on $\{t < \mathbf{e}_q\}$, $\underline{X}_{\mathbf{e}_q} \le X_t + I$. If

$$\mathrm{e}^y \le K\mathbb{E}(\mathrm{e}^{\underline{X}_{\mathbf{e}_q}}), \tag{11.10}$$

then for $x \in \mathbb{R}$

$$\begin{aligned} v_y(x) &\ge \frac{\mathbb{E}(\mathbf{1}_{(t<\mathbf{e}_q)}\mathbb{E}((K\mathbb{E}(\mathrm{e}^{\underline{X}_{\mathbf{e}_q}}) - \mathrm{e}^{x+X_t+I})\mathbf{1}_{(-(X_t+I)>x-y)}|\mathcal{F}_t))}{\mathbb{E}(\mathrm{e}^{\underline{X}_{\mathbf{e}_q}})} \\ &\ge \mathbb{E}(\mathrm{e}^{-qt}v_y(x + X_t)) \end{aligned}$$

$$= \mathbb{E}_x\big(\mathrm{e}^{-qt} v_y(X_t)\big).$$

Note that the first inequality follows by virtue of the fact that the part of the outer expectation in (11.7) taken over the event $\{\mathbf{e}_q \le t\}$ is positive, thanks to (11.10). Appealing to stationary independent increments, we can now see that, for $0 \le s \le t < \infty$,

$$\mathbb{E}\big(\mathrm{e}^{-rt} v_y(X_t)|\mathcal{F}_s\big) = \mathrm{e}^{-rs}\mathbb{E}_{X_s}\big(\mathrm{e}^{-r(t-s)} v_y(X_{t-s})\big) \le \mathrm{e}^{-rs} v_y(X_s), \qquad (11.11)$$

showing that $\{\mathrm{e}^{-qt} v_y(X_t) : t \ge 0\}$ is a $\mathbb{P}_x$-supermartingale. Right-continuity of its paths follows from the right-continuity of the paths of X and right-continuity of v_y, the latter of which can be seen from (11.8).

To conclude, we see that it would be sufficient to take $y = \log K\mathbb{E}(\mathrm{e}^{\underline{X}_{\mathbf{e}_q}})$ in order to satisfy conditions (i) and (ii) of Lemma 11.1, and, thereby, establish a solution to (11.3). □

The case that X is a compound Poisson process offers us the possibility to see that the optimal stopping time is not necessarily unique. Assume further that there are two-sided jumps whose distribution has no atoms (this excludes the possibility that X can jump exactly onto a prescribed point). For this class of compound Poisson processes, we note that

$$\inf\{t > 0 : X_t < y\} = \inf\{t \ge 0 : X_t \le y\}$$

$\mathbb{P}_x$-almost surely, unless $y = x$. In that case, the stopping time on the left is strictly positive $\mathbb{P}_x$-almost surely whereas the stopping time on the right is zero $\mathbb{P}_x$-almost surely. In other words, for all $x, y \in \mathbb{R}$, the optimal stopping time on the left is $\mathbb{P}_x$-almost surely greater than or equal to (as opposed to just equal to) the optimal stopping time on the right. Suppose we redefine $\tau_y^- = \inf\{t \ge 0 : X_t \le y\}$ and take $\tau_{x^*}^-$ (under this new definition) to be the candidate optimal stopping time to the McKean optimal stopping problem, instead of the one given in Theorem 11.2. Revisiting the proof of Theorem 11.2, we find easily that the value function is the same with this new stopping time. In showing this, one needs to start by making the strict inequality in (11.6) a weak inequality and working the consequence of this change through the computations.

With either the old or the new definition of $\tau_{x^*}^-$, the value function emerges as the same. We therefore see that, although there is a unique value for the solution to the optimal stopping problem, the optimal strategy is not necessarily unique. Indeed, we have found, at least in the case of compound Poisson jumps, that there is another optimal stopping time which can be almost surely smaller than the optimal stopping time found in the proof of Theorem 11.2, depending on the value of x.

In the case that X is spectrally negative, the solution may be expressed in terms of the scale functions. This was shown by Avram et al. (2002) and Chan (2004).

Corollary 11.3 *Suppose that X is spectrally negative. Then*

$$v(x) = KZ^{(q)}(x - x^*) - \mathrm{e}^x Z_1^{(p)}(x - x^*), \quad x \in \mathbb{R},$$

where $p = q - \psi(1)$ and

$$x^* = \log\left(K \frac{q}{\Phi(q)} \frac{\Phi(q) - 1}{q - \psi(1)}\right).$$

Here, we understand the right-hand side above in the limiting sense when $q = \psi(1)$. That is to say, $x^* = \log(K\psi(1)/\psi'(1))$.

Recall that Φ_1 is the right inverse of ψ_1, which in turn is the Laplace exponent of X under the measure $\mathbb{P}^1$. Note that we have

$$\psi_1(\lambda) = \psi(\lambda + 1) - \psi(1),$$

for all $\lambda \geq -1$. Hence, as $\Phi(q) - 1 > -1$,

$$\psi_1\big(\Phi(q) - 1\big) = q - \psi(1) = p,$$

and this implies that $\Phi_1(p) = \Phi(q) - 1$, where for negative values of p, we understand

$$\Phi_1(p) = \sup\{\lambda \geq -1 : \psi_1(\lambda) = p\}.$$

The subscripts on the functions $W_1^{(p)}$ and $Z_1^{(p)}$ indicate that they are the scale functions associated with the measure $\mathbb{P}^1$.

Proof of Corollary 11.3 We know from Theorem 11.2 that $v = v_y$, for $y = x^*$. Hence, from (11.5) and the conclusion of Exercise 8.7 (ii), we may write the given expression for v as

$$\begin{aligned} v(x) = {} & K\left(Z^{(q)}(x - x^*) - W^{(q)}(x - x^*)\frac{q}{\Phi(q)}\right) \\ & - \mathrm{e}^x\left(Z_1^{(p)}(x - x^*) - W_1^{(p)}(x - x^*)\frac{p}{\Phi_1(p)}\right). \end{aligned}$$

Next, note that the general form of x^* given in Theorem 11.2, together with the expression for one of the Wiener–Hopf factors in (8.4), allows us to deduce that

$$\mathrm{e}^{x^*} = K\frac{q}{\Phi(q)}\frac{\Phi(q) - 1}{q - \psi(1)}.$$

From (8.30), we have that $\mathrm{e}^x W_1^{(p)}(x) = W^{(q)}(x)$. Hence taking into account the definition of $\Phi_1(p)$, two of the terms in the expression for v given above cancel to give the identity in the statement of the corollary. □

11.3 Smooth Fit Versus Continuous Fit

It is clear that the solution to (11.3) is bounded from below by the gain function G, and further, is equal to the gain function on the domain on which the distribution of X_{τ^*} is concentrated. It turns out that there are different ways in which the function v "fits" on to the gain function G, according to certain path properties of the underlying Lévy process. The McKean optimal stopping problem provides a good example of where a dichotomy appears in this respect. We say that there is *continuous fit* at the threshold x^* if the left and right limit points of v at x^* exist and are equal. In addition, if the left and right derivatives of v exist at the boundary x^* and are equal, then we say that there is *smooth fit* at x^*. The remainder of this section is devoted to explaining the dichotomy of smooth and continuous fit in (11.3).

Consider again the McKean optimal stopping problem. The following theorem is again taken from Alili and Kyprianou (2005).

Theorem 11.4 *The function $v(\log y)$ is convex in $y > 0$ and, in particular, there is continuous fit of v at x^*. The right derivative at x^* is given by $v'(x^*+) = -e^{x^*} + K\mathbb{P}(\underline{X}_{\mathbf{e}_q} = 0)$. Thus, the optimal stopping problem* (11.3) *exhibits smooth fit at x^* if and only if 0 is regular for $(-\infty, 0)$.*

Proof Note that, for a fixed stopping time $\tau \in \mathcal{T}$, the expression $\mathbb{E}(e^{-q\tau}(K - e^{x+X_\tau})^+)$ is convex in e^x, as the same is true of the function $(K - ce^x)^+$, where $c > 0$ is a constant. Further, since taking the supremum is a subadditive operation, it can easily be deduced that $v(\log y)$ is a convex function in y. In particular, v is continuous.

Next, we establish necessary and sufficient conditions for smooth fit. Since $v(x) = K - e^x$, for all $x < x^*$, and hence $v'(x^*-) = -e^{x^*}$, we are required to show that $v'(x^*+) = -e^{x^*}$ for smooth fit. Starting from (11.7) and recalling that $e^{x^*} = K\mathbb{E}(e^{\underline{X}_{\mathbf{e}_q}})$, we have

$$\begin{aligned}
v(x) &= -K\mathbb{E}\big(\big(e^{x-x^*+\underline{X}_{\mathbf{e}_q}} - 1\big)\mathbf{1}_{(-\underline{X}_{\mathbf{e}_q} > x-x^*)}\big) \\
&= -K\big(e^{x-x^*} - 1\big)\mathbb{E}\big(e^{\underline{X}_{\mathbf{e}_q}}\mathbf{1}_{(-\underline{X}_{\mathbf{e}_q} > x-x^*)}\big) \\
&\quad - K\mathbb{E}\big(\big(e^{\underline{X}_{\mathbf{e}_q}} - 1\big)\mathbf{1}_{(-\underline{X}_{\mathbf{e}_q} > x-x^*)}\big).
\end{aligned}$$

From the last equality, we may then write

$$\begin{aligned}
\frac{v(x) - (K - e^{x^*})}{x - x^*} &= \frac{v(x) + K(\mathbb{E}(e^{\underline{X}_{\mathbf{e}_q}}) - 1)}{x - x^*} \\
&= -K\frac{(e^{x-x^*} - 1)}{x - x^*}\mathbb{E}\big(e^{\underline{X}_{\mathbf{e}_q}}\mathbf{1}_{(-\underline{X}_{\mathbf{e}_q} > x-x^*)}\big) \\
&\quad + K\frac{\mathbb{E}((e^{\underline{X}_{\mathbf{e}_q}} - 1)\mathbf{1}_{(-\underline{X}_{\mathbf{e}_q} \le x-x^*)})}{x - x^*}.
\end{aligned}$$

To simplify notations, let us call A_x and B_x the last two terms, respectively. It is clear that

$$\lim_{x\downarrow x^*} A_x = -K\mathbb{E}\big(\mathrm{e}^{\underline{X}_{\mathbf{e}_q}}\mathbf{1}_{(-\underline{X}_{\mathbf{e}_q}>0)}\big). \tag{11.12}$$

On the other hand, using integration by parts, we have that

$$\begin{aligned} B_x &= K\frac{\mathbb{E}((\mathrm{e}^{\underline{X}_{\mathbf{e}_q}}-1)\mathbf{1}_{(0<-\underline{X}_{\mathbf{e}_q}\le x-x^*)})}{x-x^*} \\ &= K\int_{(0,x-x^*]}\frac{\mathrm{e}^{-z}-1}{x-x^*}\mathbb{P}(-\underline{X}_{\mathbf{e}_q}\in \mathrm{d}z) \\ &= K\frac{\mathrm{e}^{x^*-x}-1}{x-x^*}\mathbb{P}\big(0<-\underline{X}_{\mathbf{e}_q}\le x-x^*\big) \\ &\quad + \frac{K}{x-x^*}\int_0^{x-x^*}\mathrm{e}^{-z}\mathbb{P}(0<-\underline{X}_{\mathbf{e}_q}\le z)\mathrm{d}z, \end{aligned}$$

where in the first equality, we have removed the possible atom at zero from the expectation by noting that $\exp\{\underline{X}_{\mathbf{e}_q}\}-1=0$ on $\{\underline{X}_{\mathbf{e}_q}=0\}$. This leads to $\lim_{x\downarrow x^*} B_x = 0$. Using the expression for e^{x^*}, we see that $v'(x^*+) = -\mathrm{e}^{x^*} + K\mathbb{P}(-\underline{X}_{\mathbf{e}_q}=0)$, which equals $-\mathrm{e}^{x^*}$ if and only if $\mathbb{P}(-\underline{X}_{\mathbf{e}_q}=0)=0$, in other words, there is smooth fit if and only if 0 is regular for $(-\infty,0)$. □

Let us now discuss the dichotomy of continuous and smooth fit as a mathematical principle. In order to make the arguments more visible, we will restrict ourselves to the case that X is a spectrally negative Lévy process, in which case v and x^* are given in Corollary 11.3. We start by looking in closer detail at the analytic properties of the candidate solution, v_y, at its boundary point y. For convenience, we shall assume that $W^{(q)}$ is continuously differentiable on $(0,\infty)$ when X has paths of bounded variation.

Returning to the candidate solutions (v_y,τ_y^-), for $y\le \log K$, we have, again from Exercise 8.7, that

$$\begin{aligned} v_y(x) &= K\left(Z^{(q)}(x-y) - W^{(q)}(x-y)\frac{q}{\Phi(q)}\right) \\ &\quad - \mathrm{e}^x\left(Z_1^{(p)}(x-y) - W_1^{(p)}(x-y)\frac{p}{\Phi_1(p)}\right) \\ &= KZ^{(q)}(x-y) - \mathrm{e}^x Z_1^{(p)}(x-y) + W^{(q)}(x-y)\frac{p}{\Phi_1(p)}\left(\mathrm{e}^y - K\frac{q\Phi_1(p)}{\Phi(q)p}\right), \end{aligned}$$

where $p=q-\psi(1)$, $x\in\mathbb{R}$ and the second equality follows from the fact that $\mathrm{e}^x W_1^{(p)}(x) = W^{(q)}(x)$; see (8.30). Thanks to the analytical properties of scale functions, we observe that v_y is continuous everywhere, except possibly at y. Indeed, at the point y, we find

$$v_y(y-) = (K - e^y)$$

and

$$v_y(y+) = v_y(y-) + W^{(q)}(0)\frac{p}{\Phi_1(p)}\left(e^y - K\frac{q\Phi_1(p)}{\Phi(q)p}\right). \quad (11.13)$$

Recall that $W^{(q)}(0) = 0$ if and only if X is of unbounded variation, and otherwise, $W^{(q)}(0) = 1/\delta$, where δ is the drift in the usual decomposition of X; see (8.3) and Lemma 8.6. As X is spectrally negative, 0 is regular for $(-\infty, 0)$ if and only if X is of unbounded variation. We see that v_y is continuous whenever 0 is regular for $(-\infty, 0)$ and otherwise, with the exception of one particular value, there is a discontinuity at y. Specifically, if $y < x^*$, then there is a negative discontinuity at y. If $y > x^*$, then there is a positive discontinuity at y and if $y = x^*$, then there is continuity at y.

Next, we compute the derivative of v_y as follows. For $x < y$ we have $v'_y(x) = -e^x$. For $x > y$, again using the fact that $e^x W_1^{(p)}(x) = W^{(q)}(x)$, we have

$$\begin{aligned} v'_y(x) &= KqW^{(q)}(x-y) - e^y pW^{(q)}(x-y) \\ &\quad - e^x Z_1^{(p)}(x-y) + W^{(q)\prime}(x-y)\frac{p}{\Phi_1(p)}\left(e^y - K\frac{q\Phi_1(p)}{\Phi(q)p}\right). \end{aligned}$$

We see that

$$\begin{aligned} v'_y(y+) &= v'_y(y-) + W^{(q)}(0)\left(Kq - e^y p\right) \\ &\quad + W^{(q)\prime}(0+)\frac{p}{\Phi_1(p)}\left(e^y - K\frac{q\Phi_1(p)}{\Phi(q)p}\right). \end{aligned} \quad (11.14)$$

Recall from Exercise 8.5 (ii) that

$$W^{(q)\prime}(0+) = \begin{cases} 2/\sigma^2 & \text{if } \nu(-\infty, 0) = \infty \text{ or } \sigma > 0 \\ (\nu(-\infty, 0) + q)/\delta^2 & \text{if } \nu(-\infty, 0) < \infty \text{ and } \sigma = 0 \end{cases}$$

where σ is the Gaussian coefficient, ν is the Lévy measure of X and $\delta > 0$ is the drift in the case that X has bounded variation. Moreover, we adopt the understanding that $1/0 = \infty$. From (11.14), we note that there is a discontinuity in the left- and right-derivative of v_y with the exception of the case that $y = x^*$. Indeed, when $y > x^*$, this discontinuity is positive, when $y < x^*$ it is negative and when $y = x^*$, $v'_y(y)$ is well defined.

Figures 11.1 and 11.2 sketch what we can expect to see for the shape of v_y, by perturbing the value y about x^*, for the cases of unbounded variation and bounded variation with infinite Lévy measure. With these diagrams in mind, we may now intuitively understand the appearance of smooth or continuous fit as a principle via the following reasoning.

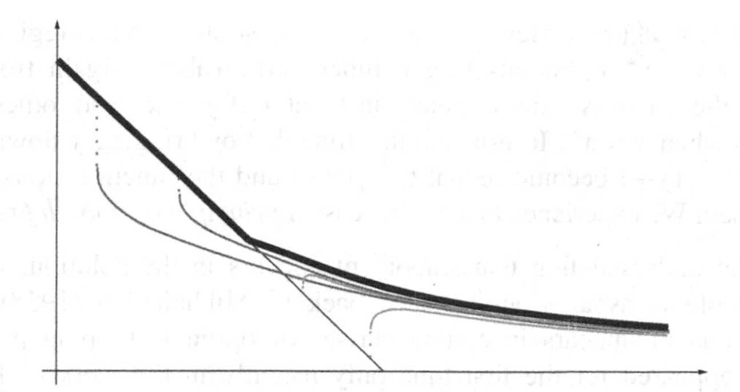

Fig. 11.1 A sketch of the functions $v_y(\log x)$ for different values of y when X is of bounded variation and $\nu(-\infty, 0) = \infty$. Curves which do not bound the function $(K - x)^+$ from above correspond to examples of $v_y(\log x)$ with $y < x^*$. Curves which are bounded from below by $(K - x)^+$ correspond to examples of $v_y(\log x)$ with $y > x^*$. The unique curve which bounds the gain from above with continuous fit corresponds to $v_{x^*}(\log x)$.

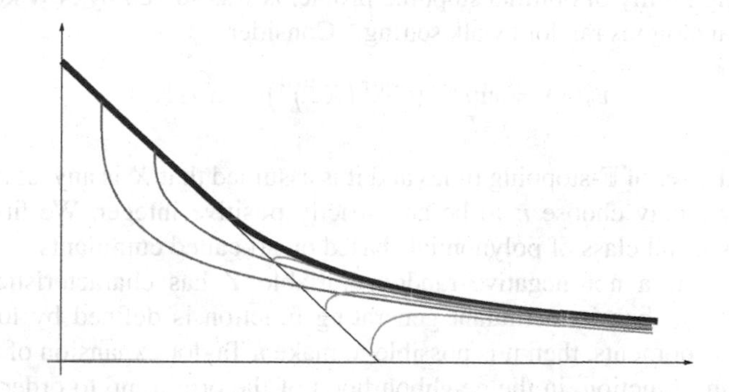

Fig. 11.2 A sketch of the functions $v_y(\log x)$ for different values of y when X is of unbounded variation and $\sigma = 0$. Curves which do not bound from above the function $(K - x)^+$ correspond to examples of $v_y(\log x)$ with $y < x^*$. Curves which are bounded from below by $(K - x)^+$ correspond to examples of $v_y(\log x)$ with $y > x^*$. The unique curve which bounds from above the gain with smooth fit corresponds to $v_{x^*}(\log x)$.

For the case 0 is irregular for $(-\infty, 0)$ for X. When $y < x^*$, thanks to the analysis of (11.13), we know that the function v_y does not bound the gain function $(K - e^x)^+$ from above due to a negative discontinuity at y. Hence τ_y^- is not a good strategy in this regime of y. On the other hand, from (11.8) and (11.9) if $y \geq x^*$, v_y bounds the gain function from above. Again from (11.13), we see that there is a positive discontinuity in v_y at y when $y > x^*$ and continuity when $y = x^*$. By bringing y down to x^* it turns out that the function v_y is pointwise optimised. Here, we experience *a principle of continuous fit* and from (11.14) it transpires there is no smooth fit.

For the case 0 is regular for $(-\infty, 0)$ for X. All curves v_y are continuous. When $y < x^*$, the function v_y cannot bound the gain function $(K - e^y)^+$ from above as $v_y'(y+) < v_y'(y-)$. Hence τ_y^- is not a good strategy in this regime of y. As before, if $y \geq x^*$, v_y bounds the gain function from above. Again, from (11.14), we see that there is a discontinuity in v_y' at y if $y > x^*$ and, otherwise, it is smooth when $y = x^*$. It turns out this time that by bringing y down to x^* the gradient $v_y'(y+)$ becomes equal to $v_y'(y-)$ and the function v_y is pointwise optimised. We experience then in this case *a principle of smooth fit* instead.

Whilst the understanding that smooth fit appears in the solutions of optimal stopping problems as a principle dates back to Mikhalevich (1958), the idea that continuous fit appears in certain classes of optimal stopping problems *as a principle* appeared for the first time only recently in the work of Peskir and Shiryaev (2000, 2002).

11.4 The Novikov–Shiryaev Optimal Stopping Problem

The following family of optimal stopping problems was solved by Novikov (2004), albeit in an analogous random walk setting.[4] Consider

$$v_n(x) = \sup_{\tau \in \mathcal{T}} \mathbb{E}_x\big(\mathrm{e}^{-q\tau}\big(X_\tau^+\big)^n\big), \quad x \in \mathbb{R}, \tag{11.15}$$

where $\mathcal{T}$ is the set of $\mathbb{F}$-stopping times and it is assumed that X is any Lévy process, $q > 0$ and we may choose n to be any strictly positive integer. We first need to introduce a special class of polynomials based on so-called cumulants.

Recall that if a non-negative random variable Y has characteristic function $\phi(\theta) = \mathbb{E}(\mathrm{e}^{\mathrm{i}\theta Y})$, then its cumulant generating function is defined by $\log \phi(\theta)$. If Y has up to n moments, then it is possible to make a Taylor expansion of the cumulant generating function, in the neighbourhood of the origin, up to order n plus an error term. Specifically,

$$\log \phi(\theta) = \sum_{j=1}^{n} \kappa_j \frac{(\mathrm{i}\theta)^j}{j!} + o\big(|\theta|^n\big) \quad \text{as } \theta \to 0.$$

In that case, the coefficients $\{\kappa_1, \ldots, \kappa_n\}$ are called the first n *cumulants*, and they may be written in terms of the first n moments. For example,

$$\begin{aligned}
\kappa_1 &= \mu_1, \\
\kappa_2 &= \mu_2 - \mu_1^2, \\
\kappa_3 &= 2\mu_1^3 - 3\mu_1\mu_2 + \mu_3,
\end{aligned}$$

and so on, where $\mu_1, \mu_2, \mu_3, \ldots$ are the first, second, third, etc. moments of Y.

[4] The continuous-time arguments are also given in Kyprianou and Surya (2005). Further work in this direction can be found in Deligiannidis and Utev (2009).

For a concise overview of cumulant generating functions, the reader is referred to Lukacs (1970) and Kendall and Stuart (1977).

Definition 11.5 (Appell Polynomials) Suppose that Y is a non-negative random variable and, for $n = 1, 2, \ldots$, its n-th cumulant is given by κ_n. Define the Appell polynomials of Y iteratively as follows. Take $Q_0(x) = 1$, $x \in \mathbb{R}$ and, assuming that $|\kappa_n| < \infty$ (equivalently, Y has an n-th moment), given $Q_{n-1}(x)$, we define $Q_n(x)$ via

$$\frac{\mathrm{d}}{\mathrm{d}x} Q_n(x) = n Q_{n-1}(x), \quad x \in \mathbb{R}. \tag{11.16}$$

This defines Q_n up to a constant. To pin this constant down, we insist that $\mathbb{E}(Q_n(Y)) = 0$. The first three Appell polynomials are given by

$$Q_0(x) = 1, \qquad Q_1(x) = x - \kappa_1, \qquad Q_2(x) = (x - \kappa_1)^2 - \kappa_2,$$

$$Q_3(x) = (x - \kappa_1)^3 - 3\kappa_2(x - \kappa_1) - \kappa_3,$$

under the assumption that $\kappa_3 < \infty$. See also Schoutens (2003) for further details of Appell polynomials.

In the following theorem, we shall work with the Appell polynomials generated by the random variable $Y = \overline{X}_{\mathbf{e}_q}$ where as usual, for each $t \in [0, \infty)$, $\overline{X}_t = \sup_{s \in [0,t]} X_s$ and $\mathbf{e}_q$ is an exponentially distributed random variable with mean $1/q$, which is independent of X.

Theorem 11.6 *Fix $n \in \{1, 2, \ldots\}$ and assume that*

$$\int_{(1,\infty)} x^n \nu(dx) < \infty. \tag{11.17}$$

Then there exists a largest root, $x_n^* \in [0, \infty)$, of the equation $Q_n(x) = 0$. Let

$$\tau_n^* = \inf\{t \geq 0 : X_t \geq x_n^*\}.$$

Then τ_n^* is an optimal strategy to (11.15). *Further,*

$$v_n(x) = \mathbb{E}_x\left(Q_n(\overline{X}_{\mathbf{e}_q})\mathbf{1}_{(\overline{X}_{\mathbf{e}_q} \geq x_n^*)}\right), \quad x \in \mathbb{R}.$$

Similarly to the McKean optimal stopping problem, we can establish a necessary and sufficient criterion for the occurrence of smooth fit. Once again, it boils down to the underlying path regularity of X.

Theorem 11.7 *For each $n = 1, 2, \ldots$, the solution to the optimal stopping problem in Theorem 9.6 is convex. In particular, there is continuous fit at x_n^*. Moreover,*

$$v_n'(x_n^*-) = v_n'(x_n^*+) - Q_n'(x_n^*)\mathbb{P}(\overline{X}_{\mathbf{e}_q} = 0).$$

Hence, there is smooth fit at x_n^* if and only if 0 *is regular for* $(0,\infty)$ *for* X.

The proofs of the last two theorems require some preliminary results, given in the following lemmas.

Lemma 11.8 (Mean value property) *Fix $n \in \{1,2,\dots\}$ and suppose that Y is a non-negative random variable satisfying $\mathbb{E}(Y^n) < \infty$. If Q_n is the n-th Appell polynomial generated by Y, then*

$$\mathbb{E}\big(Q_n(x+Y)\big) = x^n,$$

for all $x \in \mathbb{R}$.

Proof Note that the result is trivially true for $n = 1$. Next, suppose the result is true for Q_{n-1}. Then, using dominated convergence, we have from (11.16) that

$$\frac{\mathrm{d}}{\mathrm{d}x}\mathbb{E}\big(Q_n(x+Y)\big) = \mathbb{E}\bigg(\frac{\mathrm{d}}{\mathrm{d}x}Q_n(x+Y)\bigg) = n\mathbb{E}\big(Q_{n-1}(x+Y)\big) = nx^{n-1}.$$

Solving this differential equation, using $\mathbb{E}(Q_n(Y)) = 0$ to pin down the constant, we have the required result. □

Lemma 11.9 (Fluctuation identity) *Fix $n \in \{1,2,\dots\}$ and suppose that*

$$\int_{(1,\infty)} x^n \nu(\mathrm{d}x) < \infty.$$

Then, for all $a > 0$ and $x \in \mathbb{R}$,

$$\mathbb{E}_x\big(\mathrm{e}^{-qT_a^+}X^n_{T_a^+}\mathbf{1}_{(T_a^+<\infty)}\big) = \mathbb{E}_x\big(Q_n(\overline{X}_{\mathbf{e}_q})\mathbf{1}_{(\overline{X}_{\mathbf{e}_q}\geq a)}\big),$$

where $T_a^+ = \inf\{t \geq 0 : X_t \geq a\}$.

Proof On the event $\{T_a^+ < \mathbf{e}_q\}$, equivalently, on the event $\{\overline{X}_{\mathbf{e}_q} \geq a\}$, we have that $\overline{X}_{\mathbf{e}_q} = X_{T_a^+} + S$, where S is independent of $\mathcal{F}_{T_a^+}$ and has the same distribution as $\overline{X}_{\mathbf{e}_q}$. It follows that

$$\mathbb{E}_x\big(Q_n(\overline{X}_{\mathbf{e}_q})\mathbf{1}_{(\overline{X}_{\mathbf{e}_q}\geq a)}|\mathcal{F}_{T_a^+}\big) = \mathbf{1}_{(T_a^+<\mathbf{e}_q)}h(X_{T_a^+}), \quad x \in \mathbb{R},$$

where $h(x) = \mathbb{E}_x(Q_n(\overline{X}_{\mathbf{e}_q})) = x^n$, and the last equality follows from Lemma 11.8 with $Y = \overline{X}_{\mathbf{e}_q}$. Note also that, by Exercise 7.1, the integral condition on ν implies that $\mathbb{E}(\overline{X}^n_{\mathbf{e}_q}) < \infty$, which has been used in order to apply Lemma 11.8. We see, by taking expectations again in the previous calculation, that

$$\mathbb{E}_x\big(Q_n(\overline{X}_{\mathbf{e}_q})\mathbf{1}_{(\overline{X}_{\mathbf{e}_q}\geq a)}\big) = \mathbb{E}_x\big(\mathrm{e}^{-qT_a^+}X^n_{T_a^+}\mathbf{1}_{(T_a^+<\infty)}\big),$$

as required. □

Lemma 11.10 (Largest positive root) *Fix $n \in \{1, 2, \ldots\}$ and suppose that*

$$\int_{(1,\infty)} x^n \nu(\mathrm{d}x) < \infty.$$

Suppose that Q_n is generated by $\overline{X}_{\mathbf{e}_q}$. Then Q_n has a unique strictly positive root, say x_n^*, such that $Q_n(x)$ is negative on $[0, x_n^*)$ and positive and increasing on $[x_n^*, \infty)$.

Proof We start by noting that the statement of the lemma is clearly true for $Q_1(x) = x - \kappa_1$. We proceed then by induction and assume that the result is true for Q_{n-1}.

The first step is to prove that $Q_n(0) \le 0$. Let

$$\eta(a, n) = \mathbb{E}\big(\mathrm{e}^{-qT_a^+} X^n_{T_a^+} \mathbf{1}_{(T_a^+ < \infty)}\big)$$

and, for all $a \ge 0$ and $n = 1, 2, \ldots$, note that $\eta(a, n) \ge 0$. On the other hand

$$\begin{aligned} \eta(a, n) &= \mathbb{E}\big(Q_n(\overline{X}_{\mathbf{e}_q}) \mathbf{1}_{(\overline{X}_{\mathbf{e}_q} \ge a)}\big) \\ &= -\mathbb{E}\big(Q_n(\overline{X}_{\mathbf{e}_q}) \mathbf{1}_{(\overline{X}_{\mathbf{e}_q} < a)}\big) \\ &= -\mathbb{P}(\overline{X}_{\mathbf{e}_q} < a) Q_n(0) \\ &\quad + \mathbb{E}\big(\big(Q_n(0) - Q_n(\overline{X}_{\mathbf{e}_q})\big) \mathbf{1}_{(\overline{X}_{\mathbf{e}_q} < a)}\big), \end{aligned}$$

for all $n = 1, 2, \ldots$, where the first equality follows by Lemma 11.9 and the second by Lemma 11.8. Since, by definition,

$$Q_n(x) = Q_n(0) + n \int_0^x Q_{n-1}(u) \mathrm{d}y, \tag{11.18}$$

for all $x \ge 0$, we have the estimate

$$\big|\mathbb{E}_x\big(\big(Q_n(0) - Q_n(\overline{X}_{\mathbf{e}_q})\big) \mathbf{1}_{(\overline{X}_{\mathbf{e}_q} < a)}\big)\big| \le na \sup_{y \in [0,a]} \big|Q_{n-1}(y)\big| \mathbb{P}(\overline{X}_{\mathbf{e}_q} < a),$$

which tends to zero as $a \downarrow 0$. We have, in conclusion, that, as $a \downarrow 0$,

$$0 \le \eta(a, n) \le -\mathbb{P}(\overline{X}_{\mathbf{e}_q} < a)\big[Q_n(0) + O(a)\big],$$

and hence it follows that $Q_n(0) \le 0$.

Under the induction hypothesis for Q_{n-1}, we see from (11.18), together with the fact that $Q_n(0) \le 0$, that Q_n is negative and decreasing on $[0, x_{n-1}^*)$. The point x_{n-1}^* is the argument corresponding to the infimum of Q_n, thanks to the positivity and monotonicity of $Q_{n-1}(s)$ on $x > x_{n-1}^*$. In particular, $Q_n(x)$ increases to infinity from its minimum point, and hence there must be a unique strictly positive root of the equation $Q_n(x) = 0$. □

We are now ready to move to the proofs of the main theorems of this section. For the first one below, the reader is again referred to Baurdoux (2013) for an interesting alternative.

Proof of Theorem 11.6 Fix $n \in \{1, 2, \ldots\}$. As a consequence of (11.17), we have that $\mathbb{E}(X_1) \in [-\infty, \infty)$, and hence the Strong Law of Large Numbers, given in Exercise 7.2, implies that (11.2) is automatically satisfied, since $q > 0$. Indeed, $(X_t^+)^n$ grows no faster than Ct^n for some constant $C > 0$.

Define

$$v_n^a(x) = \mathbb{E}_x\big(\mathrm{e}^{-qT_a^+}\big(X_{T_a^+}^+\big)^n \mathbf{1}_{(T_a^+<\infty)}\big), \quad x \in \mathbb{R}. \tag{11.19}$$

Again, referring to the discussion following Lemma 11.1, we consider pairs (v_n^a, T_a^+), for $a > 0$, to be a class of candidate solutions to (11.15). Our goal then is to verify, with the help of Lemma 11.1, that the candidate pair (v_n^a, T_a^+) solves (11.15) for some $a > 0$.

Lower bound (i). We need to prove that $v_n^a(x) \geq (x^+)^n$ for all $x \in \mathbb{R}$. Note that this statement is obvious for $x \in (-\infty, 0) \cup (a, \infty)$, just from the definition of v_n^a. Otherwise, when $x \in (0, a)$, we have from Lemmas 11.8 and 11.9 that, for all $x \in \mathbb{R}$,

$$\begin{aligned} v_n^a(x) &= \mathbb{E}_x\big(Q_n(\overline{X}_{\mathbf{e}_q})\mathbf{1}_{(\overline{X}_{\mathbf{e}_q}\geq a)}\big) \\ &= x^n - \mathbb{E}\big(Q_n(x + \overline{X}_{\mathbf{e}_q})\mathbf{1}_{(x+\overline{X}_{\mathbf{e}_q}<a)}\big). \end{aligned} \tag{11.20}$$

Recall from Lemma 11.10 that $Q_n(x) \leq 0$ on $(0, x_n^*]$. Therefore, provided

$$a \leq x_n^*,$$

we have in (11.20) that $v_n^a(x) \geq (x^+)^n$, $x \in \mathbb{R}$.

Supermartingale property (ii). Provided

$$a \geq x_n^*,$$

we have almost surely that

$$Q_n(\overline{X}_{\mathbf{e}_q})\mathbf{1}_{(\overline{X}_{\mathbf{e}_q}\geq a)} \geq 0.$$

On the event that $\{\mathbf{e}_q > t\}$, we have that $\overline{X}_{\mathbf{e}_q}$ is equal in distribution to $(X_t + S) \vee \overline{X}_t$, where S is independent of $\mathcal{F}_t$ and equal in distribution to $\overline{X}_{\mathbf{e}_q}$. In particular $\overline{X}_{\mathbf{e}_q} \geq X_t + S$. It now follows that

$$\begin{aligned} v_n^a(x) &\geq \mathbb{E}_x\big(\mathbf{1}_{(\mathbf{e}_q>t)}Q_n(\overline{X}_{\mathbf{e}_q})\mathbf{1}_{(\overline{X}_{\mathbf{e}_q}\geq a)}\big) \\ &\geq \mathbb{E}_x\big(\mathbf{1}_{(\mathbf{e}_q>t)}\mathbb{E}_x\big(Q_n(X_t + S)\mathbf{1}_{(X_t+S\geq a)}|\mathcal{F}_t\big)\big) \\ &= \mathbb{E}_x\big(\mathrm{e}^{-qt}v_n^a(X_t)\big), \quad x \in \mathbb{R}. \end{aligned}$$

From this inequality, together with the Markov property, it is easily shown, as in the McKean optimal stopping problem, that $\{e^{-qt}v_n^a(X_t) : t \geq 0\}$ is a $\mathbb{P}_x$-supermartingale. Right-continuity follows again from the right-continuity of the paths of X, together with the right-continuity of v_n^a, which is evident from (11.20).

We now see that the unique choice $a = x_n^*$ allows all the conditions of Lemma 11.1 to be satisfied, thus giving the solution to (11.15). □

Note that the case $q = 0$ can be dealt with in essentially the same manner. In this regime it is necessary to assume that $\limsup_{t\uparrow\infty} X_t < \infty$, and if working with the gain function $(x^+)^n$, for $n = 1, 2, \ldots$, then one needs to assume that

$$\int_{(1,\infty)} x^{n+1}\nu(\mathrm{d}x) < \infty.$$

The power in the above integral is $n + 1$, and not n as one must now deal with the n-th moments of $\overline{X}_\infty$; see Exercise 7.1.

Proof of Theorem 11.7 In a similar manner to the proof of Theorem 11.4, it is straightforward to prove that v is convex and hence continuous.

To establish when there is smooth fit at x_n^*, we note that, for $x < x_n^*$,

$$\begin{aligned}\frac{v_n(x_n^*) - v_n(x^n)}{x_n^* - x} &= \frac{(x_n^*)^n - x^n}{x_n^* - x} + \frac{\mathbb{E}_x(Q_n(\overline{X}_{\mathbf{e}_q})\mathbf{1}_{(\overline{X}_{\mathbf{e}_q} < x_n^*)})}{x_n^* - x}\\ &= \frac{(x_n^*)^n - x^n}{x_n^* - x} + \frac{\mathbb{E}_x((Q_n(\overline{X}_{\mathbf{e}_q}) - Q_n(x_n^*))\mathbf{1}_{(\overline{X}_{\mathbf{e}_q} < x_n^*)})}{x_n^* - x},\end{aligned}$$

where the final equality follows because $Q_n(x_n^*) = 0$. Clearly,

$$\lim_{x\downarrow x_n^*} \frac{(x_n^*)^n - x^n}{x_n^* - x} = v_n'(x_n^*+).$$

However,

$$\begin{aligned}&\frac{\mathbb{E}_x((Q_n(\overline{X}_{\mathbf{e}_q}) - Q_n(x_n^*))\mathbf{1}_{(\overline{X}_{\mathbf{e}_q} < x_n^*)})}{x_n^* - x}\\ &\quad= \frac{\mathbb{E}_x((Q_n(\overline{X}_{\mathbf{e}_q}) - Q_n(x))\mathbf{1}_{(x<\overline{X}_{\mathbf{e}_q} < x_n^*)})}{x_n^* - x}\\ &\qquad- \frac{\mathbb{E}_x((Q_n(x_n^*) - Q_n(x))\mathbf{1}_{(\overline{X}_{\mathbf{e}_q} < x_n^*)})}{x_n^* - x},\end{aligned} \tag{11.21}$$

where, in the first term on the right-hand side, we may restrict the expectation to $\{x < \overline{X}_{\mathbf{e}_q} < x_n^*\}$ as, under $\mathbb{P}_x$, the possible atom of $\overline{X}_{\mathbf{e}_q}$ at x gives zero mass to the expectation. Denote by A_x and B_x the two expressions on the right-hand side of (11.21). We have that

$$\lim_{x\uparrow x_n^*} B_x = -Q_n'(x_n^*)\mathbb{P}(\overline{X}_{\mathbf{e}_q} = 0).$$

Integration by parts also gives

$$\begin{aligned} A_x &= \int_{(0,x_n^*-x)} \frac{Q_n(x+y)-Q_n(x)}{x_n^*-x}\mathbb{P}(\overline{X}_{\mathbf{e}_q} \in \mathrm{d}y) \\ &= \frac{Q_n(x_n^*)-Q_n(x)}{x_n^*-x}\mathbb{P}\big(\overline{X}_{\mathbf{e}_q} \in (0, x_n^*-x)\big) \\ &\quad - \frac{1}{x_n^*-x}\int_0^{x_n^*-x} \mathbb{P}\big(\overline{X}_{\mathbf{e}_q} \in (0,y]\big)Q_n'(x+y)\mathrm{d}y. \end{aligned}$$

Hence, it follows that

$$\lim_{x\uparrow x_n^*} A_x = 0.$$

In conclusion, we have that

$$\lim_{x\uparrow x_n^*} \frac{v_n(x_n^*)-v_n(x)}{x_n^*-x} = v_n'(x_n^*+) - Q_n'(x_n^*)\mathbb{P}(\overline{X}_{\mathbf{e}_q} = 0),$$

which concludes the proof. □

11.5 The Shepp–Shiryaev Optimal Stopping Problem

Consider the optimal stopping problem

$$v(x) = \sup_{\tau\in\mathcal{T}} \mathbb{E}\big(\mathrm{e}^{-q\tau+(\overline{X}_\tau \vee x)}\big), \tag{11.22}$$

where $q > 0$, $\mathcal{T}$ is the set of $\mathbb{F}$-stopping times which are almost surely finite and $x \geq 0$. This optimal stopping problem was proposed and solved by Shepp and Shiryaev (1993), for the case that X is a linear Brownian motion and q is sufficiently large. Like the McKean optimal stopping problem, (11.22) appears in the context of an option pricing problem. Specifically, it addresses the problem of the optimal time to sell a risky asset for the maximum of either e^x or its running maximum (with discounting) when the risky asset follows the dynamics of an exponential linear Brownian motion. This lies at the heart of the pricing problem of so-called *Russian options*.

In Avram et al. (2004), a solution was given to (11.22) in the case that X is a general spectrally negative Lévy process (and q is sufficiently large). In order to keep to the mathematics that has been covered earlier on in this text, we give an account of a special case of that solution here. Specifically, we deal with the case that X has bounded variation, in which case, we shall write X in the usual form

$$X_t = \delta t - S_t, \quad t \geq 0, \tag{11.23}$$

where $\delta > 0$ and S is a driftless subordinator with Lévy measure ν (concentrated on $(0, \infty)$). We shall further assume that ν has no atoms. As we shall use scale functions in our solution, this last condition will ensure that they are continuously differentiable on $(0, \infty)$; see Exercise 8.4. Our objective is the theorem below. Note that we use standard notation from Chap. 8.

Theorem 11.11 *Suppose that X has paths of bounded variation with Laplace exponent ψ satisfying $q > \psi(1)$. Define*

$$x^* = \inf\{x \geq 0 : Z^{(q)}(x) \leq q W^{(q)}(x)\}, \quad x \geq 0.$$

Then for each $x \geq 0$, a solution to (11.22) *is given by the pair*

$$v(x) = \mathrm{e}^x Z^{(q)}(x^* - x)$$

and

$$\tau^* = \inf\{t > 0 : Y_t^x > x^*\},$$

where $Y^x = \{Y_t^x : t \geq 0\}$ is the process X reflected in its supremum, when issued from $x \geq 0$, so that $Y_t^x = (x \vee \overline{X}_t) - X_t$.

Before moving to the proof of Theorem 11.11, let us consider the nature of (11.22) and its solution.

It needs to be pointed out that, due to the involvement of the running supremum in the formulation of the problem, one may consider (11.22) as an optimal stopping problem which concerns the three-dimensional Markov process $\{(t, X_t, \overline{X}_t) : t \geq 0\}$. Nonetheless, it is possible to reduce (11.22) to an optimal stopping problem driven by the two-dimensional Markov process $\{(t, Y_t^x) : t \geq 0\}$.

The way to do this was noted by Shepp and Shiryaev (1994) in a follow-up article to their original contribution. Recalling the method of change of measure described in Sects. 3.3 and 8.1 (see specifically Corollary 3.11), for each $\tau \in \mathcal{T}$, we may write

$$\mathbb{E}\big(\mathrm{e}^{-q\tau + (\overline{X}_\tau \vee x)}\big) = \mathbb{E}^1\big(\mathrm{e}^{-\alpha\tau + Y_\tau^x} \mathbf{1}_{(\tau < \infty)}\big),$$

where

$$\alpha = q - \psi(1).$$

Hence, our objective is to solve the optimal stopping problem

$$\sup_{\tau \in \mathcal{T}} \mathbb{E}^1\big(\mathrm{e}^{-\alpha\tau + Y_\tau^x} \mathbf{1}_{(\tau < \infty)}\big), \tag{11.24}$$

which is based on the two-dimensional Markov process $\{(t, Y_t^x) : t \geq 0\}$. Note that (11.24) takes a different form from the type of optimal stopping problems we have considered earlier, in that it does not conform to the description given in (11.1). We shall return to this point shortly. In the meantime, we can note that arguments along the lines of those in the paragraphs following Lemma 11.1, suggest that a suitable class of candidate solutions to (11.24) is pairs of the form $(v_a(x), \overline{\sigma}_a^x)$, $a \geq 0$, where for each $x \geq 0$,

$$v_a(x) = \mathbb{E}^1\big(\mathrm{e}^{-\alpha\overline{\sigma}_a^x + Y_{\overline{\sigma}_a^x}^x}\mathbf{1}_{(\overline{\sigma}_a^x < \infty)}\big) = \mathbb{E}^1\big(\mathrm{e}^{-\alpha\overline{\sigma}_a^x + Y_{\overline{\sigma}_a^x}^x}\big) \tag{11.25}$$

and

$$\overline{\sigma}_a^x = \inf\{t > 0 : Y_t^x \geq a\}. \tag{11.26}$$

Indeed, the times at which Y^x is zero correspond to times at which $\overline{X}$, and hence the gain in (11.22), is increasing. The times at which Y^x takes large values correspond to the times at which X is far from its running supremum. At such moments, one must wait for the process to return to its maximum before there is an increase in the gain. Exponential discounting puts a time penalty on waiting too long, suggesting that we should look for a threshold strategy in which one should stop if X moves too far from its maximum. Note from Exercise 11.1 that the stopping time (11.26) is $\mathbb{P}$-almost surely finite, which places it in the class $\mathcal{T}$. By the same token, it is also $\mathbb{P}^1$-almost surely finite, which explains the removal of the indicator in the second equality of (11.25). We see at this point that the original problem (11.22) has now been reduced to an optimal stopping problem concerning only the one-dimensional process $\{Y_t^x : t \geq 0\}$.

Next, let us consider the optimal threshold x^*. Define the function $f(x) = Z^{(q)}(x) - qW^{(q)}(x)$, $x \geq 0$. Differentiating, we have, for $x > 0$,

$$\begin{aligned} f'(x) &= q\big(W^{(q)}(x) - W^{(q)\prime}(x)\big) \\ &= q\mathrm{e}^{\Phi(q)x}\big(\big(1 - \Phi(q)\big)W_{\Phi(q)}(x) - W'_{\Phi(q)}(x)\big), \end{aligned} \tag{11.27}$$

where in the second equality, we have used (8.30). We know that $W_{\Phi(q)}(x)$ is monotone increasing (cf. Theorem 8.1). In particular $W'_{\Phi(q)}(x) > 0$ for all $x > 0$. On the other hand, the assumption that $q > \psi(1)$ implies that $\Phi(q) > 1$. Hence, the right-hand side of (11.27) is strictly negative, for all $x > 0$. Further, we may check, with the help of Exercise 8.5 (i), that

$$\lim_{x\uparrow\infty} \frac{f(x)}{qW^{(q)}(x)} = \frac{1}{\Phi(q)} - 1 < 0.$$

From (8.30) again, it is clear that the denominator on the left-hand side above tends to infinity.

Note that $f(0+) = 1 - qW^{(q)}(0)$. Since X has bounded variation, we know that $W^{(q)}(0) = \delta^{-1}$. If $q \geq \delta$, then $f(0+) < 0$ and $x^* = 0$, which corresponds to stopping immediately. Note that it is already clear that one should stop immediately for this regime of q as the gain process $\exp\{-qt + (\overline{X}_t \vee x)\}$, $t \geq 0$, is decreasing. When $q < \delta$, we have that $f(0+) > 0$, $f'(x) < 0$ for all $x > 0$ and $f(\infty) = -\infty$. It follows that there is a unique solution in $(0, \infty)$ to $f(x) = 0$, which we denote by x^*.

Note also that, as the solution can be expressed in terms of functions whose analytic properties are sufficiently well-understood, we may easily investigate the situation with regard to smooth or continuous fit. For each $x > 0$, the process Y^x has the same small-time path behaviour as $-X$, and hence $\mathbb{P}(\overline{\sigma}_x^x = 0) = 0$ as $\mathbb{P}(\tau_0^- = 0) = 0$. This property is independent of $x > 0$ (the point $x = 0$ needs to be considered as a special case on account of reflection).

Corollary 11.12 *When $q < \delta$, the value function in* (11.22) *is convex and hence exhibits continuous fit at x^*. Further, it satisfies*

$$v'(x^*-) = v'(x^*+) - \frac{q}{\delta}\mathrm{e}^{x^*},$$

showing that there is no smooth fit at x^*.

Proof The first part follows in a similar manner to the proof of convexity in the previous two optimal stopping problems. After a straightforward differentiation of the value function of (11.22), recalling that $W(0+) = 1/\delta$, the last part of the corollary follows. □

Proof of Theorem 11.11 As indicated above, we consider candidate solutions of the form $(v_a, \overline{\sigma}_a^x)$. We can develop the right-hand side of v_a in terms of scale functions with the help of Theorem 8.10 (i). However, before doing so, let us note, with the help of Lemma 8.4, that the analogue of the scale function $W_{-1}^{(q)}$, when working under the measure $\mathbb{P}^1$, can be calculated as equal to

$$\left[W_1^{(q)}\right]_{-1}(x) = \mathrm{e}^x W_1^{(q+\psi_1(-1))}(x).$$

However, we also know that $\psi_1(-1) = \psi(1-1) - \psi(1) = -\psi(1)$, and hence, applying Lemma 8.4 again, we have further that

$$\left[W_1^{(q)}\right]_{-1}(x) = \mathrm{e}^x \mathrm{e}^{-x} W^{(q)}(x) = W^{(q)}(x).$$

We may therefore read out of Theorem 8.10 (i) the identity

$$v_a(x) = \mathrm{e}^x \left(Z^{(q)}(a-x) - W^{(q)}(a-x) \frac{q W^{(q)}(a) - Z^{(q)}(a)}{W^{(q)\prime}(a) - W^{(q)}(a)} \right), \quad x \geq 0. \quad (11.28)$$

It would be convenient at this point if we could apply Lemma 11.1 to the pair $(v_a, \overline{\sigma}_a^x)$ for an appropriate choice of a. As alluded to previously, this lemma is not directly applicable to the optimal stopping problem (11.24) on two counts. Firstly, because the process Y^x is not a Lévy process and, secondly, because of the inclusion of the indicator of the event $\{\tau < \infty\}$ in the expected gain. This is not a serious obstruction however. Indeed, we leave it as an exercise to the reader to show that, following closely the proof of Lemma 11.1, the following statement is true. If, for some $\tau^* \in \mathcal{T}$, $v^*(x) := \mathbb{E}^1(\exp\{-\alpha\tau^* + Y^x_{\tau^*}\}\mathbf{1}_{(\tau^*<\infty)})$ satisfies (i) $v^*(x) \geq \mathrm{e}^x$ and (ii) $\{\mathrm{e}^{-\alpha t} v^*(Y^x_t) : t \geq 0\}$ is a right-continuous supermartingale, then the pair (v^*, τ^*) solves (11.24).

Let us continue then to check the conditions of the aforementioned modified version of Lemma 11.1 for the pair $(\overline{\sigma}^x_a, v_a)$, with an appropriate choice of a.

Lower bound (i). We need to show that $v_a(s) \geq \mathrm{e}^x$. The assumption $q > \psi(1)$ implies that $\Phi(q) > 1$, and hence

$$W^{(q)\prime}(a) - W^{(q)}(a) > W^{(q)\prime}(a) - \Phi(q)W^{(q)}(a).$$

On the other hand, from (8.30), we may compute

$$W^{(q)\prime}(a) - \Phi(q)W^{(q)}(a) = \mathrm{e}^{\Phi(q)a} W'_{\Phi(q)}(a) > 0,$$

where the inequality is due to (8.22). Together with the properties of $Z^{(q)}(a) - qW^{(q)}(a)$ as a function of a, we see that the coefficient

$$\frac{qW^{(q)}(a) - Z^{(q)}(a)}{W^{(q)\prime}(a) - W^{(q)}(a)}$$

is strictly positive when $a > x^*$ and non-positive when $a \in [0, x^*]$.

Recalling that $Z^{(q)}(x) \geq 1$, we conclude that $v_a(x) \geq \mathrm{e}^x$ when $a \in [0, x^*]$. On the other hand, if $a > x^*$, then

$$v_a(a-) = \mathrm{e}^a - W^{(q)}(0)\frac{qW^{(q)}(a) - Z^{(q)}(a)}{W^{(q)\prime}(a) - W^{(q)}(a)} < \mathrm{e}^a, \tag{11.29}$$

showing that, in order to respect the lower bound, we must take $a \leq x^*$.

Supermartingale property (ii). We know that the function $v_a(x)$ is differentiable with continuous first derivative on $(0, a)$. Further, the right-derivative at zero exists and is equal to zero. To see this, simply compute

$$\begin{aligned} v'_a(0+) &= \Bigg(Z^{(q)}(a) - W^{(q)}(a)\frac{qW^{(q)}(a) - Z^{(q)}(a)}{W^{(q)\prime}(a) - W^{(q)}(a)} \\ &\qquad - qW^{(q)}(a) + W^{(q)\prime}(a)\frac{qW^{(q)}(a) - Z^{(q)}(a)}{W^{(q)\prime}(a) - W^{(q)}(a)}\Bigg) \\ &= 0. \end{aligned}$$

Next, recall from Sect. 8.1 that, under $\mathbb{P}^1$, X remains a Lévy process of bounded variation with the same drift, but now with exponentially tilted Lévy measure, so that in the form (11.23), $\nu(\mathrm{d}y)$ becomes $\mathrm{e}^{-y}\nu(\mathrm{d}y)$. Applying the change of variable formula in the spirit of Exercises 4.2 and 4.1, we see that, despite the fact that the first derivative of v_a is not well defined at a, for $t \geq 0$,

$$\begin{aligned}
e^{-\alpha t} v_a(Y_t^x) = v_a(x) - \alpha \int_0^t e^{-\alpha s} v_a(Y_s^x) ds \\
+ \int_0^t e^{-\alpha s} (v_a(a-) - v_a(a+)) dL_t^a \\
- \delta \int_0^t e^{-\alpha s} v_a'(Y_s^x) ds + \int_0^t e^{-\alpha s} v_a'(Y_s^x) d(x \vee \overline{X}_s) \\
+ \int_0^t \int_{(0,\infty)} e^{-\alpha s} (v_a(Y_{s-}^x + y) - v_a(Y_{s-}^x)) e^{-y} \nu(dy) ds \\
+ M_t \qquad (11.30)
\end{aligned}$$

$\mathbb{P}$-almost surely, where $\{L_t^a : t \geq 0\}$ counts the number of crossings of the process Y^x over the level a. Further, for $t \geq 0$,

$$\begin{aligned}
M_t = \int_{[0,t]} \int_{(0,\infty)} e^{-\alpha s} (v_a(Y_{s-}^x + y) - v_a(Y_{s-}^x)) N^1(ds \times dy) \\
- \int_0^t \int_{(0,\infty)} e^{-\alpha s} (v_a(Y_{s-}^x + y) - v_a(Y_{s-}^x)) e^{-y} \nu(dy) ds,
\end{aligned}$$

where N^1 is the counting measure associated with the jumps of S in the decomposition (11.23) of X under $\mathbb{P}^1$. In the fourth integral of (11.30), the process $x \vee \overline{X}_s$ increases only when $Y_s^x = 0$. Since $v_a'(0+) = 0$, the fourth integral is equal to zero. Note also that it can be proved in a straightforward way, with the help of the compensation formula in Theorem 4.4, that the process $M := \{M_t : t \geq 0\}$ is a martingale. In the first and third integrals of (11.30), we can freely interchange the roles of Y_{s-}^x and Y_s^x on account of the set of jump times of X (and hence Y^x) having zero Lebesgue measure.

Recalling that $\mathbb{P}^1(\overline{\sigma}_a^x < \infty) = 1$, the Markov property and (11.25) imply that

$$\mathbb{E}^1(e^{-\alpha \overline{\sigma}_a^x + Y_{\overline{\sigma}_a^x}^x} | \mathcal{F}_t) = e^{-\alpha(\overline{\sigma}_a^x \wedge t)} v_a(Y_{\overline{\sigma}_a^x \wedge t}^x), \quad t \geq 0.$$

In particular, we have a martingale. Hence, considering the left-hand side of (11.30), we deduce that, for all $0 \leq x \leq a$ and stopping times τ for the process Y^x,

$$\mathbb{E}\left(\int_0^{\overline{\sigma}_a^x \wedge \tau} e^{-\alpha s} \mathcal{L}^1 v_a(Y_s^x) ds\right) = 0, \qquad (11.31)$$

where

$$\mathcal{L}^1 v_a(x) := \int_{(0,\infty)} (v_a(x+y) - v_a(x)) e^{-y} \nu(dy) - \delta v_a'(x) - \alpha v_a(x) = 0, \quad (11.32)$$

for all $x \in (0, a)$. It is a straightforward exercise to show that $x \mapsto \mathcal{L}^1 v_a(x)$ is a continuous function on $(0, a)$. Let us consider the particular the case that, in (11.31),

$$\tau = \inf\{t > 0 : Y_t^x \notin (u, v)\},$$

for $0 < u < x < v \leq a$, so that $\overline{\sigma}_a^x \wedge \tau = \tau$. The equality in (11.31) can be rewritten using the resolvent density, $u^{(\alpha)}(\cdot, \cdot, \cdot)$, described in Theorem 8.7. Specifically, using Fubini's Theorem, we have

$$\begin{aligned}
\mathbb{E}\left(\int_0^\tau \mathrm{e}^{-\alpha s} \mathcal{L}^1 v_a(Y_s^x) \mathrm{d}s \right) &= \int_0^\infty \mathrm{e}^{-\alpha s} \int_{(u,v)} \mathbb{P}(Y_s^x \in \mathrm{d}y,\, s < \tau) \mathcal{L}^1 v_a(y) \mathrm{d}s \\
&= \int_u^v u^{(\alpha)}(v - u, v - x, v - y) \mathcal{L}^1 v_a(y) \mathrm{d}y \\
&= 0.
\end{aligned}$$

Since we may choose $0 < u < x < v \leq a$ arbitrarily and the density $u^{(\alpha)}$ is strictly positive, it can be shown that $\mathcal{L}^1 v_a(y)$ is Lebesgue-almost everywhere zero on $(0, a)$, and hence, by continuity, $\mathcal{L}^1 v_a(y) = 0$ for all $y \in (0, a)$.

Since $v_a(x) = \mathrm{e}^x$ for all $x > a$, we know that the expression on the left-hand side of (11.32) satisfies

$$\begin{aligned}
\mathcal{L}^1 v_a(x) &= \mathrm{e}^x \int_{(0,\infty)} (1 - \mathrm{e}^{-y}) \nu(\mathrm{d}y) - \delta \mathrm{e}^x - \alpha \mathrm{e}^x \\
&= -\mathrm{e}^x (\psi(1) + \alpha) \\
&= -q\mathrm{e}^x < 0,
\end{aligned}$$

for $x > a$. In conclusion, we have shown that $\mathcal{L}^1 v_a(x) \leq 0$ on $x \in (0, \infty) \backslash \{a\}$.

If $a \geq x^*$, then from (11.29) $v_a(a-) - v_a(a+) \leq 0$. In the notation of Theorem 8.11,

$$\int_0^\infty \mathrm{e}^{-\alpha s} \mathbb{P}(Y_s^x \in \{a\}) \mathrm{d}s = \lim_{z \uparrow \infty} \overline{U}^{(\alpha)}(z, x, \{a\}) = 0.$$

This implies that we can rewrite (11.30) in the form

$$\begin{aligned}
\mathrm{e}^{-\alpha t} v_a(Y_t^x) = v_a(x) &+ \int_0^t \mathrm{e}^{-\alpha s} \mathcal{L}^1 v_a(Y_s^x) \mathrm{d}s \\
&+ \int_0^t \mathrm{e}^{-\alpha s} (v_a(a-) - v_a(a+)) \mathrm{d}L_t^a + M_t,
\end{aligned}$$

without having to worry about the value of $\mathcal{L}^1 v_a$ at a. It follows that the process $\{\mathrm{e}^{-\alpha t} v_a(X_t) : t \geq 0\}$ is the difference of a martingale M and a non-decreasing adapted process, thus making it a supermartingale. It is also clear from (11.30) that this supermartingale must be right-continuous.

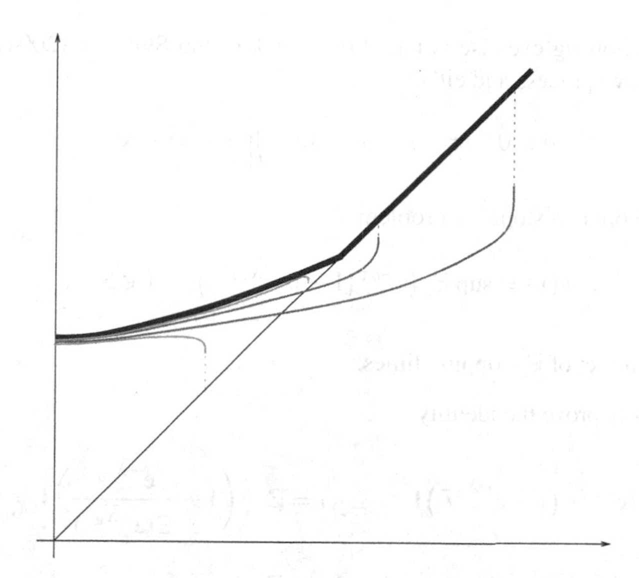

Fig. 11.3 A sketch of the functions $v_a(\log x)$ for different values of a, when X is of bounded variation and $\nu(-\infty, 0) = \infty$. Curves which do not bound from above the diagonal correspond to $v_a(\log x)$ for $a > x^*$. Curves which are bounded from below by the diagonal correspond to $v_a(\log x)$ for $0 < a < x^*$. The unique curve which bounds from above the diagonal with continuous fit corresponds to $v_a(\log x)$ with $y = x^*$.

In conclusion, all properties of Lemma 11.1 are satisfied uniquely when $a = x^*$, thus concluding the proof. □

The semi-explicit nature of the functions $\{v_a : a \geq 0\}$, once again, gives us the opportunity to show graphically how continuous fit occurs, by perturbing the function v_a about the value $a = x^*$. See Fig. 11.3.

Exercises

11.1 Suppose that X is a spectrally negative Lévy process. Consider the process $Y^x = \{Y_t^x : t \geq 0\}$, where $Y_t^x = (x \vee \overline{X}_t) - X_t$. Recall the definition

$$\overline{\sigma}_a^x = \inf\{t > 0 : Y_t^x > a\},$$

for $0 \leq x \leq a$. Use the Poisson point process of excursions, described in Theorem 6.14, to show that

$$\mathbb{P}(\overline{\sigma}_a^x = \infty) = \lim_{t \uparrow \infty} \frac{W(a-x)}{W(a)} \exp\{-tW'(a)/W(a)\},$$

and hence $\mathbb{P}(\overline{\sigma}_a^x < \infty) = 1$, for all $0 \leq x \leq a$.

11.2 The following exercise is based on Novikov and Shiryaev (2004). Suppose that X is a Lévy process and either:

$$q>0 \quad \text{or} \quad q=0 \quad \text{and} \quad \lim_{t\uparrow\infty} X_t = -\infty.$$

Consider the optimal stopping problem

$$v(x) = \sup_{\tau\in\mathcal{T}} \mathbb{E}_x\big(\mathrm{e}^{-q\tau}\big(1-\mathrm{e}^{-(X_\tau)^+}\big)\big), \quad x\in\mathbb{R}, \tag{11.33}$$

where $\mathcal{T}$ is the set of $\mathbb{F}$-stopping times.

(i) For $a>0$, prove the identity

$$\mathbb{E}_x\big(\mathrm{e}^{-qT_a^+}\big(1-\mathrm{e}^{-X_{T_a^+}}\big)\mathbf{1}_{(T_a^+<\infty)}\big) = \mathbb{E}_x\left(\left(1-\frac{\mathrm{e}^{-\overline{X}_{\mathrm{e}_q}}}{\mathbb{E}(\mathrm{e}^{-\overline{X}_{\mathrm{e}_q}})}\right)\mathbf{1}_{(\overline{X}_{\mathrm{e}_q}\geq a)}\right),$$

where $T_a^+ = \inf\{t\geq 0 : X_t \geq a\}$ and $x\in\mathbb{R}$.

(ii) Show that a solution to (11.33) is given by the pair $(v_{x^*}, T_{x^*}^+)$, where $v_{x^*}(x)$ is equal to the left-hand side of the identity in part (i) with

$$a = x^* := -\log \mathbb{E}\big(\mathrm{e}^{-\overline{X}_{\mathrm{e}_q}}\big).$$

(iii) Show that there is smooth fit at x^* if and only if 0 is regular for $(0,\infty)$ for X, and otherwise there is continuous fit.

11.3 This exercise is taken from Baurdoux (2007) and is based on a method of Beibel and Lerche (1997). Suppose that X is a spectrally negative α-stable Lévy process, with $\alpha\in(1,2)$ and probabilities $\{\mathbb{P}_x : x\in\mathbb{R}\}$. Let $\eta>0$ and define for, $x\in\mathbb{R}$,

$$H(x) = \int_0^\infty \mathrm{e}^{ux-u^\alpha} u^{\alpha\eta-1}\mathrm{d}u.$$

Now suppose that h is a function on $\mathbb{R}$ such that there exists some x^* satisfying

$$x^* = \mathrm{argmax}_{x\in\mathbb{R}} \frac{h(x)}{H(x)}.$$

(i) Show that, for all $x\in\mathbb{R}$,

$$\frac{H((t+1)^{-1/\alpha}X_t)}{H(x)(t+1)^\eta}, \quad t\geq 0$$

is a martingale under $\mathbb{P}_x$.

(ii) Use the martingale in part (i) to deduce that, for any stopping time τ,

$$\mathbb{E}_x\left[\frac{h((\tau+1)^{-1/\alpha}X_\tau)}{(\tau+1)^\eta}\mathbf{1}_{(\tau<\infty)}\right] \le H(x)\frac{h(x^*)}{H(x^*)}.$$

(iii) Define

$$\tau^* = \inf\{t > 0 : (1+t)^{-1/\alpha}X_t = x^*\}.$$

Assuming that $\mathbb{P}(\tau^* < \infty) = 1$ for $x < x^*$,[5] show that τ^* is an optimal stopping time for

$$V(x) = \sup_\tau \mathbb{E}_x\left[\frac{h((\tau+1)^{-1/\alpha}X_\tau)}{(\tau+1)^\eta}\mathbf{1}_{(\tau<\infty)}\right],$$

for $x < x^*$, where the supremum is taken over all stopping times for X. Moreover deduce that $V(x) = h(x^*)H(x)/H(x^*)$.

[5]In fact it *is* the case that $\mathbb{P}(\tau^* < \infty) = 1$ thanks to the law of the iterated logarithm for X, which states that

$$\limsup_{t\uparrow\infty} \frac{X_t}{t^{1/\alpha}(2\log\log t)^{(\alpha-1)/\alpha}} = c_\alpha$$

almost surely, for some constant $c_\alpha > 0$.

Chapter 12
Continuous-State Branching Processes

Our interest in continuous-state branching processes will be in exposing their intimate relationship with spectrally positive Lévy processes. A flavour for this has already been given in Sect. 1.3.4, where it was shown that a compound Poisson process killed on exiting $(0, \infty)$ can be time changed to obtain a continuous-time Bienaymé–Galton–Watson process, and vice versa. The analogue of this path transformation in greater generality consists of time changing the path of a spectrally positive Lévy process, killed on exiting $(0, \infty)$, to obtain a process equal in law to a Markov process which observes the so-called *branching property* (defined in more detail later) and vice versa. The latter process is what we refer to as the continuous-state branching process. The time change binding the two processes together is called the Lamperti transform, following the foundational work of Lamperti (1967a, 1967b).[1]

Having looked closely at the Lamperti transform, we shall give an account of a number of observations concerning the long-term behaviour, as well as conditioning on survival, of continuous-state branching processes. Thanks to some of the results in Chap. 8, semi-explicit results can be obtained.

12.1 The Lamperti Transform

A $[0, \infty]$-valued strong Markov process $Y = \{Y_t : t \geq 0\}$ with probabilities $\{P_x : x \geq 0\}$ is called a *continuous-state branching process* if it has paths that are right-continuous with left limits and its law observes the branching property given in Definition 1.14.[2] Another way of phrasing the branching property is that, for all $\theta \geq 0$ and $x, y \geq 0$,

$$E_{x+y}\big(\mathrm{e}^{-\theta Y_t}\big) = E_x\big(\mathrm{e}^{-\theta Y_t}\big) E_y\big(\mathrm{e}^{-\theta Y_t}\big). \qquad (12.1)$$

[1] See also Silverstein (1968).

[2] As usual, for $x \geq 0$, the measure P_x satisfies the property $P_x(Y_0 = x) = 1$.

A.E. Kyprianou, *Fluctuations of Lévy Processes with Applications*, Universitext,
DOI 10.1007/978-3-642-37632-0_12, © Springer-Verlag Berlin Heidelberg 2014

Note from the above equality that, by iterating, we may always write, for each $x > 0$,

$$E_x(\mathrm{e}^{-\theta Y_t}) = E_{x/n}(\mathrm{e}^{-\theta Y_t})^n, \tag{12.2}$$

showing that Y_t is infinitely divisible for each $t > 0$. If we define for, $\theta, t \geq 0$,

$$g(t,\theta,x) = -\log E_x(\mathrm{e}^{-\theta Y_t}),$$

then (12.2) implies that, for any positive integer m,

$$g(t,\theta,m) = ng(t,\theta,m/n) \quad \text{and} \quad g(t,\theta,m) = mg(t,\theta,1),$$

showing that for $x \in \mathbb{Q} \cap [0,\infty)$,

$$g(t,\theta,x) = xu_t(\theta), \tag{12.3}$$

where $u_t(\theta) = g(t,\theta,1) \geq 0$. From (12.1), we also see that, for $0 \leq z < y$, $g(t,\theta,z) \leq g(t,\theta,y)$, which implies that $g(t,\theta,x-)$ exists as a left limit and is less than or equal to $g(t,\theta,x+)$, which exists as a right limit. Thanks to (12.3), both left and right limits are the same, so that, for all $x > 0$

$$E_x(\mathrm{e}^{-\theta Y_t}) = \mathrm{e}^{-xu_t(\theta)}. \tag{12.4}$$

The Markov property in conjunction with (12.4) implies that, for all $x > 0$ and $t, s, \theta \geq 0$,

$$\mathrm{e}^{-xu_{t+s}(\theta)} = E_x\big(E(\mathrm{e}^{-\theta Y_{t+s}}|Y_t)\big) = E_x(\mathrm{e}^{-Y_t u_s(\theta)}) = \mathrm{e}^{-xu_t(u_s(\theta))}.$$

In other words, the Laplace exponent of Y obeys the semi-group property

$$u_{t+s}(\theta) = u_t\big(u_s(\theta)\big). \tag{12.5}$$

The first significant glimpse one gets of Lévy processes, in relation to the above definition of a continuous-state branching process, comes with the following result, for which we offer no proof on account of the associated technicalities (see, however, Exercise 1.11 for intuitive motivation or Chap. II of Le Gall (1999), Silverstein (1968), Caballero et al. (2009) for a proof).

Theorem 12.1 *For $t, \theta \geq 0$, suppose that $u_t(\theta)$ is the Laplace functional given by (12.4) of some continuous-state branching process. Then it is differentiable in t and satisfies*

$$\frac{\partial u_t}{\partial t}(\theta) + \psi\big(u_t(\theta)\big) = 0, \tag{12.6}$$

with initial condition $u_0(\theta) = \theta$, where for $\lambda \geq 0$,

$$\psi(\lambda) = -q - a\lambda + \frac{1}{2}\sigma^2\lambda^2 + \int_{(0,\infty)} \big(\mathrm{e}^{-\lambda x} - 1 + \lambda x \mathbf{1}_{(x<1)}\big)\Pi(\mathrm{d}x), \tag{12.7}$$

with $q \geq 0$, $a \in \mathbb{R}$, $\sigma \geq 0$ *and* Π *is a measure supported in* $(0, \infty)$ *satisfying* $\int_{(0,\infty)}(1 \wedge x^2)\Pi(\mathrm{d}x) < \infty$.

Note that, for $\lambda \geq 0$, $\psi(\lambda) = \log \mathbb{E}(\mathrm{e}^{-\lambda X_1})$, where X is either a spectrally positive Lévy process[3] or a subordinator, killed independently at rate $q \geq 0$, with cemetery state $+\infty$.[4] From Sects. 8.1 and 5.5, respectively, we know that ψ is convex, infinitely differentiable on $(0, \infty)$, $\psi(0) = q$ and $\psi'(0+) \in [-\infty, \infty)$. Further, if X is a (killed) subordinator, then $\psi(\infty) < 0$ and otherwise, we have that $\psi(\infty) = \infty$.

For each $\theta > 0$ the solution to (12.6) can be uniquely identified by the relation

$$-\int_{\theta}^{u_t(\theta)} \frac{1}{\psi(\xi)} \mathrm{d}\xi = t. \tag{12.8}$$

This is easily confirmed by elementary differentiation. Note, by letting $t \downarrow 0$, we notice that $u_0(\theta) = \theta$.

From the discussion above, we see that *if a continuous-state branching process exists*, then it is associated with a particular function $\psi : [0, \infty) \mapsto \mathbb{R}$ given by (12.7). Formally speaking, we shall refer to all ψ which respect the definition (12.7) as *branching mechanisms*. We will now state without proof the Lamperti transform which, amongst other things, shows that every branching mechanism ψ can be associated with a continuous-state branching process.

Theorem 12.2 (The Lamperti transform) *Let* ψ *be any given branching mechanism.*

(i) *Suppose that* $X = \{X_t : t \geq 0\}$ *is a Lévy process with no negative jumps, killed (with cemetery state* $+\infty$*) at an independent and exponentially distributed time with parameter* $q \geq 0$. *Further,* $\psi(\lambda) = \log \mathbb{E}(\mathrm{e}^{-\lambda X_1})$. *Define, for* $t \geq 0$,

$$Y_t = X_{\theta_t \wedge \tau_0^-},$$

where $\tau_0^- = \inf\{t > 0 : X_t < 0\}$ *and*

$$\theta_t = \inf\left\{s > 0 : \int_0^s \frac{\mathrm{d}u}{X_u} > t\right\}.$$

Then under $\mathbb{P}_x$, $x \geq 0$, $Y = \{Y_t : t \geq 0\}$ *is a continuous-state branching process with branching mechanism* ψ *and initial value* $Y_0 = x$.

[3] Recall that our definition of spectrally positive processes excludes subordinators. See the discussion following Lemma 2.14.

[4] As usual, we understand the process X killed at rate q to mean that it is sent to $+\infty$ after an independent and exponentially distributed time with parameter q. Further $q = 0$ means there is no killing.

(ii) *Conversely, suppose that $Y = \{Y_t : t \geq 0\}$ is a continuous-state branching process with branching mechanism ψ, such that $Y_0 = x \geq 0$. Define for $t \geq 0$,*

$$X_t = Y_{\varphi_t},$$

where

$$\varphi_t = \inf\left\{s > 0 : \int_0^s Y_u \mathrm{d}u > t\right\}.$$

Then $X = \{X_t : t \geq 0\}$ is a Lévy process with no negative jumps and initial position $X_0 = x$, which is stopped on first entry into $(-\infty, 0)$ and killed (with cemetery state $+\infty$) at an independent and exponentially distributed time with parameter $q \geq 0$. If $\mathbb{P}$ is the law of X conditional on $X_0 = 0$, then $\psi(\lambda) = \log \mathbb{E}(\mathrm{e}^{-\lambda X_1})$, $\lambda \geq 0$.

It can be shown that a general continuous-state branching process appears as the result of an asymptotic rescaling (in time and space) of the continuous-time Bienaymé–Galton–Watson process discussed in Sect. 1.3.4; see Jirina (1958). Roughly speaking, the Lamperti transform for continuous-state branching processes then follows as a consequence of the analogous construction being valid for the continuous-time Bienaymé–Galton–Watson process.

12.2 Long-Term Behaviour

For the forthcoming discussion, it will be useful to recall the definition of a Bienaymé–Galton–Watson process. This process is a discrete-time Markov chain $Z = \{Z_n : n = 0, 1, 2, \ldots\}$ with state space $\{0, 1, 2, \ldots\}$. The quantity Z_n is to be thought of as the size of the n-th generation of some asexually reproducing population. The process Z has probabilities $\{P_x : x = 0, 1, 2, \ldots\}$ such that, under P_x, $Z_0 = x$ and

$$Z_n = \sum_{i=1}^{Z_{n-1}} \xi_i^{(n)}, \tag{12.9}$$

for $n = 1, 2, \ldots$, where, for each $n \geq 1$, $\{\xi_i^{(n)} : i = 1, 2, \ldots\}$ are independent and identically distributed on $\{0, 1, 2, \ldots\}$.

Without specifying anything further about the common distribution of the offspring, there are two events which are of immediate concern for the Markov chain Z, *explosion* and *absorption*. In the first case it is not clear whether or not the event $\{Z_n = \infty\}$ has positive probability for some $n \geq 1$ (the latter could happen if, for example, the offspring distribution has no moments). When $P_x(Z_n < \infty) = 1$, for all $n \geq 1$, we say the process is *conservative* (in other words there is no explosion). In the second case, we note from the definition of Z that if $Z_n = 0$ for some $n \geq 1$, then $Z_{n+m} = 0$ for all $m \geq 0$, which makes 0 an absorbing state. As Z_n is to be thought of as the size of the n-th generation of some asexually reproducing population, the event $\{Z_n = 0 \text{ for some } n > 0\}$ is referred to as *extinction.*

In this section, we consider the analogues of conservative behaviour and extinction within the setting of continuous-state branching processes. In addition, we shall examine the laws of the supremum and total progeny process of continuous-state branching processes. These are the analogues of

$$\sup_{n\geq 0} Z_n \quad \text{and} \quad \left\{\sum_{0\leq k\leq n} Z_k : n \geq 0\right\}$$

for the Bienaymé–Galton–Watson process. Note, in the latter case, total progeny is interpreted as the total number of offspring to date.

12.2.1 Conservative Processes

A continuous-state branching process, $Y = \{Y_t : t \geq 0\}$, is said to be *conservative* if, for all $t > 0$, $P(Y_t < \infty) = 1$. The following result is taken from Grey (1974).

Theorem 12.3 *A continuous-state branching process with branching mechanism ψ is conservative if and only if*

$$\int_{0+} \frac{1}{|\psi(\xi)|} \mathrm{d}\xi = \infty.$$

Therefore, a necessary condition is that $\psi(0) = 0$ and a sufficient condition is that $\psi(0) = 0$ and $|\psi'(0+)| < \infty$ (equivalently $q = 0$ and $\int_{[1,\infty)} x\Pi(\mathrm{d}x) < \infty$).

Proof From the definition of $u_t(\theta)$, a continuous-state branching process is conservative if and only if $\lim_{\theta\downarrow 0} u_t(\theta) = 0$, since, for each $x > 0$,

$$P_x(Y_t < \infty) = \lim_{\theta\downarrow 0} E_x\left(\mathrm{e}^{-\theta Y_t}\right) = \exp\left\{-x \lim_{\theta\downarrow 0} u_t(\theta)\right\},$$

where the limits are justified by monotonicity. However, note from (12.8) that as $\theta \downarrow 0$,

$$t = -\int_0^\delta \frac{1}{\psi(\xi)} \mathrm{d}\xi + \int_{u_t(\theta)}^\delta \frac{1}{\psi(\xi)} \mathrm{d}\xi,$$

where $\delta > 0$ is sufficiently small. However, as the left-hand side is independent of θ, we are forced to conclude that $\lim_{\theta\downarrow 0} u_t(\theta) = 0$ if and only if

$$\int_{0+} \frac{1}{|\psi(\xi)|} \mathrm{d}\xi = \infty.$$

Note that $\psi(\theta)$ may be negative in the neighbourhood of the origin, and hence the absolute value is taken in the integral.

If ψ is bounded away from zero in the neighbourhood of the origin, then $1/|\psi|$ is locally integrable there. Hence, a necessary condition to be conservative is that $\psi(0)=0$. On the other hand, if $\psi(0)=0$ and ψ is locally linear in the neighbourhood of the origin, then $1/|\psi|$ is not locally integrable there. Hence, recalling that ψ is a smooth function on $[0,\infty)$, a sufficient condition to be conservative is that $\psi(0)=0$ and $|\psi'(0+)|<\infty$. □

Henceforth, we shall assume that there is no explosion (and in particular that $q=0$*).*

12.2.2 Extinction Probabilities

Thanks to the representation of continuous-state branching processes given in Theorem 12.2 (i), it is clear that they observe the fundamental property that if $Y_t=0$ for some $t>0$, then $Y_{t+s}=0$ for $s\geq 0$. This can also be seen from the branching property (12.1). By taking $y=0$ in (12.1), we see that P_0 must be the measure that assigns probability one to the process which is identically zero. Hence by the Markov property, once in state zero, the process remains in state zero.

Let $\zeta=\inf\{t>0:Y_t=0\}$. The event $\{\zeta<\infty\}=\{Y_t=0 \text{ for some } t>0\}$ is thus referred to as *extinction*, in line with the same terminology used for the Bienaymé–Galton–Watson process.

Note from (12.4) that $u_t(\theta)$ is continuously differentiable in θ on $(0,\infty)$ (since, by dominated convergence, the same is true for the left-hand side of the aforementioned equality). Differentiating (12.4) in θ, we find that, for each $\theta, x, t>0$,

$$E_x\big(Y_t \mathrm{e}^{-\theta Y_t}\big)=x\frac{\partial u_t}{\partial\theta}(\theta)\mathrm{e}^{-xu_t(\theta)}. \tag{12.10}$$

Hence taking limits as $\theta\downarrow 0$, we obtain

$$E_x(Y_t)=x\frac{\partial u_t}{\partial\theta}(0+), \tag{12.11}$$

so that both sides of the equality are infinite at the same time. Differentiating (12.6) in θ, we also find that, for $\theta>0$,

$$\frac{\partial}{\partial t}\frac{\partial u_t}{\partial\theta}(\theta)+\psi'\big(u_t(\theta)\big)\frac{\partial u_t}{\partial\theta}(\theta)=0.$$

Standard techniques for first-order differential equations then imply that

$$\frac{\partial u_t}{\partial\theta}(\theta)=c\exp\bigg\{-\int_0^t\psi'\big(u_s(\theta)\big)\mathrm{d}s\bigg\}, \tag{12.12}$$

for some constant $c > 0$. Inspecting (12.10) as $t \downarrow 0$, we see that $c = 1$. Now taking limits as $\theta \downarrow 0$ and recalling that, for each fixed $s > 0$, $u_s(\theta) \downarrow 0$ (thanks to the exclusion of explosive behaviour), it is straightforward to deduce from (12.11) and (12.12) that

$$E_x(Y_t) = x\mathrm{e}^{-\psi'(0+)t}, \tag{12.13}$$

where we understand the left-hand side to be infinite whenever $\psi'(0+) = -\infty$. Note that, from its definition, we know that ψ is convex and $\psi'(0+) \in [-\infty, \infty)$ (cf. Sects. 5.5 and 8.1). Hence, to obtain (12.13), we have used dominated convergence in the integral in (12.12) when $|\psi'(0+)| < \infty$ and monotone convergence when $\psi'(0+) = -\infty$.

This leads to the following classification of continuous-state branching processes.

Definition 12.4 A continuous-state branching process with branching mechanism ψ is called:

(i) subcritical, if $\psi'(0+) > 0$,
(ii) critical, if $\psi'(0+) = 0$ and
(iii) supercritical, if $\psi'(0+) < 0$.

The use of the terminology "criticality" refers then to whether the process will, on average, decrease, remain constant or increase. The same terminology is employed for Bienaymé–Galton–Watson processes where now the three cases in Definition 12.4 correspond to the mean of the offspring distribution being strictly less than, equal to and strictly greater than unity, respectively. A classic result, attributed to the scientists after which the latter process is named, states that there is extinction with probability 1 if and only if the mean offspring size is less than or equal to unity (see Chap. I of Athreya and Ney (1972) for example). The analogous result for continuous-state branching processes might therefore say that there is extinction with probability one if and only if $\psi'(0+) \geq 0$. However, here we encounter a subtle difference for continuous-state branching processes as the following simple example shows. In the representation given by Theorem 12.2, take $X_t = 1 - t$ corresponding to $Y_t = \mathrm{e}^{-t}$. Clearly $\psi(\lambda) = \lambda$ so that $\psi'(0+) = 1 > 0$ and yet $Y_t > 0$ for all $t > 0$.

Extinction is characterised by the following result, due to Grey (1974); see also Bingham (1976).

Theorem 12.5 *Suppose that Y is a continuous-state branching process with branching mechanism ψ. For all $x \geq 0$, let $p(x) = P_x(\zeta < \infty)$.*

(i) *If $\psi(\infty) < 0$, then for all $x > 0$, $p(x) = 0$.*
(ii) *When $\psi(\infty) = \infty$, $p(x) > 0$ for some (and then for all) $x > 0$ if and only if*

$$\int^{\infty} \frac{1}{\psi(\xi)} \mathrm{d}\xi < \infty, \tag{12.14}$$

in which case $p(x) = \mathrm{e}^{-\Phi(0)x}$, *where* $\Phi(0) = \sup\{\lambda \geq 0 : \psi(\lambda) = 0\}$. *Otherwise* $p(x) = 0$ *for all* $x > 0$.

Proof (i) If $\psi(\lambda) = \log \mathbb{E}(\mathrm{e}^{-\lambda X_1})$, $\lambda \geq 0$, where X is a subordinator, then clearly, from the path representation given in Theorem 12.2 (i), extinction occurs with probability zero. From the discussion following Theorem 12.1, the case that X is a subordinator is equivalent to $\psi(\lambda) < 0$ for all $\lambda > 0$.

(ii) Since $\{Y_t = 0\} \subseteq \{Y_{t+s} = 0\}$, for $s, t > 0$, we have by monotonicity that, for each $x > 0$,

$$P_x(Y_t = 0) \uparrow p(x) \tag{12.15}$$

as $t \uparrow \infty$. Hence $p(x) > 0$ if and only if $P_x(Y_t = 0) > 0$ for some $t > 0$. Since $P_x(Y_t = 0) = \mathrm{e}^{-xu_t(\infty)}$, we see that $p(x) > 0$ for some (and then all) $x > 0$ if and only if $u_t(\infty) < \infty$ for some $t > 0$.

Fix $t > 0$. Taking limits in (12.8) as $\theta \uparrow \infty$, we see that, if $u_t(\infty) < \infty$, then

$$\int^{\infty} \frac{1}{\psi(\xi)} \mathrm{d}\xi < \infty. \tag{12.16}$$

Conversely, if the above integral condition holds, then, again taking limits in (12.8) as $\theta \uparrow \infty$, it must necessarily hold that $u_t(\infty) < \infty$.

Finally, assuming (12.16), we have learnt that

$$\int_{u_t(\infty)}^{\infty} \frac{1}{\psi(\xi)} \mathrm{d}\xi = t. \tag{12.17}$$

From (12.15) and the fact that $u_t(\infty) = -x^{-1} \log P_x(Y_t = 0)$, we see that $u_t(\infty)$ decreases as $t \uparrow \infty$ to the largest constant, $c \geq 0$, such that $\int_c^\infty 1/\psi(\xi)\mathrm{d}\xi$ becomes infinite. Appealing to the convexity and smoothness of ψ, the constant c must necessarily correspond to a root of ψ in $[0, \infty)$, at which point it will behave linearly and thus cause $\int_c^\infty 1/\psi(\xi)\mathrm{d}\xi$ to blow up. There are at most two such points, and the largest of these is described precisely by $c = \Phi(0) \in [0, \infty)$ (see Sect. 8.1). In conclusion,

$$p(x) = \lim_{t \uparrow \infty} \mathrm{e}^{-xu_t(\infty)} = \mathrm{e}^{-\Phi(0)x},$$

as required. □

On account of the convexity of ψ, we also recover the following corollary to part (ii) of the above theorem.

Corollary 12.6 *For a continuous-state branching process with branching mechanism* ψ *satisfying* $\psi(\infty) = \infty$ *and*

$$\int^{\infty} \frac{1}{\psi(\xi)} \mathrm{d}\xi < \infty,$$

we have $p(x) < 1$ *for some (and then for all)* $x > 0$ *if and only if* $\psi'(0+) < 0$.

To summarise the conclusions of Theorem 12.5 and Corollary 12.6, we have the following cases for the extinction probability $p(x)$:

Condition	$p(x)$
$\psi(\infty) < 0$	0
$\psi(\infty) = \infty,\ \int^{\infty} \psi(\xi)^{-1}\mathrm{d}\xi = \infty$	0
$\psi(\infty) = \infty,\ \psi'(0+) < 0,\ \int^{\infty} \psi(\xi)^{-1}\mathrm{d}\xi < \infty$	$\mathrm{e}^{-\Phi(0)x} \in (0, 1)$
$\psi(\infty) = \infty,\ \psi'(0+) \geq 0,\ \int^{\infty} \psi(\xi)^{-1}\mathrm{d}\xi < \infty$	1

Let us return to the example that $\psi(\lambda) = \lambda$, resulting in a continuous-state branching process $Y_t = \mathrm{e}^{-t}$ for $t \geq 0$. We see that, in the above table, this example falls into the second category. Despite the fact that this is a process which does not become extinct, it does become *extinguished*. That is to say, $\lim_{t\uparrow\infty} Y_t = 0$ with positive probability; in fact with probability one.

What is not clear from the above table is whether, like the aforesaid example, all continuous-state branching processes which fall into the second category necessarily become *extinguished*. Exercises 12.1 and 12.3 explore this eventuality in more detail. Despite the fact that extinction is impossible within the second category, the probability of becoming extinguished is again related to the largest root of the branching mechanism. Specifically, the following result is established in the aforementioned exercises.

Theorem 12.7 *Suppose that $\psi(\infty) = \infty$. Then for all $x \geq 0$,*

$$P_x\Big(\lim_{t\uparrow\infty} Y_t = 0\Big) = \mathrm{e}^{-\Phi(0)x}.$$

This means that (sub)critical processes which do not become extinct necessarily become extinguished with probability one. Moreover, supercritical processes which do not become extinct can still become extinguished with positive probability. The only exception in the latter case is if the underlying Lévy process is a subordinator, in which case there is neither extinction nor the possibility of becoming extinguished.

12.2.3 Total Progeny and the Supremum

Thinking of a continuous-state branching process, $\{Y_t : t \geq 0\}$, as the continuous-time, continuous-state analogue of the Bienaymé–Galton–Watson process, it is reasonable to refer to

$$J_t := \int_0^t Y_u \mathrm{d}u$$

as the total progeny until time $t \geq 0$.

In this section our main goal, given in the theorem below, is to provide distributional identities for $J_{T_a^+}$, where

$$T_a^+ = \inf\{t > 0 : Y_t > a\},$$

and $\sup_{s \le \zeta} Y_s$. To facilitate the statement of the main result, let us first recall the following notation. As remarked above, for any branching mechanism ψ, when $\psi(\infty) = \infty$ (in other words when the Lévy process associated with ψ is not a subordinator), we have that ψ is the Laplace exponent of a spectrally negative Lévy process. Let $\Phi(q) = \sup\{\theta \ge 0 : \psi(\theta) = q\}$ (cf. Sect. 8.1). Associated with ψ are the scale functions $W^{(q)}$ and $Z^{(q)}$ (cf. Sect. 8.2). In particular,

$$\int_0^\infty \mathrm{e}^{-\beta x} W^{(q)}(x)\mathrm{d}x = \frac{1}{\psi(\beta) - q} \quad \text{for } \beta > \Phi(q),$$

and $Z^{(q)}(x) = 1 + q\int_0^x W^{(q)}(y)\mathrm{d}y$. Following the notational protocol of Chap. 8, we write W in place of $W^{(0)}$.

Theorem 12.8 *Let $Y = \{Y_t : t \ge 0\}$ be a continuous-state branching process with branching mechanism ψ satisfying $\psi(\infty) = \infty$.*

(i) *For each $a \ge x > 0$ and $q \ge 0$,*

$$E_x\big(\mathrm{e}^{-q\int_0^{T_a^+} Y_s \mathrm{d}s} \mathbf{1}_{(T_a^+ < \zeta)}\big) = Z^{(q)}(a - x) - W^{(q)}(a - x)\frac{Z^{(q)}(a)}{W^{(q)}(a)}.$$

(ii) *For each $a \ge x > 0$ and $q \ge 0$,*

$$E_x\big(\mathrm{e}^{-q\int_0^{\zeta} Y_s \mathrm{d}s} \mathbf{1}_{(\zeta < T_a^+)}\big) = \frac{W^{(q)}(a - x)}{W^{(q)}(a)}.$$

Proof Suppose now that X is the Lévy process mentioned in Theorem 12.2 (ii). Write in the usual way $\tau_a^+ = \inf\{t > 0 : X_t > a\}$ and $\tau_0^- = \inf\{t > 0 : X_t < 0\}$. Then the Lamperti transform implies that

$$\tau_a^+ = \int_0^{T_a^+} Y_s \mathrm{d}s \quad \text{and} \quad \tau_0^- = \int_0^{\zeta} Y_s \mathrm{d}s. \tag{12.18}$$

The proof is now completed by invoking Theorem 8.1 (iii) for the process X. Note that X is a spectrally positive Lévy process and, hence, to implement the aforementioned result, which applies to spectrally negative processes, one must consider the problem of two-sided exit from $[0, a]$ of $-X$ with initial condition $X_0 = a - x$. □

We conclude this section by noting the following two corollaries of Theorem 12.8 which, in their original form, are due to Bingham (1976).

Corollary 12.9 *Under the assumptions of Theorem* 12.8, *we have for each* $a \geq x > 0$,

$$P_x\Big(\sup_{s\geq 0} Y_s \leq a\Big) = \frac{W(a-x)}{W(a)}.$$

In particular,

$$P_x\Big(\sup_{s\geq 0} Y_s < \infty\Big) = \mathrm{e}^{-\Phi(0)x}, \quad x \geq 0,$$

and the right-hand side is equal to unity if and only if Y is not supercritical.

Proof The first part is obvious by taking limits as $q \downarrow 0$ in Theorem 12.8 (i). The second part follows by taking limits as $a \uparrow \infty$ and making use of Exercise 8.5 (i). Recall that $\Phi(0) > 0$ if and only if $\psi'(0+) < 0$. □

Corollary 12.10 *Under the assumptions of Theorem* 12.8, *we have for each* $x > 0$ *and* $q \geq 0$,

$$E_x\big(\mathrm{e}^{-q\int_0^\xi Y_s \mathrm{d}s}\big) = \mathrm{e}^{-\Phi(q)x}.$$

Proof The proof is, again, a straightforward consequence of Theorem 12.8 (ii) by taking limits as $a \uparrow \infty$ and then applying the conclusion of Exercise 8.5 (i). □

12.3 Conditioned Processes and Immigration

In the classical theory of Bienaymé–Galton–Watson processes where the offspring distribution is assumed to have finite mean, it is well understood that by taking a critical or subcritical process (for which extinction occurs with probability one) and conditioning it in the long term to remain positive, one uncovers a beautiful relationship between a martingale change of measure and processes with immigration; cf. Athreya and Ney (1972) and Lyons et al. (1995).

Let us be a little more specific. A Bienaymé–Galton–Watson process with immigration is defined as the Markov chain $Z^* = \{Z_n^* : n = 0, 1, \ldots\}$ where $Z_0^* = z \in \{0, 1, 2, \ldots\}$ and, for $n = 1, 2, \ldots$,

$$Z_n^* = Z_n + \sum_{k=1}^{n} Z_{n-k}^{(k)}, \tag{12.19}$$

where now, $Z = \{Z_n : n \geq 0\}$ has law P_z. Moreover, for each $k = 1, 2, \ldots, n$, $Z_{n-k}^{(k)}$ is independent and equal in distribution to numbers in the $(n-k)$-th generation of (Z, P_{η_k}), where it is assumed that the initial numbers, η_k, are, independently of everything else, randomly distributed according to the probabilities $\{p_i^* : i = 0, 1, 2, \ldots\}$. Intuitively speaking, one may see the process Z^* as a variant of the Bienaymé–Galton–Watson process, Z, in which, from the first and subsequent generations, there is a stream of immigrants, $\{\eta_1, \eta_2, \ldots\}$ each of whom initiates an independent copy of (Z, P_1).

Suppose now that Z is a Bienaymé–Galton–Watson process with probabilities $\{P_x : x = 1, 2, \ldots\}$ as described above. For any event A which belongs to the sigma-algebra generated by the first n generations, it turns out that for each $x = 0, 1, 2, \ldots,$

$$P_x^*(A) := \lim_{m \uparrow \infty} P_x(A | Z_k > 0 \text{ for } k = 0, 1, \ldots, n + m)$$

is well defined, and further,

$$P_x^*(A) = E_x(\mathbf{1}_A M_n),$$

where $M_n = m^{-n} Z_n / Z_0$ and $m = E_1(Z_1)$, which is assumed to be finite. It is not difficult to show, using the iteration (12.9), that $E_x(Z_n) = xm^n$ and that $\{M_n : n \geq 0\}$ is a martingale. What is perhaps more intriguing is that the new process, (Z, P_x^*), can be identified in two different ways:

1. The process $\{Z_n - 1 : n \geq 0\}$ under P_x^* can be shown to have the same law as a Bienaymé–Galton–Watson process with immigration having $x - 1$ initial ancestors. The immigration probabilities satisfy $p_i^* = (i + 1)p_{i+1}/m$, for $i = 0, 1, 2, \ldots,$ where $\{p_i : i = 0, 1, 2, \ldots\}$ is the offspring distribution of the original Bienaymé–Galton–Watson process. Moreover, immigrants initiate independent copies of Z.
2. The process Z under P_x^* has the same law as $x - 1$ initial individuals, each one independently initiating a Bienaymé–Galton–Watson process with law P_1, together with one individual initiating an independent immortal genealogical line of descent, *the spine*, along which individuals reproduce with the tilted distribution $\{ip_i/m : i = 1, 2, \ldots\}$. The siblings of individuals on the spine initiate copies of a Bienaymé–Galton–Watson process under P_1. By subtracting individuals on the spine from the aggregate population, one observes a Bienaymé–Galton–Watson process with immigration as described above.

Taking the second interpretation above, one sees that the change of measure has adjusted the statistics on just one genealogical line of descent to ensure that it, and hence the whole process itself, is immortal. See Fig. 12.1.

Our aim in this section is to establish the analogue of these ideas for critical or subcritical continuous-state branching processes. This is done in Sect. 12.3.2. However, we first address the issue of how to condition a spectrally positive Lévy process to stay positive. Apart from being a useful comparison for the case of conditioning a continuous-state branching process, there are reasons to believe that the two classes of conditioned processes might be connected through a Lamperti-type transform because of the relationship given in Theorem 12.2. This is the very last point we address in Sect. 12.3.2.

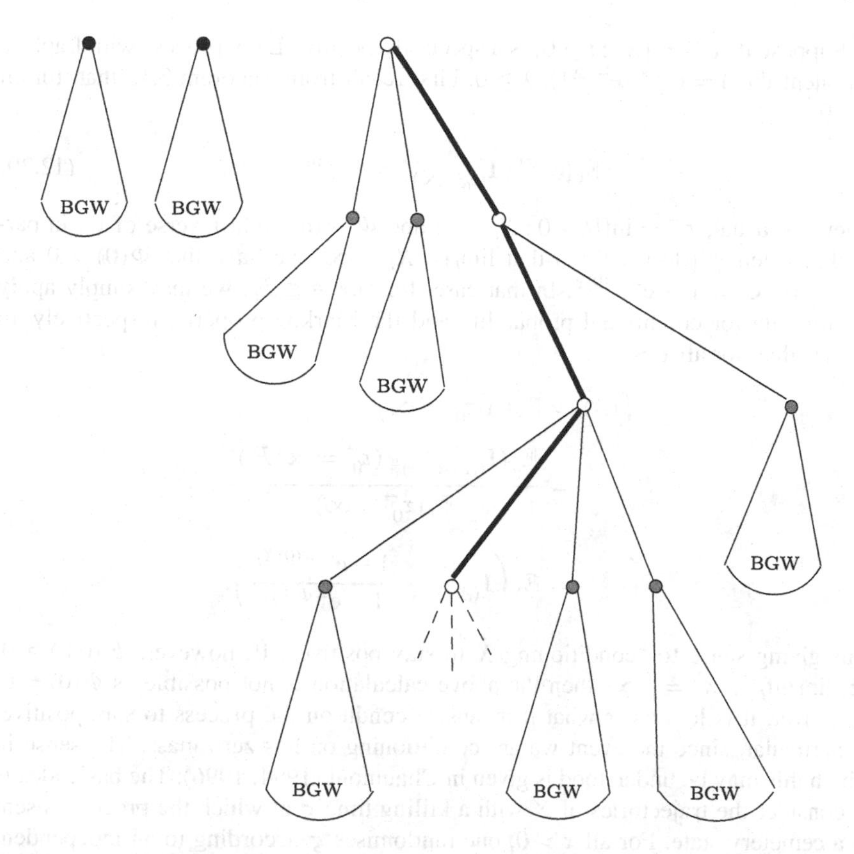

Fig. 12.1 Nodes shaded in *black* initiate Bienaymé–Galton–Watson processes under P_1. Nodes in *white* are individuals belonging to the immortal genealogical line of descent known as the spine. Nodes shaded in *grey* represent the offspring of individuals on the spine who are not themselves members of the spine. These individuals may also be considered as "immigrants".

12.3.1 Conditioning a Spectrally Positive Lévy Process to Stay Positive

It is possible to talk of conditioning any Lévy process to stay positive and this is now a well-documented phenomenon (also for the case of random walks). See Bertoin (1993), Bertoin and Doney (1994b), Chaumont (1994, 1996), Konstantopoulos and Richardson (2002), Duquesne (2003), Bryn-Jones and Doney (2006) and Chaumont and Doney (2005), to name but some of the most recent additions to the literature; see also Lambert (2000) who considers conditioning a spectrally negative Lévy process to stay in a strip. We restrict our attention to the case of conditioning a spectrally positive Lévy process to stay positive, since this is what is required for the forthcoming discussion. This also facilitates the mathematics. A more general treatment of conditioning a Lévy process to stay positive is given in Chap. 13.

Suppose that $X = \{X_t : t \geq 0\}$ is a spectrally positive Lévy process with Laplace exponent $\psi(\lambda) = \log \mathbb{E}(\mathrm{e}^{-\lambda X_1})$, $\lambda \geq 0$. First recall from Theorem 3.12 that, for all $x > 0$,

$$\mathbb{E}_x\big(\mathrm{e}^{-q\tau_0^-}\mathbf{1}_{(\tau_0^- < \infty)}\big) = \mathrm{e}^{-\Phi(q)x}, \tag{12.20}$$

where, as usual, $\tau_0^- = \inf\{t > 0 : X_t < 0\}$ and Φ is the right inverse of ψ. In particular, when $\psi'(0+) < 0$, so that $\lim_{t\uparrow\infty} X_t = \infty$, we have that $\Phi(0) > 0$ and $\mathbb{P}_x(\tau_0^- = \infty) = 1 - \mathrm{e}^{-\Phi(0)x}$. In that case, for any $A \in \mathcal{F}_t$, we may simply apply the formula for conditional probability and the Markov property, respectively, to deduce that, for all $x > 0$,

$$\begin{aligned}
\mathbb{P}_x^{\uparrow}(A) &:= \mathbb{P}_x\big(A|\tau_0^- = \infty\big) \\
&= \frac{\mathbb{E}_x(\mathbf{1}_{(A, t<\tau_0^-)}\mathbb{P}(\tau_0^- = \infty|\mathcal{F}_t))}{\mathbb{P}_x(\tau_0^- = \infty)} \\
&= \mathbb{E}_x\left(\mathbf{1}_{(A, t<\tau_0^-)}\frac{1 - \mathrm{e}^{-\Phi(0)X_t}}{1 - \mathrm{e}^{-\Phi(0)x}}\right),
\end{aligned}$$

thus giving sense to "conditioning X to stay positive". If, however, $\psi'(0+) \geq 0$, i.e. $\liminf_{t\uparrow\infty} X_t = -\infty$, then the above calculation is not possible as $\Phi(0) = 0$. Moreover, it is less clear what it means to condition the process to stay positive, in particular, since the event we are conditioning on has zero mass. The sense in which this may be understood is given in Chaumont (1994, 1996). The basic idea is to consider the trajectories of X with a killing time ς at which the process is sent to a cemetery state. For all $x > 0$, one randomises ς according to an independent exponential distribution with rate $q > 0$ under $\mathbb{P}_x$. Conditioning to stay positive (an event which now has positive probability) is then performed up to this killing time, followed by taking limits as $q \downarrow 0$.

Theorem 12.11 [5] *Suppose that* $\mathbf{e}_q$ *is an exponentially distributed random variable, with parameter* q*, that is independent of* X*. Suppose that* $\psi'(0+) \geq 0$*. For all* $x, t > 0$ *and* $A \in \mathcal{F}_t$*,*

$$\mathbb{P}_x^{\uparrow}(A, t < \varsigma) := \lim_{q\downarrow 0} \mathbb{P}_x\big(A, t < \mathbf{e}_q|\tau_0^- > \mathbf{e}_q\big) \tag{12.21}$$

exists and satisfies

$$\mathbb{P}_x^{\uparrow}(A, t < \varsigma) = \mathbb{E}_x\left(\mathbf{1}_{(A, t<\tau_0^-)}\frac{X_t}{x}\right).$$

[5] Here and later on, we make an abuse of notation in working with the independent and exponentially distributed random variables $\mathbf{e}_q$. In taking limits as $q \downarrow 0$ in, for example, (12.21), we appear to be working with an uncountable sequence of independent exponential random variables on the same probability space. This is possible, on account of the fact that, for each $q > 0$, we may write $\mathbf{e}_q = q^{-1}\mathbf{e}_1$.

Proof Appealing to the formula for conditional probability, the Markov property, the lack-of-memory property and (12.20), we have

$$\begin{aligned}\mathbb{P}_x\left(A, t < \mathbf{e}_q | \tau_0^- > \mathbf{e}_q\right) &= \frac{\mathbb{P}_x(A,\ t < \mathbf{e}_q,\ \tau_0^- > \mathbf{e}_q)}{\mathbb{P}_x(\tau_0^- > \mathbf{e}_q)} \\ &= \frac{\mathbb{E}_x(\mathbf{1}_{(A, t<\mathbf{e}_q \wedge \tau_0^-)}\mathbb{E}(\tau_0^- > \mathbf{e}_q | \mathcal{F}_t))}{\mathbb{E}_x(1 - \mathrm{e}^{-q\tau_0^-})} \\ &= \mathbb{E}_x\left(\mathbf{1}_{(A, t<\tau_0^-)}\mathrm{e}^{-qt}\frac{1 - \mathrm{e}^{-\Phi(q)X_t}}{1 - \mathrm{e}^{-\Phi(q)x}}\right). \qquad (12.22)\end{aligned}$$

Under the assumption $\psi'(0+) \geq 0$, we know that $\Phi(0) = 0$ and, hence, by l'Hôpital's Rule

$$\lim_{q \downarrow 0} \frac{1 - \mathrm{e}^{-\Phi(q)X_t}}{1 - \mathrm{e}^{-\Phi(q)x}} = \frac{X_t}{x}. \qquad (12.23)$$

Note also that, for all q sufficiently small,

$$\frac{1 - \mathrm{e}^{-\Phi(q)X_t}}{1 - \mathrm{e}^{-\Phi(q)x}} \leq \frac{\Phi(q)X_t}{1 - \mathrm{e}^{-\Phi(q)x}} \leq C\frac{X_t}{x},$$

where $C > 1$ is a constant. The condition $\psi'(0+) \geq 0$ also implies that, for all $t > 0$, $\mathbb{E}(|X_t|) < \infty$ (see Sect. 8.1) and hence, by dominated convergence, we may take limits in (12.22) as $q \downarrow 0$ and then apply (12.23) to deduce the result. □

When $\psi'(0+) < 0$, for each $x > 0$, $\mathbb{P}_x^\uparrow$ is a probability measure. This is not necessarily the case when $\psi'(0+) \geq 0$. The following lemma gives a more precise account.

Lemma 12.12 *Fix $x > 0$. When $\psi'(0+) = 0$, $\mathbb{P}_x^\uparrow$ is a probability measure and when $\psi'(0+) > 0$, $\mathbb{P}_x^\uparrow$ is a sub-probability measure.*

Proof All that is required to be shown is that, for each $t > 0$, $\mathbb{E}_x(\mathbf{1}_{(t<\tau_0^-)}X_t) = x$ for $\mathbb{P}_x^\uparrow$ to be a probability measure, and $\mathbb{E}_x(\mathbf{1}_{(t<\tau_0^-)}X_t) < x$ for it to be a sub-probability measure. To this end, recall from the proof of Theorem 12.11 that

$$\begin{aligned}\mathbb{E}_x\left(\mathbf{1}_{(t<\tau_0^-)}\frac{X_t}{x}\right) &= \lim_{q \downarrow 0}\mathbb{P}_x\left(t < \mathbf{e}_q | \tau_0^- > \mathbf{e}_q\right) \\ &= 1 - \lim_{q \downarrow 0}\mathbb{P}_x\left(\mathbf{e}_q \leq t | \tau_0^- > \mathbf{e}_q\right) \\ &= 1 - \lim_{q \downarrow 0}\int_0^t \frac{q\mathrm{e}^{-qu}}{1 - \mathrm{e}^{-\Phi(q)x}}\mathbb{P}_x\left(\tau_0^- > u\right)\mathrm{d}u\end{aligned}$$

$$= 1 - \lim_{q \downarrow 0} \frac{q}{\Phi(q)x} \int_0^t e^{-qu} \mathbb{P}_x(\tau_0^- > u) du$$

$$= 1 - \lim_{q \downarrow 0} \frac{\psi'(0+)}{x} \int_0^t e^{-qu} \mathbb{P}_x(\tau_0^- > u) du.$$

When $\psi'(0+) = 0$, it is clear that the right-hand side above is equal to unity and, otherwise, it is strictly less than unity, thereby distinguishing the case of a probability measure from a sub-probability measure. □

Note that when $\mathbb{E}_x(\mathbf{1}_{(t<\tau_0^-)} X_t) = x$, an easy application of the Markov property implies that $\{\mathbf{1}_{(t<\tau_0^-)} X_t/x : t \geq 0\}$ is a unit mean $\mathbb{P}_x$-martingale, so that $\mathbb{P}_x^{\uparrow}$ is obtained by a martingale change of measure. Similarly, when $\mathbb{E}_x(\mathbf{1}_{(t<\tau_0^-)} X_t) \leq x$, $\{\mathbf{1}_{(t<\tau_0^-)} X_t/x : t \geq 0\}$ is a supermartingale.

On a final note, the reader may be curious as to how one characterises spectrally positive Lévy processes (and indeed a general Lévy process) conditioned to stay positive when the initial value $x = 0$. In general, this is a non-trivial issue, but possible by considering the weak limit of $\mathbb{P}_x^{\uparrow}$ as a measure on the space of paths that are right-continuous with left limits. The interested reader should consult Chaumont and Doney (2005) for the most recent account as well as the commentary in Chap. 13.

12.3.2 Conditioning a (sub)Critical Continuous-State Branching Process to Stay Positive

Let us now progress to the issue of conditioning continuous-state branching processes to stay positive, following closely Chap. 3 of Lambert (2001) and Lambert (2007). We continue to adopt the notation of Sect. 12.1. Our interest is restricted to the case that there is extinction with probability one, for all initial values $x > 0$. According to Corollary 12.6, this corresponds to $\psi(\infty) = \infty$, $\psi'(0+) \geq 0$ and

$$\int^{\infty} \frac{1}{\psi(\xi)} d\xi < \infty,$$

and, henceforth, we assume that these conditions are in force. For notational convenience, we also set

$$\rho = \psi'(0+).$$

Theorem 12.13 *Suppose that $Y = \{Y_t : t \geq 0\}$ is a continuous-state branching process with branching mechanism ψ satisfying the above conditions. For each event $A \in \sigma(Y_s : s \leq t)$ and $x > 0$,*

$$P_x^{\uparrow}(A) := \lim_{s \uparrow \infty} P_x(A | \zeta > t + s)$$

is well defined as a probability measure and satisfies

$$P_x^{\uparrow}(A) = E_x\left(\mathbf{1}_A \mathrm{e}^{\rho t}\frac{Y_t}{x}\right).$$

In particular, $P_x^{\uparrow}(\zeta < \infty) = 0$ *and* $\{\mathrm{e}^{\rho t} Y_t : t \geq 0\}$ *is a* P_x*-martingale.*

Proof From the proof of Theorem 12.5, we have seen that, for $x > 0$,

$$P_x(\zeta \leq t) = P_x(Y_t = 0) = \mathrm{e}^{-x u_t(\infty)},$$

where $u_t(\theta)$ satisfies (12.17). Crucial to the proof will be the convergence

$$\lim_{s\uparrow\infty} \frac{u_s(\infty)}{u_{t+s}(\infty)} = \mathrm{e}^{\rho t}, \tag{12.24}$$

for each $t > 0$, and hence, we first show that this result holds.

To this end, note from (12.17) that

$$\int_{u_{t+s}(\infty)}^{u_s(\infty)} \frac{1}{\psi(\xi)} \mathrm{d}\xi = t.$$

On the other hand, recall from the proof of Theorem 12.5 that $u_t(\infty)$ is decreasing to $\Phi(0) = 0$, as $t \downarrow 0$. Hence, since $\lim_{\xi\downarrow 0} \psi(\xi)/\xi = \psi'(0+) = \rho$, it follows that

$$\log \frac{u_s(\infty)}{u_{t+s}(\infty)} = \int_{u_{t+s}(\infty)}^{u_s(\infty)} \frac{1}{\xi} \mathrm{d}\xi = \int_{u_{t+s}(\infty)}^{u_s(\infty)} \frac{\psi(\xi)}{\xi} \frac{1}{\psi(\xi)} \mathrm{d}\xi \to \rho t,$$

as $s \uparrow \infty$, thus proving the claim.

With (12.24) in hand, we may now proceed to note that

$$\lim_{s\uparrow\infty} \frac{1 - \mathrm{e}^{-Y_t u_s(\infty)}}{1 - \mathrm{e}^{-x u_{t+s}(\infty)}} = \frac{Y_t}{x} \mathrm{e}^{\rho t}.$$

In addition, for s sufficiently large,

$$\frac{1 - \mathrm{e}^{-Y_t u_s(\infty)}}{1 - \mathrm{e}^{-x u_{t+s}(\infty)}} \leq \frac{Y_t u_s(\infty)}{1 - \mathrm{e}^{-x u_{t+s}(\infty)}} \leq C \frac{Y_t}{x} \mathrm{e}^{\rho t},$$

for some $C > 1$. Hence, we may now apply the Markov property and then the Dominated Convergence Theorem to deduce that

$$\begin{aligned}
\lim_{s\uparrow\infty} P_x(A | \zeta > t + s) &= \lim_{s\uparrow\infty} E_x\left(\mathbf{1}_{(A,\zeta>t)} \frac{P_{Y_t}(\zeta > s)}{P_x(\zeta > t + s)}\right) \\
&= \lim_{s\uparrow\infty} E_x\left(\mathbf{1}_{(A,\zeta>t)} \frac{1 - \mathrm{e}^{-Y_t u_s(\infty)}}{1 - \mathrm{e}^{-x u_{t+s}(\infty)}}\right) \\
&= E_x\left(\mathbf{1}_{(A,\zeta>t)} \frac{Y_t}{x} \mathrm{e}^{\rho t}\right).
\end{aligned}$$

Note that we may remove the requirement that $\{t < \zeta\}$ from the indicator on the right-hand side above, as $Y_t = 0$ on $\{t \geq \zeta\}$. To show that $P_x^\uparrow$ is a probability measure, it suffices to show that, for each $x, t > 0$, $E_x(Y_t) = \mathrm{e}^{-\rho t}x$. However, this was already proved in (12.13). A direct consequence of this is that $P_x^\uparrow(\zeta > t) = 1$, for all $t \geq 0$, which implies that $P_x^\uparrow(\zeta < \infty) = 0$.

The fact that $\{\mathrm{e}^{\rho t} Y_t : t \geq 0\}$ is a martingale follows in the usual way from the necessary consistency of Radon–Nikodym densities. Alternatively, it follows directly from (12.13) by applying the Markov property as follows. For $0 \leq s \leq t$,

$$E_x\big(\mathrm{e}^{\rho t} Y_t | \sigma(Y_u : u \leq s)\big) = \mathrm{e}^{\rho s} E_{Y_s}\big(\mathrm{e}^{\rho(t-s)} Y_{t-s}\big) = \mathrm{e}^{\rho s} Y_s,$$

which establishes the martingale property. □

In older literature, the process $(Y, P_x^\uparrow)$ is called the Q-process. See for example Athreya and Ney (1972). It is also straightforward to show that $(Y, P_x^\uparrow)$ is a Markov process, using the formula for conditional expectation under change of measure.

We have thus far seen that conditioning a (sub)critical continuous-state branching process to stay positive can be performed mathematically in a similar way to conditioning a spectrally positive Lévy processes to stay positive. Our next objective is to show that, in an analogous sense to what has been discussed for Bienaymé–Galton–Watson processes, the conditioned process has the same law as a continuous-state branching process with immigration. Let us spend a little time to give a mathematical description of the latter.

In general, we define a Markov process $Y^* = \{Y_t^* : t \geq 0\}$, with probabilities $\{\mathbf{P}_x : x \geq 0\}$, to be a continuous-state branching process with branching mechanism ψ and immigration mechanism ϕ if the following hold: It is $[0, \infty)$-valued, has paths that are right-continuous with left limits and, for all $x, t > 0$ and $\theta \geq 0$,

$$\mathbf{E}_x\big(\mathrm{e}^{-\theta Y_t^*}\big) = \exp\left\{-x u_t(\theta) - \int_0^t \phi\big(u_{t-s}(\theta)\big)\mathrm{d}s\right\}, \tag{12.25}$$

where $u_t(\theta)$ is the unique solution to (12.6) and ϕ is the Laplace exponent of any subordinator. Specifically, for $\theta \geq 0$,

$$\phi(\theta) = \delta\theta + \int_{(0,\infty)} \big(1 - \mathrm{e}^{-\theta x}\big)\Upsilon(\mathrm{d}x),$$

where Υ is a measure concentrated on $(0, \infty)$ satisfying $\int_{(0,\infty)}(1 \wedge x)\Upsilon(\mathrm{d}x) < \infty$.

It is possible to see how the above definition plays an analogous role to (12.19) by considering the following sample calculations (which also show the existence of continuous-state branching processes with immigration). Suppose that $S = \{S_t : t \geq 0\}$, under $\mathbb{P}$, is a pure jump subordinator[6] with Laplace exponent $\phi(\theta)$ (hence

[6]Examples of such processes when $\phi(\lambda) = c\lambda^\alpha$ for $\alpha \in (0, 1)$ and $c > 0$ are considered by Etheridge and Williams (2003).

$\delta = 0$). Now define a process

$$Y_t^* = Y_t^{(y)} + \int_{[0,t]} \int_{(0,\infty)} Y_{t-s}^{(x)} N(\mathrm{d}s \times \mathrm{d}x), \quad t \geq 0,$$

where N is the Poisson random measure associated with the jumps of S, $\{Y_t^{(y)} : t \geq 0\}$ is a continuous-state branching process with initial value $y > 0$ and, for each (s, x) in the support of N, $Y_{t-s}^{(x)}$ is an independent copy of the process (Y, P_x) at time $t - s$. Note that, since S has a countable number of jumps, the integral above is well defined. Moreover, for the forthcoming calculations, it will be more convenient to write the expression for Y_t^* in the form

$$Y_t^* = Y_t^{(y)} + \sum_{u \leq t} Y_{t-u}^{(\Delta S_u)}, \quad t \geq 0,$$

where $\Delta S_u := S_u - S_{u-}$, so that $\Delta S_u = 0$ at all but a countable number of $u \in [0, t]$. We immediately see that $Y^* = \{Y_t^* : t \geq 0\}$ is a natural analogue of (12.19), where now the subordinator S_t plays the role of $\sum_{i=1}^n \eta_i$, the total number of immigrants in Z^* up to and including generation n. It is also straightforward to see, from its pathwise definition, that Y^* is Markovian. Indeed $Y_{t+s}^* = \widetilde{Y}_s^{(Y_t^*)} + \sum_{t<u\leq t+s} Y_{t+s-u}^{(\Delta S_u)}$, where $\widetilde{Y}_s^{(y)}$ is an independent copy of $Y_s^{(y)}$, showing that the only dependency on $\{Y_u^* : u \leq t\}$ comes through Y_t^*, in the first term on the right-hand side of the last equality.

Let us proceed further to compute the Laplace exponent of Y^*. If $\mathbf{P}_x$ is the law of Y^* when $Y_0^* = Y_0 = x$, then, with $\mathbf{E}_x$ as the associated expectation operator, for all $\theta \geq 0$,

$$\mathbf{E}_x(\mathrm{e}^{-\theta Y_t^*}) = \mathbf{E}_x\left(\mathrm{e}^{-\theta Y_t} \prod_{v \leq t} E(\mathrm{e}^{-\theta Y_{t-v}^{\Delta S_v}} | S)\right),$$

where the interchange of the product and the conditional expectation is a consequence of monotone convergence.[7] decomposition tells us that

$$\mathbf{E}\left(1 - \prod_{u \leq t} \mathbf{1}_{(\Delta S_u > \varepsilon)} \mathrm{e}^{-\theta Y_{t-u}^{(\Delta S_u)}} | S\right) = 1 - \prod_{u \leq t} \mathbf{1}_{(\Delta S_u > \varepsilon)} \mathbf{E}(\mathrm{e}^{-\theta Y_{t-u}^{(\Delta S_u)}} | S)$$

due to there being a finite number of independent jumps greater than ε. Now take limits as $\varepsilon \downarrow 0$ and apply monotone convergence. Continuing this calculation, we have

$$\mathbf{E}_x(\mathrm{e}^{-\theta Y_t^*}) = E_x(\mathrm{e}^{-\theta Y_t}) \mathbb{E}\left(\prod_{v \leq t} E_{\Delta S_v}(\mathrm{e}^{-\theta Y_{t-v}})\right)$$

[7]Note that for each $\varepsilon > 0$, the Lévy–Itô

$$
\begin{aligned}
&= \mathrm{e}^{-xu_t(\theta)}\mathbb{E}\left(\prod_{v\le t}\mathrm{e}^{-\Delta S_v u_{t-v}(\theta)}\right)\\
&= \mathrm{e}^{-xu_t(\theta)}\mathbb{E}\left(\mathrm{e}^{-\sum_{v\le t}\Delta S_v u_{t-v}(\theta)}\right)\\
&= \mathrm{e}^{-xu_t(\theta)}\mathbb{E}\left(\mathrm{e}^{-\int_{[0,t]}\int_{(0,\infty)} xu_{t-s}(\theta)N(\mathrm{d}s\times\mathrm{d}x)}\right)\\
&= \exp\left\{-xu_t(\theta) - \int_{[0,t]}\int_{(0,\infty)}\left(1-\mathrm{e}^{-xu_{t-s}(\theta)}\right)\mathrm{d}s\,\Upsilon(\mathrm{d}x)\right\}\\
&= \exp\left\{-xu_t(\theta) - \int_0^t \phi\left(u_{t-s}(\theta)\right)\mathrm{d}s\right\},
\end{aligned}
$$

where the penultimate equality follows from Theorem 2.7 (ii).

Allowing a drift component in ϕ introduces some lack of clarity with regard to a path-wise construction of Y^* in the manner shown above. Intuitively speaking, if δ is the drift of the underlying subordinator, then the term $\delta\int_0^t u_{t-s}(\theta)\mathrm{d}s$, which appears in the Laplace exponent of (12.25), may be thought of as due to a "continuum immigration" where, with rate δ, in each $\mathrm{d}t$ an independent copy of $(Y, P_\cdot)$ immigrates with infinitesimally small initial value. The problem with this heuristic is that there is an uncountable number of immigrating processes, which creates measurability problems. Nonetheless, Lambert (2002), Kyprianou et al. (2012a) and Chen and Delmas (2012) all give different pathwise constructions, using techniques that go beyond the scope of this text. Returning to the relationship between processes with immigration and conditioned processes, we see that the existence of a process Y^* with an immigration mechanism containing drift can otherwise be seen from the following lemma.

Lemma 12.14 *Fix $x > 0$. Suppose that (Y, P_x) is a continuous-state branching process with branching mechanism ψ such that $\rho \ge 0$. Then $(Y, P_x^\uparrow)$ has the same law as a continuous-state branching process with branching mechanism ψ and immigration mechanism ϕ, where for $\theta \ge 0$,*

$$\phi(\theta) = \psi'(\theta) - \rho.$$

Proof Fix $x > 0$. Clearly $(Y, P_x^\uparrow)$ has paths that are right-continuous with left limits as, for each $t > 0$, when restricted to $\sigma(Y_s : s \le t)$, we have $P_x^\uparrow \ll P_x$. Next, we compute the Laplace exponent of Y_t under $P^\uparrow$, making use of (12.4):

$$
\begin{aligned}
E_x^\uparrow\left(\mathrm{e}^{-\theta Y_t}\right) &= E_x\left(\mathrm{e}^{\rho t}\frac{Y_t}{x}\mathrm{e}^{-\theta Y_t}\right)\\
&= -\frac{\mathrm{e}^{\rho t}}{x}\frac{\partial}{\partial\theta}E_x\left(\mathrm{e}^{-\theta Y_t}\right)\\
&= -\frac{\mathrm{e}^{\rho t}}{x}\frac{\partial}{\partial\theta}\mathrm{e}^{-xu_t(\theta)}
\end{aligned}
$$

$$= \mathrm{e}^{\rho t}\mathrm{e}^{-xu_t(\theta)}\frac{\partial u_t}{\partial \theta}(\theta). \tag{12.26}$$

Recall from (12.12) that

$$\frac{\partial u_t}{\partial \theta}(\theta) = \mathrm{e}^{-\int_0^t \psi'(u_s(\theta))\mathrm{d}s} = \mathrm{e}^{-\int_0^t \psi'(u_{t-s}(\theta))\mathrm{d}s},$$

in which case, we may identify with the help of (12.7),

$$\begin{aligned}\phi(\theta) &= \psi'(\theta) - \rho \\ &= \sigma^2\theta + \int_{(0,\infty)} \left(1 - \mathrm{e}^{-\theta x}\right)x\Pi(\mathrm{d}x).\end{aligned}$$

The latter is the Laplace exponent of a subordinator with drift σ^2 and Lévy measure $x\Pi(\mathrm{d}x)$, $x > 0$. □

Looking again to the analogy with conditioned Bienaymé–Galton–Watson processes, it is natural to ask whether there is any way to decompose the conditioned process in some way so as to identify the analogue of the genealogical line of descent, earlier referred to as the spine, along which copies of the original process immigrate. This is possible, but again somewhat beyond the scope of this text. We refer the reader instead to Duquesne (2009), Duquesne and Winkel (2007), Lambert (2002) and Kyprianou et al. (2012a). See also Berestycki et al. (2011b).

Finally, as promised earlier, we show the connection between $(X, \mathbb{P}_x^\uparrow)$ and $(Y, P_x^\uparrow)$ for each $x > 0$. We are only able to make a statement for the case that $\psi'(0+) = 0$.

Lemma 12.15 *Suppose that $Y = \{Y_t : t \geq 0\}$ is a continuous-state branching process with branching mechanism ψ. Suppose further that $X = \{X_t : t \geq 0\}$ is a spectrally positive Lévy process with Laplace exponent $\psi(\theta) = \log \mathbb{E}(\mathrm{e}^{-\theta X_1})$, for $\theta \geq 0$. Fix $x > 0$. If $\psi'(0+) = 0$ and*

$$\int^\infty \frac{1}{\psi(\xi)}\mathrm{d}\xi < \infty,$$

then

(i) *the process $\{X_{\theta_t} : t \geq 0\}$ under $\mathbb{P}_x^\uparrow$ has the same law as $(Y, P_x^\uparrow)$, where*

$$\theta_t = \inf\left\{s > 0 : \int_0^s \frac{1}{X_u}\mathrm{d}u > t\right\},$$

(ii) *the process $\{Y_{\varphi_t} : t \geq 0\}$ under $P_x^\uparrow$ has the same law as $(X, \mathbb{P}_x^\uparrow)$, where*

$$\varphi_t = \inf\left\{s > 0 : \int_0^s Y_u\mathrm{d}u > t\right\}.$$

Proof Note that the condition $\psi'(0+) = 0$ necessarily excludes the case that X is a subordinator.

(i) It is easy to show that θ_t is a stopping time with respect to the filtration $\{\mathcal{F}_t : t \geq 0\}$ of X. Using Theorem 12.11 and the Lamperti transform, we have that, if $F(X_{\theta_s} : s \leq t)$ is a non-negative measurable functional of X, then, for each $x > 0$,

$$\begin{aligned}
\mathbb{E}^{\uparrow}_x\big(F(X_{\theta_s} : s \leq t)\mathbf{1}_{(\theta_t<\infty)}\big) &= \mathbb{E}_x\left(\frac{X_{\theta_t}}{x}F(X_{\theta_s} : s \leq t)\mathbf{1}_{(\theta_t<\tau_0^-)}\right) \\
&= E_x\left(\frac{Y_t}{x}F(Y_s : s \leq t)\mathbf{1}_{(t<\zeta)}\right) \\
&= E^{\uparrow}_x\big(F(Y_s : s \leq t)\big).
\end{aligned}$$

(ii) The proof of the second part is a similar argument and left to the reader. □

12.4 Concluding Remarks

It would be impossible to complete this chapter without mentioning that the material presented above is but the tip of the iceberg of a much grander theory of branching processes. If in the continuous-time Bienaymé–Galton–Watson process we allowed individuals to independently move around according to some Markov process, then we would have an example of a *spatial Markov branching particle process*. If continuous-state branching processes are the continuous-state analogue of continuous-time Bienaymé–Galton–Watson processes, then what is the analogue of a spatial Markov branching particle process? The answer to this question opens the door to the world of measure-valued processes, or *superprocesses*. Apart from their implicit probabilistic and mathematical interest, superprocesses have many applications from the point of view of mathematical biology, genetics, statistical physics and PDE theory. The interested reader is referred to the monographs of Etheridge (2000), Le Gall (1999), Duquesne and Le Gall (2002) and Dynkin (2002) for an introduction. In the direction of genetics, the extended article of Lambert (2008) gives an excellent overview.

Exercises

12.1 [8] Suppose that $Y = \{Y_t : t \geq 0\}$ is a continuous-state branching process with branching mechanism

$$\psi(\theta) = c\theta - \int_{(0,\infty)} \big(1 - \mathrm{e}^{\theta x}\big)\lambda F(\mathrm{d}x), \quad \theta \geq 0,$$

where $c, \lambda > 0$ and F is a probability distribution concentrated on $(0, \infty)$. Assume further that $\psi'(0+) > 0$ (hence Y is subcritical).

[8] This exercise is due to Prof. A.G. Pakes.

(i) Show that Y does not become extinct with probability one.
(ii) Show that for all t sufficiently large, $Y_t = \mathrm{e}^{-ct}\Delta$ where Δ is a positive random variable.

12.2 Fix $x > 0$. Suppose that (Y, P_x) is a non-explosive continuous-state branching process with branching mechanism ψ. Write, as usual, X for the Lévy process associated with Y.

(i) Use the Kella–Whitt martingale[9] for X (described in Sect. 4.4) and the Lamperti transformation to show that, for $x \geq 0$ and $\lambda > 0$,

$$M_t^\lambda := \mathrm{e}^{-\lambda Y_t} - \psi(\lambda) \int_0^t Y_s \mathrm{e}^{-\lambda Y_s} \mathrm{d}s, \quad t \geq 0,$$

is a P_x-martingale.
(ii) Recall from Theorem 12.1 that, for $x \geq 0$, $E_x(\mathrm{e}^{-\lambda Y_t}) = \mathrm{e}^{-x u_t(\lambda)}$, where $u_t(\theta)$ solves (12.6). Use the above facts to deduce directly (without using the Kella–Whitt martingale) that $E_x(M_t^\lambda) = \mathrm{e}^{-\lambda x}$, for all $x, t \geq 0$. Hence, using the Markov property, give a different proof that $\{M_t^\lambda : t \geq 0\}$ is a martingale.

12.3 (Proof of Theorem 12.7) This exercise elaborates further on the phenomena exposed in Exercise 12.1, with the help of the martingale in Exercise 12.2. In doing so, it provides the proof of Theorem 12.7. We suppose that (Y, P_x) is a continuous-state branching process, issued from $x > 0$, with branching mechanism ψ which satisfies $\psi(\infty) = \infty$ and

$$\int^\infty \frac{1}{\psi(\xi)} \mathrm{d}\xi = \infty.$$

(i) Using (12.18) and the long-term behaviour of X, show that if $\psi'(0+) \geq 0$, then, for all $x > 0$,

$$P_x\left(\lim_{t\uparrow\infty} Y_t = 0\right) = 1.$$

(ii) Now suppose that $\psi'(0+) < 0$, so that $\Phi(0) > 0$. Note from part (i) of Exercise 12.2 that $\{\mathrm{e}^{-\Phi(0)Y_t} : t \geq 0\}$ is a martingale. For all $x > 0$, show that

$$P_x\left(\lim_{t\uparrow\infty} Y_t \in \{0, \infty\}\right) = 1,$$

and, moreover, that

$$P_x\left(\lim_{t\uparrow\infty} Y_t = 0\right) = \mathrm{e}^{-\Phi(0)x}.$$

[9] In Theorem 4.7, the Kella–Whitt martingale was only introduced for Lévy processes with bounded variation paths. Thereafter it was noted that, in fact, the conclusion of this theorem is still valid when the Lévy process has paths of unbounded variation.

(iii) Show that all supercritical branching mechanisms ψ which correspond to spectrally positive Lévy processes with bounded variation paths survive with probability one, but become extinguished with positive probability.

12.4 Suppose that ψ is a branching mechanism associated with the continuous-state branching process (Y, P_x), where $x > 0$. Assume that $\psi'(0+) < 0$ and $\psi(\infty) = \infty$. Define a new probability measure, P^*, that satisfies

$$P_x^*(A) = P_x\big(A \big| \lim_{t\uparrow\infty} Y_t = 0\big)$$

for each $A \in \sigma(Y_u : u \leq t)$.

(i) Show that (Y, P_x^*) is a Markov process and, for all $\theta, t \geq 0$,

$$E_x^*\big(\mathrm{e}^{-\theta Y_t}\big) = \mathrm{e}^{-x(u_t(\theta+\Phi(0))-\Phi(0))}.$$

(ii) Using (12.6), show that (Y, P_x^*) is a continuous-state branching process with branching mechanism

$$\psi^*(\lambda) = \psi\big(\lambda + \Phi(0)\big), \quad \lambda \geq 0,$$

explaining, in particular, why ψ^* agrees with the definition of a branching mechanism.

12.5 This exercise is based in part on Chaumont (1994). Suppose that X is a spectrally positive Lévy process with Laplace exponent $\psi(\theta) = \log \mathbb{E}(\mathrm{e}^{-\theta X_1})$, for $\theta \geq 0$. Assume that $\psi'(0+) \geq 0$.

(i) Show, using the Wiener–Hopf factorisation, that, for each $x, q > 0$ and continuous, compactly supported $f : [0, \infty) \to [0, \infty)$,

$$\begin{aligned}
&\mathbb{E}_x^{\uparrow}\left(\int_0^\infty \mathrm{e}^{-qt} f(X_t)\mathrm{d}t\right) \\
&= \frac{\Phi(q)}{qx}\int_0^\infty \mathrm{d}y\, \mathrm{e}^{-\Phi(q)y}\mathbf{1}_{(y<x)}\int_{[0,\infty)} \mathbb{P}(\overline{X}_{\mathbf{e}_q} \in \mathrm{d}z) f(x+z-y)(x+z-y).
\end{aligned}$$

(ii) Hence, show the following identity holds for the potential density of the process conditioned to stay positive:

$$\int_0^\infty \mathrm{d}t \cdot \mathbb{P}_x^{\uparrow}(X_t \in \mathrm{d}y) = \frac{y}{x}\big\{W(y) - W(y-x)\big\}\mathrm{d}y, \quad y \geq 0,$$

where W is the scale function defined in Theorem 8.1.

(iii) Show that when $\psi'(0+) = 0$ (in which case it follows from Lemma 12.12 that $\mathbb{P}_x^{\uparrow}$ is a probability measure for each $x > 0$), we have

$$\mathbb{P}_x^{\uparrow}\big(\tau_z^+ < \tau_y^-\big) = 1 - \frac{y}{x}\frac{W(x-y)}{W(z-y)},$$

where $0 \leq y < x < z < \infty$ and

$$\tau_z^+ = \inf\{t > 0 : X_t > z\} \quad \text{and} \quad \tau_y^- = \inf\{t > 0 : X_t < y\}.$$

Hence deduce that, for all $x > 0$,

$$\mathbb{P}_x^\uparrow\Big(\liminf_{t\uparrow\infty} X_t = \infty\Big) = 1.$$

12.6 This exercise is taken from Lambert (2001). Suppose that Y is a conservative continuous-state branching process with branching mechanism ψ (we shall adopt the same notation as the main text in this chapter). Suppose that $\psi(\infty) = \infty$ (and hence the underlying Lévy process is not a subordinator), $\int^\infty \psi(\xi)^{-1} \mathrm{d}\xi < \infty$ and $\rho := \psi'(0+) \geq 0$.

(i) Using (12.8) show that one may write, for each $t, x > 0$ and $\theta \geq 0$,

$$E_x^\uparrow\big(\mathrm{e}^{-\theta Y_t}\big) = \mathrm{e}^{-x u_t(\theta) + \rho t} \frac{\psi(u_t(\theta))}{\psi(\theta)},$$

which is a slightly different representation from (12.25).

(ii) Assume that $\rho = 0$. Show that, for each $x > 0$,

$$P_x^\uparrow\Big(\lim_{t\uparrow\infty} Y_t = \infty\Big) = 1.$$

Hint: you may use the conclusion of Exercise 12.5 (iii).

(iii) Now assume that $\rho > 0$ so that the convexity of ψ implies, in addition, that $\rho < \infty$ and hence $\int_{(1,\infty)} x \Pi(\mathrm{d}x) < \infty$ (cf. Sect. 8.1). Show that

$$0 \leq \int_0^\theta \frac{\psi(\xi) - \rho\xi}{\xi^2} \mathrm{d}\xi = \frac{1}{2}\sigma^2\theta + \int_{(0,\infty)} x\Pi(\mathrm{d}x) \cdot \int_0^{\theta x} \left(\frac{\mathrm{e}^{-\lambda} - 1 + \lambda}{\lambda^2}\right) \mathrm{d}\lambda.$$

Hence, using the fact that $\psi(\xi) \sim \rho\xi$ as $\xi \downarrow 0$, show that

$$\int^\infty x \log x \Pi(\mathrm{d}x) < \infty$$

if and only if

$$0 \leq \int_{0+} \left(\frac{1}{\rho\xi} - \frac{1}{\psi(\xi)}\right) \mathrm{d}\xi < \infty.$$

(iv) Keeping with the assumption that $\rho > 0$ and $x > 0$, show that

$$Y_t \stackrel{P_x^\uparrow}{\to} \infty \quad \text{as } t \uparrow \infty$$

if $\int^{\infty} x \log x \Pi(\mathrm{d}x) = \infty$ and, otherwise, Y_t converges in distribution under $P_x^{\uparrow}$ as $t \uparrow \infty$ to a non-negative random variable, Y_∞, with Laplace transform

$$E_x^{\uparrow}\left(\mathrm{e}^{-\theta Y_\infty}\right) = \frac{\rho\theta}{\psi(\theta)} \exp\left\{-\rho \int_0^\theta \left(\frac{1}{\rho\xi} - \frac{1}{\psi(\xi)}\right)\mathrm{d}\xi\right\}.$$

12.7 This exercise deals with the so-called Seneta–Heyde norming for continuous-state branching processes and is based on Grey (1974). Suppose that ψ is a branching mechanism for the continuous-state branching process (Y, P_x), where $x > 0$. Assume that $\rho := \psi'(0+) \in (-\infty, 0)$.

(i) Fix $t > 0$. Show that, as a function of θ, $u_t(\theta)$ is strictly increasing from 0 to $q_t := -\log P_1(Y_t = 0)$. Hence, deduce that its inverse, say $\eta_t : [0, q_t) \to [0, \infty)$ satisfies

$$-\int_{\eta_t(\lambda)}^{\lambda} \frac{1}{\psi(\xi)}\mathrm{d}\xi = t,$$

for $\lambda \in [0, q_t)$, and, moreover, that

$$\eta_t\left(\eta_s(\lambda)\right) = \eta_{t+s}(\lambda),$$

for $\lambda \in [0, q_{t+s})$.

(ii) Show that q_t is either equal to ∞ for all $t \geq 0$ or it decreases to $\Phi(0)$.

(iii) Now fix $\lambda \in (0, \Phi(0))$. Show that

$$\left\{\mathrm{e}^{-\eta_t(\lambda)Y_t} : t \geq 0\right\}$$

is a P_x-martingale, which converges almost surely and in mean. Deduce accordingly that

$$\Xi := \lim_{t\uparrow\infty} \eta_t(\lambda) Y_t$$

exists almost surely and is valued in $[0, \infty)$.

(iv) Suppose that, for $\theta > 0$, we write $\phi(\theta) = -\log E_1(\exp\{-\theta\Xi\})$. Show that

$$\int_\lambda^{u_t(\theta\eta_t(\lambda))} \frac{1}{\psi(\xi)}\mathrm{d}\xi = \int_{\eta_t(\lambda)}^{\theta\eta_t(\lambda)} \frac{1}{\psi(\xi)}\mathrm{d}\xi \text{ and } \int_\lambda^{\phi(\theta)} \frac{1}{\psi(\xi)}\mathrm{d}\xi = \frac{1}{-\rho}\log\theta.$$

By taking limits as θ tends to 0 and ∞ respectively, deduce that for all $x > 0$, $P_x(\Xi < \infty) = 1$ and $\{\Xi = 0\} = \{\lim_{t\uparrow\infty} Z_t = 0\}$ P_x-almost surely.[10]

[10] There is a minor error in Grey (1974). In the current setting, Theorem 3 (ii) of this paper states that $P_1(\Xi = 0) = P_1(\zeta < \infty)$, which cannot be true for all supercritical continuous-state branching processes. Indeed, suppose that $\int^\infty 1/\psi(\xi)\mathrm{d}\xi = \infty$, so that $P_1(\zeta < \infty) = 0$. In that case, Theorem 12.7 tells us that $P_1(\lim_{t\uparrow\infty} Y_t = 0) = \exp\{-\Phi(0)\}$. However, since $\lambda \in (0, \Phi(0))$, $\eta_t(\lambda) \to 0$ as $t \uparrow \infty$ and we see that $\eta_t(\lambda)Y_t \to 0$ on $\{\lim_{t\uparrow\infty} Y_t = 0\}$. This also implies that $\exp\{-\Phi(0)\} \leq P_1(\Xi = 0)$, which is a contradiction. The error occurs on line 11 of p. 675 where it is claimed that "$\phi(\theta) \to -\log q$ (which may be $+\infty$) as $\theta \to \infty$".

12.8 In contrast to the previous exercise, note that

$$\{e^{\rho t} Y_t : t \geq 0\}$$

is also a non-negative P_x-martingale, and hence has an almost sure limit, which is not guaranteed to be non-trivial, however. Our aim in this exercise is to establish when the aforesaid martingale limit agrees, up to a multiplicative constant, with the random variable Ξ in the previous exercise. Assume, as before, that $\rho := \psi'(0+) \in (-\infty, 0)$ and fix $\lambda \in (0, \Phi(0))$.

(i) Use an argument similar to the one used to derive (12.24) to prove that

$$\lim_{s \uparrow \infty} \frac{\eta_{t+s}(\lambda)}{\eta_s(\lambda)} = e^{\rho t},$$

for all $t > 0$. Note that, as a consequence, there exists a slowly varying function at zero, say L, such that

$$\Xi = \lim_{t \uparrow \infty} e^{\rho t} L(e^{\rho t}) Y_t$$

P_x-almost surely.

(ii) We are interested to find out when L may be asymptotically replaced by a constant in $(0, \infty)$. To this end, prove that

$$\int_{\eta_t(\lambda)}^{\lambda} \left(\frac{1}{\rho \xi} - \frac{1}{\psi(\xi)} \right) d\xi = -\frac{1}{\rho} \log\left(\frac{\eta_t(\lambda) e^{-\rho t}}{\lambda} \right).$$

(iii) Show that $\eta_t(\lambda)$ is decreasing to zero as $t \uparrow \infty$. Hence $\eta_t(\lambda) e^{-\rho t}$ is increasing and has a finite limit if and only if

$$\int_0^{\lambda} \left(\frac{1}{\rho \xi} - \frac{1}{\psi(\xi)} \right) d\xi < \infty.$$

(iv) Finally, appealing to analysis that is similar in spirit to part (iii) of Exercise 12.6, prove that L may be asymptotically replaced by a strictly positive constant in $(0, \infty)$ if and only if

$$\int^{\infty} x \log x \, \Pi(dx) < \infty.$$

Chapter 13
Positive Self-similar Markov Processes

In this chapter, our objective is to explore in detail the general class of so-called *positive self-similar Markov processes*. Emphasis will be placed on the bijection between this class and the class of Lévy processes which are killed at an independent and exponentially distributed time. This bijection, which can be expressed through a straightforward space-time transformation, is due to Lamperti (1972).[1] Somewhat confusingly, on account of the theory presented in Sect. 12.1 for continuous-state branching processes, is also known as the Lamperti transform. To distinguish the two cases, we therefore refer to the bijection discussed in this chapter as the *second Lamperti transform.*

Through the second Lamperti transform, we are able to explore a number of specific examples of positive self-similar Markov processes which illuminate a variety of explicit and semi-explicit fluctuation identities for Lévy processes. Our first such family of examples will be positive self-similar Markov processes that are obtained when considering path transformations of stable processes and conditioned stable processes. Here, the underlying associated Lévy processes are known as *Lamperti-stable processes*. Known properties of stable processes, when transferred through the second Lamperti transform, will give us explicit fluctuation identities for Lamperti-stable processes; in particular, we will obtain their Wiener–Hopf factorisation. Another family of examples we will consider is continuous-state branching processes and continuous-state branching processes with immigration, which are also self-similar. (Note that they are automatically Markovian and positive.) Here, we shall see an interesting interplay between the first Lamperti transform, described in Chap. 12, and the second Lamperti transform.

Whilst our exposition of general positive self-similar Markov processes will, in the beginning, insist that their initial value lies in $(0, \infty)$, we will also look at the more complicated case that the point of issue is the origin. This discussion leads us to the concept of recurrent extensions of positive self-similar Markov processes. With this theory in hand, we will conclude the chapter by looking at elements of

[1] What we call here "positive self-similar Markov processes", Lamperti (1972) called "semi-stable Markov processes".

A.E. Kyprianou, *Fluctuations of Lévy Processes with Applications*, Universitext,
DOI 10.1007/978-3-642-37632-0_13, © Springer-Verlag Berlin Heidelberg 2014

fluctuation theory for positive self-similar Markov processes associated with spectrally negative Lévy processes.

13.1 Definition and Examples

A $[0, \infty)$-valued strong Markov process $X = \{X_t : t \geq 0\}$ which has paths that are almost surely right-continuous with left limits, as well as quasi-left-continous,[2] is called a *positive self-similar Markov process* if there exists a constant $\alpha > 0$ such that, for any $x > 0$ and $c > 0$,

$$\text{the law of } \{cX_{c^{-\alpha}t} : t \geq 0\} \text{ under } P_x \text{ is } P_{cx}, \tag{13.1}$$

where P_x is the law of X when issued from x.[3] In that case, we refer to α as the *index of self-similarity*. Let us turn immediately to some examples that are easily found through path transformations of a familiar class of Lévy processes.

13.1.1 Stable Subordinators

With an eye on the definition above, recall that the class of α-stable processes defined in Sect. 1.2.6 enjoys the scaling property (13.1), for $\alpha \in (0, 2]$, as well as being Markovian with the desired path properties. We understand the extreme parameter value $\alpha = 2$ as corresponding to the case of Brownian motion. Not all α-stable processes are positive, however. If $\alpha \in (0, 1)$ and we agree to restrict ourselves to the case of subordinators, then positivity is guaranteed and we find our first examples of positive self-similar Markov processes.

Although α-stable processes for $\alpha \in [1, 2]$ fail to meet the definition given above, we shall use them extensively to construct other examples of positive self-similar Markov processes by considering appropriate path transformations.

13.1.2 Modulus of a Symmetric Stable Process

We wish to consider $|Y| := \{|Y_t| : t \geq 0\}$, where $Y := \{Y_t : t \geq 0\}$ is a symmetric α-stable process with $\alpha \in (0, 2]$. Note that when $\alpha = 1$, in definition (1.13) for the characteristic exponent of Y, we must necessarily have $\eta = 0$. We understand the

---

[2]Recall that X is quasi-left-continuous if it has the following property: For each $\mathbb{F}$-stopping time T, if there exists an increasing sequence of $\mathbb{F}$-stopping times, $\{T_n : n \geq 1\}$, satisfying $\lim_{n\uparrow\infty} T_n = T$ almost surely, then $\lim_{n\uparrow\infty} X_{T_n} = X_T$ almost surely on $\{T < \infty\}$.

[3]It is important to note that our definition of a positive self-similar Markov process differs slightly from what one normally finds in the literature. Where we have assumed that it is a strong Markov process with paths that are right-continuous with left limits and quasi-left-continuous, a more usual assumption would be that it is a regular Markov process that satisfies the so-called Feller property. The latter assumption implies the former assumption.

word "symmetric" here to mean that the constants c_1 and c_2, in the definition of its Lévy measure (1.11), are equal if $\alpha \in (0,2)$. Equivalently, we can say that we require the parameter β in (1.13) to be equal to zero. When $\alpha = 2$, then Y is a Brownian motion which is clearly symmetric. In all cases, symmetry implies that Y has the same law as $-Y$.

It is clear that the process $|Y|$ is positive, as well as inheriting from Y the property that its paths are right-continuous and quasi-left-continuous. We must establish that it is both a strong Markov process as well as respecting the self-similar scaling property. To this end, rather than indicating the initial value of Y through its law, let us enhance our notation so that $Y_t^{(x)}$ is now understood as the position of Y at time $t \geq 0$ when issued from $x \in \mathbb{R}$. The process $Y^{(x)} := \{Y_t^{(x)} : t \geq 0\}$ is thus a symmetric stable process issued from $x \in \mathbb{R}$. The scaling property (13.1) for symmetric, and indeed non-symmetric, stable processes may now be written

$$\{cY^{(x)}_{c^{-\alpha}t} : t \geq 0\} \stackrel{d}{=} \{Y_t^{(cx)} : t \geq 0\}, \quad c > 0,$$

where $\stackrel{d}{=}$ means equality in distribution. With our new notation, the Markov property for stable processes can also be phrased as follows. For $s, t \geq 0$ and $x \in \mathbb{R}$,

$$Y_{t+s}^{(x)} \stackrel{d}{=} \widetilde{Y}_s^{(Y_t^{(x)})},$$

where, for each $y \in \mathbb{R}$, $\widetilde{Y}^{(y)} := \{\widetilde{Y}_s^{(y)} : s \geq 0\}$ is an independent copy of $Y^{(y)}$. A little thought now reveals that in the symmetric case, thanks to the fact that $Y^{(x)}$ has the same law as $-Y^{(-x)}$, we additionally have

$$\left|Y_{t+s}^{(x)}\right| \stackrel{d}{=} \left|\widetilde{Y}_s^{(|Y_t^{(x)}|)}\right|,$$

which is the Markov property for $|Y^{(x)}|$. It is now not difficult to derive the strong Markov property for $|Y^{(x)}|$ in the spirit of Exercise 3.2. Moreover, for $c > 0$ and $x \in \mathbb{R}$,

$$\left\{c\left|Y^{(x)}_{c^{-\alpha}t}\right| : t \geq 0\right\} \stackrel{d}{=} \left\{\left|Y_t^{(cx)}\right| : t \geq 0\right\}.$$

In conclusion $|Y^{(x)}|$ is a positive self-similar Markov process with index α. These processes have been studied in more detail in Caballero et al. (2011). In particular, the aforementioned paper more generally considers the radial part of an isotropic $\mathbb{R}^d$-valued α-stable process.

13.1.3 Bessel Processes

In the spirit of the previous example, suppose we take the radial part of an n-dimensional Brownian motion. In a similar spirit to the previous example, its radial part is also a strong Markov process thanks to radial symmetry. Clearly the radial part has continuous paths. Moreover, the scaling property of n-dimensional

Brownian motion, together with symmetry considerations, can be used to show that the radial part is self-similar with stability index equal to 2.

In conclusion, the radial part of an n-dimensional Brownian motion is a positive self-similar Markov process. Indeed, the resulting diffusion belongs to the family of so-called *Bessel processes*. The general class of Bessel processes are continuous-path strong Markov processes on $[0,\infty)$, parameterised by a constant, $\mathtt{d}\in(0,\infty)$, known as its dimension, having transition semi-group density (with respect to Lebesgue measure) given by

$$p^{\nu}(t,x,y)=\frac{1}{t}\left(\frac{y}{x}\right)^{\nu}y\exp\left\{-\frac{x^2+y^2}{2t}\right\}J_{\nu}\left(\frac{xy}{t}\right),$$

for $t,x,y>0$, where J_ν is the Bessel function of the first kind[4] with index $\nu:=\mathtt{d}/2-1$, and

$$p^{\nu}(t,0,y)=2^{-\nu}t^{-(\nu+1)}\Gamma(\nu+1)^{-1}y^{2\nu+1}\exp\left\{-\frac{y^2}{2t}\right\}.$$

To be precise, the radial part of an n-dimensional Brownian motion is a Bessel process of dimension $\mathtt{d}=n$.

It is straightforward to verify, from the expressions given above for their transition semigroup densities, that Bessel processes of all dimensions $\mathtt{d}\in(0,\infty)$ respect the scaling property (13.1) with $\alpha=2$, and hence are positive self-similar Markov processes. Indeed, one easily verifies that

$$p(t,cx,y)=\frac{1}{c}p\left(c^{-2}t,x,y/c\right),$$

for all $x,y,t\geq 0$ and $c>0$, which is equivalent to (13.1).

If we denote by $R=\{R_t:t\geq 0\}$ a $\mathtt{d}$-dimensional Bessel process, then it is also the case that, for any $q>0$, the process $\{(R_t)^q:t\geq 0\}$ is a positive strong Markov process which also possesses the scaling property (13.1).

The detailed justification of all of the above facts concerning Bessel processes can be found in Chap. XI of Revuz and Yor (2004). See also Pitman and Yor (1981). Alternatively, see Exercise 13.10.

13.1.4 Reflected Stable Processes

Keeping with the notational convention of Sect. 13.1.2, let us look at the (not necessarily symmetric) α-stable process $Y:=Y^{(0)}$, started from zero. Consider the process

$$Z_t^{(x)}:=(x\vee\overline{Y}_t)-Y_t,\quad x,t\geq 0,$$

where, as usual, $\overline{Y}_t:=\sup_{s\leq t}Y_s$. This is the reflection of Y in its supremum. As

[4] See Lebedev (1972) for further background on Bessel functions.

we have seen at earlier stages of this book, the process $Z^{(x)} := \{Z_t^{(x)} : t \geq 0\}$ is a $[0, \infty)$-valued strong Markov process (cf. Exercise 3.2), with paths that are right-continuous. Quasi-left continuity is inherited from the paths of Y. To check that it respects the scaling property (13.1), note that, for all $c > 0$,

$$\begin{pmatrix} c\overline{Y}_{c^{-\alpha}t} \\ cY_{c^{-\alpha}t} \end{pmatrix} = \begin{pmatrix} c\sup_{s\leq c^{-\alpha}t} Y_s \\ cY_{c^{-\alpha}t} \end{pmatrix} = \begin{pmatrix} \sup_{u\leq t} cY_{c^{-\alpha}u} \\ cY_{c^{-\alpha}t} \end{pmatrix}, \quad t \geq 0. \tag{13.2}$$

Using the scaling property of the α-stable process Y, the last pair in (13.2), as a process, is equal in law to $\{(\overline{Y}_t, Y_t) : t \geq 0\}$. In that case, we have, for $x, t \geq 0$ and $c > 0$,

$$cZ^{(x)}_{c^{-\alpha}t} = (cx \vee c\overline{Y}_{c^{-\alpha}t}) - cY_{c^{-\alpha}t}, \quad t \geq 0,$$

which is equal in law to the process $\{(cx \vee \overline{Y}_t) - Y_t : t \geq 0\}$, that is to say, the process $\{Z_t^{(cx)} : t \geq 0\}$. It follows that $\{Z_t^{(x)} : t \geq 0\}$ is a positive self-similar Markov process.

13.1.5 Killed Stable Processes

Excluding the case of subordinators, a general α-stable process is not a positive-valued process (albeit strong Markov and self-similar). However, by absorbing an α-stable process at the origin as it enters $(-\infty, 0)$, we can preserve the strong Markov and self-similarity properties whilst introducing the property of positivity. Again appealing to our previous notation, let us define, for $x, t > 0$,

$$X_t^{(x)} = Y_t^{(x)} \mathbf{1}_{(\underline{Y}_t^{(x)} \geq 0)}.$$

Taking account of a similar computation to (13.2), it is now straightforward to see that, for $x, c > 0$,

$$cX^{(x)}_{c^{-\alpha}t} = cY^{(x)}_{c^{-\alpha}t} \mathbf{1}_{(\underline{Y}^{(x)}_{c^{-\alpha}t} \geq 0)}, \quad t \geq 0,$$

and as a process, this is equal in law to

$$Y_t^{(cx)} \mathbf{1}_{(\underline{Y}_t^{(cx)} \geq 0)} = X_t^{(cx)}, \quad t \geq 0.$$

As with previous examples, the requirement that paths are right-continuous and quasi-left-continuous is trivially fulfilled.

13.2 Conditioned Processes and Self-similarity

Another family of positive self-similar Markov processes consists of conditioned stable processes. In order to describe them in detail, it is worth spending a little time investigating the type of conditioning we are interested in for the setting of general Lévy processes.

13.2.1 Lévy Processes Conditioned to Stay Positive

Recall that in Sect. 12.3.1, we introduced spectrally positive Lévy processes conditioned to stay positive. It is possible to treat the general class of Lévy processes with this kind of conditioning. In this respect, we briefly outline the work of Chaumont (1994, 1996) and Chaumont and Doney (2005).

Let us start with a simple fluctuation identity. In order to state it, we need to recall some notation. In what follows, we shall understand X to be a general Lévy process. In Chap. 6, it was shown that there exists a local time process, $\widehat{L}$, for $X - \underline{X}$ at zero. We defined $\widehat{H}_t = -X_{\widehat{L}_t^{-1}}$, when $t < \widehat{L}_\infty$, and otherwise $\widehat{H}_t := \infty$. It turned out that the pair $(\widehat{L}^{-1}, \widehat{H})$, also known as the descending ladder process, has the law of a (possibly-killed) two-dimensional subordinator whose range corresponds precisely to the time-space points of increase of $-\underline{X}$. The Laplace exponent of this bivariate subordinator was defined by

$$\widehat{\kappa}(\alpha, \beta) := -\frac{1}{t} \log \mathbb{E}\big(\mathrm{e}^{-\alpha \widehat{L}_t^{-1} - \beta \widehat{H}_t}\big), \quad \alpha, \beta, t \geq 0.$$

The exponent $\widehat{\kappa}$ also appears in the expression for the Laplace exponent of the running infimum sampled at $\mathbf{e}_q$, an independent and exponentially distributed time with rate $q > 0$,

$$\mathbb{E}\big(\mathrm{e}^{\beta \underline{X}_{\mathbf{e}_q}}\big) = \frac{\widehat{\kappa}(q,0)}{\widehat{\kappa}(q,\beta)}, \quad \beta \geq 0. \tag{13.3}$$

If we define the potential function

$$\widehat{U}_q(x) = \mathbb{E}\left[\int_0^\infty \mathrm{e}^{-q \widehat{L}_t^{-1}} \mathbf{1}_{(\widehat{H}_t \leq x)} \mathrm{d}t\right], \quad q, x \geq 0, \tag{13.4}$$

then a straightforward computation shows that

$$\int_{[0,\infty)} \mathrm{e}^{-\beta x} \widehat{U}_q(\mathrm{d}x) = \frac{1}{\widehat{\kappa}(q,\beta)}, \quad \beta \geq 0. \tag{13.5}$$

Using (13.3) and (13.5), and noting that $\mathbb{P}_x(\tau_0^- > \mathbf{e}_q) = \mathbb{P}(-\underline{X}_{\mathbf{e}_q} \leq x)$, we can easily deduce the following useful fluctuation identity. For $q > 0$ and $x \geq 0$,

$$\mathbb{P}_x\big(\tau_0^- > \mathbf{e}_q\big) = \widehat{U}_q(x) \widehat{\kappa}(q,0), \tag{13.6}$$

where, as usual, $\tau_0^- = \inf\{t > 0 : X_t < 0\}$. We will use (13.6) to show the existence of a martingale change of measure for stable processes, which, in turn, will lead to the law of a new strong Markov process that can be identified as positive self-similar.

To this end, let us start by remarking that, from (13.6), the ratio $\mathbb{P}_x(\tau_0^- > \mathbf{e}_q)/\widehat{\kappa}(q,0)$ is monotone decreasing in q. Moreover, its limit is clearly identifiable as $\widehat{U}(x) = \widehat{U}_0(x)$, the potential function for the descending ladder height process $\widehat{H}$.

We may now appeal to monotone convergence, together with the Markov property and the lack-of-memory property, to deduce that

$$
\begin{aligned}
&\mathbb{E}_x\big(\widehat{U}(X_t)\mathbf{1}_{(t<\tau_0^-)}\big) \\
&\quad = \lim_{q\downarrow 0}\mathbb{E}_x\left(\frac{\mathbb{P}_{X_t}(\tau_0^- > \mathbf{e}_q)}{\widehat{\kappa}(q,0)}\mathbf{1}_{(t<\tau_0^-)}\right) \\
&\quad = \lim_{q\downarrow 0}\frac{1}{\widehat{\kappa}(q,0)}\mathbb{P}_x\big(\tau_0^- > t+\mathbf{e}_q\big) \\
&\quad = \lim_{q\downarrow 0}\frac{1}{\widehat{\kappa}(q,0)}\mathbb{P}_x\big(\tau_0^- > \mathbf{e}_q | \mathbf{e}_q > t\big) \\
&\quad = \lim_{q\downarrow 0}\frac{1}{\widehat{\kappa}(q,0)}\left\{\mathrm{e}^{qt}\mathbb{P}_x\big(\tau_0^- > \mathbf{e}_q\big) - \mathrm{e}^{qt}\int_0^t q\mathrm{e}^{-qs}\mathbb{P}_x\big(\tau_0^- > s\big)\mathrm{d}s\right\} \\
&\quad = \widehat{U}(x) - \lim_{q\downarrow 0}\frac{q\mathrm{e}^{qt}}{\widehat{\kappa}(q,0)}\int_0^t \mathrm{e}^{-qs}\mathbb{P}_x\big(\tau_0^- > s\big)\mathrm{d}s. \qquad (13.7)
\end{aligned}
$$

Note that $\widehat{\kappa}'(0+,0) = \lim_{q\downarrow 0}\widehat{\kappa}(q,0)/q$ exists and is equal to $\mathbb{E}(\widehat{L}_1^{-1}) \in (0,\infty]$. We claim that $\widehat{\kappa}'(0+,0) < \infty$ if and only if $\lim_{t\uparrow\infty} X_t = -\infty$. To see why, recall from (6.30) that, up to a multiplicative constant, $q = \kappa(q,0)\widehat{\kappa}(q,0)$. Hence, when $\limsup_{t\uparrow\infty} X_t = \infty$, we have $\kappa(0,0) = 0$ and, since

$$1 = \lim_{q\downarrow 0}\kappa(q,0)\frac{\widehat{\kappa}(q,0)}{q}, \qquad (13.8)$$

it follows that $\widehat{\kappa}'(0+,0) = \infty$. On the other hand, if $\limsup_{t\uparrow\infty} X_t < \infty$, equivalently $\lim_{t\uparrow\infty} X_t = -\infty$, then $\kappa(0,0) > 0$ and (13.8) forces us to conclude that $\widehat{\kappa}'(0+,0) < \infty$.

We may now return to (13.7) and observe that if $\limsup_{t\uparrow\infty} X_t = \infty$, then, for all $x, t \geq 0$,

$$\mathbb{E}_x\big(\widehat{U}(X_t)\mathbf{1}_{(t<\tau_0^-)}\big) = \widehat{U}(x). \qquad (13.9)$$

Hence, again thanks to the Markov property, we have that, for all $x, s, t \geq 0$,

$$
\begin{aligned}
\mathbb{E}_x\big(\widehat{U}(X_{t+s})\mathbf{1}_{(t+s<\tau_0^-)}|\mathcal{F}_t\big) &= \mathbb{E}_y\big(\widehat{U}(X_s)\mathbf{1}_{(s<\tau_0^-)}\big)\big|_{y=X_t}\mathbf{1}_{(t<\tau_0^-)} \\
&= \widehat{U}(X_t)\mathbf{1}_{(t<\tau_0^-)},
\end{aligned}
$$

showing that $\{\widehat{U}(X_t)\mathbf{1}_{(t<\tau_0^-)} : t \geq 0\}$ is a martingale. Here, $\{\mathcal{F}_t : t \geq 0\}$ denotes the usual filtration associated with X. In the case that $\lim_{t\uparrow\infty} X_t = -\infty$, the equality in (13.9) is replaced by

$$\mathbb{E}_x\big(\widehat{U}(X_t)\mathbf{1}_{(t<\tau_0^-)}\big) \leq \widehat{U}(x), \qquad (13.10)$$

and we get that $\{\widehat{U}(X_t)\mathbf{1}_{(t<\tau_0^-)} : t \geq 0\}$ is a supermartingale.

We may now use the above (super)martingale to define a change of measure to a (sub-)probability measure. To accommodate for the case that $\{\widehat{U}(X_t)\mathbf{1}_{(t<\tau_0^-)} : t \geq 0\}$ is a supermartingale but not a martingale, the change of measure must take place on the space of processes which are killed at some time, say ς, and sent to a cemetery state. For each $x > 0$, under $\mathbb{P}_x$, the time ς is randomised according to the stopping time τ_0^-. In that case, we have on $\{t < \varsigma\}$

$$\left.\frac{\mathrm{d}\mathbb{P}_x^{\uparrow}}{\mathrm{d}\mathbb{P}_x}\right|_{\mathcal{F}_t} = \frac{\widehat{U}(X_t)}{\widehat{U}(x)}\mathbf{1}_{(t<\tau_0^-)}, \quad t \geq 0, \tag{13.11}$$

for all $x > 0$. It is a straightforward exercise to check that X, together with the new family of probabilities $\{\mathbb{P}_x^{\uparrow} : x > 0\}$, defines a Markov process (which is killed in the case that the change of measure induces a sub-probability measure). Indeed, for all non-negative, measurable f and $s, t, x > 0$,

$$\begin{aligned}
\mathbb{E}_x^{\uparrow}\big(f(X_{t+s})\mathbf{1}_{(t+s<\varsigma)}|\mathcal{F}_t\big) &= \mathbf{1}_{(s<\varsigma)}\mathbb{E}_x\left(f(X_{t+s})\frac{\widehat{U}(X_{t+s})}{\widehat{U}(x)}\mathbf{1}_{(t+s<\tau_0^-)}\middle|\mathcal{F}_s\right) \\
&= \mathbf{1}_{(t<\varsigma)}\mathbb{E}_y\left(f(X_t)\frac{\widehat{U}(X_t)}{\widehat{U}(x)}\right)\bigg|_{y=X_s} \\
&= \mathbb{E}_y^{\uparrow}\big(f(X_t)\big)\big|_{y=X_s} \quad \text{on } s < \varsigma,
\end{aligned} \tag{13.12}$$

where $\mathbb{E}_x^{\uparrow}$ denotes expectation with respect to $\mathbb{P}_x^{\uparrow}$. When $\limsup_{t\uparrow\infty} X_t = \infty$, we have $\mathbb{P}_x^{\uparrow}(t < \varsigma) = 1$ and the final qualification on the right-hand side of (13.12) is unnecessary. With a little further work, it can also be shown that under the change of measure (13.11), the process X remains in the class of strong Markov processes. We omit the details for the sake of brevity.

The choice of notation $\mathbb{P}_x^{\uparrow}$ was already used in Sect. 12.3.1 to denote the law of a class of spectrally positive Lévy processes conditioned to stay positive. However, there is no conflict with the use of this notation here. Indeed, we can also show that the change of measure (13.11) corresponds to the same conditioning of the process X to stay positive as found in Theorem 12.11. To see why, note that, for each $A \in \mathcal{F}_t$ and $x, t > 0$, we have

$$\begin{aligned}
\mathbb{P}_x^{\uparrow}(A, t < \varsigma) &= \lim_{q\downarrow 0}\mathbb{P}_x\big(A, t < \mathbf{e}_q|\tau_0^- > \mathbf{e}_q\big) \\
&= \lim_{q\downarrow 0}\mathbb{E}_x\left(\mathbf{1}_{(A\cap\{t<\mathbf{e}_q\wedge\tau_0^-\})}\frac{\mathbb{P}_{X_t}(\tau_0^- > \mathbf{e}_q)}{\mathbb{P}_x(\tau_0^- > \mathbf{e}_q)}\right) \\
&= \lim_{q\downarrow 0}\mathbb{E}_x\left(\mathbf{1}_{(A\cap\{t<\mathbf{e}_q\wedge\tau_0^-\})}\frac{\widehat{U}_q(X_t)}{\widehat{U}_q(x)}\right) \\
&= \mathbb{E}_x\left(\mathbf{1}_{(A\cap\{t<\tau_0^-\})}\frac{\widehat{U}(X_t)}{\widehat{U}(x)}\right).
\end{aligned} \tag{13.13}$$

In the third equality, we have used (13.6) and in the final equality, we have used that, from (13.4), $\widehat{U}_q(x) \uparrow \widehat{U}(x)$ as $q \downarrow 0$ and that $\widehat{U}_q(X_t) \leq \widehat{U}(X_t)$, so that thanks to (13.9) or (13.10), as appropriate, we may apply dominated convergence. When X is a spectrally positive Lévy process with Laplace exponent $\psi(\lambda) = \log \mathbb{E}(\mathrm{e}^{-\lambda X_1})$, $\lambda \geq 0$, its descending ladder height process is a unit rate linear drift that is killed at rate $\Phi(0)$, where Φ is the right inverse of ψ. If $\Phi(0) > 0$, equivalently $\psi'(0+) < 0$, then this gives us $\widehat{U}(x) = (1 - \mathrm{e}^{-\Phi(0)x})/\Phi(0)$. If $\Phi(0) = 0$, equivalently $\psi'(0+) \geq 0$, then $\widehat{U}(x) = x$.

Note further that $\mathbb{P}_x^{\uparrow}$ is a probability measure if and only if $\limsup_{t\uparrow\infty} X_t = \infty$. Otherwise, it is a sub-probability measure. In the case that X is a stable process, we know from Sect. 6.5.3 that its ascending ladder height is a stable subordinator and hence the condition $\limsup_{t\uparrow\infty} X_t = \infty$ is automatically satisfied.

On a final note, when considering $\{\mathbb{P}_x^{\uparrow} : x > 0\}$ as a family of probability laws on an appropriate measurable space, Chaumont and Doney (2005) show that, in an appropriate sense of weak convergence, the limiting law $\mathbb{P}_0^{\uparrow} := \lim_{x\downarrow 0} \mathbb{P}_x^{\uparrow}$ exists. The details are complicated and we refrain from giving them here as $\mathbb{P}_0^{\uparrow}$ does not appear in any of our computations below.

13.2.2 Conditioned Stable Processes

Let us return to the objective at hand, which is to illustrate another family of positive self-similar Markov processes. In the case that X is equal to Y, an α-stable process, we know from (6.37) that $\kappa(0,\beta) \propto \beta^{\alpha\rho}$, where $\rho = \mathbb{P}(Y_t \geq 0) \in (0,1)$. In other words, the ascending ladder height process is a stable subordinator (and in particular has no killing) which implies that $\limsup_{t\uparrow\infty} Y_t = \infty$. Hence, we may apply the change of measure (13.11) to generate the positive strong Markov process, which we can now identify as the α-stable process conditioned to stay positive. Moreover, thanks to (6.37), we can compute $\widehat{U}(x) \propto x^{\alpha(1-\rho)}$ for $x \geq 0$; see for example Exercise 5.8 (ii).

With regard to self-similarity, we note the following. For $c, x > 0$, $t \geq 0$ and appropriately bounded, measurable and non-negative f, we can write, with the help of (13.1),

$$\begin{aligned}
\mathbb{E}_x^{\uparrow}\left[f\left(\{cY_{c^{-\alpha}s} : s \leq t\}\right)\right] &= \mathbb{E}\left[f\left(\{cY^{(x)}_{c^{-\alpha}s} : s \leq t\}\right)\frac{(Y^{(x)}_{c^{-\alpha}t})^{\alpha(1-\rho)}}{x^{\alpha(1-\rho)}}\mathbf{1}_{(\underline{Y}^{(x)}_{c^{-\alpha}t} \geq 0)}\right] \\
&= \mathbb{E}\left[f\left(\{Y^{(cx)}_s : s \leq t\}\right)\frac{(Y^{(cx)}_t)^{\alpha(1-\rho)}}{(cx)^{\alpha(1-\rho)}}\mathbf{1}_{(\underline{Y}^{(cx)}_t \geq 0)}\right] \\
&= \mathbb{E}_{cx}^{\uparrow}\left[f\left(\{Y_s : s \leq t\}\right)\right].
\end{aligned} \tag{13.14}$$

In conclusion, any (non-monotone) α-stable process conditioned to stay positive is also a positive self-similar Markov process.

There is another type of conditioning for Lévy processes, which also boils down to a change of measure in the spirit of (13.11), and which can be used to identify positive self-similar Markov processes when applied to the special case of α-stable processes.

The increasing function $\widehat{U}$ is differentiable Lebesgue almost everywhere. If its density with respect to Lebesgue measure is denoted by $\widehat{u}$, then Chaumont (1996) notes that, under appropriate conditions, for a general Lévy process, X, $\mathbb{E}_x(\widehat{u}(X_t)\mathbf{1}_{(t<\tau_0^-)}) = \widehat{u}(x)$, for all $x, t > 0$. One may then proceed to a martingale change of measure as in (13.11), with the potential function $\widehat{U}$ replaced by its density $\widehat{u}$. Accordingly, one may define a new family of Markovian measures, say $\{\mathbb{P}_x^{\downarrow} : x > 0\}$, on the space of processes killed at some random time ς. As before, for each $x > 0$, under $\mathbb{P}_x$, ς is randomised according to the stopping time τ_0^-. We have on $t < \varsigma$

$$\left.\frac{d\mathbb{P}_x^{\downarrow}}{d\mathbb{P}_x}\right|_{\mathcal{F}_t} = \frac{\widehat{u}(X_t)}{\widehat{u}(x)}\mathbf{1}_{(t<\tau_0^-)}, \quad t \geq 0.$$

Chaumont (1996) goes on to show that this family corresponds to the law of the underlying Lévy process conditioned to be absorbed continuously at the origin before entering $(-\infty, 0)$. In particular, for each $A \in \mathcal{F}_t$ and $x, t, \eta > 0$,

$$\mathbb{P}_x^{\downarrow}(A,\, t < \tau_\eta^-) = \lim_{\varepsilon \downarrow 0} \mathbb{P}_x(A,\, t < \tau_\eta^- | \underline{X}_{\tau_0^- -} \leq \varepsilon).$$

Further details are explored in Exercise 13.5 for the case that X is a stable process. In that case, we have $\widehat{u}(x) \propto x^{\alpha(1-\rho)-1}$. Moreover, checking the self-similarity property of the positive strong Markov process $(X, \mathbb{P}_x^{\downarrow})$, $x > 0$, is as straightforward as the computation in (13.14). In the case that X is a spectrally positive stable process, from the discussion in Sect. 6.5.3, we know that $\alpha(1-\rho) - 1 = 0$ and hence, in effect, $\mathbb{P}_x^{\downarrow}$, $x > 0$, does not constitute a change of measure. Intuitively this is obvious on account of the fact that a spectrally positive Lévy process will hit the origin continuously with probability one.

13.3 The Second Lamperti Transform

In this section, we shall look at a second transformation of Lamperti (1972), which provides a bijection between the class of exponentially killed Lévy processes and positive self-similar Markov processes, up to the first moment that they hit zero. We are guided, in part, by the presentation in Chaumont (2007), as well as the original contribution of Lamperti (1972).

To this end, let us introduce some more notation. Throughout this section, we shall use $\xi := \{\xi_t : t \geq 0\}$ to denote a Lévy process which is killed and sent to the cemetery state $-\infty$ at an independent and exponentially distributed random time, $\mathbf{e} = \inf\{t > 0 : \xi_t = -\infty\}$, with rate in $[0, \infty)$. As usual, we understand $\mathbf{e}$ in the *broader sense* of an exponential distribution, so that if its rate is 0, then $\mathbf{e} = \infty$ with probability one, in which case there is no killing.

We will be interested in applying a time change to the process ξ by using its integrated exponential process, $I := \{I_t : t \geq 0\}$, where

$$I_t = \int_0^t e^{\alpha \xi_s} ds, \quad t \geq 0. \tag{13.15}$$

As the process I is increasing, we may define its limit, $I_\infty := \lim_{t\uparrow\infty} I_t$. We are also interested in the inverse process of I:

$$\varphi(t) = \inf\{s > 0 : I_s > t\}, \quad t \geq 0. \tag{13.16}$$

As usual, we work with the convention $\inf \emptyset = \infty$.

In the spirit of the examples given in the previous section, many of the arguments we shall use in the remainder of this section are of a pathwise nature. Therefore, we shall often prefer to use the notation $X^{(x)} := \{X_t^{(x)} : t \geq 0\}$ to denote a positive self-similar Markov process with initial value $x > 0$. Its lifetime until hitting zero will be denoted by $\zeta^{(x)} = \inf\{t > 0 : X_t^{(x)} = 0\}$. We shall also write ζ when the initial value of X is expressed through P_x.

Theorem 13.1 (The second Lamperti transform) *Fix $\alpha > 0$.*

(i) *If $X^{(x)}$, $x > 0$, is a positive self-similar Markov process with index of self-similarity α, then up to absorption at the origin, it can be represented as follows. For $x > 0$,*

$$X_t^{(x)} \mathbf{1}_{(t<\zeta^{(x)})} = x \exp\{\xi_{\varphi(x^{-\alpha}t)}\}, \quad t \geq 0, \tag{13.17}$$

and either

(1) *$\zeta^{(x)} = \infty$ almost surely for all $x > 0$, in which case ξ is a Lévy process satisfying $\limsup_{t\uparrow\infty} \xi_t = \infty$,*
(2) *$\zeta^{(x)} < \infty$ and $X_{\zeta^{(x)}-}^{(x)} = 0$ almost surely for all $x > 0$, in which case ξ is a Lévy process satisfying $\lim_{t\uparrow\infty} \xi_t = -\infty$, or*
(3) *$\zeta^{(x)} < \infty$ and $X_{\zeta^{(x)}-}^{(x)} > 0$ almost surely for all $x > 0$, in which case ξ is a Lévy process killed at an independent and exponentially distributed random time.*

In all cases, we may identify $\zeta^{(x)} = x^\alpha I_\infty$.

(ii) *Conversely, suppose that ξ is a given (killed) Lévy process. For each $x > 0$, define*

$$X_t^{(x)} = x \exp\{\xi_{\varphi(x^{-\alpha}t)}\}\mathbf{1}_{(t<x^\alpha I_\infty)}, \quad t \geq 0.$$

Then $X^{(x)}$ defines a positive self-similar Markov process, up to its absorption time $\zeta^{(x)} = x^\alpha I_\infty$, with index α.

Before moving to the proof of this theorem, let us point out that there is another way of connecting positive self-similar Markov processes to an underlying Lévy process. This is done through the use of stochastic differential equations. Such ideas have been pursued in Barczy and Döring (2011) and Berestycki et al. (2011a).

13.3.1 Proof of Theorem 13.1 (i)

We break the proof up into a series of lemmas. The first lemma shows that only three types of positive self-similar Markov processes can exist when categorised according to the absorption time ζ. These three cases correspond precisely to the cases (1), (2) and (3) in Theorem 13.1.

Lemma 13.2 *Simultaneously for all $x > 0$, either $P_x(\zeta = \infty) = 1$, $P_x(\zeta < \infty, X_{\zeta-} = 0) = 1$ or $P_x(\zeta < \infty, X_{\zeta-} > 0) = 1$.*

Proof We claim that the probabilities $P_x(\zeta < \infty)$ are independent of $x > 0$. To see why this is true, we can appeal to the scaling property (13.1) and write, for all $c > 0$,

$$\begin{aligned}\zeta^{(cx)} &= \inf\{t > 0 : X_t^{(cx)} = 0\}\\ &\overset{d}{=} c^\alpha \inf\{c^{-\alpha}t > 0 : cX^{(x)}_{c^{-\alpha}t} = 0\}\\ &= c^\alpha \zeta^{(x)},\end{aligned} \tag{13.18}$$

showing that $P_x(\zeta < \infty) = P_{cx}(\zeta < \infty)$, for all $x, c > 0$, as claimed. Note that this also shows that $x^{-\alpha}\zeta^{(x)}$ is independent of the value of x.

Denote by $p \in [0, 1]$ the common value of the probabilities $P_x(\zeta < \infty)$, $x > 0$. We shall now show that either $p = 0$ or $p = 1$. Thanks to the Markov property, we can now write, for all $x, t > 0$,

$$P_x(t < \zeta < \infty) = E_x\big(\mathbf{1}_{(t<\zeta)} P_{X_t}(\zeta < \infty)\big) = p P_x(t < \zeta),$$

and, hence,

$$\begin{aligned}p &= P_x(\zeta \le t) + P_x(t < \zeta < \infty)\\ &= P_x(\zeta \le t) + p\big(1 - P_x(\zeta \le t)\big)\\ &= p + (1 - p)P_x(\zeta \le t).\end{aligned}$$

This forces us to conclude that either $p = 1$ or $P_x(\zeta \le t) = 0$, for all $x, t > 0$. In other words, $p = 1$ or $p = 0$.

Next, let us assume that $P_x(\zeta < \infty) = 1$ for all $x > 0$. We are interested in the probabilities $P_x(X_{\zeta-} = 0)$, $x > 0$. In fact, similarly to the computations above, we can argue that, thanks to self-similarity, this probability does not depend on the initial value of X (the details are left as an exercise for the reader). Henceforth, we shall denote the common value of these probabilities by $\mathtt{p}$. Let us introduce the stopping times $\kappa_y^- = \inf\{t > 0 : X_t < y\}$, where $y > 0$. Note that, for a fixed $y > 0$, the events $\{\kappa_y^- = \zeta\}$ and $\{X_{\zeta-} = 0\}$ are disjoint. Together with the strong Markov property, this implies that for all $x > y > 0$,

$$\mathtt{p} = P_x(X_{\zeta-} = 0) = E_x\left(\mathbf{1}_{(\kappa_y^- < \zeta)} P_z(X_{\zeta-} = 0)|_{z = X_{\kappa_y^-}}\right) = \mathtt{p} P_x\left(\kappa_y^- < \zeta\right).$$

We are therefore forced to conclude that either $\mathtt{p} = 0$ or, for all $0 < y < x$,

$$P_x\left(\kappa_y^- < \zeta\right) = 1.$$

In other words, if $\mathtt{p}$ is not equal to 0, then X visits every $(0, y)$-neighbourhood of the origin, for $y < x$, which is another way of saying $\mathtt{p} = 1$. □

For the next lemma, we need more notation. For $x > 0$, let us write

$$\varphi(t) = \int_0^{x^\alpha t} \left(X_s^{(x)}\right)^{-\alpha} \mathrm{d}s, \quad t < x^{-\alpha}\zeta^{(x)}. \tag{13.19}$$

This choice of notation is preemptive as we shall show in due course that it agrees with (13.16). We also claim that the distribution of $\varphi(t)$ does not depend on x. Indeed, let us momentarily indicate any dependence on x by writing $\varphi^{(x)}(t)$ in place of $\varphi(t)$, for $t \geq 0$. For each $x, c > 0$ and $t < (cx)^{-\alpha}\zeta^{(cx)} \overset{d}{=} x^{-\alpha}\zeta^{(x)}$,

$$\begin{aligned}\varphi^{(cx)}(t) &\overset{d}{=} \int_0^{(cx)^\alpha t} c^{-\alpha}\left(X_{c^{-\alpha}s}^{(x)}\right)^{-\alpha} \mathrm{d}s \\ &= \int_0^{x^\alpha t} \left(X_u^{(x)}\right)^{-\alpha} \mathrm{d}u \\ &= \varphi^{(x)}(t).\end{aligned} \tag{13.20}$$

For technical reasons, it is important that we understand the behaviour of $\varphi(x^{-\alpha}\zeta-) := \lim_{t\uparrow\zeta} \varphi(x^{-\alpha}t)$. A similar argument to the one given in (13.20) also shows that, for $x > 0$, $\varphi(x^{-\alpha}\zeta-)$ does not depend on x. This is also intuitively clear as both $x^{-\alpha}\zeta^{(x)}$ and $\varphi^{(x)}$ are independent of the value of x. The next lemma says a little more about the distribution of $\varphi(x^{-\alpha}\zeta-)$.

Lemma 13.3 *In the cases that $\zeta = \infty$ or that $\{\zeta < \infty$ and $X_{\zeta-} = 0\}$, we have $P_x(\varphi(x^{-\alpha}\zeta-) = \infty) = 1$, for all $x > 0$. In the case that $\zeta < \infty$ and $X_{\zeta-} > 0$, we have that, under P_x, $\varphi(x^{-\alpha}\zeta-)$ is exponentially distributed with a rate that does not depend on the value of x.*

Proof We consider each of the three cases individually. First, we look at the case that $\zeta = \infty$. Using the Markov property, we have

$$\begin{aligned}\varphi(\infty) &= \int_0^1 \big(X_s^{(x)}\big)^{-\alpha} \mathrm{d}s + \int_1^\infty \big(X_s^{(x)}\big)^{-\alpha} \mathrm{d}s \\ &\stackrel{d}{=} \int_0^1 \big(X_s^{(x)}\big)^{-\alpha} \mathrm{d}s + \int_0^\infty \big(\widetilde{X}_s^{(z)}\big)^{-\alpha} \mathrm{d}s \\ &= \int_0^1 \big(X_s^{(x)}\big)^{-\alpha} \mathrm{d}s + \widetilde{\varphi}(\infty), \end{aligned} \tag{13.21}$$

where $z = X_1^{(x)}$ and $\widetilde{X}$ is an independent copy of X with $\widetilde{\varphi}$ defined in the obvious way. Since the integral in the final equality of (13.21) is strictly positive, we are forced to deduce that $\varphi(\infty) = \infty$, P_x-almost surely, for all $x > 0$.

Next, in the case that $\zeta < \infty$ and $X_{\zeta-} = 0$, we cannot appeal to the above method as it is not true that $1 < \zeta$ almost surely. It is, however, true that $P_x(\kappa_y^- < \zeta) = 1$ for all $x, y > 0$. We may therefore use the strong Markov property in a similar fashion to (13.21), splitting the integral in the definition of $\varphi(x^{-\alpha}\zeta^{(x)}-)$ at κ_y^-, where $y < x$, to recover the required result. The details are again left to the reader.

Finally, we consider the case that $\zeta < \infty$ and $X_{\zeta-} > 0$. Fix $x > 0$. By right-continuity and quasi-left-continuity of paths, it is trivial to note that the trajectory of $X^{(x)}$ is bounded away from zero and infinity on the time horizon $[0, \zeta^{(x)})$. It follows that $\varphi(x^{-\alpha}\zeta^{(x)}-)$ is almost surely finite. Now define the inverse of φ,

$$I_u = \inf\{0 < t < x^{-\alpha}\zeta^{(x)} : \varphi(t) > u\}, \quad u > 0, \tag{13.22}$$

which is also a stopping time for $X^{(x)}$, and moreover does not depend on the initial value x, thanks to the same being true of φ. As usual, we insist on the standard convention $\inf \emptyset = \infty$. In particular, for $u \geq \varphi(x^{-\alpha}\zeta^{(x)}-)$, $I_u = \infty$.

From (13.19), we have that $x^\alpha I_\cdot$ is the inverse of the process $\int_0^\cdot (X_s^{(x)})^{-\alpha} ds$. It follows that, for each $u > 0$, $x^\alpha I_u$ is a stopping time for X. For each $u > 0$, on the event $\{\varphi(x^{-\alpha}\zeta^{(x)}-) > u\}$, using (13.18) and the strong Markov property, we have that

$$\begin{aligned}\varphi\big(x^{-\alpha}\zeta^{(x)}-\big) &= \int_0^{x^\alpha I_u} \big(X_s^{(x)}\big)^{-\alpha} \mathrm{d}s + \int_{x^\alpha I_u}^{\zeta^{(x)}} \big(X_s^{(x)}\big)^{-\alpha} \mathrm{d}s \\ &\stackrel{d}{=} u + \int_0^{\widetilde{\zeta}^{(z)}} \big(\widetilde{X}_s^{(z)}\big)^{-\alpha} \mathrm{d}s \\ &= u + \widetilde{\varphi}\big(z^{-\alpha}\widetilde{\zeta}^{(z)}-\big), \end{aligned} \tag{13.23}$$

where $z = X_{x^\alpha I_u}^{(x)}$ and, as before, $\widetilde{X}$ is an independent copy of X with the corresponding quantities $\widetilde{\zeta}$ and $\widetilde{\varphi}$ defined in the obvious way. Let us now write, for $x, u > 0$, $\mathcal{E}(u) = P_x(\varphi(x^{-\alpha}\zeta-) > u)$, recalling that this quantity does not depend on the value of x. Thanks to (13.23), it is now clear that, for all $u, s > 0$,

$$\frac{\mathcal{E}(u+s)}{\mathcal{E}(u)} = P_x\big(\varphi(x^{-\alpha}\zeta-) > u+s|\varphi(x^{-\alpha}\zeta-) > u\big)$$
$$= P_z\big(\varphi(z^{-\alpha}\zeta) > s\big)$$
$$= \mathcal{E}(s),$$

whenever $\mathcal{E}(u) > 0$. The value of z in the above computation is irrelevant on account of the fact that $\varphi(z^{-\alpha}\zeta^{(z)}-)$ is independent of z.

We know that there must exist some $u_0 > 0$ such that $\mathcal{E}(u) > 0$ for all $u \le u_0$. In that case, $\mathcal{E}(u_0+s) = \mathcal{E}(u_0)\mathcal{E}(s)$ for all $s \le u_0$. Iterating this argument, we find that $\mathcal{E}(u) > 0$ for all $u > 0$. Right-continuity of $\mathcal{E}(u)$ is also evident. Classical theory now allows us to conclude that $\mathcal{E}(u)$ is an exponential function or identically equal to one. However the latter case can be excluded on account of the fact that $\varphi(x^{-\alpha}\zeta^{(x)}-)$ is almost surely finite. In conclusion, irrespective of the value of x, $\varphi(x^{-\alpha}\zeta^{(x)}-)$ is exponentially distributed, as required. □

With the previous two preparatory lemmas in hand, we may now give the proof of Theorem 13.1 (i). The classification of positive self-similar Markov processes into three categories has already been established by Lemma 13.2. It was proved just before the statement of Lemma 13.3 that $\varphi(x^{-\alpha}\zeta^{(x)}-)$ is independent of the value of x. Let us rename this quantity $\mathbf{e}$. The same lemma also tells us that $\mathbf{e}$ is exponentially distributed in the broader sense (i.e. $\mathbf{e} = \infty$ almost surely is interpreted as meaning that the exponential parameter is zero).

Using the same notation as above, we now define the process $\xi = \{\xi_t : t \ge 0\}$ by setting, for $x, t > 0$,

$$\xi_t = \log\big(X^{(x)}_{x^\alpha I_t}/x\big). \tag{13.24}$$

It is a straightforward consequence of the scaling arguments, which have been repeatedly used above, that the law of ξ does not depend on the value of x. For the sake of brevity, and since the arguments are now familiar, the details are, yet again, left to the reader. It is also apparent from (13.22) that $\xi_t > -\infty$ for all $t < \mathbf{e}$, and $\xi_t = -\infty$ for all $t \ge \mathbf{e}$ (in the case that $\mathbf{e} < \infty$).

The process ξ has right-continuous paths with left limits. Moreover, since $\varphi(I_t) = t$ for $t < \mathbf{e}$, we may use straightforward calculus to deduce that $\varphi'(I_t)\mathrm{d}I_t/\mathrm{d}t = 1$,

and hence

$$\frac{\mathrm{d}I_t}{\mathrm{d}t} = x^{-\alpha}\big(X^{(x)}_{x^\alpha I_t}\big)^\alpha = \mathrm{e}^{\alpha\xi_t}.$$

Recalling that $x^{-\alpha}\zeta^{(x)}$ does not depend on the value of x, we may therefore write, for each $x > 0$,

$$I_t = \int_0^t \mathrm{e}^{\alpha\xi_s}\mathrm{d}s, \quad t < \mathbf{e}.$$

Hence, as soon as we can establish that ξ is a killed Lévy process, where the time at which it is sent to the cemetery state $-\infty$ (if at all) is $\mathbf{e}$, our proof of Theorem 13.1 (i) is complete.

To this end, fix $x > 0$ and consider the event

$$\{\xi_t > -\infty\} = \{X^{(x)}_{x^\alpha I_t} > 0\} = \{x^\alpha I_t < \zeta^{(x)}\} = \{t < \varphi(x^{-\alpha}\zeta^{(x)}-)\} = \{t < \mathbf{e}\}.$$

We claim that, on $\{t < \mathbf{e}\}$, for $h > 0$,

$$\exp(\xi_{t+h} - \xi_t) = \frac{X^{(x)}_{x^\alpha I_{t+h}}}{X^{(x)}_{x^\alpha I_t}} \overset{d}{=} z^{-1}\widetilde{X}^{(z)}_{z^\alpha \widetilde{I}_h}, \tag{13.25}$$

where $z = X^{(x)}_{x^\alpha I_t}$ and, as before, $\widetilde{X}$ is an independent copy of X, with an obvious associated definition for $\widetilde{I}$. To see where the last equality in (13.25) comes from, apply the strong Markov property to deduce that, on $\{t < \mathbf{e}\}$,

$$\begin{aligned}
x^\alpha I_{t+h} &= x^\alpha I_t + x^\alpha \inf\left\{s > 0 : \int_0^{x^\alpha s} \left(X^{(x)}_{x^\alpha I_t + u}\right)^{-\alpha} \mathrm{d}u > h\right\} \\
&\overset{d}{=} x^\alpha I_t + x^\alpha \inf\left\{s > 0 : \int_0^{x^\alpha s} \left(\widetilde{X}^{(z)}_u\right)^{-\alpha} \mathrm{d}u > h\right\} \\
&= x^\alpha I_t + z^\alpha \inf\left\{r > 0 : \int_0^{z^\alpha r} \left(\widetilde{X}^{(z)}_u\right)^{-\alpha} \mathrm{d}u > h\right\} \\
&= x^\alpha I_t + z^\alpha \widetilde{I}_h.
\end{aligned}$$

However, $\widetilde{I}_h$ is independent of the value z and hence, using the scaling property (13.1) in (13.25), we deduce that $\exp(\xi_{t+h} - \xi_t)$ is independent of $\{\xi_s : s \le t\}$ and has the same distribution as $\exp(\xi_h)$.

In the case that $\mathbf{e} = \infty$, we see that ξ is a Lévy process (no killing). Moreover, referring to (13.24), we see that $\zeta^{(x)} = x^\alpha I_\infty$. Hence in case (1), we necessarily have that $I_\infty = \infty$ almost surely and in case (2), $I_\infty < \infty$ almost surely. The following lemma is dealt with in Exercise 13.6.

Lemma 13.4 *Suppose that ξ is an unkilled Lévy process. Then $\mathbb{P}(I_\infty < \infty) = 1$ if and only if $\limsup_{t\uparrow\infty} \xi_t < \infty$, and otherwise $\mathbb{P}(I_\infty = \infty) = 1$.*

It is now clear that in case (1), we have $\limsup_{t\uparrow\infty} \xi_t = \infty$ and in case (2), $\limsup_{t\uparrow\infty} \xi_t < \infty$, which is to say that $\lim_{t\uparrow\infty} \xi_t = -\infty$.

We now prove for case (3) that ξ and $\mathbf{e}$ are independent. Recall from Lemma 13.3 that, irrespective of the initial value of $X^{(x)}$, $\mathbf{e}$ is exponentially distributed. Moreover, from (13.23), again, irrespective of the value of x, for all $t > 0$, the event $\{\mathbf{e} > t\}$ is independent of $\{X^{(x)}_s : s \le x^\alpha I_t\}$, and hence is independent of $\{\xi_s : s \le t\}$.

Taking account of our earlier observation that, conditional on $\{\mathbf{e} > t\}$, the increment $\xi_{t+h} - \xi_t \stackrel{d}{=} \xi_h$, for all $t, h > 0$, we conclude that ξ must be a Lévy process killed at an independent and exponentially distributed time. □

13.3.2 Proof of Theorem 13.1 (ii)

Whilst it is clear that $X^{(x)}$ is positive and has paths that are right-continuous with left limits, we need to check that it is self-similar as well as a strong Markov process. Self-similarity is very easy to show. Indeed, for all $c, t > 0$,

$$cX^{(x)}_{c^{-\alpha}t} = cx\exp\{\xi_{\varphi((cx)^{-\alpha}t)}\}\mathbf{1}_{(t<(cx)^{\alpha}I_\infty)} = X^{(cx)}_t.$$

The remainder of the proof is thus concerned with establishing the strong Markov property.

To this end, let us write $\mathbb{H} := \{\mathcal{H}_t : t \geq 0\}$ for a right-continuous version of the natural filtration generated by the process ξ. Note in particular that $\varphi(x^{-\alpha}t)$ is a stopping time for $\mathbb{H}$ and $\mathcal{G}_t := \mathcal{H}_{\varphi(x^{-\alpha}t)}$, $t \geq 0$ is the natural right-continuous filtration to which $X^{(x)}$ is adapted. See, for example, Proposition 7.9 of Kallenberg (2002). Now suppose that τ is a stopping time with respect to $\mathbb{G} := \{\mathcal{G}_t : t \geq 0\}$. We claim that $\varphi(x^{-\alpha}\tau)$ is a stopping time with respect to $\mathbb{H}$. To see this, write for each $s > 0$,

$$\{\varphi(x^{-\alpha}\tau) < s\} = \{\tau < x^\alpha I_s\} = \bigcup_{u\in\mathbb{Q}\cap(0,\infty)} \{\tau < u < x^\alpha I_s\}. \tag{13.26}$$

For each $s, u > 0$, the set $\{\tau < u < x^\alpha I_s\}$ can be written $\{\tau < u,\ \varphi(x^{-\alpha}u) < s\}$. Since τ is a stopping time for $\mathbb{G}$, the last event belongs to $\mathcal{G}_u \cap \{\varphi(x^{-\alpha}u) < s\} = \mathcal{H}_{\varphi(x^{-\alpha}u)} \cap \{\varphi(x^{-\alpha}u) < s\} \subseteq \mathcal{H}_s$. The final inclusion uses the fact that $\varphi(x^{-\alpha}u)$ is a stopping time. In conclusion, we have from (13.26) that $\{\varphi(x^{-\alpha}\tau) < s\} \in \mathcal{H}_s$, for all $s > 0$. Recalling the discussion in Sect. 3.1, this establishes the claim that $\varphi(x^{-\alpha}\tau)$ is a stopping time for $\mathbb{H}$ since we have assumed that $\mathbb{H}$ is a right-continuous filtration.

We can now say that, from the strong Markov property for Lévy processes and the lack-of-memory property for exponential distributions, on $\{\varphi(x^{-\alpha}\tau) < \mathbf{e}\}$, the process $\widetilde{\xi} := \{\widetilde{\xi}_t : t \geq 0\}$, where $\widetilde{\xi}_t := \xi_{\varphi(x^{-\alpha}\tau)+t} - \xi_{\varphi(x^{-\alpha}\tau)}$, $t \geq 0$, is a (killed) Lévy process, which is independent of $\mathcal{G}_\tau = \mathcal{H}_{\varphi(x^{-\alpha}\tau)}$ and has the same law as ξ.

Next note that $\tau < x^\alpha I_\infty$ if and only if $\varphi(x^{-\alpha}\tau) < \mathbf{e}$, in which case $x^\alpha \int_0^{\varphi(x^{-\alpha}\tau)} \exp\{\alpha\xi_s\}\mathrm{d}s = \tau$. Moreover, we have on $\{\tau < x^\alpha I_\infty\}$,

$$\begin{aligned} x^\alpha I_\infty &= x^\alpha \int_0^{\varphi(x^{-\alpha}\tau)} \mathrm{e}^{\alpha\xi_s}\,\mathrm{d}s \\ &\quad + x^\alpha \mathrm{e}^{\alpha\xi_{\varphi(x^{-\alpha}\tau)}} \int_0^{\mathbf{e}-\varphi(x^{-\alpha}\tau)} \mathrm{e}^{\alpha(\xi_{\varphi(x^{-\alpha}\tau)+u} - \xi_{\varphi(x^{-\alpha}\tau)})}\mathrm{d}u \\ &= \tau + \big(X^{(x)}_\tau\big)^\alpha \widetilde{I}_\infty, \end{aligned}$$

where, conditional on $\mathcal{G}_\tau \cap \{\tau < x^\alpha I_\infty\}$, from the strong Markov property and lack-of-memory property, the random variable $\widetilde{I}_\infty$ has the same distribution as I_∞. We may now write

$$\mathbf{1}_{(\tau+t<x^\alpha I_\infty)} = \mathbf{1}_{(\tau<x^\alpha I_\infty)}\mathbf{1}_{(t<(X_\tau^{(x)})^\alpha \widetilde{I}_\infty)}. \tag{13.27}$$

Note also that we have on $\{\tau + t < x^\alpha I_\infty\}$,

$$I_{\varphi(x^{-\alpha}(\tau+t))} = x^{-\alpha}\tau + x^{-\alpha}t = I_{\varphi(x^{-\alpha}\tau)} + x^{-\alpha}t,$$

and hence,

$$\begin{aligned}
x^{-\alpha}t &= \int_{\varphi(x^{-\alpha}\tau)}^{\varphi(x^{-\alpha}(\tau+t))} \exp\{\alpha\xi_s\}\mathrm{d}s \\
&= \exp\{\alpha\xi_{\varphi(x^{-\alpha}\tau)}\}\int_0^{\varphi(x^{-\alpha}(\tau+t))-\varphi(x^{-\alpha}\tau)} \exp\bigl\{\alpha(\xi_s - \xi_{\varphi(x^{-\alpha}\tau)})\bigr\}\mathrm{d}s \\
&= x^{-\alpha}\bigl(X_\tau^{(x)}\bigr)^\alpha \int_0^{\varphi(x^{-\alpha}(\tau+t))-\varphi(x^{-\alpha}\tau)} \exp\{\alpha\widetilde{\xi}_s\}\mathrm{d}s.
\end{aligned}$$

It follows that if we define $\widetilde{\varphi}$ to play the role of φ for the process $\widetilde{\xi}$, then given $\mathcal{G}_\tau \cap \{\tau < x^\alpha I_\infty\}$, for $t < (X_\tau^{(x)})^\alpha \widetilde{I}_\infty$,

$$\varphi\bigl(x^{-\alpha}(\tau+t)\bigr) - \varphi\bigl(x^{-\alpha}\tau\bigr) = \widetilde{\varphi}\bigl((X_\tau^{(x)})^{-\alpha}t\bigr). \tag{13.28}$$

Finally, taking account of (13.27) and (13.28), we have, for $t \geq 0$,

$$\begin{aligned}
X_{\tau+t}^{(x)} &= x\mathrm{e}^{\xi_{\varphi(x^{-\alpha}\tau)}}\mathrm{e}^{(\xi_{\varphi(x^{-\alpha}(\tau+t))}-\xi_{\varphi(x^{-\alpha}\tau)})}\mathbf{1}_{(\tau<x^\alpha I_\infty)}\mathbf{1}_{(t<(X_\tau^{(x)})^\alpha \widetilde{I}_\infty)} \\
&= X_\tau^{(x)}\exp\{\widetilde{\xi}_{\widetilde{\varphi}((X_\tau^{(x)})^{-\alpha}t)}\}\mathbf{1}_{(t<(X_\tau^{(x)})^\alpha \widetilde{I}_\infty)},
\end{aligned}$$

from which the strong Markov property is now evident.

Finally, there is the issue of quasi-left-continuity. We recall from Lemma 3.2 that Lévy processes have the aforementioned property. In turn, quasi-left-continuity of ξ transfers through the second Lamperti transform in a straightforward way (using in particular that the time-change for ξ is a stopping time with respect to $\mathbb{H}$) and is inherited by the process X. We leave the details to the reader. □

13.4 Lamperti-Stable Processes

Let us return to some of the examples given in Sects. 13.1 and 13.2. Our objective is to compute in explicit terms the characteristics of the underlying Lévy process ξ in the second Lamperti transform. In particular, we are interested in the three cases of a stable process conditioned to stay positive, a stable process conditioned to hit the origin continuously, and a stable process killed on first entry into $(-\infty, 0)$. For each of these three cases, we shall write $\xi^{\uparrow}$, ξ^{*} and $\xi^{\downarrow}$, respectively, for the three underlying Lévy processes that appear through the second Lamperti transform. Our presentation is inspired by the original investigation of these processes found in Caballero and Chaumont (2006b) and Chaumont et al. (2009).

As we shall shortly see, $\xi^{\uparrow}$, $\xi^{\downarrow}$ and ξ^{*} all belong to the class of hypergeometric class of Lévy processes that was described in Sect. 6.6.1. However, these three processes are more precisely named Lamperti-stable processes on account of their intimate relationship with stable processes, through the second Lamperti transform. See Caballero et al. (2010).

13.4.1 The Case of $\xi^{\uparrow}$

Recall from Sect. 13.2 that, for $x > 0$, the process $(Y, \mathbb{P}_x^{\uparrow})$ is used to denote an α-stable Lévy process conditioned to stay positive. From Theorem 13.1, we know that $\xi^{\uparrow} = \{\xi_t^{\uparrow} : t \geq 0\}$, the associated Lévy process through the second Lamperti transform, is not killed and drifts to $+\infty$. Our strategy for characterising $\xi^{\uparrow}$ will revolve around the Wiener–Hopf factorisation. In particular, if we write $\Psi^{\uparrow}$ for the characteristic exponent of $\xi^{\uparrow}$, then the Wiener–Hopf factorisation tells us that, up to a multiplicative constant,

$$\Psi^{\uparrow}(\theta) = \phi^{\uparrow}(-\mathrm{i}\theta)\widehat{\phi}^{\uparrow}(\mathrm{i}\theta), \quad \theta \in \mathbb{R},$$

where $\phi^{\uparrow}$ and $\widehat{\phi}^{\uparrow}$ are the Laplace exponents of the ascending and descending ladder height processes, respectively, cf. Theorem 6.15. Therefore, in order to get our hands on $\Psi^{\uparrow}$, it suffices to try to compute closed form expressions for $\phi^{\uparrow}$ and $\widehat{\phi}^{\uparrow}$. An obvious place to look for information concerning these quantities will be in the overshoot distribution of $\xi^{\uparrow}$, when it crosses thresholds both upwards and downwards. As we shall shortly see, these overshoot distributions conveniently turn out to be easily recovered from overshoot distributions of the associated stable process, thanks to the second Lamperti transform.

Let us start by computing the Laplace exponent $\phi^{\uparrow}$. Fix $\alpha \in (0, 2)$ and, for convenience, assume that Y has positive jumps. That is to say, we rule out the case of a spectrally negative stable process. This means that, in particular, we are excluding the extreme case of a Brownian motion.[5] The case that Y is a Brownian motion is treated in Exercise 13.3. More generally, the spectrally negative case is dealt with in Exercise 13.8.

As usual, write

$$\tau_u^+ = \inf\{t > 0 : Y_t > u\} \quad \text{and} \quad \tau_a^- = \inf\{t > 0 : Y_t < a\}, \tag{13.29}$$

for any $a \in \mathbb{R}$. Now take $y > 1$. In the light of the change of measure (13.11), we have, for all $z > 0$,

[5] Rather obviously, we also rule out the case that $-Y$ is a subordinator.

$$\begin{aligned}
&\mathbb{P}_1^{\uparrow}\left(\frac{Y_{\tau_y^+}-y}{y}\in \mathrm{d}z\right)\\
&=\mathbb{P}_1\left((Y_{\tau_y^+})^{\alpha(1-\rho)};\ \frac{Y_{\tau_y^+}-y}{y}\in \mathrm{d}z,\ \tau_y^+<\tau_0^-\right)\\
&=\big(y(1+z)\big)^{\alpha(1-\rho)}\mathbb{P}_1\left(\frac{Y_{\tau_y^+}}{y}-1\in \mathrm{d}z,\ \tau_y^+<\tau_0^-\right), \qquad (13.30)
\end{aligned}$$

where $\mathbb{P}_x$ is the law of an α-stable process issued from $x\in\mathbb{R}$. Using scaling arguments similar to those that were used repeatedly in Sect. 13.3, it is straightforward to show that

$$\mathbb{P}_1\left(\frac{Y_{\tau_y^+}}{y}-1\in \mathrm{d}z,\ \tau_y^+<\tau_0^-\right)=\mathbb{P}_{1/y}\big(Y_{\tau_1^+}-1\in \mathrm{d}z,\ \tau_1^+<\tau_0^-\big). \qquad (13.31)$$

On the one hand, thanks to the second Lamperti transform, if we write $\tau^{+,\uparrow}_{\log y}=\inf\{t>0:\xi_t^{\uparrow}>\log y\}$, then, under $\mathbb{P}_1^{\uparrow}$,

$$\frac{Y_{\tau_y^+}}{y}=\exp\{\xi^{\uparrow}_{\tau^{+,\uparrow}_{\log y}}-\log y\}.$$

On the other hand, the probability on the right-hand side of (13.31) is known explicitly and has been derived in Exercise 7.7. Taking account of these facts, back in (13.30), we have, after a little algebra, that, for $z>0$,

$$\begin{aligned}
&\mathbf{P}^{\uparrow}\left(\exp\{\xi^{\uparrow}_{\tau^{+,\uparrow}_{\log y}}-\log y\}-1\in \mathrm{d}z\right)\\
&=\frac{\sin\pi\alpha\rho}{\pi}\left(1-\frac{1}{y}\right)^{\alpha\rho}z^{-\alpha\rho}\left(z+1-\frac{1}{y}\right)^{-1}\mathrm{d}z, \qquad (13.32)
\end{aligned}$$

where $\mathbf{P}^{\uparrow}$ is the law of $\xi^{\uparrow}$ (with associated expectation operator $\mathbf{E}^{\uparrow}$).

Before we are in a position to extract information about $\phi^{\uparrow}$ from this last identity, we must first take limits as $y\uparrow\infty$. To see why, recall that the overshoot $\xi^{\uparrow}_{\tau^{+,\uparrow}_{\log y}}-\log y$ agrees precisely with the overshoot of the ascending ladder height process of $\xi^{\uparrow}$. Moreover, thanks to Theorem 5.7, providing the ascending ladder height process has finite mean, we should expect to see a non-degenerate limiting distribution in (13.32). This will give us, up to a multiplicative constant, the tail of the Lévy measure of the ascending ladder height process. If the ascending ladder height process of $\xi^{\uparrow}$ does not have finite mean, then it is straightforward to check from the proof of Theorem 5.7, using Corollary 5.3, that the limiting overshoot distribution should be degenerate. In conclusion, by taking limits as $y\uparrow\infty$ in (13.32), we deduce that if $\Upsilon^{\uparrow}$ is the Lévy measure of the ascending ladder height process of $\xi^{\uparrow}$, then

$$\Upsilon^{\uparrow}(x,\infty)\mathrm{d}x\propto\mathbf{P}^{\uparrow}\big(\Delta^{\uparrow}\in \mathrm{d}x\big),$$

where the random variable $\Delta^{\uparrow}$ satisfies

$$\mathbf{P}^{\uparrow}\left(\mathrm{e}^{\Delta^{\uparrow}}-1\in\mathrm{d}z\right)=\frac{\sin\pi\alpha\rho}{\pi}z^{-\alpha\rho}(z+1)^{-1}\mathrm{d}z.$$

This tells us that

$$\Upsilon^{\uparrow}(x,\infty)\propto\frac{(\mathrm{e}^{x}-1)^{-\alpha\rho}}{\Gamma(1-\alpha\rho)\Gamma(\alpha\rho)},\tag{13.33}$$

where we have used Euler's reflection formula[6] for gamma functions.

Note that $\xi^{\uparrow}$ (and hence its ascending ladder height process) cannot creep upwards on account of the fact that the same is true of Y. Therefore, there is no drift component in the Laplace exponent $\phi^{\uparrow}$ and hence, recalling Exercise 2.11, for all $\lambda\geq 0$,

$$\begin{aligned}\phi^{\uparrow}(\lambda)&\propto\lambda\int_{0}^{\infty}\mathrm{e}^{-\lambda x}\Upsilon^{\uparrow}(x,\infty)\mathrm{d}x\\&=\frac{\lambda}{\Gamma(1-\alpha\rho)\Gamma(\alpha\rho)}\int_{0}^{\infty}\mathrm{e}^{-\lambda x}\left(\mathrm{e}^{x}-1\right)^{-\alpha\rho}\mathrm{d}x\\&=\frac{\lambda}{\Gamma(1-\alpha\rho)\Gamma(\alpha\rho)}\int_{0}^{1}u^{\lambda+\alpha\rho-1}(1-u)^{-\alpha\rho}\mathrm{d}u.\end{aligned}$$

The right-hand side above is a beta integral and hence, for all $\lambda\geq 0$, up to a multiplicative constant, we have

$$\phi^{\uparrow}(\lambda)=\lambda\frac{\Gamma(\lambda+\alpha\rho)\Gamma(1-\alpha\rho)}{\Gamma(1-\alpha\rho)\Gamma(\alpha\rho)\Gamma(\lambda+1)}=\frac{\Gamma(\alpha\rho+\lambda)}{\Gamma(\alpha\rho)\Gamma(\lambda)}.\tag{13.34}$$

Note that we could have equivalently derived this expression from (13.33) using Example 5.26.

Next, we turn our attention to the derivation of $\widehat{\phi}^{\uparrow}$. We cannot apply the same technique as above since $\mathbf{P}^{\uparrow}(\liminf_{t\uparrow\infty}\xi_t^{\uparrow}>-\infty)=1$, and hence an asymptotic overshoot in the downwards direction would not make sense. Moreover, the descending ladder height process of $\xi^{\uparrow}$ is exponentially killed, in which case $\widehat{\phi}^{\uparrow}(0)>0$. This means that we need to find more than just the Lévy measure of the descending ladder height process in order to construct $\widehat{\phi}^{\uparrow}$. Note that, as before, there will be no drift term in $\widehat{\phi}^{\uparrow}$ on account of the fact that Y does not creep downwards.

Instead, we can look at the law of the global infimum of $\xi^{\uparrow}$, which, again thanks to the second Lamperti transform, can easily be derived from the global infimum of Y under $\mathbb{P}_1^{\uparrow}$. Indeed, we have from (6.29) that, for $\lambda\geq 0$,

$$\frac{\widehat{\phi}^{\uparrow}(0)}{\widehat{\phi}^{\uparrow}(\lambda)}=\mathbf{E}^{\uparrow}\left(\mathrm{e}^{\lambda\underline{\xi}^{\uparrow}_{\infty}}\right)=\mathbb{E}_1^{\uparrow}\left((\underline{Y}_{\infty})^{\lambda}\right),\tag{13.35}$$

[6] Recall again (see the first footnote in Sect. 5.6) that Euler's reflection formula for gamma functions says that $\Gamma(1-u)\Gamma(u)=\pi/\sin\pi u$ for $u\in\mathbb{C}\backslash\mathbb{Z}$.

where $\underline{\xi}^{\uparrow}_{\infty} = \inf_{s\geq 0} \xi^{\uparrow}_s$ and $\underline{Y}_{\infty} = \inf_{s\geq 0} Y_s$. Moreover, referring to Exercise 7.7 (i), we have, for $0 < y < 1$,

$$
\begin{aligned}
&\mathbb{P}^{\uparrow}_1(\underline{Y}_{\infty} \leq y)\\
&= \mathbb{P}^{\uparrow}_1(\tau^-_y < \infty)\\
&= \mathbb{E}_1\big((Y_{\tau^-_y})^{\alpha(1-\rho)};\ \tau^-_y < \tau^-_0\big)\\
&= \mathbb{E}_{1-y}\big((y + Y_{\tau^-_0})^{\alpha(1-\rho)};\ -Y_{\tau^-_0} \leq y\big)\\
&= \frac{\sin \pi\alpha(1-\rho)}{\pi} \int_0^y (y-z)^{\alpha(1-\rho)} \left(\frac{z}{1-y}\right)^{-\alpha(1-\rho)} \left(1 + \frac{z}{1-y}\right)^{-1} \frac{1}{1-y} \mathrm{d}z\\
&= \frac{\sin \pi\alpha(1-\rho)}{\pi} (1-y)^{\alpha(1-\rho)}\\
&\quad \times \int_0^y \frac{1}{y}\left(1 - \frac{z}{y}\right)^{\alpha(1-\rho)} \left(\frac{z}{y}\right)^{-\alpha(1-\rho)} \left(\frac{1}{y} - 1 + \frac{z}{y}\right)^{-1} \mathrm{d}z. \qquad (13.36)
\end{aligned}
$$

The expression on the right-hand side reduces to $1 - (1-y)^{\alpha(1-\rho)}$. However, this requires one to first spot some straightforward, but nonetheless non-obvious, manipulations. Set $1 - z/y = (v+1)^{-1}$ and note that the integral on the right-hand side of (13.36) can be developed as follows:

$$
\begin{aligned}
&\int_0^{\infty} \frac{y}{(v+1-y)(v+1)v^{\alpha(1-\rho)}} \mathrm{d}v\\
&= \int_0^{\infty} \frac{1}{v^{\alpha(1-\rho)}(v+1-y)} \mathrm{d}v - \int_0^{\infty} \frac{1}{v^{\alpha(1-\rho)}(v+1)} \mathrm{d}v\\
&= \big[(1-y)^{-\alpha(1-\rho)} - 1\big] \int_0^{\infty} \frac{1}{z^{\alpha(1-\rho)}(z+1)} \mathrm{d}z, \qquad (13.37)
\end{aligned}
$$

where we have used the change of variable $v = (1-y)z$ to deal with the first integral in the first equality. Note, moreover, that by setting $w = (1+z)^{-1}$, we get a familiar beta integral,

$$
\begin{aligned}
\int_0^{\infty} \frac{1}{z^{\alpha(1-\rho)}(z+1)} \mathrm{d}z &= \int_0^1 (1-w)^{-\alpha(1-\rho)} w^{\alpha(1-\rho)} \mathrm{d}w\\
&= \Gamma\big(1 - \alpha(1-\rho)\big)\Gamma\big(\alpha(1-\rho)\big)\\
&= \frac{\pi}{\sin \pi\alpha(1-\rho)}, \qquad (13.38)
\end{aligned}
$$

where we have used the reflection formula again in the final equality.

Now plugging (13.37) and (13.38) back into (13.36), we get, as promised,

$$
\mathbb{P}^{\uparrow}_1(\underline{Y}_{\infty} \leq y) = 1 - (1-y)^{\alpha(1-\rho)},
$$

for $0 < y < 1$. Returning to (13.35), we now have, up to a multiplicative constant,

$$\begin{aligned}\widehat{\phi}^{\uparrow}(\lambda) &= \left(\alpha(1-\rho)\int_0^1 y^{\lambda}(1-y)^{\alpha(1-\rho)-1}\mathrm{d}y\right)^{-1} \\ &= \frac{\Gamma(1+\lambda+\alpha(1-\rho))}{\alpha(1-\rho)\Gamma(1+\lambda)\Gamma(\alpha(1-\rho))} \\ &= \frac{\Gamma(1+\lambda+\alpha(1-\rho))}{\Gamma(1+\lambda)\Gamma(\alpha(1-\rho)+1)},\end{aligned} \tag{13.39}$$

for $\lambda \geq 0$.

Through the Wiener–Hopf factorisation, we may now write down the characteristic exponent of $\xi^{\uparrow}$ up to a multiplicative constant:

$$\Psi^{\uparrow}(\theta) = \frac{\Gamma(\alpha\rho - \mathrm{i}\theta)}{\Gamma(-\mathrm{i}\theta)}\frac{\Gamma(1+\mathrm{i}\theta+\alpha(1-\rho))}{\Gamma(1+\mathrm{i}\theta)}. \tag{13.40}$$

This clearly puts the process $\xi^{\uparrow}$ in the class of hypergeometric Lévy processes. See Sect. 6.6.1.

13.4.2 The Case of $\xi^{\downarrow}$

In Sect. 13.2, we also introduced the process $(Y, \mathbb{P}_x^{\downarrow})$, for $x > 0$, in other words, an α-stable process conditioned to be absorbed continuously at the origin before entering $(-\infty, 0)$. From Theorem 13.1, we know that $\xi^{\downarrow} = \{\xi_t^{\downarrow} : t \geq 0\}$, the associated Lévy process through the second Lamperti transform, is not killed and drifts to $-\infty$.

As in the previous section, we can again try to reconstruct the characteristic exponent of $\xi^{\downarrow}$ by piecing together its Wiener–Hopf factorisation. This time, we must take account of the fact that $\overline{\xi}_{\infty}^{\downarrow} := \sup_{s\geq 0}\xi_s^{\downarrow} < \infty$. If $\Psi^{\downarrow}$ is the characteristic exponent of $\xi^{\downarrow}$, then its factorisation will be written

$$\Psi^{\downarrow}(\theta) = \phi^{\downarrow}(-\mathrm{i}\theta)\widehat{\phi}^{\downarrow}(\mathrm{i}\theta), \quad \theta \in \mathbb{R},$$

where $\phi^{\downarrow}$ and $\widehat{\phi}^{\downarrow}$ are the Laplace exponents of the ascending and descending ladder height processes, respectively. On account of the fact that $\lim_{t\uparrow\infty}\xi_t^{\downarrow} = -\infty$, we must have $\phi^{\downarrow}(0) > 0$. That is to say, the ascending ladder height process is exponentially killed.

To compute $\phi^{\downarrow}$, we again appeal to (6.29) to deduce that, for $\lambda \geq 0$,

$$\frac{\phi^{\downarrow}(0)}{\phi^{\downarrow}(\lambda)} = \mathbf{E}^{\downarrow}(\mathrm{e}^{-\lambda\overline{\xi}_{\infty}^{\downarrow}}) = \mathbb{E}_1^{\downarrow}((\overline{Y}_{\infty})^{-\lambda}), \tag{13.41}$$

where $\mathbf{P}^{\downarrow}$ is the law of $\xi^{\downarrow}$ (with associated expectation operator $\mathbf{E}^{\downarrow}$) and $\overline{Y}_{\infty} = \sup_{s\geq 0} Y_s$. Appealing to the change of measure discussed at the end of Sect. 13.2, we have, for $z > 1$,

$$\begin{aligned}
&\mathbb{P}_1^{\downarrow}(\overline{Y}_\infty > z)\\
&\quad = \mathbb{P}_1^{\downarrow}(\tau_z^+ < \infty)\\
&\quad = \mathbb{E}_1\big((Y_{\tau_z^+})^{\alpha(1-\rho)-1};\ \tau_z^+ < \tau_0^-\big)\\
&\quad = z^{\alpha(1-\rho)-1}\mathbb{E}_{1/z}\big((Y_{\tau_1^+})^{\alpha(1-\rho)-1};\ \tau_1^+ < \tau_0^-\big)\\
&\quad = \frac{\sin\pi\alpha\rho}{\pi}\left(1-\frac{1}{z}\right)^{\alpha\rho}\int_0^\infty \frac{1/z}{(1+y)y^{\alpha\rho}(y+1-1/z)}\mathrm{d}y\\
&\quad = 1-\left(1-\frac{1}{z}\right)^{\alpha\rho}, \qquad (13.42)
\end{aligned}$$

where in the third equality, we have used the scaling property, in the fourth equality, we have used Exercise 7.7 and for the fifth equality, we have used (13.37) and (13.38), writing $1-\rho$ in place of ρ. With (13.42) in hand, one may return to (13.41) and easily show that, up to a multiplicative constant,

$$\phi^{\downarrow}(\lambda) = \frac{\Gamma(1+\lambda+\alpha\rho)}{\Gamma(1+\lambda)\Gamma(\alpha\rho+1)},$$

for $\lambda \geq 0$. The computation is left to the reader.

In order to deal with $\widehat{\phi}^{\downarrow}$, write $\tau_{\log y}^{-,\downarrow} = \inf\{t > 0 : \xi_t^{\downarrow} < \log y\}$ and use the second Lamperti transform to deduce that, for $0 < z, y < 1$,

$$\begin{aligned}
\mathbf{P}^{\downarrow}\big(1-\exp\{\xi^{\downarrow}_{\tau^{-,\downarrow}_{\log y}} - \log y\} \leq z\big) &= \mathbb{P}_1^{\downarrow}\left(\frac{y - Y_{\tau_y^-}}{y} \leq z\right)\\
&= \mathbb{E}_1\left((Y_{\tau_y^-})^{\alpha(1-\rho)-1};\ \frac{y-Y_{\tau_y^-}}{y} \leq z\right)\\
&= \mathbb{E}_{1-y}\left((y+Y_{\tau_0^-})^{\alpha(1-\rho)-1};\ \frac{-Y_{\tau_0^-}}{y} \leq z\right).
\end{aligned}$$

Again, making use of Exercise 7.7, we deduce that

$$\begin{aligned}
&\mathbf{P}^{\downarrow}\big(1-\exp\{\xi^{\downarrow}_{\tau^{-,\downarrow}_{\log y}} - \log y\} \in \mathrm{d}z\big)\\
&\quad = \frac{\sin\pi\alpha(1-\rho)}{\pi}(1-y)^{\alpha(1-\rho)}(1-z)^{\alpha(1-\rho)-1}z^{-\alpha(1-\rho)}(1-y+zy)^{-1}\mathrm{d}z.
\end{aligned}$$

Suppose we write $\widehat{\Upsilon}^{\downarrow}$ for the Lévy measure of $\widehat{\phi}^{\downarrow}$. Taking limits as $y \downarrow 0$, we may again appeal to Corollary 5.3 to deduce that, for $0 < z < 1$ and $x > 0$,

$$\widehat{\Upsilon}^{\downarrow}(x,\infty)\mathrm{d}x \propto \mathbf{P}^{\downarrow}(\Delta^{\downarrow} \in \mathrm{d}x),$$

where

$$\mathbf{P}^{\downarrow}\left(1-\exp\{-\Delta^{\downarrow}\}\in \mathrm{d}z\right)=\frac{\sin\pi\alpha(1-\rho)}{\pi}(1-z)^{\alpha(1-\rho)-1}z^{-\alpha(1-\rho)}\mathrm{d}z.$$

Said another way, for $x>0$, we have

$$\widehat{\Upsilon}^{\downarrow}(x,\infty)\propto\frac{\mathrm{e}^{-\alpha(1-\rho)x}(1-\mathrm{e}^{-x})^{-\alpha(1-\rho)}}{\Gamma(1-\alpha(1-\rho))\Gamma(\alpha(1-\rho))},$$

where again, we have used Euler's reflection formula for gamma functions.

Using similar reasoning to previously, it is now easy to verify that

$$\widehat{\phi}^{\downarrow}(\lambda)\propto\lambda\int_0^\infty \mathrm{e}^{-\lambda x}\widehat{\Upsilon}^{\downarrow}(x,\infty)\mathrm{d}x=\frac{\Gamma(\lambda+\alpha(1-\rho))}{\Gamma(\alpha(1-\rho))\Gamma(\lambda)},$$

for $\lambda\geq 0$.

In conclusion, we see that, up to a multiplicative constant,

$$\Psi^{\downarrow}(\theta)=\frac{\Gamma(1-\mathrm{i}\theta+\alpha\rho)}{\Gamma(1-\mathrm{i}\theta)}\frac{\Gamma(\mathrm{i}\theta+\alpha(1-\rho))}{\Gamma(\mathrm{i}\theta)},\quad \theta\in\mathbb{R}. \tag{13.43}$$

Again, we see that $\xi^{\downarrow}$ also belongs to the class of hypergeometric Lévy processes.

13.4.3 The Case of ξ^*

The Lévy process ξ^* comes about by applying the second Lamperti transform to an α-stable process killed on first entering $(-\infty,0)$. In terms of the three categories described in Theorem 13.1 (i), if we exclude the case of a spectrally positive stable process, then, since all other stable processes do not creep downwards, we are forced to conclude that ξ^* belongs to the third category. Specifically, it is killed at some independent and exponentially distributed random time. A little thought also reveals that the case of a spectrally positive stable process has already been covered through our study of $\xi^{\downarrow}$.

We are interested in computing Ψ^*, the characteristic exponent of ξ^*. Suppose that $\mathbf{e}$ is the independent exponentially distributed time at which ξ is killed. Since we only aim to compute Ψ^* up to a multiplicative constant, we may assume that the rate associated with $\mathbf{e}$ is unity and appeal to the Wiener–Hopf factorisation to write

$$\mathbf{E}^*\left(\mathrm{e}^{\mathrm{i}\theta\xi^*_{\mathbf{e}}}\right)=\mathbf{E}^*\left(\mathrm{e}^{\mathrm{i}\theta\overline{\xi}^*_{\mathbf{e}}}\right)\mathbf{E}^*\left(\mathrm{e}^{\mathrm{i}\theta\underline{\xi}^*_{\mathbf{e}}}\right)=\frac{1}{\Psi^*(\theta)},\quad \theta\subset\mathbb{R},$$

where $\mathbf{P}^*$ (with associated expectation operator $\mathbf{E}^*$) is the law of ξ^*. Written another way, we have

$$\Psi^*(\theta)=\frac{1}{\mathbf{E}^*(\mathrm{e}^{\mathrm{i}\theta\overline{\xi}^*_{\mathbf{e}}})\mathbf{E}^*(\mathrm{e}^{\mathrm{i}\theta\underline{\xi}^*_{\mathbf{e}}})},\quad \theta\in\mathbb{R}. \tag{13.44}$$

Note that

$$\mathbf{E}^*\left(\mathrm{e}^{\mathrm{i}\theta\overline{\xi}^*_{\mathbf{e}}}\right) = \mathbb{E}_1\left((\overline{Y}_{\tau_0^-})^{\mathrm{i}\theta}\right) \quad \text{and} \quad \mathbf{E}^*\left(\mathrm{e}^{\mathrm{i}\theta\underline{\xi}^*_{\mathbf{e}}}\right) = \mathbb{E}_1\left((\underline{Y}_{\tau_0^- -})^{\mathrm{i}\theta}\right), \tag{13.45}$$

where $\overline{Y}_{\tau_0^-} = \sup_{s\leq\tau_0^-} Y_s$, $\underline{Y}_{\tau_0^- -} = \inf_{s<\tau_0^-} Y_s$ and $\mathbb{P}_1$ is the law of an α-stable process with initial value 1 (with associated expectation operator $\mathbb{E}_1$). To obtain the first equality in (13.45), note that, for all $z > 1$,

$$\begin{aligned}
\mathbb{P}_1(\overline{Y}_{\tau_0^-} \leq z) &= \mathbb{P}_1\left(\tau_0^- < \tau_z^+\right) \\
&= \mathbb{P}_{1/z}\left(\tau_0^- < \tau_1^+\right) \\
&= \frac{\Gamma(\alpha)}{\Gamma(\alpha\rho)\Gamma(\alpha(1-\rho))}\int_0^{1-1/z} u^{\alpha\rho-1}(1-u)^{\alpha(1-\rho)-1}\mathrm{d}u,
\end{aligned}$$

where the second equality is the result of scaling and the third equality uses Exercise 7.7. We can now compute, for $\theta \in \mathbb{R}$,

$$\begin{aligned}
\mathbb{E}_1\left((\overline{Y}_{\tau_0^-})^{\mathrm{i}\theta}\right) &= \frac{\Gamma(\alpha)}{\Gamma(\alpha\rho)\Gamma(\alpha(1-\rho))}\int_1^\infty z^{\mathrm{i}\theta}\left(1-\frac{1}{z}\right)^{\alpha\rho-1}\left(\frac{1}{z}\right)^{\alpha(1-\rho)-1}\frac{\mathrm{d}z}{z^2} \\
&= \frac{\Gamma(\alpha)}{\Gamma(\alpha\rho)\Gamma(\alpha(1-\rho))}\int_0^1 u^{\alpha(1-\rho)-\mathrm{i}\theta-1}(1-u)^{\alpha\rho-1}\mathrm{d}u \\
&= \frac{\Gamma(\alpha)\Gamma(\alpha(\rho-1)-\mathrm{i}\theta)}{\Gamma(\alpha\rho)\Gamma(\alpha-\mathrm{i}\theta)}.
\end{aligned}$$

In order to deal with the second equality in (13.45), observe that $\underline{Y}_{\tau_0^- -}$ is equal to the position of the descending ladder height process of Y immediately before entering into $(-\infty, 0)$. Recall from (6.37) that the descending ladder height process is a stable subordinator with index $\alpha(1-\rho)$. Using this fact, together with Theorem 5.6, or more conveniently Exercise 5.8, a straightforward computation gives, for $0 < z < 1$,

$$\begin{aligned}
&\mathbb{P}_1(\underline{Y}_{\tau_0^- -} \in \mathrm{d}z) \\
&\quad = \mathbb{P}(1-\widehat{H}_{\widehat{T}_1^+ -} \in \mathrm{d}z) \\
&\quad = \frac{\sin\pi\alpha(1-\rho)}{\pi}(1-z)^{\alpha(1-\rho)-1}z^{-\alpha(1-\rho)}\mathrm{d}z,
\end{aligned}$$

where $\{\widehat{H}_t : t \geq 0\}$ is the descending ladder height of Y and $\widehat{T}_1^+ = \inf\{t > 0 : \widehat{H}_t > 1\}$. It follows that, for $\theta \in \mathbb{R}$,

$$\mathbb{E}_1\left((\underline{Y}_{\tau_0^- -})^{\mathrm{i}\theta}\right) = \frac{\Gamma(\mathrm{i}\theta+1-\alpha(1-\rho))}{\Gamma(1-\alpha(1-\rho))\Gamma(\mathrm{i}\theta+1)}.$$

Putting everything together in (13.44), we find that, up to a multiplicative constant,

$$\Psi^*(\theta) = \frac{\Gamma(\alpha - \mathrm{i}\theta)}{\Gamma(\alpha(1-\rho) - \mathrm{i}\theta)} \frac{\Gamma(\mathrm{i}\theta + 1)}{\Gamma(\mathrm{i}\theta + 1 - \alpha(1-\rho))}, \quad \theta \in \mathbb{R}. \tag{13.46}$$

Again, we see that ξ^* belongs to the class of hypergeometric Lévy processes.

13.4.4 The Relation Between $\xi^\uparrow$, $\xi^\downarrow$ and ξ^*

Careful inspection of the three characteristic exponents $\Psi^\uparrow$, $\Psi^\downarrow$ and Ψ^* reveals that, for $\theta \in \mathbb{R}$,

$$\Psi^\uparrow(\theta) = \Psi^\downarrow(\theta - \mathrm{i}) = \Psi^*\big(\theta - \mathrm{i}\alpha(1-\rho)\big). \tag{13.47}$$

We shall now offer a very straightforward explanation of this connection.

Note that both $\mathbb{P}_1^\uparrow$ and $\mathbb{P}_1^\downarrow$ are absolutely continuous with respect to $\mathbb{P}_1^*$, the law of a stable process issued from 1 and killed on entering $(-\infty, 0)$. Taking account of their respective densities, we may write, for any stopping time T with respect to the filtration of Y,

$$\mathbb{E}_1^\uparrow\big(f(Y_T)\mathbf{1}_{(T<\infty)}\big) = \mathbb{E}_1^\downarrow\big(Y_T f(Y_T)\mathbf{1}_{(T<\infty)}\big) = \mathbb{E}_1^*\big(Y_T^{\alpha(1-\rho)} f(Y_T)\mathbf{1}_{(T<\infty)}\big),$$

where f is any bounded, measurable function and $\mathbb{E}_1^*$ is expectation with respect to $\mathbb{P}_1^*$. Glancing back to (13.24) and noting, in particular, that the quantity $x^\alpha I_t$ there is a stopping time, we see that, for all $t > 0$,

$$\mathbf{E}^\uparrow\big(f(\mathrm{e}^{\xi_t^\uparrow})\big) = \mathbf{E}^\downarrow\big(\mathrm{e}^{\xi_t^\downarrow} f(\mathrm{e}^{\xi_t^\downarrow})\big) = \mathbf{E}^*\big(\mathrm{e}^{\alpha(1-\rho)\xi_t^*} f(\mathrm{e}^{\xi_t^*})\big). \tag{13.48}$$

This last identity can be easily extended to complex-valued functions with bounded, measurable real and imaginary components. In that case, taking $f(z) = z^{\mathrm{i}}$, (13.47) follows immediately.

Taking $f = 1$ in (13.48) shows, with the help of the Markov property, that $\{\exp\{\xi_t^\downarrow\} : t \geq 0\}$ and $\{\exp\{\alpha(1-\rho)\xi_t^*\} : t \geq 0\}$ are martingales with respect to $\mathbb{P}^\downarrow$ and $\mathbb{P}^*$, respectively. Moreover, these martingales describe Esscher transforms in the spirit of (8.5), allowing one to transform between $\mathbb{P}^\downarrow$ and $\mathbb{P}^\uparrow$ and $\mathbb{P}^*$ and $\mathbb{P}^\uparrow$ respectively.[7]

Naturally, we could have used this observation to give shorter proofs of (13.43) and (13.46). However, the proofs given in Sects. 13.4.2 and 13.4.3 expose a number of interesting identities for $\xi^\downarrow$ and ξ^*, some of which are exploited in the exercises at the end of this chapter.

---

[7]Formally the case of Esscher transforms for killed Lévy processes was not discussed in (8.5). However, it is not difficult to check that one may similarly change measure in this way when the underlying Lévy process has independent exponential killing.

13.5 Self-similar Continuous-State Branching Processes

Fix $x > 0$. Suppose now that $(Y, \mathbb{P}_x)$ is a spectrally positive α-stable process with index $\alpha \in (1, 2)$, starting from $x > 0$. Let (Z, P_x) be the associated continuous-state branching process. That is to say, Z is the continuous-state branching process whose branching mechanism is given by $\psi(\lambda) = \lambda^\alpha$, $\lambda \geq 0$. Recall from Chap. 12 that the processes Y and Z are connected through the first Lamperti transform, given in Theorem 12.2. Specifically,

$$Z_t = Y_{\theta_t \wedge \tau_0^-},$$

with $\tau_0^- = \inf\{t > 0 : Y_t < 0\}$ and $\theta. = \theta.(Y)$, where for any positive stochastic process $X = \{X_t : t \geq 0\}$, we define

$$\theta_t(X) = \inf\left\{ s > 0 : \int_0^s \frac{1}{X_u} \mathrm{d}u > t \right\}. \tag{13.49}$$

Recall, moreover, that, from Lemma 12.15, if $P_x^\uparrow$ is the law of Z conditioned to stay positive with initial value $x > 0$, then $(Z, P_x^\uparrow)$ has the same law as $(Y_{\theta.}, \mathbb{P}_x^\uparrow)$.

The question we would like to address in this section is whether the process Z is a positive self-similar Markov process under either of the measures P_x or $P_x^\uparrow$. This is a natural question to ask. Indeed, we have already shown in Sect. 13.4 that a spectrally positive α-stable process up until its first entry in $(-\infty, 0)$ and a spectrally positive α-stable process conditioned to stay positive are both positive self-similar Markov processes. Therefore (Z, P_x) and $(Z, P_x^\uparrow)$ are both time-changed positive self-similar Markov processes. Our question thus boils down to whether self-similarity is preserved through the time change in the first Lamperti transform.

Remarkably, we can show something a little stronger. Namely that the class of positive self-similar Markov processes remains closed under the operation of taking the first Lamperti transform. The following result is due to Patie (2009a) and Kyprianou and Pardo (2008).

Proposition 13.5 *Suppose that X is any positive self-similar Markov process with initial value $x > 0$ and self-similarity index $\alpha > 1$. Then, recalling the definition (13.49), $\{X_{\theta_t} \mathbf{1}_{(\theta_t < \zeta)} : t \geq 0\}$ is a positive self-similar Markov process with initial value x, self-similarity index $\alpha - 1$ and the same underlying Lévy process appearing in the second Lamperti transform.*

Proof Suppose that ξ is the Lévy process associated with X through the second Lamperti transform. Write $I_t^{(\alpha)} = \int_0^t \exp\{\alpha \xi_s\} \mathrm{d}s$, $t \geq 0$, indicating the index of self-similarity, where previously, in (13.15), we have just written I_t. Accordingly, we shall additionally modify our notation and write φ_α for the right-continuous inverse of $I^{(\alpha)}$. Define

$$A_t = \int_0^t \frac{\mathrm{d}s}{X_s}, \quad t \geq 0.$$

Appealing to the second Lamperti transform, we have that, for $t < \varphi_\alpha(x^{-\alpha}\zeta)$,

$$A_{x^\alpha I_t^{(\alpha)}} = \int_0^{x^\alpha I_t^{(\alpha)}} \frac{1}{x} \exp\{-\xi_{\varphi_\alpha(x^{-\alpha}s)}\}ds$$
$$= x^{\alpha-1}\int_0^t e^{(\alpha-1)\xi_u}du$$
$$= x^{\alpha-1} I_t^{(\alpha-1)}, \tag{13.50}$$

where in the second equality we have changed variables using $s = x^\alpha I_u^{(\alpha)}$. On the other hand, for any $0 \le t < x^{\alpha-1} I_\infty^{(\alpha-1)}$,

$$\varphi_{\alpha-1}\big(x^{-(\alpha-1)}t\big) = \inf\{s \ge 0 : I_s^{(\alpha-1)} > x^{-(\alpha-1)}t\}$$
$$= \inf\{s \ge 0 : A_{x^\alpha I_s^{(\alpha)}} > t\}$$
$$= \inf\{\varphi_\alpha\big(x^{-\alpha}u\big) \ge 0 : A_u > t\}$$
$$= \varphi_\alpha\big(x^{-\alpha}\theta(t)\big), \tag{13.51}$$

where in the final inequality we have used the monotonicity and continuity of φ_α.

Using the fact that $\zeta = x^\alpha I_\infty$, and that A and θ are mutually inverse to one another, (13.50) gives us that

$$\inf\{t \ge 0 : X_{\theta(t)} = 0\} = A_\zeta = A_{x^\alpha I_\infty^{(\alpha)}} = x^{\alpha-1} I_\infty^{(\alpha-1)},$$

and, for all $0 \le t < x^{\alpha-1} I_\infty^{(\alpha-1)}$,

$$X_{\theta(t)} = x\exp\{\xi_{\varphi_\alpha(x^{-\alpha}\theta(t))}\} = x\exp\{\xi_{\varphi_{\alpha-1}(x^{-(\alpha-1)}t)}\},$$

thus completing the proof. □

We may now return to our original question concerning the processes (Z, P_x) and $(Z, P_x^\uparrow)$. From Sect. 13.4, we know that the underlying Lévy process that drives $(Y, \mathbb{P}_x)$ through the second Lamperti transform is ξ^* (in the case that it is spectrally positive). Said another way, it is the Lévy process with characteristic exponent given by (13.46) such that $\alpha(1-\rho) = 1$. The computations in Sect. 13.4 assumed that there were jumps in both directions. However, given the simpler form of the Wiener–Hopf factorisation in the spectrally one-sided case, the reader can readily check that a straightforward modification of the arguments given there can be adapted to handle the spectrally one-sided case as well. Similarly, one easily verifies that $(Y, \mathbb{P}_x^\uparrow)$ is associated, through the second Lamperti transform, with $\xi^\uparrow$ (in the case that it is spectrally positive), i.e. the Lévy process with characteristic exponent (13.40) such that $\alpha(1-\rho) = 1$. In both cases, we also recall that $(Y, \mathbb{P}_x)$ and $(Y, \mathbb{P}_x^\uparrow)$ have index of self-similarity $\alpha \in (1, 2)$.

It now follows from Proposition 13.5 that (Z, P_x) and $(Z, P_x^{\uparrow})$ are positive self-similar Markov processes, associated through the second Lamperti transform with ξ^* and $\xi^{\uparrow}$ respectively, but now with index of self-similarity $\alpha - 1$.

We can summarise the above conclusions with the schematic below, which also indicates the relevant functions that are used to construct changes of measure between the laws $(\mathbb{P}_x, P_x, \mathbf{P}^*)$ and $(\mathbb{P}_x^{\uparrow}, P_x^{\uparrow}, \mathbf{P}^{\uparrow})$. For convenience, we shall write Y^* for the process $\{Y_t \mathbf{1}_{(t<\tau_0^-)} : t \geq 0\}$.

$$
\begin{array}{ccccccc}
(\xi^*, \mathbf{P}^*) & \overset{\text{Lamperti 2}}{\longleftarrow - - \longrightarrow} & (Y^*, \mathbb{P}_x) & \overset{\text{Lamperti 1}}{\longleftarrow - - \longrightarrow} & (Z, P_x) & \overset{\text{Lamperti 2}}{\longleftarrow - - \longrightarrow} & (\xi^*, \mathbf{P}^*) \\
\uparrow \mathrm{e}^{-y} & & y^{-1} \uparrow & & \uparrow y^{-1} & & \mathrm{e}^{-y} \uparrow \\
| & \alpha & | & & | & \alpha - 1 & | \\
\downarrow \mathrm{e}^{y} & & y \downarrow & & \downarrow y & & \mathrm{e}^{y} \downarrow \\
(\xi^{\uparrow}, \mathbf{P}^{\uparrow}) & \overset{\text{Lamperti 2}}{\longleftarrow - - \longrightarrow} & (Y, \mathbb{P}_x^{\uparrow}) & \overset{\text{Lamperti 1}}{\longleftarrow - - \longrightarrow} & (Z, P_x^{\uparrow}) & \overset{\text{Lamperti 2}}{\longleftarrow - - \longrightarrow} & (\xi^{\uparrow}, \mathbf{P}^{\uparrow})
\end{array}
$$

13.6 Entrance Laws and Recurrent Extensions

The second Lamperti transform in Theorem 13.1 requires that the initial value of the underlying positive self-similar Markov process, X, is strictly positive. Taking limits as the initial value, x, tends to zero in the representation (13.17) does not offer any insight with regard to the following question: For a given positive self-similar Markov process, is it possible to give a construction of the process issued from the origin in terms of the underlying Lévy process ξ? Said another way, can we make sense of "P_0"?

We know from Sect. 13.1 that the point 0 may be included in the state space as an initial value, for at least some positive self-similar Markov processes. In this respect, one may take, for example, stable subordinators, the modulus of symmetric stable processes, Bessel processes and reflected stable processes. It turns out that addressing this issue in general is highly non-trivial. Our objective in this section is to summarise what is known in this direction. However, on account of the mathematical complexity involved, we shall offer no proofs, presenting instead the relevant intuition where possible.

Theorem 13.1 (i) indicates that positive self-similar Markov processes naturally divide into two classes. Firstly, *conservative processes*, for which $\zeta = \infty$ almost surely, and, secondly, *non-conservative processes*, for which $\zeta < \infty$ almost surely. It turns out that the way to deal with the first case is to construct an *entrance law* and the way to deal with the second case is to construct a *recurrent extension*.

13.6.1 Entrance Law

Suppose that X is a conservative positive self-similar Markov process. We want to find a way to give a meaning to "$P_0 := \lim_{x\downarrow 0} P_x$". One way to do this is to look at the behaviour of the transition semigroup of X as its initial value tends to zero. That is to say, to consider whether the weak limit

$$P_0(X_t \in \mathrm{d}y) := \lim_{x\downarrow 0} P_x(X_t \in \mathrm{d}y), \quad t, y > 0, \tag{13.52}$$

exists. In that case, for any sequence of times $0 < t_1 \leq t_2 \leq \cdots \leq t_n < \infty$ and $y_1, \ldots, y_n \in (0, \infty)$, $n \in \mathbb{N}$, the Markov property gives us

$$\begin{aligned}
&P_0(X_{t_1} \in \mathrm{d}y_1, \ldots, X_{t_n} \in \mathrm{d}y_n) \\
&\quad := \lim_{x\downarrow 0} P_x(X_{t_1} \in \mathrm{d}y_1, \ldots, X_{t_n} \in \mathrm{d}y_n) \\
&\quad = \lim_{x\downarrow 0} P_x(X_{t_1} \in \mathrm{d}y_1) P_{y_1}(X_{t_2-t_1} \in \mathrm{d}y_2, \ldots, X_{t_n-t_2} \in \mathrm{d}y_n) \\
&\quad = P_0(X_{t_1} \in \mathrm{d}y_1) P_{y_1}(X_{t_2-t_1} \in \mathrm{d}y_2, \ldots, X_{t_n-t_2} \in \mathrm{d}y_n).
\end{aligned}$$

The limit (13.52), when it exists, thus implies the existence of P_0 as limit of P_x as $x \downarrow 0$, in the sense of convergence of finite-dimensional distributions.

The existence of an entrance law was first investigated by Bertoin and Caballero (2002) for the case of positive self-similar processes with monotone increasing paths and processes with no positive jumps. The general case was treated in a series of papers: Bertoin and Yor (2002b), Caballero and Chaumont (2006a), Chaumont et al. (2012) and Bertoin and Savov (2011). In particular, amongst other things, one can find in these papers the following result.

Theorem 13.6 *Assume that X is a conservative positive self-similar Markov process. Moreover, suppose that the Lévy process $(\xi, \mathbb{P})$, associated with X through the second Lamperti transform, is not a compound Poisson process and has an ascending ladder height process H which satisfies $\mathbb{E}(H_1) < \infty$. Then $P_0 := \lim_{x\downarrow 0} P_x$ exists in the sense of convergence of finite-dimensional distributions. Conversely, if $\mathbb{E}(H_1) = \infty$, then this limit does not exist.*

The contents of the above theorem understates the actual contribution found in the aforementioned literature. This is partly due to the fact that we have not developed all the appropriate tools here in order to state the strongest available form of this theorem. Indeed, what has been shown by Bertoin and Savov (2011) and Chaumont et al. (2012) is that, when considering $\{P_x : x > 0\}$ as a family of probability laws on the measurable space of trajectories that are right-continuous with left limits, accompanied by the sigma-algebra generated by the so-called Skorokhod topology,[8] then, under the same assumption as Theorem 13.6, there exists $P_0 := \lim_{x\downarrow 0} P_x$, in the sense of weak convergence on the aforesaid measurable space. Moreover, when this assumption fails, the limit does not exist.

Under the additional assumption that $\mathbb{E}(\xi_1) > 0$, Bertoin and Yor (2002b) are also able to characterise transitions from the origin under the measure P_0. They showed that, for any positive measurable function f and $t > 0$,

$$E_0\big(f(X_t)\big) = \frac{1}{\alpha\mathbb{E}(\xi_1)}\mathbb{E}\bigg(\frac{1}{I_\infty^-} f\bigg(\bigg(\frac{t}{I_\infty^-}\bigg)^{1/\alpha}\bigg)\bigg),$$

where $I_\infty^- = \int_0^\infty \exp\{-\alpha\xi_s\}\mathrm{d}s$.

Bertoin and Savov (2011) go further and give a pathwise construction of the process $X^{(0)}$, using the forthcoming intuition, which, itself, was developed earlier in Caballero and Chaumont (2006a). Recall from the definition of self-similarity (13.1) that, for fixed $t, y > 0$,

$$P_1\big(t^{-1/\alpha}X_t \in \mathrm{d}y\big) = P_{t^{-1/\alpha}}(X_1 \in \mathrm{d}y).$$

This suggests that, in terms of the underlying Lévy process $\xi = \{\xi_t : t \geq 0\}$, one needs to sample ξ over a longer and longer time horizon in order to construct the law of $X^{(x)}$ as $x \downarrow 0$. Ultimately, one would therefore expect that a pathwise construction of the limiting process $X^{(0)}$ over any period $[0, t]$ would necessarily require one to sample ξ over an infinite time horizon. However, appealing to the Markov property, one would still need to further sample from an independent copy of ξ, again over an infinite time horizon, to construct the path of $X^{(0)}$ over the period (t, ∞).

Appealing to this logic, Bertoin and Savov (2011) define the law of the Lévy process ξ with time index running over $\mathbb{R}$, and then give a pathwise construction of $X^{(0)}$ with this extended definition of ξ. They reason that a Lévy process indexed by $\mathbb{R}$, now written $\xi^\circ = \{\xi_t^\circ : t \in \mathbb{R}\}$, must have the stationarity property that $(\xi^\circ_{\sigma_x^+} - x, x - \xi^\circ_{\sigma_x^+ -})$ is independent of $x \in \mathbb{R}$, where $\sigma_x^+ = \inf\{t > -\infty : \xi_t^\circ > x\}$. Moreover, the distribution of $(\xi^\circ_{\sigma_x^+} - x, x - \xi^\circ_{\sigma_x^+ -})$ must be equal to that of the joint law of the overshoot and undershoot of ξ at first passage over a level as this level tends to ∞. Suppose that ξ has ascending ladder height process H and the potential measure of its descending ladder height is denoted by $\widehat{U}$. Assume that $\mathbb{E}(H_1) < \infty$ (which necessarily implies that H has no killing, and hence $\limsup_{t\uparrow\infty} \xi_t = \infty$) and write $\gamma \geq 0$ for its drift coefficient. Then referring back to Exercise 7.9, the aforementioned joint law is given by

$$\chi(\mathrm{d}y, \mathrm{d}z) = \frac{1}{\mathbb{E}(H_1)}\big(\widehat{U}(z)\Pi(z + \mathrm{d}y)\mathrm{d}z + \gamma\delta_0(\mathrm{d}y)\delta_0(\mathrm{d}z)\big), \quad y, z \geq 0.$$

[8]Let $\mathbb{D}$ be the space of mappings from $[0, \infty)$ to $\mathbb{R}$ which are right-continuous with left limits. The Skorokhod topology is generated by an appropriate metric on the space $\mathbb{D}$, which has the property that most events of interest belong to the sigma-algebra generated by its open sets. The details are far too involved to provide a concise overview here. The reader is instead referred to Chap. VI of Jacod and Shiryaev (1987), or indeed Billingsley (1999).

Next, write $\mathbb{P}_x$ (resp. $\mathbb{P}_x^{\uparrow}$) for the law of ξ (resp. ξ conditioned to stay positive) when $\xi_0 = x \geq 0$.[9] Let us also write $\xi^{\uparrow} = \{\xi_t^{\uparrow} : t \geq 0\}$ for an independent copy of the process ξ conditioned to stay positive.

The process ξ° can now be described as follows. We suppose that the two-dimensional random variable $(\Delta, \Delta^{\uparrow})$ has distribution χ. Then

$$\xi_t^{\circ} := \begin{cases} \xi_t & \text{under } \mathbb{P}_{\Delta} \text{ if } t \geq 0, \\ -\xi_{|t|-}^{\uparrow} & \text{under } \mathbb{P}_{\Delta^{\uparrow}}^{\uparrow} \text{ if } t < 0. \end{cases}$$

Now, define

$$I_t^{\circ} = \int_{-\infty}^{t} \mathrm{e}^{\alpha \xi_s^{\circ}} \mathrm{d}s$$

and let $\varphi^{\circ}(t) = \inf\{s > 0 : I_s^{\circ} \geq t\}$. Then Bertoin and Savov (2011) showed, under the same assumptions as Theorem 13.6, that $I_{\infty}^{\circ} = \infty$ almost surely and the process

$$X_t^{(0)} = \exp\{\xi_{\varphi^{\circ}(t)}^{\circ}\}, \quad t \geq 0,$$

has the same law as P_0.

A convenient feature of this construction, and indeed the earlier inspiration for this construction in Caballero and Chaumont (2006a), is that it offers transparency with regard to the need for the assumption that $\mathbb{E}(H_1) < \infty$. This condition is crucial to the definition of the process ξ°, around which the whole construction pivots.

13.6.2 Recurrent Extension

Suppose now that X is a non-conservative positive self-similar Markov process. If there is a way to describe how X can be issued from the origin then, in principle, one should be able to reissue it from the origin at all subsequent hitting times of this point, in such a way that the resulting process remains strong Markov, thereby generating what is known as a recurrent extension. To be more precise, we say that the strong Markov process, $\overset{\leadsto}{X} := \{\overset{\leadsto}{X}_t : t \geq 0\}$, possessing paths that are right-continuous with left limits and probabilities $\{\overset{\leadsto}{P}_x,\ x \geq 0\}$, is a *recurrent extension* of X if, for each $x > 0$, the origin is not an absorbing state $\overset{\leadsto}{P}_x$-almost surely and $\{\overset{\leadsto}{X}_{t \wedge \overset{\leadsto}{\zeta}} : t \geq 0\}$ under $\overset{\leadsto}{P}_x$ has the same law as (X, P_x), where

$$\overset{\leadsto}{\zeta} = \inf\{t > 0 : \overset{\leadsto}{X}_t = 0\}.$$

[9] Our lack of willingness to give a precise description of $\mathbb{P}_0^{\uparrow}$ at the end of Sect. 13.2.1 comes at the price of lack of clarity at this point of our informal discussion. On the other hand, in the case that $\gamma = 0$, that is to say, there is no upward creeping in ξ, there is no need for us to be clear about the meaning of $\mathbb{P}_0^{\uparrow}$ as then it is not used in this construction.

Showing that a recurrent extension exists is a very technical task and revolves around the theory of excursions. Roughly speaking, instead of constructing an entrance law for the family $\{P_x : x > 0\}$, it turns out that the correct mathematical procedure is to use $\{P_x : x > 0\}$ to construct an entrance law for an excursion measure that will describe the sojourns of $\widetilde{X}$ away from zero. Then with the help of what is known as *Itô synthesis*, one may piece together excursions end to end in an appropriate way to generate the desired recurrent extension.

In theory, one may approach the problem of constructing an excursion entrance law, and hence the problem of constructing a recurrent extension, in two different ways. Either the excursion may start by leaving the origin with a jump, or it leaves the origin continuously. Necessary and sufficient conditions are given by Rivero (2005) in the first case and by Rivero (2007) and Fitzsimmons (2006) in the second case. We focus on the case of recurrent extensions which leave the origin continuously, on account of the fact that the construction is unique. Otherwise, in the case of processes which leave the origin with a jump, there is no unique construction.

Theorem 13.7 *Assume that X is a non-conservative positive self-similar Markov process. Suppose that $(\xi, \mathbb{P})$ is the (killed) Lévy process associated with X through the second Lamperti transform and, moreover, it is not a compound Poisson process. Then there exists a unique recurrent extension of X which leaves 0 continuously if and only if there exists a $\beta \in (0, \alpha)$ such*

$$\mathbb{E}\big(e^{\beta\xi_1}\big) = 1. \tag{13.53}$$

Here, as usual, α is the index of self-similarity.

Condition (13.53) is also known as the Cramér condition.

13.7 Spectrally Negative Processes

For any given positive self-similar Markov process, X, if the (killed) Lévy process, ξ, associated with X through the second Lamperti transform, is spectrally negative, then we say that X is a *positive self-similar Markov process of the spectrally negative type*. In this final section, we shall briefly introduce some fluctuation theory for positive self-similar Markov processes of the spectrally negative type.

Conforming to the notation in Chap. 8, we shall write

$$\psi(\lambda) = \log \mathbb{E}\big(\exp\{\lambda\xi_1\}\big), \quad \lambda \geq 0,$$

where

$$\psi(\lambda) = -q - a\lambda + \frac{1}{2}\sigma^2\lambda^2 + \int_{(-\infty,0)} \big(e^{\lambda x} - 1 - \lambda x \mathbf{1}_{(x>-1)}\big)\Pi(dx),$$

such that $q \geq 0$, $a \in \mathbb{R}$, $\sigma^2 \geq 0$ and Π is a measure concentrated on $(-\infty, 0)$ such that $\int_{(-\infty,0)}(1 \wedge x^2)\Pi(dx) < \infty$. In the case that ξ is killed, we have $q = -\psi(0) > 0$.

As usual, we write X, with probabilities $\{P_x : x > 0\}$, for the positive self-similar Markov process associated with ξ by the second Lamperti transform. When $\psi(0) = 0$ and $\psi'(0+) \geq 0$, the state 0 is never visited at strictly positive times and Theorem 13.6 gives us the existence of an entrance law P_0. In the case that the boundary state 0 is reached continuously in an almost surely finite time, we have $\psi(0) = 0$ and $\psi'(0+) < 0$. Otherwise, when $-\psi(0) > 0$, the state zero it is reached in an almost surely finite time by a jump. Moreover, for these last two cases, Theorem 13.7 tells us that, providing there exists a $\Phi(0) \in (0, \alpha)$, a unique recurrent extension of X exists which leaves 0 continuously, thereby giving meaning to P_0. As usual, $\alpha > 0$ is the index of self-similarity.

13.7.1 Patie's Scale Functions

Recall from Chap. 8 that, for any spectrally negative Lévy process, ξ, there exists a family of so-called scale functions. These functions play a fundamental role in many fluctuation identities for spectrally negative Lévy processes. Patie (2009b) introduced a family of functions which, just like scale functions for spectrally negative Lévy processes, play a similarly natural role for many fluctuation identities of positive self-similar Markov processes of the spectrally negative type, as well as possessing related martingale properties (cf. Exercise 8.12). Unlike scale functions for spectrally negative Lévy processes, they can be explicitly identified through a power series representation with coefficients written in terms of the Laplace exponent ψ. Defined immediately below, we henceforth refer to them as *Patie's scale functions.*

Definition 13.8 (Patie's scale functions) Fix $\alpha > 0$. For a given (killed) spectrally negative Lévy process with Laplace exponent ψ, let

$$\mathsf{a}_0(\psi;\alpha) = 1 \text{ and } \mathsf{a}_n(\psi;\alpha) = \left(\prod_{k=1}^{n} \psi(\alpha k)\right)^{-1}, \quad n \in \mathbb{N} \tag{13.54}$$

and define the function

$$\mathcal{I}_{\psi,\alpha}(x) = \sum_{n=0}^{\infty} \mathsf{a}_n(\psi;\alpha) x^n, \quad x \geq 0.$$

Let us make some immediate observations regarding basic analytical properties of $\mathcal{I}_{\psi,\alpha}$. Firstly, note that, since $\lim_{\lambda\uparrow\infty} \psi(\lambda) = \infty$, we have

$$\lim_{n\uparrow\infty} \frac{|\mathsf{a}_{n+1}(\psi;\alpha)|}{|\mathsf{a}_n(\psi;\alpha)|} = \lim_{n\uparrow\infty} \frac{1}{|\psi(\alpha(n+1))|} = 0.$$

Hence for all $x \geq 0$, $\mathcal{I}_{\psi,\alpha}(x)$ is a convergent series. In fact this argument shows that if we consider $\mathcal{I}_{\psi,\alpha}$ as a mapping on $\mathbb{C}$, then it is an entire function. We also see that, whenever $\Phi(0) < \alpha$, all of the coefficients $\mathtt{a}_n(\psi;\alpha)$ are strictly positive and $\mathcal{I}_{\psi,\alpha}$ is a positive and strictly increasing function.

13.7.2 Exit Problems

Our aim here is to use Patie's scale functions to address some simple exit problems in the spirit of what we have seen in Chap. 8 for spectrally negative Lévy processes. Recall that $\zeta = \inf\{t > 0 : X_t = 0\}$. As a first step, we examine a martingale property, similar to the martingale property observed in Exercise 8.12 for scale functions of spectrally negative Lévy processes.

Theorem 13.9 *Suppose that $\psi(0) = 0$ and $\psi'(0+) \geq 0$ (equivalently $\Phi(0) = 0$) and fix $q > 0$. The process*

$$\mathrm{e}^{-qt}\mathcal{I}_{\psi,\alpha}\big(qX_t^\alpha\big), \quad t \geq 0$$

is a martingale.

Proof Let us start by first proving the following claim, lifted from Bertoin and Yor (2002a). For $x, q, t > 0$,

$$E_x\big(X_t^{\alpha n}\big) = \sum_{k=0}^{n} \frac{\mathtt{a}_{n-k}(\psi;\alpha)}{\mathtt{a}_n(\psi;\alpha)} x^{\alpha(n-k)} \frac{t^k}{k!}. \tag{13.55}$$

To this end, let us define $u_n(x,t) = x^{-\alpha n}E_x(X_t^{\alpha n})$ for $x, t > 0$ and $n \in \mathbb{N}$. In the notation of Sect. 13.3,

$$u_n(x,t) = \mathbb{E}\big(\exp\{\alpha n \xi_{\varphi(x^{-\alpha}t)}\}\big) = \mathbb{E}^{\alpha n}\big(\mathrm{e}^{\psi(\alpha n)\varphi(x^{-\alpha n}t)}\big),$$

where we have used the fact that $\varphi(x^{-\alpha}t)$ is a stopping time in the filtration of ξ and $\mathbb{E}^{\alpha n}$ means expectation with respect to the probability measure $\mathbb{P}^{\alpha n}$, which is defined through the exponential change of measure in (8.5).[10]

Next, recall that, for all $t \geq 0$,

$$\int_0^{\varphi(x^{-\alpha}t)} \mathrm{e}^{\alpha\xi_s}\mathrm{d}s = x^{-\alpha}t, \tag{13.56}$$

so that

$$\frac{\mathrm{d}}{\mathrm{d}t}\varphi\big(x^{-\alpha}t\big) = x^{-\alpha}\mathrm{e}^{-\alpha\xi_{\varphi(x^{-\alpha}t)}}.$$

[10] As remarked upon earlier in this chapter, although the exponential change of measure has only been defined for processes ξ with no killing, the reader can easily verify that it is equally applicable to spectrally negative Lévy processes with killing.

Accordingly, we have

$$\frac{\mathrm{d}}{\mathrm{d}t}\mathrm{e}^{\psi(\alpha n)\varphi(x^{-\alpha}t)} = x^{-\alpha}\psi(\alpha n)\mathrm{e}^{\psi(\alpha n)\varphi(x^{-\alpha}t)}\mathrm{e}^{-\alpha\xi_{\varphi(x^{-\alpha}t)}},$$

so that

$$\mathrm{e}^{\psi(\alpha n)\varphi(x^{-\alpha}t)} = 1 + x^{-\alpha}\psi(\alpha n)\int_0^t \mathrm{e}^{\psi(\alpha n)\varphi(x^{-\alpha}s)}\mathrm{e}^{-\alpha\xi_{\varphi(x^{-\alpha}s)}}\mathrm{d}s.$$

Now taking expectations above with respect to $\mathbb{P}^{\alpha n}$ and reverting back to the original measure $\mathbb{P}$, we find, for $x, t > 0$ and $n \geq 1$,

$$u_n(x,t) = 1 + x^{-\alpha}\psi(\alpha n)E_x\left(\int_0^t u_{n-1}(x,s)\mathrm{d}s\right),$$

where $u_0(x,t) = 1$. It is now a straightforward exercise, left to the reader, to show by induction that

$$u_n(x,t) = \sum_{k=0}^{n}\frac{\mathsf{a}_{n-k}(\psi;\alpha)}{\mathsf{a}_n(\psi;\alpha)}x^{-\alpha k}\frac{t^k}{k!},$$

from which the claim (13.55) follows.

We can now compute for all $q, t, x > 0$,

$$\begin{aligned}
E_x\big(\mathrm{e}^{-qt}\mathcal{I}_{\psi,\alpha}\big(qX_t^\alpha\big)\big) &= \mathrm{e}^{-qt}\sum_{n=0}^{\infty}\mathsf{a}_n(\psi;\alpha)q^n E_x\big(X_t^{\alpha n}\big) \\
&= \mathrm{e}^{-qt}\sum_{n=0}^{\infty}\mathsf{a}_n(\psi;\alpha)q^n\sum_{k=0}^{n}\frac{\mathsf{a}_{n-k}(\psi;\alpha)}{\mathsf{a}_n(\psi;\alpha)}x^{\alpha(n-k)}\frac{t^k}{k!} \\
&= \mathrm{e}^{-qt}\sum_{k=0}^{\infty}q^k\frac{t^k}{k!}\sum_{n=k}^{\infty}\mathsf{a}_{n-k}(\psi;\alpha)q^{n-k}x^{\alpha(n-k)} \\
&= \mathcal{I}_{\psi,\alpha}\big(qx^\alpha\big),
\end{aligned} \tag{13.57}$$

where we have used the fact that $\Phi(0) = 0$ (which implies that the coefficients $\mathsf{a}_n(\psi;\alpha)$ are all positive) and Fubini's Theorem in the first equality. The case that $x = 0$ can be obtained by taking limits as $x \downarrow 0$ in (13.57) making use of Theorem 13.6.

Finally, the Markov property together with the identity (13.57) gives us the required martingale property. Indeed, if $\{\mathcal{G}_t : t \geq 0\}$ is the natural filtration generated by X, then for $s, t > 0$,

$$\begin{aligned}
E_x\big(\mathrm{e}^{-q(t+s)}\mathcal{I}_{\psi,\alpha}\big(qX_{t+s}^\alpha\big)|\mathcal{G}_s\big) &= \mathrm{e}^{-qs}E_y\big(\mathrm{e}^{-qt}\mathcal{I}_{\psi,\alpha}\big(qX_t^\alpha\big)\big)\big|_{y=X_s} \\
&= \mathrm{e}^{-qs}\mathcal{I}_{\psi,\alpha}\big(qX_s^\alpha\big).
\end{aligned}$$

The proof is now complete. □

The next theorem deals with the promised exit problems for our class of positive self-similar Markov processes of the spectrally negative type.

Theorem 13.10 *Fix $q \geq 0$ and, for all $a > 0$, let $\kappa_a^+ = \inf\{t > 0 : X_t > a\}$.*

(i) *Suppose that $\Phi(0) < \alpha$. For $0 \leq x \leq a$, we have*

$$E_x\left(e^{-q\kappa_a^+}\right) = \frac{\mathcal{I}_{\psi,\alpha}(qx^\alpha)}{\mathcal{I}_{\psi,\alpha}(qa^\alpha)}.$$

(ii) *Moreover, for $0 < x \leq a$, we have*

$$E_x\left(e^{-q\kappa_a^+}\mathbf{1}_{(\kappa_a^+ < \zeta)}\right) = \left(\frac{x}{a}\right)^{\Phi(0)} \frac{\mathcal{I}_{\psi_{\Phi(0)},\alpha}(qx^\alpha)}{\mathcal{I}_{\psi_{\Phi(0)},\alpha}(qa^\alpha)},$$

where $\psi_{\Phi(0)}(\lambda) = \psi(\lambda + \Phi(0))$.

Proof (i) We start by giving the proof in the case that $\psi(0) = 0$ and $\psi'(0+) \geq 0$ (equivalently $\Phi(0) = 0$) and hence $\zeta = \infty$ almost surely. Applying Doob's Optional Sampling Theorem at the bounded stopping time $t \wedge \kappa_a^+$, we have, for all $q \geq 0$ and $0 \leq x \leq a$,

$$E_x\left(e^{-q(t\wedge\kappa_a^+)}\mathcal{I}_{\psi,\alpha}\left(qX^\alpha_{t\wedge\kappa_a^+}\right)\right) = \mathcal{I}_{\psi,\alpha}\left(qx^\alpha\right).$$

Noting that $X_{t\wedge\kappa_a^+} \leq a$, we may apply bounded convergence to conclude that as $t \uparrow \infty$,

$$\mathcal{I}_{\psi,\alpha}\left(qx^\alpha\right) = E_x\left(e^{-q\kappa_a^+}\mathcal{I}_{\psi,\alpha}\left(qX^\alpha_{\kappa_a^+}\right)\right) = E_x\left(e^{-q\kappa_a^+}\right)\mathcal{I}_{\psi,\alpha}\left(qa^\alpha\right),$$

where in the second equality, we have used the fact that $X_{\kappa_a^+} = a$, which follows as a consequence of spectral negativity.

The remaining cases, when there is a recurrent extension, are somewhat more complicated and therefore omitted. The reader is referred instead to Patie (2009b) for further details. See also the comments in Kyprianou and Patie (2011).

(ii) When $\psi(0) = 0$ and $\psi'(0+) \geq 0$ there is nothing to prove as $\Phi(0) = 0$ and hence $\zeta < \infty$ almost surely. We therefore concentrate on the case that either $\psi(0) = 0$ and $\psi'(0+) < 0$ or $-\psi(0) > 0$.

The process X with probabilities $\{P_x : x \geq 0\}$ corresponds, through the second Lamperti transform, to the spectrally negative Lévy process $(\xi, \mathbb{P})$. Suppose that we consider instead the positive self-similar Markov process of the spectrally negative type, X, with probabilities $\{P_x^{\Phi(0)} : x \geq 0\}$, corresponding, through the second Lamperti transform, to the Lévy process $(\xi, \mathbb{P}^{\Phi(0)})$. From the discussion around the change of measure (8.5), under $\mathbb{P}^{\Phi(0)}$, the Laplace exponent of ξ is equal to $\psi_{\Phi(0)}(\lambda)$ for $\lambda \geq 0$. Note also that since $\psi_{\Phi(0)}(0) = 0$ and $\psi'_{\Phi(0)}(0+) = \psi'(\Phi(0)) > 0$, the process $(\xi, \mathbb{P}^{\Phi(0)})$ has no killing and drifts to ∞. This implies that $P_x^{\Phi(0)}(\zeta = \infty) = 1$ for all $x \geq 0$.

Next, we are going to use the fact that, for $x > 0$, under P_x,

$$a = X_{\kappa_a^+} = x \exp\{\xi_{\tau^+_{\log(a/x)}}\},$$

and accordingly $\tau^+_{\log(a/x)} = \varphi(x^{-\alpha}\kappa_a^+)$, whenever the stopping times κ_a^+ and $\tau^+_{\log(a/x)}$ are finite.

Hence, from (13.16),

$$\kappa_a^+ = x^\alpha \int_0^{\tau^+_{\log(a/x)}} \exp\{\xi_s\} \mathrm{d}s.$$

We may now compute, for $0 < x \le a$,

$$\begin{aligned}
E_x^{\Phi(0)}\big(\mathrm{e}^{-q\kappa_a^+}\big) &= \mathbb{E}^{\Phi(0)}\left(\exp\left\{-qx^\alpha \int_0^{\tau^+_{\log(a/x)}} \mathrm{e}^{\xi_s}\mathrm{d}s\right\}\mathbf{1}_{(\tau^+_{\log(a/x)}<\infty)}\right) \\
&= \mathbb{E}\left(\exp\left\{-qx^\alpha \int_0^{\tau^+_{\log(a/x)}} \mathrm{e}^{\xi_s}\mathrm{d}s + \Phi(0)\xi_{\tau^+_{\log(a/x)}}\right\}\mathbf{1}_{(\tau^+_{\log(a/x)}<\mathbf{e})}\right) \\
&= \left(\frac{a}{x}\right)^{\Phi(0)} \mathbb{E}\left(\exp\left\{-qx^\alpha \int_0^{\tau^+_{\log(a/x)}} \mathrm{e}^{\xi_s}\mathrm{d}s\right\}\mathbf{1}_{(\tau^+_{\log(a/x)}<\mathbf{e})}\right) \\
&= \left(\frac{a}{x}\right)^{\Phi(0)} E_x\big(\mathrm{e}^{-q\kappa_a^+}\mathbf{1}_{(\kappa_a^+<\zeta)}\big).
\end{aligned}$$

Note that in the second equality, we have simply applied the change of measure (8.5), noting that $\mathbf{e}$ is the killing time of $(\xi, \mathbb{P})$, which is almost surely infinite in the case that $\psi(0) = 0$. The desired result now follows by using the expression derived in part (i). □

13.7.3 Ciesielski–Taylor Identity

Recall the definition of Bessel processes in Sect. 13.1.3. Suppose that $(X, Q^{(\mathrm{d})})$ is a Bessel process starting from 0, with dimension $\mathrm{d} > 0$. For these processes, Ciesielski and Taylor (1962) observed that the following curious identity holds in distribution. For $a > 0$ and *integer* $\mathrm{d} > 0$,

$$\big(\kappa_a^+, Q^{(\mathrm{d})}\big) \stackrel{(d)}{=} \left(\int_0^\infty \mathbf{1}_{(X_s \le a)} ds,\ Q^{(\mathrm{d}+2)}\right), \tag{13.58}$$

where we recall that $\kappa_a^+ = \inf\{s \ge 0;\ X_s > a\}$. They proved this relationship by showing that the densities of both random variables coincide. Getoor and Sharpe (1979) extended this identity to any dimension $\mathrm{d} > 0$ by means of Laplace transforms and recurrence relationships for Bessel functions. Other generalisations of this identity have been explored for a variety of other one-dimensional Markov processes. See for example Biane (1985), Carmona et al. (1998) and Bertoin (1992). In particular Yor (1991) gives a pathwise explanation for this identity for certain parameter regimes. We shall follow the exposition of Kyprianou and Patie (2011) here and look at the generalisation of the Ciesielski–Taylor identity for the class of positive self-similar Markov processes of the spectrally negative type. This setting captures all of the aforementioned results with the exception of those of Biane (1985), who considers the case of one-dimensional diffusions.

In order to state the main result, we need to establish some appropriate notation. Recall that we are working in a setting where P_0 is well defined for our class of positive self-similar Markov processes of the spectrally negative type, either as an entrance law or the law of a recurrent extension issued from the origin. It will be convenient to adjust our notation to include some information about the underlying spectrally negative Lévy process, ξ. We shall therefore prefer to write P_0^ψ from now on, where ψ is the Laplace exponent of ξ. Accordingly, we shall also write $\mathbb{P}^\psi$ for the law of ξ.

Next, recall from Exercise 9.8 that for any given Laplace exponent ψ of a spectrally negative Lévy process and any $\beta > 0$,

$$\mathcal{T}_\beta\psi(\lambda) = \frac{\lambda}{\lambda+\beta}\psi(\lambda+\beta), \quad \lambda \geq -\beta, \tag{13.59}$$

is also the Laplace exponent of another spectrally negative Lévy process. Although the case of killed processes was not covered by Exercise 9.8, we may easily verify that the same conclusion still holds. In particular, when $-\psi(0) > 0$, $\mathcal{T}_\beta\psi(0) = 0$ and, hence, the spectrally negative Lévy process with Laplace exponent $\mathcal{T}_\beta\psi$ has no killing.

We are now ready to state our generalisation of the Ciesielski–Taylor identity.

Theorem 13.11 *Suppose that ψ is the Laplace exponent of a (possibly-killed) spectrally negative Lévy process. Assume that $\Phi(0) < \alpha$. Then, for any $a > 0$, the following Ciesielski–Taylor type identity holds:*

$$\left(\kappa_a^+, P_0^\psi\right) \stackrel{(d)}{=} \left(\int_0^\infty \mathbf{1}_{(X_s \leq a)} \mathrm{d}s, P_0^{\mathcal{T}_\alpha\psi}\right). \tag{13.60}$$

Proof Recall from Theorem 13.10 that, for $0 \leq x \leq a$ and $q \geq 0$, we have

$$E_0^\psi\left[\mathrm{e}^{-q\kappa_a^+}\right] = \frac{1}{\mathcal{I}_{\psi,\alpha}(qa^\alpha)}. \tag{13.61}$$

The proof is thus complete as soon as we can show the right-hand side above is equal to $O_q^{\mathcal{T}_\alpha\psi}(0;a)$, where, for any $q \geq 0$, $a > 0$ and $0 \leq x \leq a$,

$$O_q^{\mathcal{T}_\alpha\psi}(x;a) = E_x^{\mathcal{T}_\alpha\psi}\left[\mathrm{e}^{-q\int_0^\infty \mathbf{1}_{(X_s \leq a)}\mathrm{d}s}\right].$$

To this end, observe that a straightforward computation, using the self-similarity property (13.1), implies that

$$O_q^{\mathcal{T}_\alpha\psi}(x;a) = O_{qa^\alpha}^{\mathcal{T}_\alpha\psi}(x/a;1).$$

Hence, without loss of generality, we may henceforth assume that $a = 1$. Note moreover that with the help of the strong Markov property, we have, for $q \geq 0$,

$$O_q^{\mathcal{T}_\alpha\psi}(0;1) = E_0^{\mathcal{T}_\alpha\psi}\left(e^{-q\kappa_1^+}\right)O_q^{\mathcal{T}_\alpha\psi}(1;1).$$

Therefore, again appealing to Theorem 13.10, the theorem is proved as soon as we can show that, for all $q \geq 0$,

$$O_q^{\mathcal{T}_\alpha\psi}(1;1) = \frac{\mathcal{I}_{\mathcal{T}_\alpha\psi,\alpha}(q)}{\mathcal{I}_{\psi,\alpha}(q)}.$$

We start by computing $O_q^{\mathcal{T}_\alpha\psi}(1;1)$. Let

$$\kappa_1^- = \inf\{s > 0;\ X_s < 1\}$$

be the first-passage time of X below the level 1. Fixing $y > 1$, we may make use of the strong Markov property and spectral negativity to deduce that

$$O_q^{\mathcal{T}_\alpha\psi}(1;1) = E_1^{\mathcal{T}_\alpha\psi}\left[e^{-q\int_0^{\kappa_y^+}\mathbf{1}_{(X_s\leq 1)}ds}\right]E_y^{\mathcal{T}_\alpha\psi}\left[e^{-q\int_0^\infty \mathbf{1}_{(X_s\leq 1)}ds}\right]$$

and

$$\begin{aligned}E_y^{\mathcal{T}_\alpha\psi}&\left[e^{-q\int_0^\infty \mathbf{1}_{(X_s\leq 1)}ds}\right]\\ &= E_y^{\mathcal{T}_\alpha\psi}\left[\mathbf{1}_{(\kappa_1^-<\infty)}E_{X_{\kappa_1^-}}^{\mathcal{T}_\alpha\psi}\left[e^{-q\kappa_1^+}\right]\right]O_q^{\mathcal{T}_\alpha\psi}(1;1) + P_y^{\mathcal{T}_\alpha\psi}\left[\kappa_1^- = \infty\right].\end{aligned}$$

Solving for $O_q^{\mathcal{T}_\alpha\psi}(1;1)$, we get for all $y > 1$,

$$\begin{aligned}O_q^{\mathcal{T}_\alpha\psi}&(1;1)^{-1}\\ &= \frac{\{E_1^{\mathcal{T}_\alpha\psi}[e^{-q\int_0^{\kappa_y^+}\mathbf{1}_{(X_s\leq 1)}ds}]\}^{-1} - E_y^{\mathcal{T}_\alpha\psi}[\mathbf{1}_{(\kappa_1^-<\infty)}E_{X_{\kappa_1^-}}^{\mathcal{T}_\alpha\psi}[e^{-q\kappa_1^+}]]}{P_y^{\mathcal{T}_\alpha\psi}[\kappa_1^- = \infty]}. \qquad (13.62)\end{aligned}$$

Now we evaluate some of the expressions on the right-hand side above. First, we deal with the denominator. We write as usual $\tau_0^- = \inf\{t > 0 : \xi_t < 0\}$. Note that

$$X_{\kappa_1^-}^{(y)} = y\exp\{\xi_{\varphi(y^{-\alpha}\kappa_1^-)}\} = y\exp\{\xi_{\tau_0^-}\}.$$

Hence, from (13.16), under $P_y^{\mathcal{T}_\alpha\psi}$, we have

$$y^{-\alpha}\kappa_1^- = \int_0^{\tau_0^-} \mathrm{e}^{\alpha\xi_s}\mathrm{d}s, \tag{13.63}$$

with the understanding that both left- and right-hand sides are infinite in value at the same time. Next, recall that $(\mathcal{T}_\alpha\psi)'(0+) = \psi(\alpha)/\alpha > 0$ and hence, using (8.10) in Theorem 8.1 together with (13.63), we have

$$\begin{aligned} P_y^{\mathcal{T}_\alpha\psi}\left[\kappa_1^- = \infty\right] &= \mathbb{E}_{\log y}^{\mathcal{T}_\alpha\psi}\left(\int_0^{\tau_0^-} \mathrm{e}^{\alpha\xi_s}\mathrm{d}s = \infty\right) \\ &= \mathbb{P}_{\log y}^{\mathcal{T}_\alpha\psi}\left(\tau_0^- = \infty\right) \\ &= \frac{\psi(\alpha)}{\alpha} W_{\mathcal{T}_\alpha\psi}(\log y) \end{aligned} \tag{13.64}$$

where $W_{\mathcal{T}_\alpha\psi}$ is the 0-scale function of $(\xi, \mathbb{P}^{\mathcal{T}_\alpha\psi})$.

For the second term in the numerator of (13.62), we may use Fubini's Theorem, together with Theorem 13.10 and the second Lamperti transform, to get

$$\begin{aligned} &E_y^{\mathcal{T}_\alpha\psi}\left[\mathbf{1}_{(\kappa_1^-<\infty)} E_{X_{\kappa_1^-}}^{\mathcal{T}_\alpha\psi}\left[\mathrm{e}^{-q\kappa_1^+}\right]\right] \\ &= E_y^{\mathcal{T}_\alpha\psi}\left[\frac{\mathcal{I}_{\mathcal{T}_\alpha\psi,\alpha}(qX_{\kappa_1^-}^\alpha)\mathbf{1}_{(\kappa_1^-<\infty)}}{\mathcal{I}_{\mathcal{T}_\alpha\psi,\alpha}(q)}\right] \\ &= \frac{1}{\mathcal{I}_{\mathcal{T}_\alpha\psi,\alpha}(q)} E_y^{\mathcal{T}_\alpha\psi}\left[\sum_{n=0}^\infty \mathrm{a}_n(\mathcal{T}_\alpha\psi;\alpha)q^n X_{\kappa_1^-}^{\alpha n}\mathbf{1}_{(\kappa_1^-<\infty)}\right] \\ &= \frac{1}{\mathcal{I}_{\mathcal{T}_\alpha\psi,\alpha}(q)} \sum_{n=0}^\infty \mathrm{a}_n(\mathcal{T}_\alpha\psi;\alpha)q^n \mathbb{E}_{\log y}^{\mathcal{T}_\alpha\psi}\left[\mathrm{e}^{\alpha n\xi_{\tau_0^-}}\mathbf{1}_{(\tau_0^-<\infty)}\right]. \end{aligned} \tag{13.65}$$

Using the fluctuation identity found in Exercise 8.7 (ii), and again recalling that $(\mathcal{T}_\alpha\psi)'(0^+) > 0$, we have, for $x \geq 0$ and $u \geq 0$,

$$\begin{aligned} &\mathbb{E}_x^{\mathcal{T}_\alpha\psi}\left(\mathrm{e}^{u\xi_{\tau_0^-}}\mathbf{1}_{(\tau_0^-<\infty)}\right) \\ &= \mathrm{e}^{ux} - \mathcal{T}_\alpha\psi(u)\mathrm{e}^{ux}\int_0^x \mathrm{e}^{-uz}W_{\mathcal{T}_\alpha\psi}(z)\mathrm{d}z - \frac{\mathcal{T}_\alpha\psi(u)}{u}W_{\mathcal{T}_\alpha\psi}(x), \end{aligned} \tag{13.66}$$

where $\mathcal{T}_\alpha\psi(u)/u$ is understood to be $(\mathcal{T}_\alpha\psi)'(0^+)$ when $u = 0$. Hence incorporating (13.64), (13.65) and (13.66) into (13.62), recalling the identity (13.61), and then taking limits as $y \downarrow 1$, we have

$$
\begin{aligned}
&\frac{1}{O_q^{\mathcal{T}_\alpha\psi}(1;1)}\\
&= \lim_{y\downarrow 1}\sum_{n=0}^{\infty}\frac{\mathtt{a}_n(\mathcal{T}_\alpha\psi;\alpha)}{\psi(\alpha)\mathcal{I}_{\mathcal{T}_\alpha\psi,\alpha}(q)}\alpha q^n\left\{\mathcal{T}_\alpha\psi(\alpha n)y^{\alpha n}\int_0^{\log y}\mathrm{e}^{-\alpha n z}\frac{W_{\mathcal{T}_\alpha\psi}(z)}{W_{\mathcal{T}_\alpha\psi}(\log y)}\mathrm{d}z\right\}\\
&\quad+\sum_{n=0}^{\infty}\frac{\mathtt{a}_n(\mathcal{T}_\alpha\psi;\alpha)}{\psi(\alpha)\mathcal{I}_{\mathcal{T}_\alpha\psi,\alpha}(q)}\alpha q^n\frac{\mathcal{T}_\alpha\psi(\alpha n)}{\alpha n}\\
&\quad-\lim_{y\downarrow 1}\frac{\alpha\mathcal{I}_{\mathcal{T}_\alpha\psi,\alpha}(qy^\alpha)\{E_1^{\mathcal{T}_\alpha\psi}[\mathrm{e}^{-q\int_0^{\kappa_y^+}\mathbf{1}_{(X_s\le 1)}\mathrm{d}s}]-E_1^{\mathcal{T}_\alpha\psi}[\mathrm{e}^{-q\kappa_y^+}]\}}{\psi(\alpha)\mathcal{I}_{\mathcal{T}_\alpha\psi,\alpha}(q)W_{\mathcal{T}_\alpha\psi}(\log y)E_1^{\mathcal{T}_\alpha\psi}[\mathrm{e}^{-q\int_0^{\kappa_y^+}\mathbf{1}_{(X_s\le 1)}\mathrm{d}s}]}. \qquad (13.67)
\end{aligned}
$$

Now using the simple estimate that, for $\theta \ge \epsilon \ge 0$, $\mathrm{e}^{-(\theta-\epsilon)} - \mathrm{e}^{-\theta} \le \epsilon$, we have that

$$
\begin{aligned}
E_1^{\mathcal{T}_\alpha\psi}\left[\mathrm{e}^{-q\int_0^{\kappa_y^+}\mathbf{1}_{(X_s\le 1)}\mathrm{d}s}\right]-E_1^{\mathcal{T}_\alpha\psi}\left[\mathrm{e}^{-q\kappa_y^+}\right] &\le qE_1^{\mathcal{T}_\alpha\psi}\left[\int_0^{\kappa_y^+}\mathbf{1}_{(X_s>1)}\mathrm{d}s\right]\\
&= q\mathbb{E}^{\mathcal{T}_\alpha\psi}\left[\int_0^{\tau_{\log y}^+}\mathrm{e}^{\alpha\xi_s}\mathbf{1}_{(\xi_s>0)}\mathrm{d}s\right]\\
&= q\int_0^{\log y}\mathrm{e}^{\alpha z}u(\log y,z)\mathrm{d}z\\
&\le qy^\alpha\int_0^{\log y}u(\log y,z)\mathrm{d}z,
\end{aligned}
$$

where $\tau_{\log y}^+ = \inf\{s>0;\ \xi_s > \log y\}$ and $u(\log y, z)$ is the potential density of $(\xi, \mathbb{P}^{\mathcal{T}_\alpha\psi})$ when killed on exiting $(-\infty, \log y]$. It can easily be deduced from Corollary 8.8 that, for $z > 0$, $u(\log y, z) = W_{\mathcal{T}_\alpha\psi}(\log y - z)$. Since $W_{\mathcal{T}_\alpha\psi}$ is monotone increasing it now follows that

$$
\begin{aligned}
&\lim_{y\downarrow 1}\frac{1}{W_{\mathcal{T}_\alpha\psi}(\log y)}\{E_1^{\mathcal{T}_\alpha\psi}[\mathrm{e}^{-q\int_0^{\kappa_y^+}\mathbf{1}_{(X_s\le 1)}\mathrm{d}s}]-E_1^{\mathcal{T}_\alpha\psi}[\mathrm{e}^{-q\kappa_y^+}]\}\\
&\le \lim_{y\downarrow 1} qy^\alpha\int_0^{\log y}\frac{W_{\mathcal{T}_\alpha\psi}(\log y - z)}{W_{\mathcal{T}_\alpha\psi}(\log y)}\mathrm{d}z\\
&= 0. \qquad (13.68)
\end{aligned}
$$

Similarly, we have that

$$
\begin{aligned}
&\lim_{y\downarrow 1}\sum_{n=0}^{\infty}\mathtt{a}_n(\mathcal{T}_\alpha\psi;\alpha)q^n\mathcal{T}_\alpha\psi(\alpha n)y^{\alpha n}\int_0^{\log y}\mathrm{e}^{-\alpha n z}\frac{W_{\mathcal{T}_\alpha\psi}(z)}{W_{\mathcal{T}_\alpha\psi}(\log y)}\mathrm{d}z\\
&\le \lim_{y\downarrow 1}\mathcal{I}_{\mathcal{T}_\alpha\psi,\alpha}(qy^\alpha)\log y\\
&= 0. \qquad (13.69)
\end{aligned}
$$

Using (13.68) and (13.69) together with the trivial observation that

$$\lim_{y\downarrow 1} E_1^{\mathcal{T}_\alpha\psi}\left[\mathrm{e}^{-q\int_0^{\kappa_y^+}\mathbf{1}_{(X_s\le 1)}\mathrm{d}s}\right]=1,$$

we may now return to (13.67) to identify the limit and find that

$$O_q^{\mathcal{T}_\alpha\psi}(1;1)=\frac{\psi(\alpha)\mathcal{I}_{\mathcal{T}_\alpha\psi,\alpha}(q)}{\sum_{n=0}^\infty \mathfrak{a}_n(\mathcal{T}_\alpha\psi;\alpha)\alpha q^n\frac{\mathcal{T}_\alpha\psi(\alpha n)}{\alpha n}}.$$

Appealing to the definitions (13.54) and (13.59), we have that for any $n\in\mathbb{N}$,

$$\frac{\alpha\mathcal{T}_\alpha\psi(\alpha n)}{\psi(\alpha)\alpha n}\mathfrak{a}_n(\mathcal{T}_\alpha\psi;\alpha)=\left(\prod_{k=1}^n\psi(\alpha k)\right)^{-1}. \tag{13.70}$$

Moreover, with the interpretation that $\mathcal{T}_\alpha\psi(u)/u=(\mathcal{T}_\alpha\psi)'(0^+)$ when $u=0$, we also see that the left-hand side of (13.70) is also equal to 1 when $n=0$. We finally come to rest at the identity

$$O_q^{\mathcal{T}_\alpha\psi}(1;1)=\frac{\mathcal{I}_{\mathcal{T}_\alpha\psi,\alpha}(q)}{\sum_{n=0}^\infty \mathfrak{a}_n(\psi;\alpha)q^n}=\frac{\mathcal{I}_{\mathcal{T}_\alpha\psi,\alpha}(q)}{\mathcal{I}_{\psi,\alpha}(q)},$$

as required. □

See Exercise 13.10 to check that the conclusion of Theorem 13.11 agrees with the original statement of the Ciesielski–Taylor identity in the setting of Bessel processes.

Exercises

13.1 Suppose that $Y=\{Y_t:t\ge 0\}$ is an α-stable process. Define the occupation time of $(0,\infty)$,

$$A_t=\int_0^t \mathbf{1}_{(X_s>0)}\mathrm{d}s,\quad t\ge 0,$$

and let $\gamma(t):=\inf\{s\ge 0: A_s>t\}$ be its right-continuous inverse. Show that $Y_{\gamma(t)}\mathbf{1}_{(t<T_0)}$, $t\ge 0$, is a positive self-similar Markov process, where $T_0=\inf\{t>0:Y_{\gamma(t)}=0\}$.

13.2 Suppose that $X=\{X_t:t\ge 0\}$ is an α-stable subordinator, so that (necessarily) $\alpha\in(0,1)$. Use the method of looking at the asymptotic overshoot of X at first entry into (y,∞), as $y\uparrow\infty$, to deduce that the Lévy process that appears in the second

Lamperti transform is a subordinator with no killing, no drift and jump measure ν satisfying

$$\nu(\mathrm{d}x) = c\frac{\mathrm{e}^x}{(\mathrm{e}^x - 1)^{\alpha+1}}\mathrm{d}x, \quad x > 0,$$

where $c > 0$ is a constant.

13.3 Consider the case that $B = \{B_t : t \geq 0\}$ is a standard Brownian motion. Use first-passage problems to deduce the following facts.

(i) Set $\tau_0^- = \inf\{t > 0 : B_t < 0\}$. The process $\{B_t \mathbf{1}_{(t<\tau_0^-)} : t \geq 0\}$ is a positive self-similar Markov process driven through the second Lamperti transform by a constant multiple of a standard Brownian motion with drift $-1/2$.

(ii) Fix $x > 0$. Suppose that $\mathbb{P}_x^\uparrow$ is the law of B conditioned to stay positive. Show that the process $(B, \mathbb{P}_x^\uparrow)$ is a positive self-similar Markov process driven through the second Lamperti transform by a constant multiple of standard Brownian motion with drift $1/2$.

In both cases, how might one deduce that the unspecified constant is equal to 1?

13.4 Suppose that $(X, \mathbb{P})$ is a Lévy process which satisfies $\limsup_{t\uparrow\infty} X_t = \infty$. We exclude the case that X is a compound Poisson process. Fix $0 < y \leq x$. Write, as usual, $\tau_y^- = \inf\{t > 0 : X_t < y\}$. By considering the event $\{\tau_y^- < \infty\}$ under $\mathbb{P}_x^\uparrow$ and the computation in (13.13), show that

$$\mathbb{P}_x^\uparrow\left(\inf_{s\geq 0} X_s \geq y\right) = \frac{\widehat{U}(x-y)}{\widehat{U}(x)}, \quad x, y \geq 0.$$

Deduce that the law of the global minimum of a standard Brownian motion conditioned to stay positive, with initial value $x > 0$, is uniformly distributed on $(0, x)$.

13.5 In this exercise, our aim is to follow Chaumont (1996) in constructing the law of a stable process conditioned to be absorbed continuously at the origin. To this end, we suppose that $Y = \{Y_t : t \geq 0\}$ is an α-stable process and that, for all $x \in \mathbb{R}$, $\tau_x^- = \inf\{t > 0 : Y_t < x\}$. Following our usual notation, we shall also write $\underline{Y}_t = \inf_{s\leq t} Y_s$.

(i) Show that, for any $\varepsilon, x > 0$,

$$\mathbb{P}_x(\underline{Y}_{\tau_0^- -} \leq \varepsilon) \propto \int_0^\varepsilon \frac{(x-u)^{\alpha(1-\rho)-1}}{u^{\alpha(1-\rho)}}\mathrm{d}u.$$

(ii) Deduce that, for $x, y > 0$,

$$\lim_{\varepsilon\downarrow 0} \frac{\mathbb{P}_x(\underline{Y}_{\tau_0^- -} \leq \varepsilon)}{\mathbb{P}_y(\underline{Y}_{\tau_0^- -} \leq \varepsilon)} = \left(\frac{x}{y}\right)^{\alpha(1-\rho)-1}.$$

(iii) Now suppose that A belongs to the sigma-algebra generated by $\{Y_s : s \le t\}$. Show, using the Markov property, that, for all $x, t, \eta > 0$ and $0 < \varepsilon < \eta$,

$$\mathbb{P}_x\big(A,\, t < \tau_\eta^- \big| \underline{Y}_{\tau_0^- -} \le \varepsilon\big) = \mathbb{E}_x\left(\mathbf{1}_{(A,\, t<\tau_{(0,\eta)})} \frac{\mathbb{P}_{Y_{\tau_{(0,\eta)}}}(\underline{Y}_{\tau_0^- -} \le \varepsilon)}{\mathbb{P}_x(\underline{Y}_{\tau_0^- -} \le \varepsilon)}\right),$$

where $\tau_{(0,\eta)} = \inf\{t > 0 : Y_t \in (0,\eta)\}$.

(iv) Now assume that, for all $x, t > 0$, $\mathbb{E}_x(Y_t^{\alpha(1-\rho)-1}\mathbf{1}_{(t<\tau_0^-)}) = x^{\alpha(1-\rho)-1}$. Show that

$$\lim_{\varepsilon \downarrow 0} \mathbb{P}_x\big(A,\, t < \tau_\eta^- \big| \underline{Y}_{\tau_0^- -} \le \varepsilon\big) = \mathbb{E}_x\left(\mathbf{1}_{(A,\, t<\tau_{(0,\eta)})} \frac{X_t^{\alpha(1-\rho)-1}}{x^{\alpha(1-\rho)-1}}\right).$$

13.6 This exercise is concerned with the proof of Lemma 13.4. Suppose that ξ is a Lévy process which is killed at rate $q \ge 0$.

(i) Suppose that $q > 0$. Using pathwise arguments, explain why it is trivial that $\mathbb{P}(I_\infty < \infty) = 1$.
(ii) Now suppose that $q = 0$ and $\limsup_{t\uparrow\infty} \xi_t < \infty$. Use Theorem 7.2 to deduce that $\mathbb{P}(I_\infty < \infty) = 1$.
(iii) Keeping with the case that $q = 0$, suppose that $\lim_{t\uparrow\infty} \xi_t = \infty$. Use the same hint as in part (ii) to deduce that $\mathbb{P}(I_\infty < \infty) = 0$.
(iv) Finally, in the case that $q = 0$ and ξ oscillates, define the sequence of stopping times

$$T_1 = \inf\{t > 0 : \xi_t > 2\} \quad \text{and} \quad S_1 = \inf\{t > T_1 : \xi_t < 1\}$$

and for $n \ge 2$

$$T_n = \inf\{t > S_{n-1} : \xi_t > 2\} \quad \text{and} \quad S_n = \inf\{t > T_n : \xi_t < 1\}.$$

Show that

$$I_\infty \ge \mathrm{e}^\alpha \sum_{n\ge1}(S_n - T_n)$$

and hence, by comparing the random variable $T_1 - S_1$ to $\tau_0^- = \inf\{t > 0 : X_t < 0\}$ under $\mathbb{P}_1$, show that $\mathbb{P}(I_\infty = \infty) = 1$.

13.7 In the notation of Sect. 13.4.2, let $\tau_x^{+,\downarrow} = \inf\{t > 0 : \xi_t^\downarrow > x\}$ and $\tau_x^{-,\downarrow} = \inf\{t > 0 : \xi_t^\downarrow < x\}$ for any $x \in \mathbb{R}$. Fix $-\infty < v < 0 < u < \infty$.

(i) Show that, for $\theta \ge 0$,

$$\begin{aligned}
&\mathbf{P}^\downarrow\big(\xi^\downarrow_{\tau_u^{+,\downarrow}} - u \in \mathrm{d}\theta;\, \tau_u^{+,\downarrow} < \tau_v^{-,\downarrow}\big)\\
&= \frac{\sin \pi\alpha(1-\rho)}{\pi}(\mathrm{e}^u - 1)^{\alpha(1-\rho)}(1 - \mathrm{e}^v)^{\alpha\rho}\\
&\quad\times \big(\mathrm{e}^{u+\theta}\big)^{\alpha\rho}\big(\mathrm{e}^{u+\theta} - \mathrm{e}^u\big)^{-\alpha(1-\rho)}\big(\mathrm{e}^{u+\theta} - \mathrm{e}^v\big)^{-\alpha\rho}\big(\mathrm{e}^{u+\theta} - 1\big)^{-1}\mathrm{d}\theta.
\end{aligned}$$

(ii) Show moreover that, for $\theta \geq 0$,

$$\mathbf{P}^{\downarrow}\left(v - \xi^{\downarrow}_{\tau_v^{-,\downarrow}} \in \mathrm{d}\theta;\, \tau_u^{+,\downarrow} > \tau_v^{-,\downarrow}\right)$$
$$= \frac{\sin \pi\alpha(1-\rho)}{\pi}\left(\mathrm{e}^u - 1\right)^{\alpha(1-\rho)}\left(1-\mathrm{e}^v\right)^{\alpha\rho}$$
$$\times \left(\mathrm{e}^{v-\theta}\right)^{\alpha\rho}\left(\mathrm{e}^v - \mathrm{e}^{v-\theta}\right)^{-\alpha\rho}\left(\mathrm{e}^u - \mathrm{e}^{v-\theta}\right)^{-\alpha(1-\rho)}\left(1-\mathrm{e}^{v-\theta}\right)^{-1}\mathrm{d}\theta.$$

13.8 We use here the notation of Sect. 13.4 and consider the case of scale functions for spectrally negative Lamperti-stable processes; see Patie (2009a) and Chaumont et al. (2009).

(i) Show that, for $z \geq 0$, $\mathbf{P}^{\uparrow}(-\underline{\xi}^{\uparrow}_{\infty} \leq z) = (1-\mathrm{e}^{-z})^{\alpha(1-\rho)}$. Hence deduce that there exists a spectrally negative Lévy process with Laplace exponent

$$\psi^{\uparrow}(\theta) = \frac{\Gamma(\theta+\alpha)}{\Gamma(\theta)}, \quad \theta \geq 0,$$

whose associated 0-scale function, say $W^{\uparrow}$, is given by

$$W^{\uparrow}(x) = \frac{1}{\Gamma(\alpha)}\left(1-\mathrm{e}^{-x}\right)^{\alpha-1}, \quad x \geq 0.$$

(ii) Show that, for $z \geq 0$, $\mathbf{P}^{\downarrow}(\overline{\xi}^{\downarrow}_{\infty} \leq z) = (1-\mathrm{e}^{-z})^{\alpha\rho}$. Hence deduce that there exists a spectrally negative Lévy process with Laplace exponent

$$\psi^{\downarrow}(\theta) = \frac{\Gamma(\theta-1+\alpha)}{\Gamma(\theta-1)}, \quad \theta \geq 0,$$

whose associated 0-scale function, say $W^{\downarrow}$, is written

$$W^{\downarrow}(x) = \frac{1}{\Gamma(\alpha)}\left(1-\mathrm{e}^{-x}\right)^{\alpha-1}\mathrm{e}^x, \quad x \geq 0.$$

13.9 Consider the case of an α-stable process conditioned to stay positive, as discussed in Sect. 13.4.1. As usual, it is denoted by $\{Y_t : t \geq 0\}$ with probabilities $\{\mathbb{P}^{\uparrow}_x : x > 0\}$.

(i) Let $b > x > 0$. Use the quintuple law applied to the Lévy process $\xi^{\uparrow}$ to deduce that for $u \in [0, b-x]$, $v \in [u, b)$ and $y > 0$,

$$\mathbb{P}^{\uparrow}_x(b - \overline{Y}_{\tau_b^+ -} \in \mathrm{d}u,\, b - Y_{\tau_b^+ -} \in \mathrm{d}v,\, Y_{\tau_b^+} - b \in \mathrm{d}y)$$
$$= \frac{\sin(\pi\alpha\rho)}{\pi}\frac{\Gamma(\alpha+1)}{\Gamma(\alpha\rho)\Gamma(\alpha(1-\rho))}$$
$$\times \frac{(b-x-u)^{\alpha\rho-1}(v-u)^{\alpha(1-\rho)-1}(b-v)^{\alpha\rho}(y+b)^{\alpha(1-\rho)}}{(b-u)^{\alpha}(y+v)^{\alpha+1}}\mathrm{d}u\,\mathrm{d}v\,\mathrm{d}y.$$

(ii) Deduce from the previous part of the question that, for $u \in [0, b-x]$, $v \in [u, b)$ and $y > 0$,

$$\mathbb{P}_x\big(b - \overline{Y}_{\tau_b^+ -} \in \mathrm{d}u,\, b - Y_{\tau_b^+ -} \in \mathrm{d}v,\, Y_{\tau_b^+} - b \in \mathrm{d}y,\, \tau_b^+ < \tau_0^-\big)$$
$$= \frac{\sin(\pi\alpha\rho)}{\pi} \frac{\Gamma(\alpha+1)}{\Gamma(\alpha\rho)\Gamma(\alpha(1-\rho))}$$
$$\times \frac{x^{\alpha\rho}(b-x-u)^{\alpha\rho-1}(v-u)^{\alpha(1-\rho)-1}(b-v)^{\alpha\rho}(y+b)^{\alpha(1-2\rho)}}{(b-u)^{\alpha}(y+v)^{\alpha+1}} \mathrm{d}u\, \mathrm{d}v\, \mathrm{d}y.$$

13.10 An alternative definition of the Bessel process uses the second Lamperti transform. Specifically, we define a Bessel process of dimension $\mathrm{d} > 0$, say $R = \{R_t : t \geq 0\}$, to be a positive self-similar Markov process with index of self-similarity 2, whose driving Lévy process, $\xi = \{\xi_t : t \geq 0\}$, is given by

$$\xi_t := B_t + \left(\frac{\mathrm{d}}{2} - 1\right)t, \quad t \geq 0,$$

where $\{B_t : t \geq 0\}$ is a Brownian motion. Note in particular that the resulting process must have continuous paths.

In the case $\mathrm{d} \geq 2$, we have that $\limsup_{t \geq 0} \xi_t = \infty$. Hence R never hits the origin and has an entrance law at the origin. When $\mathrm{d} \in (0, 2)$ then the process R visits the origin in an almost surely finite time. Moreover, since

$$\mathbb{E}(\mathrm{e}^{\lambda\xi_1}) = \exp\left\{\lambda^2/2 + \left(\frac{\mathrm{d}}{2} - 1\right)\lambda\right\}, \quad \lambda \in \mathbb{R}$$

it follows that $\Phi(0) = (2 - \mathrm{d}) < 2$ and hence there exists a recurrent extension which leaves the origin continuously.

(i) Verify that the generalised Ciesielski–Taylor identity proved in Theorem 13.11 confirms the original result for Bessel processes.
(ii) Now suppose that $q > 0$ is a constant. Use the second Lamperti transform or otherwise to show that $\{(R_t)^q : t \geq 0\}$ is a positive self-similar Markov process. In particular, show that its index of self-similarity is $2/q$.

Epilogue

The applications featured in this book have been chosen specifically because they exemplify, utilise and have stimulated many different aspects of the mathematical subtleties which are commonly referred to as the fluctuation theory of Lévy processes. There are, of course, many other applications of Lévy processes which we have not touched upon. The literature in this respect is vast.

None the less, let us mention a few topics, with a few key references for the interested reader. The list is by no means exhaustive but merely a selection of current research activities at the time of writing.

Stable and stable-like processes Stable processes and variants thereof are a core class of Lévy processes which offer the luxury of a higher degree of mathematical tractability in a wide variety of problems. This is in part due to their inherent scaling properties. Samorodnitsky and Taqqu (1994) provides an excellent starting point for further reading.

Stochastic control The step from optimal stopping problems driven by diffusions to optimal stopping problems driven by processes with jumps comes hand in hand with the movement to stochastic control problems driven by jump processes. Recent progress is summarised in Øksendal and Sulem (2004).

Financial mathematics In Sect. 2.7.3, we made some brief remarks concerning how properties of Lévy processes may be used to one's advantage when modelling risky assets. This picture is far from complete as, at the very least, we have made no reference to the more substantial and effective stochastic volatility models. The use of such models increases the mathematical demands on subtle financial issues such as hedging, completeness, exact analytic pricing, measures of risk and so on. Whilst solving some problems in mathematical finance, the use of Lévy processes also creates many problems. The degree of complexity of the latter now supports a large and vibrant community of researchers engaged in many new and interesting forms of mathematical theories. The reader is again referred to Boyarchenko and Levendorskii (2002a), Schoutens (2003), Bingham and Kiesel (2004), Cont and

A.E. Kyprianou, *Fluctuations of Lévy Processes with Applications*, Universitext,
DOI 10.1007/978-3-642-37632-0, © Springer-Verlag Berlin Heidelberg 2014

Tankov (2004), Kyprianou et al. (2005), Bingham (2006), Schoutens and Cariboni (2009) and Bingham et al. (2010).

Regenerative sets and combinatorics By sampling values independently from an exponential distribution and grouping them in a way that is determined by a prespecified regenerative set on $[0, \infty)$ (for example the range of a subordinator) one may describe certain combinatorial sampling formulae. This is but part of a much bigger theory which studies the relationship between stochastic processes and combinatorial structures. See Chap. 9 of Kingman (1993), Gnedin and Pitman (2005) or Pitman (2006).

Stochastic differential equations driven by Lévy processes There is a well-established theory for existence, uniqueness and characterisation of the solution to stochastic differential equations driven by Brownian motion which crop up in countless scenarios within the physical and engineering sciences (cf. Øksendal 2003). It is natural to consider analogues of these equations where now the driving source of randomness is a Lévy process. Applebaum (2004) offers a recent treatment. See also Bass (2004) and Situ (2005).

Continuous-time time series models Lévy processes are the continuous-time analogue of random walks. What, then, are the continuous-time analogues of time series models, particularly those that are popular in mathematical finance such as GARCH processes? The answer to this question has been addressed in recent literature such as Klüppelberg et al. (2004b, 2006) and Brockwell et al. (2006). See also the discussion in Bingham (2013). Lévy processes play an important role here.

Integrated exponential Lévy processes For any pair of (not necessarily independent) Lévy processes one may formally define the associated integrated exponential as the stochastic integral over non-negative times of the exponential of the first process with respect to the increments of the second process. The resulting random variable appears in a variety of applications, for example in the, already seen, setting of positive self-similar Markov processes. Other applications include risk theory, the theory of Brownian diffusions in random environments, mathematical finance and the theory of self-similar fragmentation; see the review paper of Bertoin and Yor (2005) and references therein. Particular issues of concern are the almost sure convergence of the above integral, as well as its moments and the tail behaviour of its distribution. See Erickson and Maller (2004), Bertoin and Yor (2005) and Maulik and Zwart (2006).

Generalised Ornstein–Uhlenbeck process In the stochastic differential equation which describes the classical Ornstein–Uhlenbeck process, if one replaces the role of Brownian motion by a general Lévy process, then one obtains the definition of a generalised Ornstein–Uhlenbeck process. See, for example, Chaps. 3 and 10 of Sato (1999). Fluctuations for such processes have been studied in the spectrally negative case; see for example Hadjiev (1985) and more recently Novikov and Shiryaev

(2004) and Patie (2004, 2005). Their stationary distributions are also closely related to the distribution of integrated exponential Lévy processes, cf. Lindner and Maller (2005). See also Bingham and Kiesel (2004) or Bingham et al. (2010) for a view on modelling with generalised Ornstein–Uhlenbeck processes.

Lévy copulas The method of using copulas to build in certain parametric dependencies in multivariate distributions from their marginals is a well-established theory. See for example the up-to-date account in Nelson (2006). Inspired by this methodology, a limited volume of recent literature has proposed to address the modelling of multi-dimensional Lévy processes by working with copulas on the Lévy measure. The foundational ideas are to be found in Tankov (2003) and Kallsen and Tankov (2006).

Lévy-type processes and pseudodifferential operators Jacob (2001, 2002, 2005) summarises the analysis of Markov processes through certain pseudodifferential operators. The latter are intimately related to the infinitesimal generator of the underlying process via complex analysis.

Fractional Lévy processes The concept of fractional Brownian motion also has its counterpart for Lévy processes; see Samorodnitsky and Taqqu (1994). Interestingly, whilst fractional Brownian motion has at least two consistent definitions in the form of stochastic integrals with respect to Brownian motion (the harmonisable representation and the moving-average representation), the analogues of these two definitions for fractional Lévy processes throw out subtle differences. See for example Benassi et al. (2002, 2004).

Quantum independent increment processes Lévy processes have also been introduced in quantum probability, where they can be thought of as an abstraction of a "noise" perturbing a quantum system. The first examples arose in models of quantum systems coupled to a heat bath and in von Waldenfels' investigations of light emission and absorption. The algebraic structure underlying the notions of increment and independence in this setting was developed by Accardi, Schürmann and von Waldenfels. For an introduction to the subject and a description of the latest research in this area, see Applebaum et al. et al. (2005) and Barndorff-Nielsen et al. (2006).

Lévy networks These systems can be thought of as multi-dimensional Lévy processes reflected on the boundary of the positive orthant of $\mathbb{R}^d$, which appear as limiting models of communication networks with traffic process of an unconventional (i.e. long-range dependent) type. See, for example, Harrison and Williams (1987), Kella (1993) and Konstantopoulos et al. (2004). The justification of these as limits can be found, for example, in Konstantopoulos and Lin (1998) and Mikosch et al. (2002). Although Brownian stochastic networks have, in some cases, stationary distributions which can be simply described, this is not the case with more general Lévy networks. The area of multi-dimensional Lévy processes is a challenging field of research.

Fragmentation and coagulation theory Closely related to classical spatial branching processes, a core class of fragmentation processes model the way in which an object of unit total mass *dislocates* in continuous time. In a way that has familiarities with the theory of Lévy processes, the construction of fragmentation processes is done with the help of Poisson point processes. Accordingly, one finds embedded probabilistic structures which are closely related to subordinators.

Conversely to fragmentation processes, coagulation processes model the coalescence of mass over time. Similarly to fragmentation processes, however, coagulation processes can be assembled in the pathwise sense, again through the use of Poisson point processes, and, again, one finds an intimate relationship with subordinators.

We refer to the monograph of Bertoin (2006) for an introduction to the state of the art for both fragmentation and coagulation processes.

Hints for Exercises

Here, we offer a terse set of hints for most parts of the exercises. There are no hints for parts of exercises which are self-explanatory.

Chapter 1

1.1 As an intermediate step, show that it suffices to check that, for all $0 \leq s \leq t \leq u < \infty$, $A \in \mathcal{F}_s$ and $\theta \in \mathbb{R}$,

$$\mathbb{E}\big[\mathbf{1}_A \mathrm{e}^{\mathrm{i}\theta(X_u - X_t)}\big] = \mathbb{P}(A)\mathbb{E}\big[\mathrm{e}^{\mathrm{i}\theta X_{t-u}}\big].$$

1.2 (i) The distribution $\boldsymbol{\Gamma}_p$ is a special example of a negative binomial distribution. A negative binomial random variable, $\boldsymbol{\Lambda}_{c,p}$, with parameter range $c > 0$ and $p \in (0, 1)$, has mass distribution function

$$\mathbb{P}(\boldsymbol{\Lambda}_{c,p} = k) = \binom{-c}{k} p^c (-q)^k = (k!)^{-1}(-c)(-c-1)\cdots(-c-k+1)p^c(-q)^k,$$

where k runs through the non-negative integers. Infinite divisibility of $\boldsymbol{\Gamma}_p$ can be seen by computing the characteristic exponent of $\boldsymbol{\Lambda}_{c,p}$. (ii) Use the infinite divisibility of $\boldsymbol{\Gamma}_p$ to write $S_{\boldsymbol{\Gamma}_p}$ as a sum of n i.i.d. random variables for any $n \in \mathbb{N}$.

1.3 (i) Integrate $\int_a^b f'(yx)\mathrm{d}y$ as a function in x and use Fubini's Theorem. (ii) One should use the convention that, for $z \in \mathbb{C}$, $1/(1 - z/\alpha)^\beta = \exp\{-\beta \log(1 - z/\alpha)\}$, where the principal value of the logarithm function is taken, thus showing that the right-hand side of (1.24) is analytic. Use a power series expansion of e^z and Fubini's Theorem to show that $\int_0^\infty (1 - \mathrm{e}^{zx})\frac{\beta}{x}\mathrm{e}^{-\alpha x}\mathrm{d}x$ is analytic on $\Re z < 0$. Now use the Identity Theorem (for analytic functions in $\mathbb{C}$) and continuity to deduce that (1.24) holds for $z \in \mathbb{C}$ such that $\Re z \leq 0$.

A.E. Kyprianou, *Fluctuations of Lévy Processes with Applications*, Universitext,
DOI 10.1007/978-3-642-37632-0, © Springer-Verlag Berlin Heidelberg 2014

1.4 (i) Use integration by parts. Analytic extension may be performed in a similar manner to the calculations in Exercise 1.3. Use (1.25) to write

$$\int_{\mathbb{R}}(1-\mathrm{e}^{\mathrm{i}\theta x})\Pi(\mathrm{d}x)=-c_1\Gamma(-\alpha)|\theta|^{\alpha}\mathrm{e}^{-\mathrm{i}\pi\alpha\,\mathrm{sgn}\,\theta/2}-c_2\Gamma(-\alpha)|\theta|^{\alpha}\mathrm{e}^{\mathrm{i}\pi\alpha\,\mathrm{sgn}\,\theta/2}, \tag{S.1}$$

the rest is algebra. The desired representation requires a particular normalisation of constants. Replace $-\Gamma(-\alpha)\cos(\pi\alpha/2)c_i$ by another constant (also called c_i) for $i=1,2$ and then set $\beta=(c_1-c_2)/(c_1+c_2)$ and $c=(c_1+c_2)$. (ii) The first part is a straightforward computation. Fourier inversion allows one to write

$$\frac{1}{2\pi}\int_{\mathbb{R}}2\left(\frac{1-\cos\theta}{\theta^2}\right)\mathrm{e}^{\mathrm{i}\theta x}\mathrm{d}\theta=1-|x|.$$

Choose $x=0$ and use symmetry for the second claim. For $z>0$, write

$$\begin{aligned}&\int_0^{\infty}(1-\mathrm{e}^{\mathrm{i}rz}+\mathrm{i}rz\mathbf{1}_{(r<1)})\frac{1}{r^2}\mathrm{d}r\\&\quad=\int_0^{\infty}(1-\cos zr)\frac{1}{r^2}-\mathrm{i}\int_0^{1/z}\frac{1}{r^2}(\sin zr-zr)\mathrm{d}r\\&\qquad-\mathrm{i}\int_{1/z}^{\infty}\frac{1}{r^2}\sin rz\mathrm{d}r+\mathrm{i}\int_{1/z}^{1}\frac{1}{r^2}zr\mathrm{d}r\end{aligned}$$

and accordingly compute the integrals to get the third claim. For $\theta\in\mathbb{R}$,

$$\begin{aligned}&\int_{\mathbb{R}}(1-\mathrm{e}^{\mathrm{i}\theta x}+\mathrm{i}\theta x)\Pi(\mathrm{d}x)\\&\quad=-c_1\Gamma(-\alpha)|\theta|^{\alpha}\mathrm{e}^{-\mathrm{i}\pi\alpha\,\mathrm{sgn}\,\theta/2}-c_2\Gamma(-\alpha)|\theta|^{\alpha}\mathrm{e}^{\mathrm{i}\pi\alpha\,\mathrm{sgn}\,\theta/2}.\end{aligned}$$

The right-hand side above is the same as (S.1) and the calculation thus proceeds in the same way as it does there.

1.5 Let $M_t=\exp\{\mathrm{i}\theta X_t+\Psi(\theta)t\}$. Clearly $\{M_t:t\geq 0\}$ is adapted to the filtration $\mathcal{F}_t=\sigma(X_s:s\leq t)$. Check that

$$\mathbb{E}(|M_t|)\leq\left\{ct\int_{\mathbb{R}}(1\wedge x^2)\Pi(\mathrm{d}x)\right\}$$

for some sufficiently large constant $c>0$. Finally, for $0\leq s\leq t<\infty$, $\mathbb{E}(M_t|\mathcal{F}_s)=M_s$ thanks to stationary and independent increments and (1.3).

1.6 (i) Similar arguments to those given in the solution to Exercise 1.5 show that $\{\exp\{\lambda B_t-\lambda^2t/2\}:t\geq 0\}$ is a martingale. Doob's Optional Sampling Theorem gives us

$$1=\mathbb{E}(\mathrm{e}^{\lambda(B_{t\wedge\tau_s}+b(t\wedge\tau_s))-(\frac{1}{2}\lambda^2+b\lambda)(t\wedge\tau_s)}).$$

Take limits as $s \uparrow \infty$, using dominated convergence and the fact that $\lim_{t\uparrow\infty} B_t + bt = \infty$, to recover the Laplace transform of τ_s. Both the left- and right-hand side of this Laplace transform can easily be shown to be analytic functions and hence, by the Identity Theorem, equal on $\{z \in \mathbb{C} : \Re z > 0\}$. The characteristic exponent is recovered by taking limits onto the imaginary axis. (ii) When $\Pi(\mathrm{d}x) = (2\pi x^3)^{-1/2}\mathrm{e}^{-xb^2/2}$ on $x > 0$, write

$$\int_0^\infty (1 - \mathrm{e}^{\mathrm{i}\theta x})\Pi(\mathrm{d}x)$$
$$= \int_0^\infty \frac{1}{\sqrt{2\pi x^3}}(1 - \mathrm{e}^{\mathrm{i}\theta - b^2x/2})\mathrm{d}x - \int_0^\infty \frac{1}{\sqrt{2\pi x^3}}(1 - \mathrm{e}^{-b^2x/2})\mathrm{d}x$$

and use Exercise 1.4 (i). (iii) Use the change of variable $sx^{-1/2} = ((2\lambda + b^2)u)^{1/2}$ in the first integral to obtain the second. Add the two integrals, writing the sum in terms of a common dummy variable x, and make a change of variable $\eta = sx^{-1/2} - \sqrt{(b^2 + 2\lambda)x}$.

1.7 Note, by definition, that $\tau = \{\tau_s : s \geq 0\}$ is also the inverse of the continuous process $\{\overline{B}_t : t \geq 0\}$, where $\overline{B}_t = \sup_{s\leq t} B_s$. One may thus deduce that τ satisfies the first two conditions of Definition 1.1. The strong Markov property and spatial homogeneity of Brownian motion implies that $\{\tau_s : s \geq 0\}$ has stationary and independent increments. Similar analysis to the solution of Exercise 1.6 shows that

$$\mathbb{E}(\mathrm{e}^{-q\tau_s}) = \mathrm{e}^{-\sqrt{2q}s}, \quad s \geq 0.$$

1.8 (ii) Let $T_0 = 0$ and define recursively, for $n = 1, 2, \ldots$, $T_n = \inf\{k > T_{n-1} : S_k > S_{T_{n-1}}\}$ and let $H_n = S_{T_n}$ if $T_n < \infty$. The indices T_n are called the strong ascending ladder times and H_n are the ladder heights. It is straightforward to prove that T_n are stopping times. Note that for each $n \geq 1$, from the Strong Markov Property, $H_n - H_{n-1}$ has the same distribution as $S_{T_0^+}$. The required identity follows by showing, for $x \geq 0$,

$$P(S_{T_0^-} \in \mathrm{d}x) = P(S_1 \in \mathrm{d}x)$$
$$+ \int_{(0,\infty)} \sum_{n\geq 1} P(S_n \in \mathrm{d}y, S_n > S_j \text{ for all } j = 0, \ldots, n-1)Q(\mathrm{d}x - y).$$

(iii) Note that $Q(z, \infty) = P(\mathbf{e}_\beta > z + \xi_1)$ and integrate out the exponential distribution. (iv) The security loading condition guarantees that $S_{T_0^+}$ has a proper distribution (first passage above the origin has probability one). The lack-of-memory property together with the fact that upward jumps in S are exponentially distributed implies that $S_{T_0^+}$ is exponentially distributed with parameter β. From this it follows that, for each $n = 1, 2, \ldots$, H_n has a gamma distribution. One may therefore compute $V(\mathrm{d}y)$ explicitly as indicated in the question.

1.9 (i) The Laplace exponent may be computed directly using the Poisson distribution of N_1 in a similar manner to the calculations in Sect. 1.2.2. With the help of the Dominated Convergence Theorem, one can show directly that $\psi''(\theta) > 0$ for all $\theta > 0$. One easily confirms that $\psi(0) = 0$ and $\psi(\infty) = \infty$. Convexity thus dictates the existence of a second root of ψ in $(0, \infty)$. (ii) The martingale properties follow from similar arguments to those given in the solution to Exercise 1.5. An argument using Doob's Optional Sampling Theorem with this martingale, similar to the one given in the solution of Exercise 1.6, allows one to deduce

$$\mathrm{e}^{\theta^* x}\mathbb{P}(\tau_x^+ < \infty) = 1,$$

from which the remaining conclusions can be drawn. (iii) Note that

$$\{W_s = 0\} = \{\overline{X}_s = X_s \text{ and } \overline{X}_s > w\}.$$

Moreover, on these events, for $s \geq 0$, $\mathrm{d}s = \mathrm{d}X_s = \mathrm{d}\overline{X}_s$. (iv) A direct computation shows that $\psi'(0+) = 1 - \lambda\mu$. Hence $\lambda\mu \leq 1$ if and only if $\theta^* = 0$. (v) When $\lambda\mu > 1$, we have that $\theta^* > 0$. Moreover, $I = 0$ on $\{\tau_w^+ = \infty\}$. This is sufficient to draw the first conclusion. On the other hand,

$$\mathbb{P}(I \in \mathrm{d}x, \tau_w^+ < \infty | W_0 = w) = \mathbb{P}(\overline{X}_\infty - w \in \mathrm{d}x, \overline{X}_\infty > w | W_0 = w).$$

The strong Markov property and the lack-of-memory property for the exponential distribution can now be used to derive the second conclusion.

1.10 (i) See Exercise 1.5. (ii) Use similar reasoning to the proof of part (i) of Exercise 1.6. (iii) Differentiate in a, taking care to note that $v(x) \geq K - \mathrm{e}^a$. Why?

1.11 (i) Use the branching property to deduce that, under P_y, Y_t is equal in law to the independent sum $\sum_{i=1}^{y} Y_t^{(i)}$, where $Y_t^{(i)}$ has the same distribution as Y_t under P_1. From here the required expression for $u_t(\phi)$ follows. Strict positivity of $u_t(\phi)$ follows by proving that $u_t(\phi) \geq \mathrm{e}^{-\phi y + \lambda t}$. (ii) Use the Markov property. (iii) Under the assumption $\pi_0 = 0$, we have, for $i = -1, 1, 2, \ldots$,

$$P_1(Y_h = 1 + i) = \lambda\pi_i h + o(h)$$

as $h \downarrow 0$, from which one may show

$$\lim_{h \downarrow 0} \frac{\mathbb{E}_1(\mathrm{e}^{-\phi Y_h}) - \mathrm{e}^{-\phi}}{h} = -\mathrm{e}^{-\phi}\psi(\phi).$$

Chapter 2

2.1 Both parts (i) and (ii) can be addressed by considering the moment-generating function of $\sum_{i=1}^{n} N_i$ and taking limits thereof as $n \uparrow \infty$.

2.2 Suppose that $\{S_i : i = 1, 2, \ldots, n\}$ are independent and exponentially distributed. Use the classical density transform to deduce that for $A \in \mathcal{B}([0, \infty)^n)$,

$$P\big((T_1, \ldots, T_n) \in A \text{ and } N_t = n\big) = \int_{(t_1,\ldots,t_n)\in A} \mathbf{1}_{(t_1 \le t_2 \le \cdots \le t_n \le t)} \lambda^n \mathrm{e}^{-\lambda t_n} \mathrm{d}t_1 \cdots \mathrm{d}t_n. \quad \text{(S.2)}$$

Parts (i) and (ii) can be deduced from (S.2).

2.3 Find upper and lower bounds of $1 - \mathrm{e}^{-\phi y}$ which are multiples of $1 \wedge y$. A diagram may help.

2.4 (i) For left-continuity, it suffices to show that, for each $x \in (0, 1]$, $f(x - \varepsilon)$ is a Cauchy sequence with respect to the distance metric $|\cdot|$ as $\varepsilon \downarrow 0$. This can be done by applying the triangle inequality to $|f(x-\varepsilon) - f(x-\eta)|$. Right-continuity can be proved directly by applying the triangle inequality to $|f(x+\varepsilon) - f(x)|$. (ii) Suppose for contradiction that, for a given $c > 0$, the set Δ_c has an accumulation point, say x. This means there exists a sequence, say $y_n \to x$, such that, for each $n \ge 1$, $y_n \in \Delta_c$. From this sequence, assume without loss of generality that there exists an *increasing* subsequence, say $x_n \uparrow x$. Noting that

$$f(x_n) - f(x_m) = \big[f(x_n) - f(x_n-)\big] + \big[f(x_n-) - f(x_m)\big],$$

one can deduce that there is a contradiction with the existence of the limit $f(x-)$.

2.5 (i) The function f^{-1} jumps for values in its argument that correspond to values in the range of f. (ii) Note that, for any $y > 0$ and $k = 1, 2, 3, \ldots$, there are either an even or odd number of jumps of magnitude 2^{-k} in $[0, y]$. This leads to a straightforward upper estimate of $|f(y)|$. Showing that f has paths of unbounded variation is straightforward as there are countable increments. Countable increments also implies that f is right-continuous with left limits.

2.6 (i) Apply Theorem 2.7 to the case that $S = [0, \infty) \times \mathbb{R}$ and the intensity measure is $\mathrm{d}t \times \Pi(\mathrm{d}x)$. One should not forget to check that $\int_{(-1,1)} |x|^n \Pi(\mathrm{d}x) < \infty$. (ii) The proof follows very closely the proof of Lemma 2.9.

2.7 (i) Consider the probability $\mathbb{P}(N([0, t] \times \{\mathbb{R}\backslash(-a, a)\}) = 0)$. (ii) Use part (i). (iii) and (iv) Piecewise linear paths have bounded variation. Moreover, consider $\lim_{a\downarrow 0} \mathbb{P}(T_a > t)$, where $T_a := \inf\{t > 0 : |X_t - X_{t-}| \ge a\}$, together with stationary and independent increments.

2.8 Suppose that N is the Poisson random measure on $[0, \infty) \times \mathbb{R}$ that describes jumps. Consider the independence that arises when restricting N to $[0, \infty) \times (0, \infty)$ and to $[0, \infty) \times (-\infty, 0)$.

2.9 (i) See Lemma 2.15. (ii) Use the factorisation

$$\left(1 - \mathrm{i}\frac{\theta c}{\alpha} + \frac{\sigma^2\theta^2}{2\alpha}\right) = \left(1 - \frac{\mathrm{i}\theta}{\alpha^{(1)}}\right) \times \left(1 - \frac{\mathrm{i}\theta}{\alpha^{(2)}}\right), \quad \theta \in \mathbb{R}.$$

2.10 Increasing the dimension of the space on which the Poisson random measure of jumps is defined has little effect on the computations we have seen for the Lévy–Itô decomposition of one-dimensional Lévy processes.

2.11 (i) Non-negativity of X_1 allows for analytic extension of the characteristic exponent to the Laplace exponent. (ii) As an intermediary step, use integration by parts to deduce that

$$\int_{(\epsilon,\infty)} (1 - \mathrm{e}^{-qx})\Pi(\mathrm{d}x) = q \int_0^\infty \mathrm{e}^{-qx}\Pi(x \vee \epsilon, \infty)\mathrm{d}x.$$

The remaining parts use the representation in (ii).

2.12 In all parts of this exercise, one can appeal to Lemma 2.15.

Chapter 3

3.1 (i) Use the martingale property from Exercise 1.5. (ii) This is a consequence of standard measure theory (see for example Theorem 4.6.3 of Durrett 2004) and part (i). (iii) Use the conclusion in part (ii) to deduce that, almost surely,

$$\mathbf{1}_A = \mathbb{P}(A|\mathcal{F}_{t+}) = \mathbb{P}(A|\mathcal{F}_t),$$

for any $A \in \mathcal{F}_{t+}$. Use the completion of $\mathbb{F}$ by null sets of $\mathbb{P}$ to deduce that $\mathcal{F}_{t+} \subseteq \mathcal{F}_t$.

3.2 It suffices to prove the result for the first process. Define for each $y \geq 0$, $Y_t^y = (y \vee \overline{X}_t) - X_t$ and, for each $\mathbb{F}$-stopping time T, let $\widetilde{X}_u = X_{T+u} - X_T$, on $\{T < \infty\}$. Show that, on $\{T < \infty\}$,

$$Y_{t+s}^y = \left[Y_t^y \vee \sup_{u\in[0,s]} \widetilde{X}_u\right] - \widetilde{X}_s.$$

3.3 The solution to this exercise is taken from Sect. 25 of Sato (1999). (i) Use submultiplicativity to show that, for integer n chosen so that $|x| \in (n-1, n]$, $g(x) \leq c^{n-1}g(x/n)^n$ and hence, since g is bounded on compacts, deduce that $g(x) \leq a_g \mathrm{e}^{b_g|x|}$, for some $a_g \in (0,\infty)$. Suppose that $\mathbb{E}(g(X_t)) < \infty$ for all $t > 0$. Using the Lévy–Itô decomposition to write

$$\mathbb{E}\big(g(X_t)\big) = \int_{\mathbb{R}}\int_{\mathbb{R}} g(x+y)\mathrm{d}F_2(y)\mathrm{d}F_{1,3}(x),$$

where F_2 is the distribution of $X_t^{(2)}$ and $F_{1,3}$ is the distribution of $X_t^{(1)} + X_t^{(3)}$, deduce that, for some $x \in \mathbb{R}$, $\mathbb{E}(g(X_t^{(2)})) \leq c a_g \mathrm{e}^{b_g|x|} \int_{\mathbb{R}} g(x+y)\mathrm{d}F_2(y) < \infty$. Conclude that $\int_{|y|\geq 1} g(y)\Pi(\mathrm{d}y) < \infty$. Conversely, suppose that $\int_{|y|\geq 1} g(y)\Pi(\mathrm{d}y) < \infty$. Use submultiplicativity, together with the method found in the proof of Theorem 3.6, to find that $\mathbb{E}(g(X_t)) < \infty$. (ii) Suppose that $h(x)$ is a positive increasing function on $\mathbb{R}$ such that, for $x \leq b$, it is constant and, for $x > b$, $\log h(x)$ is concave. Show that

$$h(|x+y|) \leq h(|x|+|y|) \leq ch(|x|)h(|y|),$$

where $c > 0$ is a constant. Now consider the discussion preceding Theorem 3.8. (iii) Apply the conclusion of Theorem 3.8 to the Lévy measure of a stable process.

3.4 It suffices to show that any Lévy process can be written as the difference of two spectrally positive processes. One should also take advantage of Theorem 3.9 in part (iii).

3.5 It is convenient to write, for $\beta > 0$,

$$\psi(\beta) = -a\beta + \frac{1}{2}\sigma^2\beta^2 + \int_{x<-1} (e^{\beta x} - 1)\Pi(dx) + \int_{0>x>-1} (e^{\beta x} - 1 - \beta x)\Pi(dx),$$

before using dominated convergence to perform multiple derivatives. See also the hints in the solution to Exercise 1.9.

3.6 (i) Use spectral negativity, stationary independent increments of the subordinator $\{\tau_x^+ : x \geq 0\}$ and that $\{\tau_x^+ < \mathbf{e}_q\} = \{\overline{X}_{\mathbf{e}_q} > 0\}$. (ii) Define the right-continuous function $f(x) = \mathbb{P}(\overline{X}_{\mathbf{e}_q} > x)$ for all $x \geq 0$. Show that, for positive integers p, q, $f(p/q) = f(1/q)^p$ and $f(1) = f(1/q)^q$ and that this leads to the desired exponential distribution. Use continuity of the process $\{\overline{X}_t : t \geq 0\}$ to infer, and then use, the finiteness of $\mathbb{E}(e^{\overline{X}_t})$ for all $t \geq 0$. (iii) Use Theorem 3.12 and Exercise 3.5, noting that, as $q \downarrow 0$, the random variable $\overline{X}_{\mathbf{e}_q}$ converges in distribution to $\overline{X}_\infty$.

3.7 For both (i) and (ii), first show that

$$\Psi(\theta) = c\big(\cos(\pi\alpha/2)\big)^{-1}(-\theta i)^\alpha, \quad \theta \in \mathbb{R},$$

and then use analytical extension, with the help of Theorem 3.6. See also Theorem 3.8.

Chapter 4

4.1 (i) Note from Lemma 4.11 that $\mathbb{P}(\tau^{\{0\}} > 0) = 1$, where $\tau^{\{0\}} = \inf\{t > 0 : X_t = 0\}$, and apply the strong Markov property at this time. (ii) Apply the change of variable formula over the time intervals $[T_n, T_{n+1})$, where $T_0 = 0$ and, for $n = 1, 2, \ldots$, T_n is the time of the n-th visit of X to 0. Note, moreover, that

$$\int_{(0,t]} (f(X_s) - f(X_{s-}))dL_t^0 = \sum_{i=1}^{n_t - 1} (f(X_{T_i}) - f(X_{T_i-})),$$

where n_t is the number of visits to 0 on the time horizon $(0, t]$.

4.2 Follow the proof of Theorem 4.2, taking care to apply regular Lebesgue–Stieltjes calculus between the jumps of the process $X^{(\varepsilon)}$ (note in particular that $\{\overline{X}_t : t \geq 0\}$ has non-decreasing paths and is continuous).

4.3 Starting with an approximation to ϕ via (4.8), one may establish this identity in a similar way to the proof of Theorem 4.4. The calculations are somewhat more technically involved, however.

4.4 (i) The Lévy–Itô decomposition tells us that we may write, for each $t \geq 0$,

$$X_t^{(\varepsilon)} = -at + \int_{[0,t]} \int_{|x|\geq 1} xN(\mathrm{d}s \times \mathrm{d}x) + \int_{[0,t]} \int_{\varepsilon \leq |x| < 1} xN(\mathrm{d}s \times \mathrm{d}x)$$

$$- t \int_{\varepsilon \leq |x| < 1} x\Pi(\mathrm{d}x),$$

where N is the Poisson random measure associated with the jumps of X. Apply the change of variable formula. (ii) Use Exercise 4.3 to analyse $||M^{(\varepsilon)} - M^{(\eta)}||$ as $\varepsilon, \eta \downarrow 0$, taking account of the boundedness of the first derivative of f in x and the necessary property that $\int_{(-1,1)} x^2 \Pi(\mathrm{d}x) < \infty$. (iii) We know from the Lévy–Itô decomposition that $X^{(\varepsilon)}$ converges uniformly on $[0, T]$, with probability one, along some deterministic subsequence, say $\{\varepsilon_n : n = 1, 2, \ldots\}$, to X. Similar reasoning also shows that, thanks to the result in part (ii), there exists a subsubsequence, say $\epsilon = \{\epsilon_n : n = 1, 2, \ldots\}$, of the latter subsequence along which $M^{(\varepsilon)}$ converges uniformly on $[0, T]$, with probability one to its limit, say M.

For the convergence of the other terms in (4.24), use the assumed continuity and boundedness conditions of f in conjunction with dominated convergence. In doing so, it will be convenient to show that, for all $0 \leq s \leq t$ and $y \in \mathbb{R}$,

$$\left| f(s, y + x) - f(s, y) - x\frac{\partial f}{\partial x}(s, y) \right| \leq Cx^2.$$

(iv) This is a standard localisation technique.

4.5 (i) The period B is equal to the time it takes for the workload to become zero again. The latter has the same distribution as $\tau_x^+ = \inf\{t > 0 : X_t > x\}$, when x is independently randomised using the distribution F. (ii) Decompose the path of X into excursions from the maximum, interlaced by intervals of time when $X = \overline{X}$. Accordingly, show that

$$\int_0^\infty \mathbf{1}_{(W_t=0)} \mathrm{d}t = \sum_{k=1}^{\Gamma_p+1} \mathbf{e}_\lambda^{(k)},$$

where Γ_p is a geometric random variable with parameter $p = 1 - \hat{F}(\Phi(0))$ and $\{\mathbf{e}_\lambda^{(k)} : k = 1, 2, \ldots\}$ is a sequence of independent random variables (also independent of Γ_p), each of which is exponentially distributed with parameter λp.

4.6 (i) Use Itô's formula for semi-martingales. (ii) Use Exercise 3.6 (iii) and consider $\mathbb{E}(\overline{X}_{\mathbf{e}_q})$. (iii) Use the positivity of the process Z and the fact that $\overline{X}$ increases, to deduce that $\mathbb{E}(\sup_{s\leq t}|M_s|)<\infty$. Use dominated convergence with a localising sequence of stopping times, to show that M is a real martingale and not just a local martingale.

4.7 (i) The proof is similar to the previous question. (ii) Show that

$$\mathbb{E}\left(\int_0^{\mathbf{e}_q} \mathrm{e}^{\mathrm{i}\alpha(X_s-\overline{X}_s)+\mathrm{i}\beta\overline{X}_s}\mathrm{d}s\right)=\frac{1}{q}\mathbb{E}\left(\mathrm{e}^{\mathrm{i}\alpha(X_{\mathbf{e}_q}-\overline{X}_{\mathbf{e}_q})+\mathrm{i}\beta\overline{X}_s}\right)$$

and that

$$\mathbb{E}\left(\int_0^{\mathbf{e}_q} \mathrm{e}^{\mathrm{i}\alpha(X_s-\overline{X}_s)+\mathrm{i}\beta\overline{X}_s}\mathrm{d}\overline{X}_s\right)=\frac{1}{\Phi(q)-\mathrm{i}\beta}.$$

(iii) Note that (4.25) factorises as follows

$$\frac{\Phi(q)}{\Phi(q)-\mathrm{i}\beta}\times\frac{q}{\Phi(q)}\frac{\Phi(q)-\mathrm{i}\alpha}{q+\Psi(\alpha)}.$$

4.8 (i) Write X as the difference of two subordinators. (ii) Use that X is stochastically bounded below by a spectrally negative Lévy process of bounded variation. (iii) Consider the Pollaczek–Khintchine formula in Sect. 4.6.

Chapter 5

5.1 (i) One reasons as for a general Lévy process, using stationary and independent increments with a minor adaptation for killing. (ii) Start by writing $\Phi(\theta)=\lim_{n\uparrow\infty} n(1-\exp\{-\Phi(\theta)/n\})$ for $\theta\geq 0$. (iii) Note that $\overline{\Pi}_n(x)\mathrm{d}x$ converges vaguely. As $\overline{\Pi}$ is monotone, atoms in the aforementioned measure can only accumulate at 0 or ∞. Uniqueness is clear. (iv) The quantity $\int_{(0,\infty)}(1\wedge x)\Pi(\mathrm{d}x)$ must be finite for $\Phi(\theta)$ to be finite. Why?

5.2 Write out $\Phi(\theta+q)-\Phi(q)$ in detail.

5.3 The proof follows verbatim the proof of Lemma 5.5 (ii), using Corollary 5.3 in place of Theorem 5.1. Note, moreover, that

$$1-\lim_{x\uparrow\infty}\mathbb{P}(X_{\tau_x^+}=x)=\frac{1}{\mu}\int_0^\infty \Pi(y,\infty)\mathrm{d}y. \tag{S.3}$$

Consider also the expression for $\Phi'(0+)$.

5.4 (i) From Sect. 4.6, we know that for the process Y, $\mathbb{P}(\sigma_0^+>0)=1$. Consider the auxiliary process, say $X=\{X_t:t\geq 0\}$, which has positive and independent jumps which are equal in distribution to $Y_{\sigma_0^+}$. (ii) Apply Theorem 5.5. (iii) When modelling a risk process by $-Y$, the previous limit is the asymptotic joint distribution of the deficit at ruin and the wealth prior to ruin, conditional on ruin occurring when starting from an arbitrary large capital.

5.5 Show, using Theorem 5.6, that

$$\lim_{x\uparrow\infty} \mathbb{P}(X_{\tau_x^+} - X_{\tau_x^+ -} \in \mathrm{d}z) = \lim_{x\uparrow\infty} \frac{1}{\mu} z\Pi(\mathrm{d}z), \quad z > 0.$$

5.6 An analogue of Theorem 4.4 is required for the case of a Poisson random measure N which is defined on $([0,\infty) \times \mathbb{R}^d, \mathcal{B}[0,\infty) \times \mathcal{B}(\mathbb{R}), \mathrm{d}t \times \mathbf{\Pi})$, where $d \in \mathbb{N}$ (although our attention will be restricted to the case that $d = 2$). Following the proof of Theorem 4.4, one deduces that, if $\phi : [0,\infty) \times \mathbb{R}^d \times \Omega \to [0,\infty)$ is a random time-space function such that

(i) as a trivariate function, $\phi = \phi(t,x)[\omega]$ is measurable,
(ii) for each $t \geq 0$, $\phi(t,x)[\omega]$ is $\mathcal{F}_t \times \mathcal{B}(\mathbb{R}^d)$-measurable, and
(iii) for each $x \in \mathbb{R}^d$, with probability one, $\{\phi(t,x)[\omega] : t \geq 0\}$ is a left-continuous process,

then for all $t \geq 0$,

$$\mathbb{E}\left(\int_{[0,t]}\int_{\mathbb{R}^d} \phi(s,x) N(\mathrm{d}s \times \mathrm{d}x)\right) = \mathbb{E}\left(\int_0^t \int_{\mathbb{R}^d} \phi(s,x) \mathrm{d}s\, \mathbf{\Pi}(\mathrm{d}x)\right). \qquad \text{(S.4)}$$

Here, we have the understanding that the right-hand side is infinite if and only if the left-hand side is. From this, the required result follows from calculations similar in nature to those found in the proof of Theorem 5.6.

5.7 (i) Show, and use, that conditionally on $\mathcal{F}_{\tau_x^+}$, on the event $\{\tau_x^+ < \mathbf{e}_\alpha\}$, the random variables $X_{\mathbf{e}_\alpha} - X_{\tau_x^+}$ and $X_{\mathbf{e}_\alpha}$ have the same distribution. (ii) Use Laplace transforms. (iii) Take limits as $\beta \uparrow \infty$ in the previous part of the question, appealing to Exercise 2.11, and recall that

$$\int_{[0,\infty)} \mathrm{e}^{-qy} U^{(\alpha)}(\mathrm{d}y) = \frac{1}{\alpha + \Phi(q)}.$$

(iv) Use Lemma 5.11.

5.8 (ii) Use the definition of the gamma function. (iii) Refer to (5.8). (iv) Check for a degenerate distribution in part (iii).

5.9 (i) Take Laplace transforms, using the conclusion of Theorem 5.6. (ii) and Theorem 5.9. (ii) Use the previous exercise. (iii) With the help of the Dominated Convergence Theorem and the assumed regular variation, consider

$$\int_0^\infty \mathrm{d}x \cdot \mathrm{e}^{-qx} \lim_{t\uparrow\infty} \mathbb{E}\left(\mathrm{e}^{-\beta(X_{\tau_{tx}^+ -}/t) - \gamma(X_{\tau_{tx}^+} - tx)/t}\right).$$

(iv) Take limits in the identity in part (ii), first considering the case that $\gamma = 0$, and then the case that $\beta = 0$ as t tends to infinity (resp. zero). Note also from the discussion in Sect. 5.5, if Φ is regularly varying with index α, then necessarily $\alpha \in [0,1]$.

5.10 (i) Introduce the auxiliary process $\widetilde{X} = \{\widetilde{X}_t : t \geq 0\}$, which is the subordinator whose Laplace exponent is given by $\Phi(q) - \eta$. Define the quantity $\widetilde{\tau}_x^+ = \inf\{t > 0 : \widetilde{X}_t > x\}$ and consider the expectation $\mathbb{E}(\int_0^\infty \mathrm{e}^{-\eta t} \mathbf{1}_{(\widetilde{X}_t > x)} \mathrm{d}t)$. (ii) Make use of (5.37). (iv) Compute $G(0,\infty)$. (v) Use the Continuity Theorem for Laplace transforms.

5.11 Use inductive differentiation.

5.12 This question requires patience in computing Laplace transforms. For part (ii), use (5.31).

5.13 (i) and (ii) Show that it sufficient to consider the case that $\eta = 0$. Use the expression for $\Phi(u)/u$ appearing in part (ii) of Exercise 2.11.

5.14 (i) Apply Laplace transforms. (ii) Use Theorem 5.6 to write down $\mathbb{P}(X_{\tau_x^+} > x)$.

Chapter 6

6.1 For the first part, consider any symmetric Lévy process. Which symmetric Lévy processes do not have the desired properties? For the second part, consider any spectrally positive Lévy process of bounded variation. Alternatively, consider the difference of two independent stable subordinators with different indices. Which one should have the larger index?

6.2 (i) By assumption, $\lim_{q\uparrow\infty} \Phi(q) = \infty$. Why? Use Exercise 2.11 to show that $\delta = \lim_{\theta\uparrow\infty} 1/\psi'(\theta)$ and hence deduce that $\delta = 0$. (ii) Use Theorem 4.11 and that $X_{\tau_x^+} = x$ on $\{\tau_x^+ < \infty\}$. (iii) From the Wiener–Hopf factorisation given in Sect. 6.5.2, we have that

$$\mathbb{E}(\mathrm{e}^{\theta \underline{X}_{\mathbf{e}_q}}) = \frac{q}{\Phi(q)} \frac{\Phi(q) - \theta}{q - \psi(\theta)}.$$

Take limits as $\theta \uparrow \infty$.

6.3 Suppose that $N = \{N_t : t \geq 0\}$ is a Poisson process with rate $\lambda\rho$ and $\{\xi_n : n = 1, 2, \ldots\}$ are independent (also of N) and identically distributed and, further, $\mathbf{e}_{\lambda(1-\rho)}$ is an independent exponentially distributed random variable with parameter $\lambda(1-\rho)$. On the other hand, suppose that $\widetilde{N} = \{\widetilde{N}_t : t \geq 0\}$ is an independent Poisson process with rate λ and $\mathbf{\Gamma}_{1-\rho}$ is a geometric distribution with parameter $1-\rho$. Compare the expectations

$$\mathbb{E}\big(\mathrm{e}^{-\theta \sum_{i=1}^{N_1} \xi_i} \mathbf{1}_{(1 < \mathbf{e}_{\lambda(1-\rho)})}\big) \quad \text{and} \quad \mathbb{E}\big(\mathrm{e}^{-\theta \sum_{i=1}^{\widetilde{N}_t} \xi_i} \mathbf{1}_{(\widetilde{N}_t \leq \mathbf{\Gamma}_{1-\rho})}\big).$$

6.4 Make the connection with path regularity.

6.5 (i) It will be helpful to show that

$$\mathbb{E}(X_1) = -a + \int_{(-\infty,-1)} x\Pi(\mathrm{d}x).$$

(ii) First explain why, for $\theta \in \mathbb{R}$, $\widehat{\kappa}(0, \mathrm{i}\theta) = -\psi(\mathrm{i}\theta)/\mathrm{i}\theta$.

6.6 (i) See Exercise 3.7, Theorem 3.12 and Corollary 3.14. See also the calculations in Sect. 6.5.3. (ii) Use (5.31).

6.7 (i) Given $\mathcal{F}_{\tau_x^+}$, consider the law of $\overline{X}_{\mathbf{e}_\alpha} - X_{\tau_x^+}$ on $\{\tau_x^+ < \mathbf{e}_\alpha\}$. (ii) Take Laplace transforms on both sides of (6.46) and use Fubini's Theorem.

6.8 (i) Recall that $\mathbb{P}(\overline{X}_{\mathbf{e}_p} = 0) = \lim_{\beta\uparrow\infty} \mathbb{E}(\mathrm{e}^{-\beta \overline{X}_{\mathbf{e}_p}})$ and make use of Theorem 6.15. (ii) Similarly, use the same theorem, together with the fact that $\lim_{\lambda\uparrow\infty} \mathbb{E}(\mathrm{e}^{-\lambda \overline{G}_{\mathbf{e}_p}}) = \mathbb{P}(\overline{G}_{\mathbf{e}_p} = 0) > 0$.

6.9 (i) For $s \in (0, 1)$ and $\theta \in \mathbb{R}$, let $q = 1 - p$ and show, with the help of Fubini's Theorem, that

$$\exp\left\{-\int_{\mathbb{R}} \sum_{n=1}^{\infty} (1 - s^n \mathrm{e}^{\mathrm{i}\theta x}) q^n \frac{1}{n} F^{*n}(\mathrm{d}x)\right\} = \frac{p}{1 - qsE(\mathrm{e}^{\mathrm{i}\theta S_1})}.$$

Exercise 2.10 will also be useful to address infinite divisibility. (ii) The path of the random walk may be broken into $\nu \in \{0, 1, 2, \ldots\}$ finite excursions from the maximum, followed by an additional excursion which straddles the random time $\mathbf{\Gamma}_p$. (iii) The independence of (G, S_G) and $(\mathbf{\Gamma}_p - G, S_{\mathbf{\Gamma}_p} - S_G)$ is immediate from the decomposition described in part (ii). Duality[1] for random walks implies that the latter pair is equal in distribution to (D, S_D). (iv) We know that $(\mathbf{\Gamma}_p, S_{\mathbf{\Gamma}_p})$ may be written as the independent sum of (G, S_G) and $(\mathbf{\Gamma}_p - G, S_G - S_{\mathbf{\Gamma}_p})$, where the latter is equal in distribution to (D, S_D). Reviewing the proof of part (ii), when the strong ladder height is replaced by a weak ladder height, we see that $(\mathbf{\Gamma}_p - G, S_{\mathbf{\Gamma}_p} - S_G)$, like (G, S_G), is infinitely divisible (for the weak ladder height, one works with the stopping time $N' = \inf\{n > 0 : S_n \leq 0\}$; note the relationship between the inequality in the definition of N' and the max in the definition of D). Further, (G, S_G) is supported on $\{1, 2, \ldots\} \times (0, \infty)$ and $(\mathbf{\Gamma}_p - G, S_{\mathbf{\Gamma}_p} - S_G)$ is supported on $\{1, 2, \ldots\} \times (-\infty, 0)$. This means that, in the variable θ, $E(s^G \mathrm{e}^{\mathrm{i}\theta S_G})$ can be analytically extended to the upper half of the complex plane and $E(s^{(\mathbf{\Gamma}_p - G)} \mathrm{e}^{\mathrm{i}\theta(S_{\mathbf{\Gamma}_p} - S_G)})$ to the lower half of the complex plane. (v) Note that the path decomposition given in part (ii) shows that

$$E\left(s^G \mathrm{e}^{\mathrm{i}\theta S_G}\right) = E\left(s^{\sum_{i=1}^{\nu} N^{(i)}} \mathrm{e}^{\mathrm{i}\theta \sum_{i=1}^{\nu} H^{(i)}}\right),$$

[1] Duality for random walks is the same concept as for Lévy processes. In other words, for any $n = 0, 1, 2, \ldots$ (which may later be randomised with any independent distribution), note that the independence and common distribution of increments implies that $\{S_{n-k} - S_n : k = 0, 1, \ldots, n\}$ has the same law as $\{-S_k : k = 0, 1, \ldots, n\}$.

where the pairs $\{(N^{(i)}, H^{(i)}) : i = 1, 2, \ldots\}$ are independent, having the same distribution as (N, S_N) conditioned on $\{N \leq \mathbf{\Gamma}_p\}$.

6.10 (i) Use Lemma 1.7 with the parameter choices $\alpha = \Phi(q)$ and $\beta = 1$. (iii) Consider the characterisation of measures through their transforms.

Chapter 7

7.1 (i) Use Exercise 3.3. (ii) First suppose that $q > 0$. The Wiener–Hopf factorisation gives us

$$\mathbb{E}\big(\mathrm{e}^{-\mathrm{i}\theta \overline{X}^K_{\mathbf{e}_q}}\big) = \mathbb{E}\big(\mathrm{e}^{\mathrm{i}\theta X^K_{\mathbf{e}_q}}\big) \frac{\widehat{\kappa}^K(q, \mathrm{i}\theta)}{\widehat{\kappa}^K(q, 0)}, \tag{S.5}$$

where $\widehat{\kappa}^K$ is the Laplace exponent of the bivariate descending ladder process of X^K and $\mathbf{e}_q$ is an independent and exponentially distributed random variable with mean $1/q$. Show, and use, that the descending ladder height process of X^K has moments of all orders. Together with the fact that $\mathbb{E}(|X^K_t|^n) < \infty$, for all $t > 0$, consider the Maclaurin expansion up to order n of (S.5). (iii) Use the Wiener–Hopf factorisation for X^K (up to a multiplicative constant) in the form

$$\kappa^K(0, -\mathrm{i}\theta) = \frac{\Psi^K(\theta)}{\widehat{\kappa}^K(0, \mathrm{i}\theta)},$$

where κ^K and Ψ^K are defined in an obvious way, and appeal to reasoning similar to that used in part (ii). Note that $\overline{X}^K_\infty$ is equal in law to the ascending ladder height process of X^K, stopped at an independent and exponentially distributed time.

7.2 (i) Show that $\mathbb{E}(Y_n) \leq \mathbb{E}(\overline{X}_1) - \mathbb{E}(\underline{X}_1)$. (iii) Let $[t]$ be the integer part of t. Write

$$\frac{X_t}{t} = \frac{\sum_{i=1}^{[t]}(X_i - X_{i-1})}{[t]} \frac{[t]}{t} + \frac{X_t - X_{[t]}}{[t]} \frac{[t]}{t}.$$

(iv) The assumption $\mathbb{E}(X_1) = \infty$ implies that $\mathbb{E}(\max\{-X_1, 0\}) < \infty$ and $\mathbb{E}(\max\{X_1, 0\}) = \infty$. Use truncation ideas from Exercise 7.1.

7.3 (i) Take limits in the appropriate Wiener–Hopf factor (see Sect. 6.5.2). (ii) Similar to part (i). (iii) The trichotomy in Theorem 7.1, together with the conclusions of parts (i) and (ii), gives the required asymptotic behaviour. (iv) The given process has Laplace exponent given by $\psi(\theta) = c\theta^\alpha$, for some $c > 0$.

7.4 (i) See Exercise 3.7 and Lemma 7.10. (ii) The measure $U(\mathrm{d}x)$ is the potential measure of the ascending ladder height process. Revisit Exercise 5.8. (iii) The potential $\widehat{U}(\mathrm{d}x)$ can be derived as in the previous part of the question. Apply the quintuple law. (iv) Use part (i), the beta integral

$$\int_0^1 u^{p-1}(1-u)^{q-1}\mathrm{d}u = \frac{\Gamma(p)\Gamma(q)}{\Gamma(p+q)},$$

for $p, q > 0$, and the reflection formula for the gamma function.

7.5 (i) Consider, in the light of Exercise 5.6, the quadruple law

$$\mathbb{P}\big(\tau_x^+ - \overline{G}_{\tau_x^+ -} \in \mathrm{d}t, \overline{G}_{\tau_x^+ -} \in \mathrm{d}s, X_{\tau_x^+} - x \in \mathrm{d}u, x - \overline{X}_{\tau_x^+ -} \in \mathrm{d}y\big),$$

for $u > 0$, $y \in [0, x]$ and $s, t \geq 0$. (ii) When X is spectrally positive, we have $\widehat{H}_t = t$ on $\{t < \widehat{L}_\infty\}$.

7.6 (i) It suffices to prove that

$$\lim_{|\theta|\uparrow\infty} \frac{1}{\theta^2} \int_{\mathbb{R}} \big(1 - \mathrm{e}^{\mathrm{i}\theta x} + \mathrm{i}\theta x \mathbf{1}_{(|x|<1)}\big)\Pi(\mathrm{d}x) = 0. \tag{S.6}$$

It will help to first prove that $|1 - \cos a| \leq 2(1 \wedge a^2)$ and $|a - \sin a| \leq 2(|a| \wedge |a|^3)$. For the second assertion, recall from Lemma 7.10 that there is creeping upwards if and only if $\lim_{\beta\uparrow\infty} \kappa(0, \beta)/\beta > 0$. (ii) Show, and then use, that L^{-1} is a subordinator which has a non-zero drift coefficient. (iii) What class of subordinators does the ladder height process H belong to in the case of irregularity? (iv) Use part (i). (v) Any symmetric process must either creep in both directions or not at all (by symmetry). Consider the integral test in Theorem 7.12.

7.7 (i) It will be convenient to use the identity:

$$\theta^{-\alpha}(1+\theta)^{-1} = \alpha \int_0^1 (1-\phi)^{\alpha-1}(\phi+\theta)^{-(\alpha+1)}\mathrm{d}\phi,$$

valid for all $\theta > 0$. By changing variables, via $(1+\theta)(1-u) = \phi + \theta$, this identity can be derived by showing that, for all $\theta > 0$,

$$\frac{\theta^{-\alpha}}{\alpha} = \int_0^{1/(1+\theta)} u^{\alpha-1}(1-u)^{-(\alpha+1)}\mathrm{d}u,$$

which, in turn, follows by differentiation. (ii) In the appropriate probabilistic setting, subtract from those paths which enter $(1, \infty)$, the paths which first enter $(-\infty, 0)$. (iii) It turns out to be easier to differentiate the equations in (ii) in the variable y first.

7.8 Show that, on the event $\{\mathcal{U}_x > u, \mathcal{V}_x > v, \mathcal{O}_x > w\}$, the interval $[x-u, x+w]$ does not belong to the range of $\overline{X}$. Show, moreover, that when $\mathcal{O}_{x-u} > u + w$ and $\mathcal{V}_{x-u} > v - u$, the interval $[x-u, x+w]$ does not belong to the range of $\overline{X}$.

7.9 Marginalise the quintuple law to the joint law of the overshoot and undershoot and then take limits, using the Renewal Theorem 5.1. Don't forget to take account of creeping.

7.10 Start by showing, for $q > 0$ and $x \geq 0$,

$$\mathbb{E}_x\left(\int_0^{\tau_0^-} \mathrm{e}^{-qt} f(X_t)\mathrm{d}t\right) = \frac{1}{q}\int_{[0,\infty)} \mathbb{P}(\overline{X}_{\mathbf{e}_q} \in \mathrm{d}y)\int_{[0,x]} \mathbb{P}(-\underline{X}_{\mathbf{e}_q} \in \mathrm{d}z) f(x+y-z).$$

Take limits as $q \downarrow 0$, making use of (7.14).

Chapter 8

8.1 Show that

$$\mathbb{P}(A|\tau_x^+ < \infty) = \mathrm{e}^{\Phi(0)x}\mathbb{P}(A, \tau_x^+ < t) + \mathbb{P}^{\Phi(0)}(A, \tau_x^+ \geq t),$$

so that it suffices to establish

$$\lim_{x\uparrow\infty} \mathrm{e}^{\Phi(0)x}\mathbb{P}(A, \tau_x^+ < t) = 0, \tag{S.7}$$

for all $t > 0$. Show instead that, for all $q > 0$, $\lim_{x\uparrow\infty} \mathrm{e}^{\Phi(0)x}\mathbb{P}(A, \tau_x^+ < \mathbf{e}_q) = 0$. Since we can choose q arbitrarily small, we can make $\mathbb{P}(\mathbf{e}_q > t) = \mathrm{e}^{-qt}$ arbitrarily close to 1, from which one can recover (S.7).

8.2 All parts can be handled through Laplace transforms.

8.3 (i) Use the assumption that $\psi'(0+) > 0$. (ii) See (8.20). (iii) Use the second assertion of part (ii), noting that, for $n \geq 1$, $\nu^{*n}(\mathrm{d}x) = ((n-1)!)^{-1}(\lambda/\delta)^n x^{n-1}\mathrm{e}^{-\mu x}\mathrm{d}x$ on $(0, \infty)$.

8.4 (ii) Use (8.26). (iii) Consider the atomic support of the Lévy measure associated to $(X, \mathbb{P}^{\Phi(q)})$ relative to the support of Π.

8.5 (i) Take limits in (8.12) as $a \uparrow \infty$. For the second assertion, use (8.11). (ii) Note that

$$W_+^{(q)\prime}(0+) = \lim_{\beta\uparrow\infty}\int_0^\infty \beta\mathrm{e}^{-\beta x} W^{(q)\prime}(x)\mathrm{d}x.$$

8.6 Use Theorem 5.9 and the fact that, for $\beta \geq 0$,

$$\int_{[0,\infty)} \mathrm{e}^{-\beta y}\widehat{U}(\mathrm{d}y) = \frac{\beta - \Phi(0)}{\psi(\beta)},$$

where $\widehat{U}$ is the potential measure of the descending ladder height process. Be careful with the normalisation of local time at the maximum, which is implicit in the definition of $\widehat{U}$.

8.7 (i) Follows by definition. (ii) Apply an exponential change of measure in (8.9). Analytic extension of the resulting identity may thereafter be necessary. (iii) Choose

$c = \Phi(p+u)$ and $q = p - \psi(\Phi(p+u)) = -u$, then take limits, first as $u \downarrow 0$, and then as $x \downarrow 0$, in the identity established in part (ii).

8.8 [2] (i) The event $\{\exists t > 0 : B_t = \overline{B}_t = t\}$ is equivalent to $\{\exists s > 0 : L_s^{-1} = H_s\}$, where (L^{-1}, H) is the ascending ladder height process. (ii) Apply the conclusion of part (iv) in Exercise 8.7.

8.9 (i) Starting by showing that

$$\mathbb{E}_y\big(\mathrm{e}^{-q\Lambda_0}\big) = q\int_0^\infty \mathrm{e}^{-qt}\mathbb{P}_y(0 \le \Lambda_0 < t)\mathrm{d}t.$$

(ii) Use (8.18) and Corollary 8.9. (iii) Note that, on $\{\Lambda_0 > 0\}$, we have $\Lambda_0 > \tau_0^-$. Condition on $\mathcal{F}_{\tau_0^-}$ and apply the strong Markov property. (iv) Make an exponential change of measure.

8.10 (i) The process Z^x will either exit from $[0, a)$ by first hitting zero, or by directly passing above a, before hitting zero. (ii) Take the limit as x tends to zero in the identity from part (i).

8.11 (i) This is a repetition of Exercise 7.7 (ii) with some simplifications. One can take advantage of the fact that $r(x, y) = r(x, 0) = \mathbb{P}(\tau_1^+ < \tau_0^-)$, which follows from spectral negativity. Recall from Exercise 6.6 (i), that $\rho = 1/\alpha$. (ii) Use Exercise 8.2 and Theorem 8.1. (iii) Starting from $x \ge 1$, the process X either first enters $(-1, 1)$ simultaneously on first entering $(-\infty, -1)$, or, at the latter time, it jumps over the interval $(-1, 1)$ and creeps in at the lower boundary at some later time.

8.12 First take $a \in (0, \infty)$. For the first martingale, justify the almost sure equality

$$\mathrm{e}^{-q(\tau_a^+ \wedge \tau_0^-)}\mathbf{1}_{(\tau_a^+ < \tau_0^-)} = \mathrm{e}^{-q(\tau_a^+ \wedge \tau_0^-)}W^{(q)}(X_{\tau_a^+ \wedge \tau_0^-})/W^{(q)}(a),$$

and take expectation conditional on $\mathcal{F}_t$. For the case $a = \infty$, use dominated convergence and the representation (8.23). Use similar reasoning for the second martingale by taking expectations of $\exp\{-q\tau_0^-\}$, conditional on $\mathcal{F}_t$.

Chapter 9

9.1 (i) and (ii) Consider using the Tauberian theorems from Sect. 5.4. (iii) Take Laplace transforms.

9.2 Visit the Laplace inversion formula on page 233 of Konstantopoulos et al. (2011).

---

[2] It is worth pointing out that this exercise can be adapted to cover the case when we replace B by any spectrally negative Lévy process in the original question.

9.3 Consider the Lévy measure of the subordinator used in Exercise 9.1 and look at the extreme case that $\alpha = 0$. Which well-known subordinator is this? (i) Use the definition of W as a potential measure.

9.4 (iii) Consider a partial fraction decomposition of $(\psi(\theta) - q)^{-1}$. (iv) Consider using (8.24). (v) See Corollary 5.24.

9.5 (i) Consider the jump density of the descending ladder height process. (ii) What is the Lévy density associated with $(X, \mathbb{P}^{\Phi(q)})$ and is it still completely monotone? (iii) Consider $W^{(q)\prime\prime\prime}$, using the fact that $W_{\Phi(q)}$ has a completely monotone density, together with Theorem 5.21.

9.6 Suppose that ϕ^*_β has associated triple $(\kappa^*_\beta, \delta^*_\beta, \Upsilon^*_\beta)$. On the one hand, note that, thanks to conjugacy,

$$W_\beta(x) = \delta^*_\beta + \int_0^x \kappa^*_\beta + \Upsilon^*_\beta(y, \infty)\mathrm{d}y,$$

where W_β is the potential function associated to ϕ_β. On the other hand, one easily verifies that $W_\beta(\mathrm{d}x) = \mathrm{e}^{-\beta x} W(\mathrm{d}x)$ for $x \geq 0$, where W is the potential function associated to ϕ.

9.7 Substitute the descending ladder height potential into formula (9.8).

9.8 (i) and (ii) It will be useful to note that

$$\mathcal{T}_\beta \psi(\theta) = \big(\psi(\theta + \beta)\big) - \psi(\beta) - \beta\big(\Phi(\theta + \beta) - \Phi(\beta)\big), \quad \theta \geq 0,$$

where $\Phi(\theta) = \psi(\theta)/\theta$. (iii) Use Laplace transforms.

Chapter 10

10.1 Note that $\mathbb{E}_x(\mathrm{e}^{-q\tau_0^-}; X_{\tau_0^-} = 0) = \mathrm{e}^{\Phi(q)x}\mathbb{P}_x^{\Phi(q)}(X_{\tau_0^-} = 0)$.

10.2 Appeal again to the quintuple law, being careful to note that the scale functions $W^{(q)}$, $q \geq 0$, have a discontinuity at the origin.

10.3 (i) Let N be a Poisson random measure associated with the jumps of X. Note that $N = \sum_{i=1}^n N^{(i)}$, where, for $i = 1, \ldots, n$, $N^{(i)}$ are the independent Poisson random measures associated with the jumps of $X^{(i)}$. For Borel $A \subseteq (-\infty, 0)$, show that

$$\begin{aligned}
&\mathbb{P}_x\big(X_{\tau_0^- -} \in \mathrm{d}y, X_{\tau_0^-} \in A, \Delta X_{\tau_0^-} = \Delta X^{(i)}_{\tau_0^-}\big) \\
&= \mathbb{E}_x\left(\int_{[0,\infty)}\int_{(-\infty,0)} \mathbf{1}_{(\underline{X}_{t-} > 0)}\mathbf{1}_{(X_{t-} \in \mathrm{d}y)}\mathbf{1}_{(y + a \in A)} N^{(i)}(\mathrm{d}t \times \mathrm{d}a)\right).
\end{aligned}$$

(ii) Use Corollary 8.8. (iii) Use that X cannot creep below the origin.

10.4 (i) Work with

$$J_t = \int_t^\infty \mathrm{e}^{-qS_t}\mathbf{1}_{(t<\mathbf{e}_\kappa)}\mathrm{d}t.$$

(ii) The subordinator S has jumps corresponding to excursion lengths conditional on the excursion heights being bounded by a. What then is the value of κ?

10.5 (iii) The equation in (i) can be solved explicitly (and hence uniquely). The equality in (ii) can be used to simplify this expression.

10.6 Repeat the computation in (8.32), taking care to include exponential discounting.

10.7 Use that

$$\begin{aligned}&\mathbb{E}_x\big(\mathrm{e}^{-q\kappa_0^-}\mathbf{1}_{\{\kappa_0^-<\kappa_a^+\}}\mathbf{1}_{\{\kappa_0^-<\infty\}}\big)\\&\quad= \mathbb{E}_x\big(\mathrm{e}^{-q\kappa_0^-}\mathbf{1}_{\{\kappa_0^-<\infty\}}\big) - \mathbb{E}_x\big(\mathrm{e}^{-q\kappa_a^+}\mathbf{1}_{\{\kappa_a^+<\kappa_0^-\}}\big)\mathbb{E}_a\big(\mathrm{e}^{-q\kappa_0^-}\mathbf{1}_{\{\kappa_0^-<\infty\}}\big).\end{aligned}$$

10.8 (i) Every sojourn below the origin must survive exponential killing. (ii) Condition on the behaviour of the process according to its first sojourn below the origin.

10.9 For the first assertion, use (10.41) and note that the running maximum of U, up to time t, will be attained at the last time that $\overline{X} - X$ is zero.

10.10 Appeal to transformations of identities found in this chapter.

10.11 (i) Show that the integral

$$\int_x^{s^*(x)} \frac{W'(\bar{\gamma}(s))}{W(\bar{\gamma}(s))}\mathrm{d}s$$

is comparable with the quantity $\log(W(x)) - \log(W(0))$. (ii) Use Corollary 10.11.

Chapter 11

11.1 For the process of excursions of $\overline{X} - X$, consider the probability that there have been no excursions with height exceeding a up to local time t, for arbitrary large t.

11.2 (i) Note that, on the event $\{\tau_a^+ \le \mathbf{e}_q\} = \{\overline{X}_{\mathbf{e}_q} \ge a\}$, we have that $\overline{X}_{\mathbf{e}_q} = X_{\tau_a^+} + S$, where stationary independent increments and the lack-of-memory property imply that S is independent of $\mathcal{F}_{\tau_a^+}$ and equal in distribution to $\overline{X}_{\mathbf{e}_q}$. (ii) Check the conditions of Lemma 11.1. For the lower bound, show, and then use, that

$$v_{x^*}(x) = \big(1-\mathrm{e}^{-x}\big) - \mathbb{E}\left(\mathbf{1}_{(\overline{X}_{\mathbf{e}_q}<x^*-x)}\left(1 - \frac{\mathrm{e}^{-x-\overline{X}_{\mathbf{e}_q}}}{\mathbb{E}(\mathrm{e}^{-\overline{X}_{\mathbf{e}_q}})}\right)\right).$$

For the supermartingale property, use that, on the event $\{\mathbf{e}_q > t\}$, $\overline{X}_{\mathbf{e}_q} = (X_t + S) \vee \overline{X}_t \geq X_t + S$, where, by stationary and independent increments and the lack-of-memory property, S is independent of $\mathcal{F}_t$ and has the same distribution as $\overline{X}_{\mathbf{e}_q}$. (iii) Work with computations in the spirit of the proof of Theorem 11.4.

11.3 (i) Make a change of variables in the integral in the definition of the function H, using $y = u(t+1)^{-1/\alpha}$, and use the fact that $\exp\{yX_t - y^\alpha t\}$, $t \geq 0$, is a martingale. (ii) In the expectation, multiply and divide by the martingale from (i). (iii) Show that the upper bound in (ii) can be attained.

Chapter 12

12.1 (i) Check the integral test in Theorem 12.5. (ii) to deduce that extinction occurs with probability zero. (ii) Show that X has an almost surely finite number of jumps on the time interval $[0, \tau_0^-]$. Let n^* be the number of jumps that X has undertaken by time τ_0^-, and denote their times, in increasing order, by $T_1, T_2, \ldots, T_{n^*}$, with $T_0 := 0$. If $\{\xi_i : i = 1, 2, 3, \ldots\}$ are the independent and identically distributed sequence of jumps of X and, for $k = 1, 2, 3, \ldots$, we let $S_k = \sum_{j=1}^k \xi_j$, show that

$$\Delta = (x + S_{n^*} - cT_{n^*}) \prod_{k=1}^{n^*} \frac{x + S_{k-1} - cT_{k-1}}{x + S_{k-1} - cT_k}.$$

12.2 (i) Sample the Kella–Whitt martingale[3] at an appropriate sequence of stopping times. (ii) Consider the formula (12.10).

12.3 (i) When $\liminf_{t\uparrow\infty} X_t = -\infty$ and $\zeta = \infty$, note that $\int_0^\infty Y_s \mathrm{d}s < \infty$. (ii) Consider (12.18) on the event $\{\tau_0^- < \infty\}$ and its complement. Use the Martingale Convergence Theorem to note that $\lim_{t\uparrow\infty} Y_t$ must exist. In particular, show that

$$\lim_{t\uparrow\infty} \mathrm{e}^{-\Phi(0)Y_t} = \mathbf{1}_{(\lim_{t\uparrow\infty} Y_t = 0)}.$$

(iii) In the light of (12.14), consider a computation in the spirit of (2.11).

12.4 (i) Let $E = \{\lim_{s\uparrow\infty} Y_s = 0\}$. From the definition of P_x^* and the Markov property, we have

$$E_x^*\big(\mathrm{e}^{-\theta Y_t} | E\big) = E^*\big(\mathrm{e}^{-\theta Y_t} P_{X_t}^*(E)\big).$$

Show that P_x^* can be written as a martingale change of measure with respect to P_x. (ii) Consider the evolution equation (12.6) with ψ replaced by ψ^*.

[3] Recall from the discussion following the proof of Theorem 4.7 that the assumption of bounded variation paths in the underlying spectrally negative Lévy process is not needed.

12.5 (i) Let $\mathbf{e}_q$ be an independent and exponentially distributed random variable with parameter $q > 0$ and set $g(x) = xf(x)$. Start by showing that

$$\mathbb{E}_x^{\uparrow}\left(\int_0^{\infty} \mathrm{e}^{-qt} f(X_t)\mathrm{d}t\right) = \frac{1}{qx}\mathbb{E}\big(g(x + X_{\mathbf{e}_q})\mathbf{1}_{(-\underline{X}_{\mathbf{e}_q} < x)}\big).$$

(ii) Use (8.24) and that $\lim_{q \downarrow 0} q/\Phi(q) = \psi'(0+)$. (iii) With the help of Theorem 12.11, deduce that

$$\mathbb{P}_x^{\uparrow}(\tau_y^- < \tau_z^+ \wedge t) = \frac{y}{x}\mathbb{P}_x(\tau_y^- < \tau_z^+ \wedge t),$$

and take $t \uparrow \infty$. Consider what happens as $y \downarrow 0$, then what happens as $z \uparrow \infty$, and finally what happens as $x \uparrow \infty$. Show, and use, that

$$\liminf_{t \uparrow \infty} X_t = \liminf_{t \uparrow \infty} X_{\tau_z^+ + t}$$

under $\mathbb{P}_x^{\uparrow}$.

12.6 (i) Considering (12.26), one needs to show that, for each $t > 0$,

$$\frac{\partial u_t}{\partial \theta}(\theta) = \frac{\psi(u_t(\theta))}{\psi(\theta)}.$$

(ii) Use part (i) to deduce that $Y_t \to \infty$ in $P_x^{\uparrow}$-probability. Consider the implications of this in Lemma 12.15 (ii), taking account of the last part of Exercise 12.5. (iii) The first part is a straightforward manipulation. It will help to understand the behaviour of

$$\int_0^{\theta x}\left(\frac{\mathrm{e}^{-\lambda} - 1 + \lambda}{\lambda^2}\right)\mathrm{d}\lambda,$$

as $x \downarrow 0$ and as $x \uparrow \infty$. (iv) Use part (i) of the question and check that, for t sufficiently large,

$$0 \le \int_{u_t(\theta)}^{\theta}\left(\frac{1}{\rho\xi} - \frac{1}{\psi(\xi)}\right)\mathrm{d}\xi = \frac{1}{\rho}\log\left(\frac{\theta}{u_t(\theta)\mathrm{e}^{\rho t}}\right).$$

12.7 (i) All of the assertions follow from the definition of $u_t(\theta)$, its semigroup property (12.5) as well as the fact that it solves (12.8). (ii) See Exercise 12.3. (iii) Use the Markov branching property and that, for $s, t > 0$, $u_t(\eta_{t+s}(\lambda)) = \eta_s(\lambda)$. (iv) The integral $\int_\lambda^{\Phi(\theta)} 1/\psi(\xi)\mathrm{d}\xi$ can only explode as $\theta \downarrow 0$ and $\theta \uparrow \infty$ if the upper delimiter tends to 0, $\Phi(0)$ or possibly ∞ (depending on whether extinction occurs or not). The latter of these three eventualities can be excluded, see the footnote in the question.

12.8 (ii) Use part (i) of Exercise 12.7.

Chapter 13

13.1 The strong Markov property may be established using piecewise stopping each time Y crosses the origin (note that there is no creeping which means that the amount of time spent on each sojourn below the origin has positive Lebsgue measure; and the same applies to sojourns above the origin). Define the rescaled process $\{\tilde{X}_t : t \geq 0\}$ by $\tilde{X}_t = cX_{c^{-\alpha}t}$, $t \geq 0$, and, correspondingly, let $\tilde{\gamma}$ be the right-inverse of $\int_0^{\cdot} \mathbf{1}_{(\tilde{X}_s>0)}\,\mathrm{d}s$. Show, and then use, that

$$c^{\alpha}\gamma\left(c^{-\alpha}t\right) = \tilde{\gamma}(t).$$

13.2 Use Exercise 5.8.

13.3 Let $(B, \mathbb{P})$ be a standard Brownian motion. (i) Consider the probability $\mathbb{P}_1(\tau_x^+ < \tau_0^-)$, for $x > 1$. If ψ is the Laplace exponent of the underlying Lévy process through the second Lamperti transform and Φ is its right inverse, show, and use, that $\Phi(0) = 1$. (ii) Show that, for $0 < \varepsilon < 1$, $\mathbb{P}_1^{\uparrow}(\tau_\varepsilon^- < \infty) = \varepsilon$ and reason further as in part (i). To pin down the multiplicative constant, one may consider using Itô's formula to match the quadratic variation of Brownian motion against the quadratic variation of the process in the second Lamperti transform.

13.4 For the second part, note that the descending ladder height process of a standard Brownian motion is a deterministic unit drift process.

13.5 (i) Use Exercise 5.8, noting that the descending ladder height process of a stable process is a stable subordinator.

13.6 (i) Use right-continuous paths and finite integral length. (ii) and (iii) Write $\xi_t = t(\xi_t/t)$. (iv) On account of the randomised position ξ_{T_n}, the time difference $T_n - S_n$ is longer than the time it takes for ξ to start at 2 and cross below 1.

13.7 Both (i) and (ii) can be recovered from the two-sided exit problem in Exercise 7.7, using (13.11) and a logarithmic change of spatial scale.

13.8 (i) Follow, for example, the reasoning in (13.36), or use Exercise 13.4. Take Laplace transforms of $\underline{\xi}_\infty^{\uparrow}$ and appeal to the Wiener–Hopf factorisation to derive $\psi^{\uparrow}$. Consider the ruin probability for $\xi^{\uparrow}$ (which involves an expression for $W^{\uparrow}$), or use the beta integral to take inverse Laplace transforms of $1/\psi^{\uparrow}$. (ii) Follow a similar programme to (i).

13.9 (i) Note that the required triple law at first-passage for $(Y, \mathbb{P}_x^{\uparrow})$ can be written in terms of a triple law for $(\xi^{\uparrow}, \mathbf{P}^{\uparrow})$, via a logarithmic change of spatial scale. In turn, the latter can be written down explicitly, using properties of the ascending and descending ladder height processes of $(\xi^{\uparrow}, \mathbf{P}^{\uparrow})$. (ii) Use (13.11).

13.10 (i) If we write ψ_{d} for the Laplace exponent of a Brownian motion with drift $\mathrm{d}/2 - 1$, then one should check that

$$\psi_{\mathrm{d}+2}(\lambda) = \frac{\theta}{\lambda+2}\psi_{\mathrm{d}}(\lambda), \quad \lambda \geq 0.$$

(ii) Using standard notation, note that

$$\left(R_t^{(x)}\right)^q = x^q \exp\{q\xi_{\varphi((x^q)^{-2/q}t)}\}, \quad t \leq \zeta^{(x)}.$$

References

Albrecher, H. and Gerber, H.U. (2011) A note on moments of dividends. *Acta Math. Appl. Sin.* **27**, 353–354.

Albrecher, H. and Hipp, C. (2007) Lundberg's risk process with tax. *Blätter DGVFM* **28**, 13–28.

Albrecher, H., Renaud, J.–F. and Zhou, X. (2008) A Lévy insurance risk process with tax. *J. Appl. Probab.* **45**, 363–375.

Alili, L. and Kyprianou, A.E. (2005) Some remarks on first passage of Lévy processes, the American put and smooth pasting. *Ann. Appl. Probab.* **15**, 2062–2080.

Andrew, P. (2006) Proof from first principles of Kesten's result for the probabilities with which a subordinator hits points. *Electron. Commun. Probab.* **11**, 58–63.

Applebaum, D. (2004) *Lévy Processes and Stochastic Calculus*. Cambridge University Press, Cambridge.

Applebaum, D., Bhat, B.V.R., Kustermans, J. and Lindsay, J.M. (2005) *Quantum Independent Increment Processes I. From Classical Probability to Quantum Stochastic Calculus*. Lecture Notes in Mathematics, Springer, Berlin.

Ash, R. and Doléans-Dade, C.A. (2000) *Probability and Measure Theory. 2nd Edition*. Harcourt, New York.

Asmussen, S. and Albrecher, H. (2010) *Ruin probabilities. 2nd Edition*. World Scientific, New Jersey.

Asmussen, S., Avram, F. and Pistorius, M. (2004) Russian and American put options under exponential phase-type Lévy models. *Stoch. Process. Appl.* **109**, 79–111.

Asmussen, S. and Klüppelberg, C. (1996) Large deviations results for subexponential tails, with applications to insurance risk. *Stoch. Process. Appl.* **64**, 103–125.

Asmussen, S. and Taksar, M. (1997) Controlled diffusion models for optimal dividend pay-out. *Insur. Math. Econ.* **20**, 1–15.

Athreya, S. and Ney, P. (1972) *Branching Processes*. Springer, Berlin.

Avram, F., Chan, T. and Usabel, M. (2002) On the valuation of constant barrier options under spectrally one-sided exponential Lévy models and Carr's approximation for American puts. *Stoch. Process. Appl.* **100**, 75–107.

Avram, F., Kyprianou, A.E. and Pistorius, M.R. (2004) Exit problems for spectrally negative Lévy processes and applications to (Canadized) Russian options. *Ann. Appl. Probab.* **14**, 215–235.

Avram, F., Palmowski, Z. and Pistorius, M.R. (2007) On the optimal dividend problem for a spectrally negative Lévy process. *Ann. Appl. Probab.* **17**, 156–180.

Azcue, P. and Muler, N. (2005) Optimal reinsurance and dividend distribution policies in the Cramér–Lundberg model. *Math. Finance* **15**, 261–308.

Bachelier, L. (1900) Théorie de la spéculation. *Ann. Sci. Éc. Norm. Super.* **17**, 21–86. [*Louis Bachelier's Theory of Speculation. The origins of modern finance*. Translated and with commentary by Mark Davis and Alison Etheridge. Foreword by Paul A. Samuelson. Princeton University Press, 2006.]

A.E. Kyprianou, *Fluctuations of Lévy Processes with Applications*, Universitext,
DOI 10.1007/978-3-642-37632-0, © Springer-Verlag Berlin Heidelberg 2014

Bachelier, L. (1901) Théorie mathematique du jeu. *Ann. Sci. Éc. Norm. Super.* **18**, 143–210.

Barczy, M. and Döring, L. (2011) A jump-type SDE approach to positive self-similar Markov processes. *Preprint.*

Barndorff-Nielsen, O.E., Franz, U., Gohm, R., Kümmerer, B. and Thorbjørnsen, S. (2006) *Quantum Independent Increment Processes II Structure of Quantum Lévy Processes, Classical Probability, and Physics.* Lecture Notes in Mathematics, Springer, Berlin.

Barndorff-Nielsen, O.E. and Shephard, N. (2001) Modelling by Lévy Processes for Financial Econometrics. In, *Lévy Processes: Theory and Applications.* O.E. Barndorff-Nielsen, T. Mikosch, and S. Resnick (Eds.), Birkhäuser, Basel, 283–318.

Bass, R.F. (2004) Stochastic differential equations with jumps. *Probab. Surv.*. **1**, 1–19.

Baurdoux, E.J. (2007) Examples of optimal stopping via measure transformation for processes with one-sided jumps. *Stochastics* **79**, 303–307.

Baurdoux, E.J. (2009) Last exit before an exponential time for spectrally negative Lévy processes. *J. Appl. Probab.* **46**, 542–558.

Baurdoux, E.J. (2013) A direct method for solving optimal stopping problems for Lévy processes. *Preprint.*

Baurdoux, E.J. and van Schaik, K. (2012) Predicting the time at which a Lévy process attains its ultimate supremum. *Preprint.*

Baxter, G. (1958) An operator identity. *Pac. J. Math.* **8**, 649–663.

Beibel, M. and Lerche, H.R. (1997) New look at optimal stopping problems related to mathematical finance. *Stat. Sin.* **7**, 93–108.

Benassi, A., Cohen, S. and Istas, J. (2002) Identification and properties of real harmonizable fractional Lévy motions. *Bernoulli* **8**, 97–115.

Benassi, A., Cohen, S. and Istas, J. (2004) On roughness indexes for fractional fields. *Bernoulli* **10**, 357–373.

Berestycki, J., Döring, L., Mytnik, L. and Zambotti, L. (2011a) Hitting properties and non-uniqueness for SDE driven by stable processes. *Preprint.*

Berestycki, J., Kyprianou, A.E. and Murillo-Salas, A. (2011b) The prolific backbone for supercritical superdiffusions. *Stoch. Process. Appl.* **121**, 1315–1331.

Bernyk, V., Dalang, R. and Peskir, G. (2008) The law of the supremum of a stable Lévy process with no negative jumps. *Ann. Probab.* **36**, 1777–1789.

Bertoin, J. (1992) An extension of Pitman's theorem for spectrally positive Lévy processes. *Ann. Probab.* **20**, 1464–1483.

Bertoin, J. (1993) Splitting at the infimum and excursions in half-lines for random walks and Lévy processes. *Stoch. Process. Appl.* **47**, 17–35.

Bertoin, J. (1996a) *Lévy Processes.* Cambridge University Press, Cambridge.

Bertoin, J. (1996b) On the first exit time of a completely asymmetric stable process from a finite interval. *Bull. Lond. Math. Soc.* **28**, 514–520.

Bertoin, J. (1997a) Regularity of the half-line for Lévy processes. *Bull. Sci. Math.* **121**, 345–354.

Bertoin, J. (1997b) Exponential decay and ergodicity of completely asymmetric Lévy processes in a finite interval. *Ann. Appl. Probab.* **7**, 156–169.

Bertoin, J. (1997c) Regenerative embedding of Markov sets. *Probab. Theory Relat. Fields.* **108**, 559–571.

Bertoin, J. (2006) *Random Fragmentation and Coagulation Processes.* Cambridge Studies in Advanced Mathematics, 102. Cambridge University Press, Cambridge.

Bertoin, J. and Caballero, M.-E. (2002) Entrance from 0+ for increasing semi-stable Markov processes. *Bernoulli* **8**, 195–205.

Bertoin, J. and Doney, R.A. (1994a) Cramér's estimate for Lévy processes. *Stat. Probab. Lett.* **21**, 363–365.

Bertoin, J. and Doney, R.A. (1994b) On conditioning a random walk to stay nonnegative. *Ann. Probab.* **22**, 2152–2167.

Bertoin, J. and Savov, M. (2011) Some applications of duality for Lévy processes in a half-line. *Bull. Lond. Math. Soc.* **43**, 97–110.

Bertoin, J., van Harn, K. and Steutel, F.W. (1999) Renewal theory and level passage by subordinators. *Stat. Probab. Lett.* **45**, 65–69.

Bertoin, J. and Yor, M. (2002a) On the entire moments of self-similar Markov processes and exponential functionals of Lévy processes. *Ann. Fac. Sci. Toulouse Math. Sér. 6* **11**, 33–45.

Bertoin, J. and Yor, M. (2002b) The entrance laws of self-similar Markov processes and exponential functionals. *Potential Anal.* **17**, 389–400.

Bertoin, J. and Yor, M. (2005) Exponential functionals of Lévy processes. *Probab. Surv.* **2**, 191–212.

Biane, Ph. (1985) Comparaison entre temps d'atteinte et temps de séjour de certaines diffusions réelles. In, *Séminaire de Probabilités, XIX*, 291–296. Lecture Notes in Mathematics, Springer, Berlin.

Bichteler, K. (2002) *Stochastic Integration with Jumps.* Cambridge University Press, Cambridge.

Biffis, E. and Morales, M. (2010) On a generalization of the Gerber–Shiu function to path-dependent penalties. *Insur. Math. Econ.* **46**, 92–97.

Billingsley, P. (1999) *Convergence of Probability Measures. 2nd Edition.* Wiley Series in Probability and Statistics: Probability and Statistics, Wiley, New York.

Bingham, N.H. (1971) Limit theorems for occupation-times of Markov processes. *Z. Wahrscheinlichkeitstheor. Verw. Geb.* **17**, 1–22.

Bingham, N.H. (1972) Limit theorems for regenerative phenomena, recurrent events and renewal theory. *Z. Wahrscheinlichkeitstheor. Verw. Geb.* **21**, 20–44.

Bingham, N.H. (1973a) Limit theorems for a class of Markov processes: some thoughts on a postcard from Kingman. In, *Stochastic Analysis*. Rollo Davidson Memorial Volume, E.F. Harding, and D.G. Kendall (Eds.), Wiley, New York, 266–293.

Bingham, N.H. (1973b) Maxima of sums of random variables and suprema of stable processes. *Z. Wahrscheinlichkeitstheor. Verw. Geb.* **4**, 273–296.

Bingham, N.H. (1975) Fluctuation theory in continuous time, *Adv. Appl. Probab.* **7**, 705–766.

Bingham, N.H. (2001) Random walk and fluctuation theory. In, *Stochastic Processes: Theory and Methods*, 19, 171–213. Handbook of Statist., North-Holland, Amsterdam.

Bingham, N.H. (1976) Continuous branching processes and spectral positivity. *Stoch. Process. Appl.* **4**, 217–242.

Bingham, N.H. (2006) Lévy processes and self-decomposability in finance. *Probab. Math. Stat.* **26**, 367–378.

Bingham, N.H. (2013) Modelling and prediction of financial time series. *To appear in Commun. Stat. Theory Methods.*

Bingham, N.H., Fry, J.M. and Kiesel, R. (2010) Multivariate elliptic processes. *Stat. Neerl.* **64**, 352–366.

Bingham, N.H., Goldie, C.M. and Teugels, J.L. (1987) *Regular Variation.* Cambridge University Press, Cambridge.

Bingham, N.H. and Kiesel, R. (2004) *Risk-Neutral Valuation. Pricing and Hedging of Financial Derivatives*. Springer, Berlin.

Black, F. and Scholes, M. (1973) The pricing of options and corporate liabilities. *J. Polit. Econ.* **81**, 637–659.

Blumenthal, R.M. (1992) *Excursions of Markov Processes*. Birkhäuser, Basel.

Blumenthal, R.M. and Getoor, R.K. (1968) *Markov Processes and Potential Theory*. Academic, New York.

Borodin, A. and Salminen, P. (2002) *Handbook of Brownian Motion—Facts and Formulae. 2nd Edition.* Birkhäuser Basel.

Borovkov, A.A. (1976) *Stochastic Processes in Queueing Theory*. Springer, Berlin.

Boyarchenko, S.I. and Levendorskii, S.Z. (2002a) Perpetual American options under Lévy processes. *SIAM J. Control Optim.* **40**, 1663–1696.

Boyarchenko, S.I. and Levendorskii, S.Z. (2002b) *Non-Gaussian Merton–Black–Scholes Theory.* World Scientific, Singapore.

Bratiychuk, N.S. and Gusak, D.V. (1991) *Boundary Problem for Processes with Independent Increments*. Naukova Dumka, Kiev (in Russian).

Bretagnolle, J. (1971) Résultats de Kesten sur les processus à accroissements indépendants. In, *Séminaire de Probabilités V*, 21–36. Lecture Notes in Mathematics, Springer, Berlin.

Brockwell, P.J., Chadraa, E. and Lindner, A. (2006) Continuous-time GARCH processes. *Ann. Appl. Probab.* **16**, 790–826.

Bryn-Jones, A. and Doney, R.A. (2006) A functional central limit theorem for random walks conditioned to stay non-negative. *J. Lond. Math. Soc.* **74**, 244–258.

Busbridge, I. W. (1960) *The Mathematics of Radiative Transfer*. Cambridge Tracts in Mathematics and Mathematical Physics, 50. Cambridge University Press, Cambridge.

Caballero, M.-E. and Chaumont, L. (2006a) Weak convergence of positive self-similar Markov processes and overshoots of Lévy processes. *Ann. Probab.* **34**, 1012–1034.

Caballero, M.-E. and Chaumont, L. (2006b) Conditioned stable Lévy processes and the Lamperti representation. *J. Appl. Probab.* **43**, 967–983.

Caballero, M.-E., Lambert, A. and Uribe Bravo, G. (2009) Proof(s) of the Lamperti representation of continuous-state branching processes. *Probab. Surv.* **6**, 62–89.

Caballero, M.-E., Pardo, J.C. and Pérez, J.L. (2010) On Lamperti stable processes. *Probab. Math. Stat.* **30**, 1–28.

Caballero, M.-E., Pardo, J.C. and Pérez, J.L. (2011) Explicit identities for Lévy processes associated to symmetric stable processes. *Bernoulli* **17**, 34–59.

Cai, J. (2004) Ruin probabilities and penalty functions with stochastic rates of interest. *Stoch. Process. Appl.* **112**, 53–78.

Cai, J. and Dickson, D.C.M. (2002) On the expected discounted penalty function at ruin of a surplus process with interest. *Insur. Math. Econ.* **30**, 389–404.

Campbell, N.R. (1909) The study of discontinuous phenomena. *Proc. Camb. Philos. Soc.* **15**, 117–136.

Campbell, N.R. (1910) Discontinuities in light emission. *Proc. Camb. Philos. Soc.* **15**, 310–328.

Carmona, Ph., Petit, F. and Yor, M. (1998) Beta-gamma random variables and intertwining relations between certain Markov processes. *Rev. Mat. Iberoam.* **14**, 311–368.

Carr, P., Geman, H., Madan, D. and Yor, M. (2003) Stochastic volatility for Lévy processes. *Math. Finance* **13**, 345–382.

Chan, T. (2004) Some applications of Lévy processes in insurance and finance. *Finance* **25**, 71–94.

Chandrasekhar, S. (1960) *Radiative Transfer*. Dover, New York.

Chaumont, L. (1994) Sur certains processus de Lévy conditionnés à rester positifs. *Stoch. Stoch. Rep.* **47**, 1–20.

Chaumont, L. (1996) Conditionings and path decomposition for Lévy processes. *Stoch. Process. Appl.* **64**, 39–54.

Chaumont, L. (2007) *Introduction aux Processus Auto-Similaires*. Lecture Notes. Unpublished.

Chaumont, L. and Doney, R.A. (2005) On Lévy processes conditioned to stay positive. *Electron. J. Probab.* **10**, 948–961.

Chaumont, L., Kyprianou, A.E. and Pardo, J.C. (2009) Some explicit identities associated with positive self-similar Markov processes, *Stoch. Process. Appl.* **119**, 980–1000.

Chaumont, L., Kyprianou, A.E., Pardo, J.C. and Rivero, V. (2012) Fluctuation theory and exit systems for positive self-similar Markov processes. *Ann. Probab.* **40**, 245–279.

Chazal, M., Kyprianou, A.E. and Patie, P. (2012) A transformation for Lévy processes with one-sided jumps and applications. *To appear in Adv. Appl. Probab.*

Chen, Y.T. and Delmas, J.-F. (2012) Smaller population size at the MRCA time for stationary branching processes. *Ann. Probab.* **40**, 2034–2068.

Chesney, M. and Jeanblanc, M. (2004) Pricing American currency options in an exponential Lévy model. *Appl. Math. Finance* **11**, 207–225.

Chistyakov, V.P. (1964) A theorem on sums of independent positive random variables and its applications to branching random processes. *Theory Probab. Appl.* **9**, 640–648.

Chiu, S.K. and Yin, C. (2005) Passage times for a spectrally negative Lévy process with applications to risk theory. *Bernoulli* **11**, 511–522.

Chow, Y.S., Robbins, H. and Siegmund, D. (1971) *Great Expectations: The Theory of Optimal Stopping*. Houghton Mifflin, Boston.

Chung, K.L. and Fuchs, W.H.J. (1951) On the distribution of values of sums of random variables. *Mem. Am. Math. Soc.* **6**, 1–12.

Ciesielski, Z. and Taylor, S.J. (1962) First-passage times and sojourn times for Brownian motion in space and the exact Hausdorff measure of the sample path. *Trans. Am. Math. Soc.* **103**, 434–450.

Cont, R. and Tankov, P. (2004) *Financial Modeling with Jump Processes*. Chapman and Hall/CRC, Boca Raton.

Cramér, H. (1994a) *Collected Works. Vol. I*. Edited and with a preface by Anders Martin-Löf. Springer, Berlin.

Cramér, H. (1994b) *Collected Works. Vol. II*. Edited and with a preface by Anders Martin-Löf. Springer, Berlin.

Darling, D.A., Liggett, T. and Taylor, H.M. (1972) Optimal stopping for partial sums. *Ann. Math. Stat.* **43**, 1363–1368.

de Finetti, B. (1929) Sulle funzioni ad incremento aleatorio. *Rend. Accad. Naz. Lincei* **10**, 163–168.

de Finetti, B. (1957) Su un'impostazion alternativa dell teoria collecttiva del rischio. In, *Transactions of the XVth International Congress of Actuaries*, 2, 433–443.

Deligiannidis, G., Le, H., and Utev, S. (2009) Optimal stopping for processes with independent increments, and applications. *J. Appl. Probab.* **46**, 1130–1145.

Dellacherie, C. and Meyer, P.A. (1975–1993) *Probabilités et Potentiel*. Hermann, Paris. Chaps. I–VI, 1975; Chaps. V–VIII, 1980; Chaps. IX–XI, 1983; Chaps. XII–XVI, 1987; Chaps. XVII–XXIV, with Maisonneuve, B., 1992.

Dickson, D.C.M. (1992) On the distribution of the surplus prior to ruin. *Insur. Math. Econ.* **11**, 191–207.

Dickson, D.C.M. (1993) On the distribution of the claim causing ruin. *Insur. Math. Econ.* **12**, 143–154.

Dickson, D.C.M. and Waters, H.R. (2004) Some optimal dividends problems. *ASTIN Bull.* **34**, 49–74.

Doney, R.A. (1987) On Wiener–Hopf factorisation and the distribution of extrema for certain stable processes. *Ann. Probab.* **15**, 1352–1362.

Doney, R.A. (1991) Hitting probabilities for spectrally positive Lévy processes. *J. Lond. Math. Soc.* **44**, 566–576.

Doney, R.A. (2005) Some excursion calculations for spectrally one-sided Lévy processes. In, *Séminaire de Probabilités XXXVIII*, 5–15. Lecture Notes in Mathematics, Springer, Berlin.

Doney, R.A. (2007) Fluctuation theory for Lévy processes. In, *École d'Eté de Saint-Flour XXXV*, 1897. Lecture Notes in Mathematics. Springer, Berlin.

Doney, R.A. and Kyprianou, A.E. (2005) Overshoots and undershoots of Lévy processes. *Ann. Appl. Probab.* **16**, 91–106.

Doney, R.A. and Savov, M.S. (2010) The asymptotic behavior of densities related to the supremum of a stable process. *Ann. Probab.* **38**, 316–326.

Dube, P., Guillemin, F. and Mazumdar, R. (2004) Scale functions of Lévy processes and busy periods of finite capacity. *J. Appl. Probab.* **41**, 1145–1156.

Duquesne, T. (2003) Path decompositions for real Lévy processes. *Ann. Inst. Henri Poincaré* **39**, 339–370.

Duquesne, T. (2009) Continuum random trees and branching processes with immigration. *Stoch. Process. Appl.* **119**, 99–129.

Duquesne, T. and Le Gall, J.-F. (2002) *Random trees, Lévy processes and spatial branching processes*. Astérisque nr. 281.

Duquesne, T. and Winkel, M. (2007) Growth of Lévy trees. *Probab. Theory Relat. Fields* **139**, 313–371.

Durrett, R. (2004) *Probability: Theory and Examples. 3rd Edition*. Duxbury.

Dynkin, E.B. (1961) Some limit theorems for sums of independent random variables with infinite mathematical expectations. In, *Selected Translations Math. Stat. Prob.* 1, 171–189. Inst. Math. Satistics Amer. Math. Soc.

Dynkin, E.B. (1963) The optimum choice of the instant for stopping a Markov process. *Sov. Math. Dokl.* **4**, 627–629.

Dynkin, E.B. (2002) *Diffusions, superdiffusions and partial differential equations.* American Mathematical Society Colloquium Publications, 50. American Mathematical Society, Providence.

Eberlein, E. (2001) Application of generalized hyperbolic Lévy motions to finance. In, *Lévy Processes: Theory and Applications*, O.E. Barndorff-Nielsen, T. Mikosch, and S. Resnick (Eds.), Birkhäuser, Basel, 319–337.

Einstein, A. (1905) On the motion of small particles suspended in a stationary Liquid, as required by the molecular kinetic theory of heat. *Ann. Phys.*. **322**, 549–560.

Embrechts, P., Goldie, C.M. and Veraverbeke, N. (1979) Subexponentiality and infinite divisibility. *Z. Wahrscheinlichkeitstheor. Verw. Geb.* **49**, 335–347.

Embrechts, P., Klüppelberg, C. and Mikosch, T. (1997) *Modelling Extremal Events for Insurance and Finance*. Springer, Berlin.

Emery, D.J. (1973) Exit problems for a spectrally positive process. *Adv. Appl. Probab.* **5**, 498–520.

Erickson, K.B. (1973) The strong law of large numbers when the mean is undefined. *Trans. Am. Math. Soc.* **185**, 371–381.

Erickson, K.B. and Maller, R.A. (2004) Generalised Ornstein–Uhlenbeck processes and the convergence of Lévy processes. In, *Séminaire de Probabilités XXXVIII*, 70–94. Lecture Notes in Mathematics, Springer, Berlin.

Etheridge, A.M. (2000) *An Introduction to Superprocesses*. American Mathematical Society, Providence.

Etheridge, A.M. and Williams, D.R.E. (2003) A decomposition of the $(1+\beta)$-superprocess conditioned on survival. *Proc. R. Soc. Edinb. A* **133**, 829–847.

Feller, W. (1971) *An Introduction to Probability Theory and Its Applications. Vol II. 2nd Edition.* Wiley, New York.

Fitzsimmons, P. J. (2006) On the existence of recurrent extensions of self-similar Markov processes. *Electron. Commun. Probab.* **11**, 230–241.

Fristedt, B.E. (1974) Sample functions of stochastic processes with stationary independent increments. In, *Adv. Probab.*, 3, 241–396. Dekker, New York.

Garrido, J. and M. Morales (2006) On the expected discounted penalty function for Lévy risk processes. *N. Am. Actuar. J.* **10**, 196–218.

Geman, H., Madan, D. and Yor, M. (2001) Asset prices are Brownian motion: only in business time. In, *Quantitative Analysis in Financial Markets*, 103–146. World Scientific, Singapore.

Gerber, H.U. (1969) Entscheidungskriterien für den zusammengesetzten Poisson-Prozess. *Mitt. - Schweiz. Ver. Versicher.math.* **69**, 185–228.

Gerber, H.U. (1972) Games of economic survival with discrete- and continuous-income processes. *Oper. Res.* **20**, 37–45.

Gerber, H.U. and B. Landry (1998) On the discounted penalty at ruin in a jump-diffusion and the perpetual put option. *Insur. Math. Econ.* **22**, 263–276.

Gerber, H.U. and Shiu, E.S.W. (1994) Martingale approach to pricing perpetual American options. *ASTIN Bull.* **24**, 195–220.

Gerber, H.U. and Shiu, E.S.W. (1997) The joint distribution of the time of ruin, the surplus immediately before ruin, and the deficit at ruin. *Insur. Math. Econ.* **21**, 129–137.

Gerber, H.U. and Shiu, E.S.W. (1998) On the time value of ruin. *N. Am. Actuar. J.* **2**, 48–78.

Gerber, H.U. and Shiu, E.S.W. (2006a) On optimal dividends: from reflection to refraction. *J. Comput. Appl. Math.* **186**, 4–22.

Gerber, H.U. and Shiu, E.S.W. (2006b) On optimal dividend strategies in the compound Poisson model. *N. Am. Actuar. J.* **10**, 76–93.

Getoor, R.K. and Sharpe, M.J. (1979) Excursions of Brownian motion and Bessel processes. *Z. Wahrscheinlichkeitstheor. Verw. Geb.* **47**, 83–106.

Gikhman, I.I. and Skorokhod, A.V. (1975) *The Theory of Stochastic Processes II*. Springer, Berlin.

Gnedin, A. (2010) Regeneration in random combinatorial structures. *Probab. Surv.* **7**, 105–156.

Gnedin, A. and Pitman, J. (2005) Regenerative composition structures. *Ann. Probab.* **33**, 445–479.

Good, I.J. (1953) The population frequencies of species and the estimation of populations parameters. *Biometrika* **40**, 260–273.

Graczyk, P. and Jakubowski, T. (2011) On Wiener–Hopf factors for stable processes. *Ann. Inst. Henri Poincaré.* **47**, 9–19.

Greenwood, P.E. and Pitman, J.W. (1980a) Fluctuation identities for Lévy processes and splitting at the maximum. *Adv. Appl. Probab.* **12**, 839–902.

Greenwood, P.E. and Pitman, J. W. (1980b) Fluctuation identities for random walk by path decomposition at the maximum. In, *Abstracts of the Ninth Conference on Stochastic Processes and Their Applications*, Evanston, Illinois, 6–10 August 1979, Adv. Appl. Probab., 12, 291–293.

Greenwood, P.E. and Pitman, J.W. (1980c) Construction of local times and Poisson processes from nested arrays. *J. Lond. Math. Soc.* **22**, 182–192.

Grey, D.R. (1974) Asymptotic behaviour of continuous time, continuous state-space branching processes. *J. Appl. Probab.* **11**, 669–677.

Grosswald, E. (1976) The student t-distribution of any degree of freedom is infinitely divisible. *Z. Wahrscheinlichkeitstheor. Verw. Geb.* **36**, 103–109.

Gusak, D.V. and Korolyuk, V.S. (1969) On the joint distribution of a process with stationary independent increments and its maximum. *Theory Probab. Appl.* **14**, 400–409.

Hadjiev, D.I. (1985) The first passage problem for generalized Ornstein–Uhlenbeck processes with nonpositive jumps. In, *Séminaire de Probabilités, XIX*, 80–90. Lecture Notes in Mathematics, Springer, Berlin.

Halgreen, C. (1979) Self-decomposability of the generalized inverse Gaussian and hyperbolic distributions. *Z. Wahrscheinlichkeitstheor. Verw. Geb.* **47**, 13–18.

Harrison, J.M. and Williams, R.J. (1987) Brownian models of open queueing networks with homogeneous customer populations. *Stoch. Stoch. Rep.* **22**, 77–115.

Heyde, C.C. and Seneta, E. (1977) I. J. Bienaymé: statistical theory anticipated. In, *Studies in the History of Mathematics and Physical Sciences*, 3. Springer, Berlin.

Hopf, E. (1934) *Mathematical Problems of Radiative Equilibrium.* Cambridge Tracts, 31. Cambridge University Press, Cambridge.

Horowitz, J. (1972) Semilinear Markov processes, subordinators and renewal theory. *Z. Wahrscheinlichkeitstheor. Verw. Geb.* **24**, 167–193.

Hougaard, P. (1986) Survival models for heterogeneous populations derived from stable distributions. *Biometrika* **73**, 386–396.

Hubalek, F. and Kyprianou, A.E. (2010) Old and new examples of scale functions for spectrally negative Lévy processes. In, *Sixth Seminar on Stochastic Analysis, Random Fields and Applications*, R. Dalang, M. Dozzi, and F. Russo (Eds.), Progress in Probability, 63. Birkhäuser, Basel, 119–146.

Huzak, M., Perman, M., Šikić, H. and Vondraček, Z. (2004a) Ruin probabilities and decompositions for general perturbed risk processes. *Ann. Appl. Probab.* **14**, 1378–1397.

Huzak, M., Perman, M., Šikić, H. and Vondraček, Z. (2004b) Ruin probabilities for competing claim processes. *J. Appl. Probab.* **41**, 679–690.

Ismail, M.E.H. (1977) Bessel functions and the infinite divisibility of the Student t-distribution. *Ann. Probab.* **5**, 582–585.

Ismail, M.E.H. and Kelker, D.H. (1979) Special functions, Stieltjes transforms and infinite divisibility. *SIAM J. Math. Anal.* **10**, 884–901.

Itô, K. (1942) On stochastic processes. I. (Infinitely divisible laws of probability). *Jpn. J. Math.* **18**, 261–301.

Itô, K. (1970) Poisson point processes attached to Markov processes. In, *Proc. 6th Berkeley Symp. Math. Stat. Probab. III*, 225–239.

Itô, K. (2004) *Stochastic Processes*. Springer, Berlin.

Jacob, N. (2001) *Pseudo-Differential Operators and Markov Processes. Vol. 1: Fourier Analysis and Semigroups*. Imperial College Press, London.

Jacob, N. (2002) *Pseudo-Differential Operators and Markov Processes. Vol. 2: Generators and Their Potential Theory*. Imperial College Press, London.

Jacob, N. (2005) *Pseudo-Differential Operators and Markov Processes. Vol. 3: Markov Processes and Applications*. Imperial College Press, London.

Jacod, J. and Shiryaev, A.N. (1987) *Limit Theorems for Stochastic Processes*. Springer, Berlin.

Jeanblanc, M. and Shiryaev, A.N. (1995) Optimization of the flow of dividends. *Russ. Math. Surv.* **50**, 257–277.

Jirina, M. (1958) Stochastic branching processes with continuous state space. *Czechoslov. Math. J.* **8**, 292–312.

Johnson, N.L. and Kotz, S. (1970) *Distributions in Statistics. Continuous Univariate Distributions. Vol 1*. Wiley, New York.

Jørgensen, B. (1982) *Statistical Properties of the Generalized Inverse Gaussian Distribution*. Lecture Notes in Statistics, 9. Springer, Berlin.

Kallenberg, O. (2002) *Foundations of Modern Probability. 2nd Edition*. Springer, Berlin.

Kallsen, J. and Tankov, P. (2006) Characterization of dependence of multidimensional Lévy processes using Lévy copulas. *J. Multivar. Anal.* **97**, 1551–1572.

Kella, O. (1993) Parallel and tandem fluid networks with dependent Lévy inputs. *Ann. Appl. Probab.* **3**, 682–695.

Kella, O. and Whitt, W. (1992) Useful martingales for stochastic storage processes with Lévy input. *J. Appl. Probab.* **29**, 396–403.

Kella, O. and Whitt, W. (1996) Stability and structural properties of stochastic storage networks. *J. Appl. Probab.* **33**, 1169–1180.

Kendall, M. and Stuart, A. (1977) *The Advanced Theory of Statistics. Vol. 1. Distribution Theory. 4th Edition*. Macmillan, New York.

Kennedy, D. (1976) Some martingales related to cumulative sum tests and single server queues. *Stoch. Process. Appl.* **4**, 261–269.

Kesten, H. (1969) Hitting probabilities of single points for processes with stationary independent increments. *Mem. Am. Math. Soc.* **93**, 1–129.

Khintchine A. (1937) A new derivation of one formula by Levy P. *Bull. Mosc. State Univ.* **I**, 1–5.

Kingman, J.F.C. (1964) Recurrence properties of processes with stationary independent increments. *J. Aust. Math. Soc.* **4**, 223–228.

Kingman, J.F.C. (1967) Completely random measures. *Pac. J. Math.* **21**, 59–78.

Kingman, J.F.C. (1993) *Poisson Processes*. Oxford University Press, Oxford.

Klüppelberg, C., and Kyprianou, A.E. (2006) On extreme ruinous behaviour of Lévy insurance risk processes. *J. Appl. Probab.* **43**, 594–598.

Klüppelberg, C., Kyprianou, A.E. and Maller, R.A. (2004a) Ruin probabilities for general Lévy insurance risk processes. *Ann. Appl. Probab.* **14**, 1766–1801.

Klüppelberg, C., Lindner, A. and Maller, R. (2004b) A continuous-time GARCH process driven by a Lévy process: stationarity and second order behaviour. *J. Appl. Probab.* **41**, 601–622.

Klüppelberg, C., Lindner, A. and Maller, R. (2006) Continuous-time volatility modelling: COGARCH versus Ornstein–Uhlenbeck models. In, Kabanov, Y., Lipster, R. and Stoyanov, J. (Eds.), *From Stochastic Calculus to Mathematical Finance. The Shiryaev Festschrift*. Springer, Berlin.

Kolmogorov, N.A. (1932) Sulla forma generale di un processo stocastico omogeneo (un problema di B. de Finetti). *Atti Accad. Naz. Lincei, Rend.* **15**, 805–808.

Konstantopoulos, T., Last, G. and Lin, S.J. (2004) Non-product form and tail asymptotics for a class of Lévy stochastic networks. *Queueing Syst.* **46**, 409–437.

Konstantopoulos, T. and Lin, S.J. (1998) Macroscopic models for long-range dependent network traffic. *Queueing Syst.* **28**, 215–243.

Konstantopoulos, T., Kyprianou, A.E. and Salminen, P. (2011) On the excursions of reflected local time processes and stochastic fluid queues. *J. Appl. Probab.* **48**, 79–98.

Konstantopoulos, T. and Richardson, G. (2002) Conditional limit theorems for spectrally positive Lévy processes. *Adv. Appl. Probab.* **34**, 158–178.

Koponen, I. (1995) Analytic approach to the problem of convergence of truncated Lévy flights towards the Gaussian stochastic process. *Phys. Rev. E.* **52**, 1197–1199.

Korolyuk, V.S. (1974) Boundary problems for a compound Poisson process. *Theory Probab. Appl.* **19**, 1–14.

Korolyuk V.S. (1975a) *Boundary Problems for Compound Poisson Processes*. Naukova Dumka, Kiev (in Russian).

Korolyuk, V.S. (1975b) On ruin problem for compound Poisson process. *Theory Probab. Appl.* **20**, 374–376.

Korolyuk, V.S. and Borovskich, Ju.V. (1981) *Analytic Problems of the Asymptotic Behaviour of Probability Distributions*. Naukova Dumka, Kiev (in Russian).

Korolyuk, V.S., Suprun, V.N. and Shurenkov, V.M. (1976) Method of potential in boundary problems for processes with independent increments and jumps of the same sign. *Theory Probab. Appl.* **21**, 243–249.

Kuznetsov, A. (2010a) Wiener–Hopf factorization and distribution of extrema for a family of Lévy processes. *Ann. Appl. Probab.* **20**, 1801–1830.

Kuznetsov, A. (2010b) Wiener–Hopf factorization for a family of Lévy processes related to theta functions. *J. Appl. Probab.* **47**, 1023–1033.

Kuznetsov, A. (2011) On extrema of stable processes. *Ann. Probab.* **39**, 1027–1060.

Kuznetsov, A., Kyprianou, A.E. and Pardo, J.C. (2012) Meromorphic Lévy processes and their fluctuation identities. *Ann. Appl. Probab.* **22**, 1101–1135.

Kuznetsov, A., Kyprianou, A.E., Pardo, J.C., and van Schaik, K. (2011) A Wiener–Hopf Monte Carlo simulation technique for Lévy process. *Ann. Appl. Probab.* **21**, 2171–2190.

Kuznetsov, A. and Pardo, J.C. (2012) Fluctuations of stable processes and exponential functionals of hypergeometric Lévy processes. *Acta Appl. Math.* **123**, 113–139.

Kyprianou, A.E., Liu, R.-L., Murillo-Salas, A. and Ren, Y.-X. (2012a) Supercritical super-Brownian motion with a general branching mechanism and travelling waves. *Ann. Inst. Henri Poincaré* **48**, 661–687.

Kyprianou, A.E. and Loeffen, R.L. (2010) Refracted Lévy processes. *Ann. Inst. Henri Poincaré.* **46**, 24–44.

Kyprianou, A.E., Loeffen, R.L. and Pérez, J.L. (2012b) Optimal control with absolutely continuous strategies for spectrally negative Lévy processes. *J. Appl. Probab.* **49**, 150–166.

Kyprianou, A.E. and Ott, C. (2012) Spectrally negative Lévy processes perturbed by functionals of their running supremum. *J. Appl. Probab.* **49**, 1005–1014.

Kyprianou, A.E. and Palmowski, Z. (2005) A martingale review of some fluctuation theory for spectrally negative Lévy processes. In, *Séminaire de Probabilités XXXVIII*, 16–29. Lecture Notes in Mathematics, Springer, Berlin.

Kyprianou A.E. and Palmowski, Z. (2007) Distributional study of De Finetti's dividend problem for a general Lévy insurance risk process. *J. Appl. Probab.* **44**, 428–443.

Kyprianou A.E. and Pardo, J.C. (2008) Continuous-state branching processes and self-similarity. *J. Appl. Probab.* **45**, 1140–1160.

Kyprianou, A.E., Pardo, J.C. and Rivero, V. (2010a) Exact and asymptotic n-tuple laws at first and last passage. *Ann. Appl. Probab.* **20**, 522–564.

Kyprianou, A.E. and Patie, P. (2011) A Ciesielski–Taylor type identity for positive self-similar Markov processes. *Ann. Inst. Henri Poincaré* **47**, 917–928.

Kyprianou, A.E. and Rivero, V. (2008) Special, conjugate and complete scale functions for spectrally negative Lévy processes. *Electron. J. Probab.* **13**, 1672–1701.

Kyprianou, A.E., Rivero, V. and Song, R. (2010b) Convexity and smoothness of scale functions and de Finetti's control problem. *J. Theor. Probab.* **23**, 547–564.

Kyprianou, A.E., Schoutens, W. and Wilmott, P. (2005) *Exotic Option Pricing and Advanced Lévy Models*. Wiley, New York.

Kyprianou, A.E. and Surya, B. (2005) On the Novikov–Shiryaev optimal stopping problem in continuous time. *Electron. Commun. Probab.* **10**, 146–154.

Kyprianou, A.E. and Zhou, X. (2009) General tax structures and the Lévy insurance risk model *J. Appl. Probab.* **46**, 1146–1156.

Lambert, A. (2000) Completely asymmetric Lévy processes confined in a finite interval. *Ann. Inst. Henri Poincaré* **36**, 251–274.

Lambert, A. (2001) *Arbres, excursions, et processus de Lévy complètement asymétriques*. Thèse, doctorat de l'Université Pierre et Marie Curie, Paris.

Lambert, A. (2002) The genealogy of continuous-state branching processes with immigration. *Probab. Theory Relat. Fields* **122**, 42–70.

Lambert, A. (2007) Quasi-stationary distributions and the continuous-state branching process conditioned to be never extinct. *Electron. J. Probab.* **12**, 420–446.

Lambert, A. (2008) Population dynamics and random genealogies. *Stoch. Models* **24**, 45–163.

Lamperti, J. (1962) An invariance principle in renewal theory. *Ann. Math. Stat.* **33**, 685–696.

Lamperti, J. (1967a) Continuous-state branching processes. *Bull. Am. Math. Soc.* **73**, 382–386.

Lamperti, J. (1967b) The limit of a sequence of branching processes. *Z. Wahrscheinlichkeitstheor. Verw. Geb.* **7**, 271–288.

Lamperti, J. (1972) Semi-stable Markov processes. *Z. Wahrscheinlichkeitstheor. Verw. Geb.* **22**, 205–225.

Lebedev, N.N. (1972) *Special Functions and Their Applications*. Dover, New York.

Le Gall, J.-F. (1999) *Spatial Branching Processes, Random Snakes and Partial Differential Equations*. Lectures in Mathematics, ETH Zürich, Birkhäuser, Basel.

Lévy, P. (1924) Théorie des erreurs. La loi de Gauss et les lois exceptionnelles. *Bull. Soc. Math. Fr.* **52**, 49–85.

Lévy, P. (1925) *Calcul des Probabilités*. Gauthier-Villars, Paris.

Lévy, P. (1934a) Sur les intégrales dont les éléments sont des variables aléatoires indépendantes. *Ann. Sc. Norm. Super. Pisa* **3**, 337–366.

Lévy, P. (1934b) Sur les intégrales dont les éléments sont des variables aléatoires indépendantes. *Ann. Sc. Norm. Super. Pisa* **4**, 217–218.

Lévy, P. (1948) *Processus Stochastiques et Mouvement Brownien*. Gauthiers–Villars, Paris (Second edition 1965).

Lévy, P. (1954) *Théorie de l'Addition des Variables Aléatoires, 2nd Edition*. Gaulthier-Villars, Paris.

Lin, X. and G. Willmot (1999) Analysis of a defective renewal equation arising in ruin theory. *Insur. Math. Econ.* **25**, 63–84.

Lindner, A. and Maller, R.A. (2005) Lévy integrals and the stationarity of generalised Ornstein–Uhlenbeck processes. *Stoch. Process. Appl.* **115**, 1701–1722.

Loeffen, R.L. (2008) On optimality of the barrier strategy in de Finetti's dividend problem for spectrally negative Lévy processes. *Ann. Appl. Probab.* **18**, 1669–1680.

Loeffen, R. L. and Renaud, J.-F. (2010) De Finetti's optimal dividends problem with an affine penalty function at ruin. *Insur. Math. Econ.* **46**, 98–108.

Lucretius, C.T. (ca. 99 BC–ca. 55 BC). *De rerum natura*.

Lukacs, E. (1970) *Characteristic Functions. 2nd Edition*, revised and enlarged. Hafner, New York.

Lundberg, F. (1903) *Approximerad framställning av sannolikhetsfunktionen. Återförsäkring av kollektivrisker*. Akad. Afhandling. Almqvist. och Wiksell, Uppsala.

Lyons, R., Pemantle, R. and Peres, Y. (1995) Conceptual proofs of $L \log L$ criteria for mean behaviour of branching processes. *Ann. Probab.*. **23**, 1125–1138.

Madan, D.P. and Seneta, E. (1990) The VG for share market returns *J. Bus.* **63**, 511–524.

Maisonneuve, B. (1975) Exit systems. *Ann. Probab.* **3**, 399–411.

Maulik, K. and Zwart, B. (2006) Tail asymptotics for exponential functionals of Lévy processes. *Stoch. Process. Appl.* **116**, 156–177.

McKean, H. (1965) Appendix: a free boundary problem for the heat equation arising from a problem of mathematical economics. *Ind. Manage. Rev.* **6**, 32–39.

Merton, R.C. (1973) Theory of rational option pricing. *Bell J. Econ. Manag. Sci.* **4**, 141–183.

Mikhalevich, V.S. (1958) Bayesian choice between two hypotheses for the mean value of a normal process. *Vìsn. Kiïv. Unìv.* **1**, 101–104.

Mikosch, T., Resnick, S.I., Rootzén, H. and Stegeman, A. (2002). Is network traffic approximated by stable Lévy motion or fractional Brownian motion? *Ann. Appl. Probab.* **12**, 23–68.

Morales, M. (2007) On the expected discounted penalty function for a perturbed risk process driven by a subordinator. *Insur. Math. Econ.* **40**, 293–301.

Moran, P.A. (1968) *An Introduction to Probability Theory* Oxford University Press, Oxford.

Mordecki, E. (2002) Optimal stopping and perpetual options for Lévy processes. *Finance Stoch.* **6**, 473–493.

Mordecki, E. (2008) Wiener–Hopf factorization for Lévy processes having negative jumps with rational transforms. *J. Appl. Probab.* **45**, 118–134.

Nelson, R.B. (2006) *An Introduction to Copulas*. Springer, Berlin.

Nguyen-Ngoc, L. and Yor, M. (2005) Some martingales associated to reflected Lévy processes. In, *Séminaire de Probabilités XXXVII*, 42–69. Lecture Notes in Mathematics, Springer, Berlin.

Noble, B. (1958) *Methods Based on the Wiener–Hopf Technique for the Solution of Partial Differential Equations.* International Series of Monographs on Pure and Applied Mathematics. 7. Pergamon Press, New York.

Novikov, A.A. (2004) Martingales and first-exit times for the Ornstein–Uhlenbeck process with jumps. *Theory Probab. Appl.* **48**, 288–303.

Novikov, A.A. and Shiryaev, A.N. (2004) On an effective solution to the optimal stopping problem for random walks. *Theory Probab. Appl.* **49**, 344–354.

Øksendal, B., (2003) *Stochastic Differential Equations. An Introduction with Applications.* Springer, Berlin.

Øksendal, B. and Sulem, A. (2004) *Applied Stochastic Control of Jump Diffusions*. Springer, Berlin.

Patie, P. (2004) *On some first passage time problems motivated by financial applications*. Ph.D. Thesis, ETH Lecture Series Zürich.

Patie, P. (2005) On a martingale associated to generalized Ornstein–Uhlenbeck processes and an application to finance. *Stoch. Process. Appl.* **115**, 593–607.

Patie, P. (2009a) Exponential functional of a new family of Lévy processes and self-similar continuous state branching processes with immigration. *Bull. Sci. Math.* **133**, 355–382.

Patie, P. (2009b) Infinite divisibility of solutions to some self-similar integro-differential equations and exponential functionals of Lévy processes. *Ann. Inst. Henri Poincaré* **45**, 667–684.

Payley, R. and Wiener, N. (1934) Fourier transforms in the complex domain. *Am. Math. Soc. Colloq. Publ.* **19**.

Percheskii, E.A. and Rogozin, B.A. (1969) On the joint distribution of random variables associated with fluctuations of a process with independent increments. *Theory Probab. Appl.* **14**, 410–423.

Peskir, G. and Shiryaev, A.N. (2000), Sequential testing problems for Poisson processes. *Ann. Stat.* **28**, 837–859.

Peskir, G. and Shiryaev, A.N. (2002) Solving the Poisson disorder problem. In, *Advances in Finance and Stochastics*. 295–312. Springer, Berlin.

Peskir, G. and Shiryaev, A.N. (2006) *Optimal Stopping and Free-Boundary Value Problems.* Lecture Notes in Mathematics, ETH Zürich, Birkhäuser, Basel.

Pistorius, M.R. (2004) On exit and ergodicity of the completely asymmetric Lévy process reflected at its infimum. *J. Theor. Probab.* **17**, 183–220.

Pistorius, M.R. (2006) On maxima and ladder processes for a dense class of Lévy processes. *J. Appl. Probab.* **43**, 208–220.

Pitman, J. (2006) Combinatorial stochastic processes. In, *Lectures from the 32nd Summer School on Probability Theory*, Saint–Flour, July 7–24, 2002. With a foreword by Jean Picard. Lecture Notes in Mathematics, 1875. Springer, Berlin.

Pitman, J. and Yor, M. (1981) Bessel processes and infinitely divisible laws. Stochastic integrals. In, *Proc. Sympos.*, Univ. Durham, Durham, 1980, 285–370. Lecture Notes in Math., 851. Springer, Berlin, 1981.

Port, S.C. (1963) An elementary probability approach to fluctuation theory. *J. Math. Anal. Appl.* **6**, 109–151.

Port, S.C. (1967) Hitting times and potentials for recurrent stable processes. *J. Anal. Math.* **20**, 371–395.

Port, S.C. and Stone, C.J. (1971a) Infinitely divisible processes and their potential theory I, II. *Ann. Inst. Fourier* **21**, 157–275.

Port, S.C. and Stone, C.J. (1971b) Infinitely divisible processes and their potential theory I, II. *Ann. Inst. Fourier* **21**, 179–265.

Prabhu, N.U. (1998) *Stochastic Storage Processes. Queues, Insurance Risk, Dams and Data Communication. 2nd Edition.* Springer, Berlin.

Protter, P. (2004) *Stochastic Integration and Differential Equations. 2nd Edition.* Springer, Berlin.

Renaud, J.–F. and Zhou, X. (2007) Moments of the expected present value of total dividends until ruin in a Lévy risk model. *J. Appl. Probab.* **44**, 420–427.

Revuz, D. and Yor, M. (2004) *Continuous Martingales and Brownian Motion. 3rd Edition.* Springer, Berlin.

Rivero, V. (2005) Recurrent extensions of self-similar Markov processes and Cramér's condition. *Bernoulli*, **11**, 471–509.

Rivero, V. (2007) Recurrent extensions of self-similar Markov processes and Cramér's condition. *Bernoulli*, **13**, 1053–1070.

Rogers, L.C.G. (1990) The two-sided exit problem for spectrally positive Lévy processes, *Adv. Appl. Probab.* **22**, 486–487.

Rogozin, B.A. (1968) The local behavior of processes with independent increments. *Theory Probab. Appl.* **13**, 507–512 (Russian)

Rogozin, B.A. (1972) The distribution of the first hit for stable and asymptotically stable random walks on an interval. *Theory Probab. Appl.* **17**, 332–338.

Rubinovitch, M. (1971) Ladder phenomena in stochastic processes with stationary independent increments. *Z. Wahrscheinlichkeitstheor. Verw. Geb.* **20**, 58–74.

Samorodnitsky, G. and Taqqu, M.S. (1994) *Stable Non-Gaussian Random Processes.* Chapman and Hall/CRC, Boca Raton.

Samuelson, P. (1965) Rational theory of warrant pricing. *Ind. Manage. Rev.* **6**, 13–32.

Sato, K. (1999) *Lévy Processes and Infinitely Divisible Distributions.* Cambridge University Press, Cambridge.

Schilling, R., Song, R., and Vondraček, Z. (2010) *Bernstein Functions. Theory and Applications.* de Gruyter Studies in Mathematics, 37. Walter de Gruyter, Berlin.

Schneider, W.R. (1996) Completely monotone generalized Mittag–Leffler functions. *Expo. Math.* **14**, 3–16.

Schoutens, W. (2003) *Lévy Processes in Finance. Pricing Finance Derivatives.* Wiley, New York.

Schoutens, W. and Cariboni, J. (2009) *Lévy Processes in Credit Risk.* Wiley, New York.

Schoutens, W. and Teugels, J.L. (1998) Lévy processes, polynomials and martingales. *Commun. Stat., Stoch. Models* **14**, 335–349.

Shepp, L. and Shiryaev, A.N. (1993) The Russian option: reduced regret. *Ann. Appl. Probab.* **3**, 603–631.

Shepp, L. and Shiryaev, A.N. (1994) A new look at the pricing of the Russian option. *Theory Probab. Appl.* **39**, 103–120.

Shiryaev, A.N. (1978) *Optimal Stopping Rules.* Springer, Berlin.

Shtatland, E.S. (1965) On local properties of processes with independent increments. *Theory Probab. Appl.* **10**, 317–322.
Silverstein, M.L. (1968) A new approach to local time. *J. Math. Mech.* **17**, 1023–1054.
Silverstein, M.L. (1980) Classification of coharmonic and coinvariant functions for a Lévy process. *Ann. Probab.* **8**, 539–575.
Situ, R. (2005) *Theory of Stochastic Differential Equations with Jumps and Applications*. Mathematical and Analytical Techniques with Applications to Engineering, Springer, Berlin.
Snell, J.L. (1952) Applications of martingale system theorems. *Trans. Am. Math. Soc.* **73**, 293–312.
Song, R. and Vondraček, Z. (2006) Potential theory of special subordinators and subordinate killed stable processes. *J. Theor. Probab.* **119**, 817–847.
Song, R. and Vondraček, Z. (2010) Some remarks on special subordinators. Rocky Mt. J. Math. **40**, 321–337.
Spitzer, F. (1956) A combinatorial lemma and its application to probability theory. *Trans. Am. Math. Soc.* **82**, 323–339.
Spitzer, F. (1957) The Wiener–Hopf equation whose kernel is a probability density. *Duke Math. J.* **24**, 327–343.
Spitzer, F. (1960a) The Wiener–Hopf equation whose kernel is a probability density. II. *Duke Math. J.* **27**, 363–372.
Spitzer, F. (1960b) A Tauberian theorem and its probability interpretation. *Trans. Am. Math. Soc.* **94**, 150–169.
Spitzer, F. (1964) *Principles of Random Walk.* Van Nostrand, New York.
Steutel, F.W. (1970) *Preservation of Infinite Divisibility under Mixing and Related Topics.* Math. Centre Tracts, 33. Math. Centrum, Amsterdam.
Steutel, F.W. (1973) Some recent results in infinite divisibility. *Stoch. Process. Appl.* **1**, 125–143.
Steutel, F.W. and van Harn, K. (1977) Generalized renewal sequences and infinitely divisible lattice distributions. *Stoch. Process. Appl.* **5**, 47–55.
Suprun, V.N. (1976) Problem of ruin and resolvent of terminating processes with independent increments. *Ukr. Math. J.* **28**, 39–45.
Takács, L. (1966) *Combinatorial Methods in the Theory of Stochastic Processes.* Wiley, New York.
Tankov, P. (2003) Dependence structure of positive multidimensional Lévy processes. http://www.math.jussieu.fr/~tankov/.
Thorin, O. (1977a) On the infinite divisibility of the Pareto distribution. *Scand. Actuar. J.* 31–40.
Thorin, O. (1977b) On the infinite divisibility of the lognormal distribution. *Scand. Actuar. J.* 121–148.
Tweedie, M.C.K. (1984) An index which distinguishes between some important exponential families. In, *Statistics: Applications and New Directions.* Calcutta, 1981, 579–604. Indian Statist. Inst., Calcutta.
Urbanik, K. (2005) Infinite divisibility of some functionals on stochastic processes. *Probab. Math. Stat.* **15**, 493–513.
Vigon, V. (2002a) *Simplifiez vos Lévy en titillant la factorisation de Wiener–Hopf.* Thèse de l'INSA, Rouen.
Vigon, V. (2002b) Votre Lévy rampe-t-il? *J. Lond. Math. Soc.* **65**, 243–256.
Watson, H.W. and Galton, F. (1874) On the probability of the extinction of families. *J. Anthropol. Inst. G. B. Irel.* **4**, 138–144.
Widder, D.V. (2010) *The Laplace Transform.* Dover, New York.
Wiener, N. and Hopf, E. (1931) Über einer Klasse singulärer Integralgleichungen. *Sitzungsb. Preuss. Akad. Wiss. Berlin Kl. Math. Phys. Tech.*, 696–706.
Yin, C. and Wang, C. (2009) Optimality of the barrier strategy in de Finetti's dividend problem for spectrally negative Lévy processes: an alternative approach. *J. Comput. Math.* **233**, 482–491.
Yor, M. (1991) Une explication du théorème de Ciesielski–Taylor. *Ann. Inst. Henri Poincaré* **27**, 201–213.
Zhou, X. (2005) On a classical risk model with a constant dividend barrier. *N. Am. Actuar. J.* **9**, 95–108.

Zolotarev, V.M. (1964) The first passage time of a level and the behaviour at infinity for a class of processes with independent increments, *Theory Probab. Appl.* **9**, 653–661.
Zolotarev, V.M. (1986) *One Dimensional Stable Distributions*. American Mathematical Society, Providence.

Index

A.E. Kyprianou, *Fluctuations of Lévy Processes with Applications*, Universitext,
DOI 10.1007/978-3-642-37632-0, © Springer-Verlag Berlin Heidelberg 2014